is a volume in

LICATIONS OF MODERN ACOUSTICS

Editors:

rd Stern
ed Research Laboratory
ylvania State University
College, Pennsylvania

Moises Levy
Department of Physics
University of Wisconsin at Milwaukee
Milwaukee, Wisconsin

Surface Acoustic Wave D
for Mobile and
Wireless Communicati

This
APPI

Serie
Richa
Appli
Penns
State

Surface Acoustic Wave Devices for Mobile and Wireless Communications

C. K. CAMPBELL

DEPARTMENT OF ELECTRICAL AND COMPUTER ENGINEERING
MCMASTER UNIVERSITY
HAMILTON, ONTARIO
CANADA

 ACADEMIC PRESS, INC.

San Diego London Boston New York
Sydney Tokyo Toronto

This book is printed on acid-free paper.

ACADEMIC PRESS, INC.
525 B Street, Suite 1900, San Diego, CA 92101-4495, USA
1300 Boylston Street, Chestnut Hill, MA 02167, USA
http://www.apnet.com

Academic Press Limited
24–28 Oval Road, London NW1 7DX, UK
http://www.hbuk.co.uk/ap/

```
Campbell, Colin, 1927-
    Surface acoustic wave devices for mobile and wireless
communications / Colin Campbell.
        p.   cm. -- (Applications of modern acoustics)
    Includes bibliographical references and index.
    ISBN 0-12-157340-0 (alk. paper)
    1. Acoustic surface wave devices.  2. Mobile communication
systems--Equipment and supplies.  3. Wireless communication systems-
-Equipment and supplies.   I. Title.  II. Series.
    TK5981.C3523  1998
    621.382'8--dc21                                    97-40551
                                                          CIP
```

Printed in the United States of America
98 99 00 CP 9 8 7 6 5 4 3 2 1

For Vivian, Gwyn,
Barry and Ian

And to the memory of
my Mother

Contents

Chapter 3

Principles of Linear-Phase SAW Filter Design

Chapter 4

Equivalent Circuit and Analytic Models for a SAW Filter

Chapter 5

Some Matching and Trade-Off Concepts for SAW Filter Design

Chapter 6

Compensation for Second-Order Effects in SAW Filters

Chapter 7

Designing SAW Filters for Arbitrary Amplitude/Phase Response

Chapter 8

Interdigital Transducers with Chirped or Slanted Fingers

Chapter 9

IDT Finger Reflections and Radiation Conductance

PART 2

Techniques, Devices and Mobile/Wireless Applications

Chapter 10

Overview of Systems and Devices

Chapter 11

SAW Reflection Gratings and Resonators

Chapter 12

Single-Phase Unidirectional Transducers For Low-Loss Filters

Chapter 13

RF and Antenna-Duplexer Filters for Mobile/Wireless Transceivers

Chapter 14

Other RF Front-End and Interstage Filters for Mobile/Wireless Transceivers

Chapter 15

SAW IF Filters for Mobile Phones and Pagers

Chapter 16

Fixed-Code SAW IDTs for Spread-Spectrum Communications

Chapter 17

Real-Time SAW Convolvers For Voice and Data Spread-Spectrum
Communications

Chapter 18

Surface Wave Oscillators and Frequency Synthesizers

Chapter 19

SAW Filters For Digital Microwave Radio, Fiber Optic, and Satellite Systems

Chapter 20

Postscript

Preface

As a reflection of my experimental training in electrical engineering and physics disciplines, I have attempted to give this text an experimental and theoretical flavour. In this way it is hoped that it will serve those in industry who are using, or plan to use, SAW devices in their communications systems. It is additionally hoped that it will be an aid to students and others, who wish to increase their theoretical understanding of the principles of operation of these fascinating devices. As a text, therefore, it progresses from basic considerations of SAW principles in Part 1 to more complex applications in Part 2. Readers with experience in the field may wish to begin with Part 2, and use Part 1 as a reference.

My deep appreciation is extended to the many individual authorities I have contacted during the compilation of this material over the past three years. These include Dr. Rodolfo Almar, Mr. Ulrich Bauernschmitt, Dr. Ji-Dong Dai, Dr. Peter Edmonson, Mr. Jim Flowers, Dr. Mitsutaka Hikita, Dr. André du Hamél, Mr. Clinton Hartmann, Associate Professor Ken-ya Hashimoto, Dr. James Heighway, Mr. Alan Holdway, Dr. Waguih Ishak, Dr. Michio Kadota, Dr. Chandra Kudsia, Dr. C. S. Lam, Dr. Jürgen Machui, Dr. David Morgan, Mr. Hideaki Nakahata, Dr. Ger Riha, Dr. Clemens Ruppel, Dr. Yoshio Satoh, Dr. John Saw, Professor Franz Seifert, Dr. Mark Suthers, Professor Maseo Takeuchi, Dr. Nobuhiko Shibagaki, Professor Yasutaka Shimizu, Professor Noburu Wakatsuki, Dr. Bert Wall, Dr. Yufeng Xu, Dr. Jun Yamada, Professor Masatsune Yamaguchi, Professor Kazuhiko Yamanouchi, Mr. Hiromi Yatsuda, Dr. Hiroyuki Odagawa, and Mr. Ben Zarlingo.

I also appreciate the illustrative material given with reproduction permission by Andersen Laboratories, Bloomfield, Connecticut; COM DEV, Ontario, Canada; Crystal Technology Inc., Palo Alto, California; Electronics Letters, Great Britain; Fujitsu Ltd., Kawasaki, Japan; Hewlett-Packard Company, Palo Alto, California; Hewlett-Packard (Canada) Ltd, Ontario, Canada; Hitachi Ltd., Central Research Laboratory, Tokyo, Japan; Institute of Electrical and Electronics Engineers, (IEEE), New York, New

York; Japan Radio Company, Ltd., Tokyo, Japan; Murata Manufacturing Co. Ltd., Kyoto, Japan; NORTEL, Northern Telecom, Ontario, Canada; Research In Motion Limited, Ontario, Canada; Sawtek Inc., Orlando, Florida; Siemens-Matsushita Components GmbH & Co. KG, Munich, Germany; Sumitomo Electric Industries, Ltd., Itami, Japan; Vectron Technologies Incorporated, Hudson, New Hampshire; and VI TELEFILTER, Teltow, Germany.

I am especially indebted to Dr. Peter Edmonson for his continued advice on mobile/wireless networks during the preparation of this text. Additionally, my thanks are expressed to Professor Franz Seifert of the Technische Universität Wien, Vienna, Austria to whose SAW convolver work I have heavily referred in the chapter on SAW convolvers.

Appreciation is also extended to the Japan Society For the Promotion of Science (JSPS) for the 1995 award of a 60-day Invitation Fellowship to Japan, for studies of *Nanometer Fabrication Technology and Application for Surface Acoustic Wave Devices.*

Most of all, I could not have completed this work without the understanding and support of my wife, Vivian.

Now I can get back to some quiet fishing.

Colin Campbell
Ancaster, Ontario
Canada
17 April 1997

PART 1

Fundamentals of Surface Acoustic Waves and Devices

"... it is proposed to investigate the behaviour of waves upon the plane free surface of an infinite homogeneous isotropic elastic solid, their character being such that the disturbance is confined to a superficial region, of thickness comparable with the wavelength..." Lord Rayleigh, 12 November 1855.

Introduction

1.1. Background

Electronic signal processing by means of the selective manipulation of surface acoustic waves on piezoelectric substrates was initiated in 1965 with the invention of the thin-film interdigital transducer (IDT) by White and Voltmer at the University of California, at Berkeley [1]. Their idea was deceptively simple but ingenious. With reference to the basic structure sketched in Fig. 1.1, their technique was to fabricate metal thin-film IDTs on the surface of a suitable piezoelectric substrate that would act as electrical input and output ports. Application of an appropriate ac voltage stimulus to the input transducer launched and manipulated a surface acoustic wave (SAW). The receiving transducer detected the incident surface acoustic wave and converted it back to a suitably filtered electrical one. This surface acoustic wave type was a Rayleigh wave[1], with motion confined to within about one acoustic wavelength under the free surface of the piezoelectric.

The electronics industry quickly recognized that such SAW signal processing could readily be applied to the design of analog electrical filters operating at selected frequencies in the range from about 10 MHz to 1 GHz. Despite this, it was some time before it was appreciated that an IDT could be considered to represent a sampled data structure to which digital sampling concepts could be applied [2].

This new technology generated two major electrical engineering product design challenges, with quite divergent applications. At one extreme, for the high-volume low-cost TV component market, the requirement was for mass-produced SAW filters to be competitive in price and performance with the inductance-capacitance (LC) filters then employed in the

---

[1] Named in honor of Lord Rayleigh, who first reported on the behaviour of surface acoustic waves in homogeneous, isotropic elastic solids at an address to the London Mathematical Society on 12 November 1855.

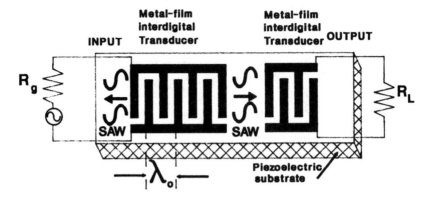

Fig. 1.1. Basic SAW/pseudo-SAW delay line fabricated on piezoelectric substrate, with metal thin-film input/output interdigital transducers. Finger period at center frequency f_o corresponds to acoustic wavelength $\lambda_o = v/f_o$, where $v =$ SAW/pseudo-SAW velocity.

intermediate-frequency (IF) circuit stages. At the opposite extreme, for low-volume high-cost components for radar signal processing, maximum emphasis was given to the efficient implementation of SAW pulse compression filters with large compression gains. Between these limits, a wide range of other SAW device configurations and applications began to receive intensive research scrutiny.

Subsequent developments quickly yielded applications to the consumer, commercial, and military markets. In 1977, Williamson [3] listed 45 different types of SAW devices that had already undergone development efforts, with varying degrees of success, including 10 major devices with exceptional performance. Development successes continued, so that by 1985 (only 20 years after the introduction of the IDT), Hartmann [4] listed 9 major consumer applications, 9 major commercial applications, and 18 major military applications of the technology.

By about 1980, however, SAW technology had reached a plateau in research, development, and manufacturing activity. This was due principally to the limitations imposed *at that time* by the inherently high insertion losses (> 6 dB) of existing SAW filter designs. In communications receiver circuitry, such high losses limited SAW filter usage to IF signal processing stages, operating with millivolt signal levels. Signal-to-noise limitations, imposed by insertion loss, rendered them unsuitable for RF filtering stages[2] involving microvolt-level inputs. As a result, many companies and

---

[2] The generic term *radio frequency* (*RF*) as used here is intended to include the VHF (30–300 MHz), UHF (300 MHz to 3 GHz), and SHF (3 GHz) frequency ranges.

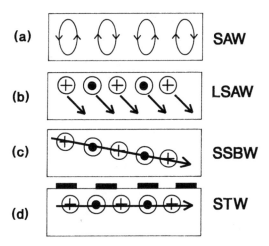

Fig. 1.2. Simplistic artistic impression of wave motions for (a): "true" SAW (Rayleigh wave); (b) leaky-SAW; (c) surface skimming bulk wave (SSBW); and (d) surface transverse wave (STW).

research laboratories around the world abandoned the SAW field and activities coalesced into strategic groups in the United States, Europe, and Japan.

The preceding slow-growth scenario has changed most dramatically in recent years, with the discovery and utilization of other types of surface acoustic waves (often referred to as *pseudo*-SAW), made possible by the discovery of new piezoelectric substrates and/or crystal cuts. To date, these include the leaky surface acoustic wave (LSAW), the surface-skimming bulk wave (SSBW)[3] and the surface transverse wave (STW). An initial, simplistic artistic impression of their respective propagation characteristics is shown in Fig 1.2. The utilization of LSAW techniques has been particularly significant, as it has opened up the gateway to the use of low-loss (e.g., 1 to 3 dB) RF front-end filters and antenna duplexers for mobile/wireless transceivers.

Along with the Rayleigh wave, these three additional forementioned wave types are also commonly referred to as *surface acoustic waves*. This all-encompassing nomenclature has also been adopted in the title of this book. For clarification in the text, however, the specific wave type will be identified as required.

---

[3] Surface skimming bulk wave (SSBW) propagation [5] is also referred to as shallow bulk acoustic wave (SBAW) propagation [6].

1.2. Merits of Rayleigh-Wave Devices

As noted by Hartmann in a 1985 review paper on the systems impact of SAW—then a strictly Rayleigh wave—technology [4], many such devices and systems exhibited several excellent features when compared to competing technologies of that time. Many of these early features are still valid, as included here:

- Surface acoustic wave devices can generally be designed to provide quite complex signal processing functions within a single package that contains only a single piezoelectric substrate with superimposed thin-film input and output interdigital transducers. Thus, for example, bandpass filters with outstanding response characteristics can now be routinely designed to achieve responses that would require several hundred inductors and capacitors in conventional LC filter designs.

- They can be mass produced using semiconductor microfabrication techniques. As a result, they can be made to be cost-competitive in mass-volume applications, with some products selling for less than $1.00.

- They can have outstanding reproducibility in performance from device to device. This is especially desirable for the design of radio frequency (RF) and IF filters for mobile phones, narrowband high-Q resonators, as well as for channelized receivers for spectral analyses or other systems requiring precision filtering.

- As they can be implemented in small, rugged, light and power-efficient modules, they are now standard components in mobile, wireless, and space-borne communications systems.

- Although SAW devices are of the analog variety, they can be employed in many digital communications systems. One example of this is in the use of Nyquist filters in quadrature-amplitude-modulation (QAM) digital radio modems.

- Surface acoustic wave filters can be made to operate in high-harmonic modes [7]. Some devices for the 2-GHz band are now being fabricated in this manner in order to meet demanding lithographic tolerances and constraints. One example is for 2.488-GHz timing-extraction filters in optical-communications links, where a long-term phase stability of 20° is required over a 20-year period [8].

1.3. Additional Merits of Pseudo-SAW Devices

Pseudo-SAW (e.g., LSAW, STW, SSBW) and Rayleigh-wave devices both employ IDTs; they also share many of the forementioned device merits in terms of operation and size. In many instances they may be visually indistinguishable from one another or from Rayleigh-wave structures. Operation in one or the other mode on quartz, for example, merely involves the choice of a different crystal axis for wave propagation. In operation, however, pseudo-SAW devices possess a number of attractive features over their SAW counterparts, as listed here:

- Velocities for pseudo-SAW devices can be much higher than for Rayleigh-wave devices, to the extent that they can be designed for operation at frequencies up to about 1.6 times higher than for a Rayleigh-wave counterpart with the same lithographic geometry. This is particularly attractive for the fabrication of RF-filters for mobile/ wireless communications operating at gigahertz frequencies.

- As will be discussed in this book, pseudo-SAW piezoelectric crystal cuts can have much higher values of electromechanical coupling efficiencies[4], with corresponding increase in operational bandwidth capability—in conjunction with lower attainable insertion loss. This is the dominant reason for their application to front-end filtering in mobile/ wireless transceivers requiring fractional bandwidths of 4 percent or more.

- Some pseudo-SAW piezoelectric crystal cuts of quartz can have superior temperature stability coefficients of delay (TCD) over their SAW counterparts [9].

- Because LSAW and SSBW propagation is beneath the piezoelectric surface, such devices can be significantly less sensitive to surface contamination than Rayleigh-wave devices. (However, the surface-fabricated IDT can still be susceptible to contamination.)

- Because the pseudo-SAW wave penetrates farther into the substrate than does the Rayleigh wave, corresponding acoustic power densities will be less. This means that pseudo-SAW devices, e.g., oscillators, are capable of handling larger powers before the onset of IDT degradation due to violent surface vibrations. They can also operate at higher powers before the onset of piezoelectric nonlinearities.

1.4. Some Device Applications

The signal processing and frequency response characteristics of a SAW device on a piezoelectric substrate are governed primarily by three interrelated factors involving: (i) the geometry of the metal-film interdigital transducers (IDTs); (ii) the piezoelectric substrate; and (iii) the wave-

---

[4] This coupling efficiency is given in terms of an electromechanical coupling coefficient K^2, as will be introduced in Chapter 2.

propagation type involved. Figure 1.1 illustrated a basic delay line structure, employing metal thin-film input/output IDTs for the launching, processing, and detection of acoustic waves passing underneath. In the period since 1965 (when White and Voltmer demonstrated the basic technique employing this configuration), a highly varied (and often bewildering looking) succession of IDT geometries has evolved, catering to a multitude of SAW and pseudo-SAW signal processing functions. Despite their considerable variety, the various SAW and pseudo-SAW transducers and device structures can be grouped under four general categories, as listed in Table 1.1.

Surface acoustic wave RF filters are currently available for mobile and wireless applications in frequency ranges from 800 MHz to 1.5 GHz, with package sizes in the order of $3.2 \times 2.5 \times 0.9 \, \text{mm}^3$ [10], and smaller packages $3 \times 3 \times 1.2 \, \text{mm}^3$ for Personal Communications Service (PCS), Wide Area

TABLE 1.1

SOME APPLICATIONS OF SAW AND PSEUDO-SAW DEVICES FOR
MOBILE/WIRELESS COMMUNICATIONS

Category 1. Bidirectional IDTs
- Fixed delay lines for oscillators, path length equalizers
- PN-coded SAW tapped-delay lines for combined CDMA-TDMA systems
- Clock-recovery filters for fiber optics communications repeater stages
- Delay lines for reduction of multipath interference
- Precision fixed-frequency reference oscillators
- VCOs for first- or second-mixing stages in mobile transceivers
- Intermediate frequency (IF) filters for mobile/wireless receivers and pagers
- Nyquist filters for digital radio

Category 2. Resonators and Resonator-Filters
- Precision fixed-frequency and tunable oscillators
- Resonators and resonator-filters for automotive keyless entry
- Resonators for garage-door transmitter control circuits
- Resonators for medical-alert transmitter circuits
- Compact resonator-filters for RF front-end and interstage filters for mobile phones
- Compact resonator-filters for RF front-end filters for wireless receivers and pagers
- High-power antenna duplexers (2 to 4 W) for mobile transceivers

Category 3. Unidirectional IDTs
- Low-loss IF filters for mobile and wireless circuits (<3 dB)
- RF front-end filters for mobile communications
- Multimode oscillators for frequency-agile spread-spectrum communications

Category 4. Nonlinear Operation
- Convolvers for spread-spectrum communications

Network (WAN), and Wireless Local Area Network (WLAN) communications, ranging from the 800-MHz to 2.4-GHz bands [11]. As one example of the size reduction and technological advances attainable with this technology, Figure 1.3 shows a hand-held two-way half-duplex *Inter@ctive* pager employing leaky-SAW RF front-end filters, for operation at 800 or 900 MHz on ARDIS or MOBITEX WAN systems, respectively. Figure 1.4 gives two examples of the packaging miniaturization available with SAW RF filters. The smaller package (3 mm), a SAW Tx-filter at a center frequency of 1880 MHz, is for a USA Personal Communications Services (PCS) mobile transceiver. The larger package (3.8 mm), a SAW Rx-filter at 1489 MHz, is for a Japan Personal Digital Cellular (PDC) transceiver. Figure 1.5 illustrates enlargements of packages frequently used for RF and IF filters in mobile circuitry. The QCC-10 ceramic package, for example, is a hermetically sealed one used for surface mounted technology, while the DIP-18 package is a plastic one. SAW-filter designs are also being increasingly incorporated into RF modules, or into Application Specific Integrated Circuits (ASICs) in mobile, wireless, and optical communications networks. Figure 1.6 illustrates some examples of SAW devices and RF modules.

FIG. 1.3.   Hand-held two-way half-duplex *Inter@ctive* pager employing leaky-SAW RF front-end filters, for operation at 800 or 900 MHz on ARDIS or MOBITEX WAN systems, respectively. (Courtesy of Research In Motion Limited, Waterloo, Ontario, Canada.)

FIG. 1.4.   Smaller package (3 mm) is a SAW Tx-filter at 1880-MHz center frequency for USA Personal Communications Services (PCS). Larger package (3.8 mm) is a SAW Rx-filter at 1489-MHz center frequency for Japan Personal Digital Cellular (PDC). (Courtesy of Fujitsu Ltd., Kawasaki, Japan.)

Examples of ceramic packages

QCC 10B   QCC8   QCC10

Examples of plastic packages

DIP18D   SIP5K

FIG. 1.5.   Enlarged illustrations of frequently used ceramic and plastic packages for SAW RF and IF filters. (Courtesy of Siemens-Matsushita Components GmbH & Co. KG, Munich, Germany.)

FIG. 1.6. Some examples of SAW devices and RF modules used in mobile and wireless communications. (Courtesy of NORTEL, Northern Telecom, Canada.)

1.5. Global Activities and Participants

The dramatic upsurge in SAW and pseudo-SAW products for mobile and wireless communications is illustrated by a report from the Quartz Industries Association of Japan (QIAJ) that the total output of quartz device production in Japan in 1994 exceeded 200 billion yen. Further, it has been indicated that there are now more SAW and pseudo-SAW device manufacturers than quartz device manufacturers in Japan. In 1993, at least 28 Japanese industries were using SAW materials and/or devices [12]. While statistics are not available, it is believed that the current SAW device output for wireless applications in Japan alone is huge [13]. An unofficial estimate obtained by the author gives the current (1997) worldwide production of SAW devices as *800 million/year.*

At this time, the major suppliers of low-loss (2 to 3 dB) RF pseudo-SAW filters are from Japan (including Fujitsu, Hitachi, Japan Energy, Japan Radio, Matsushita, Murata Manufacturing, OKI, Sanyo, Sumitomo, Toshiba, and Toyocom) and Europe (including MICRONAS Semiconductor, and Siemens-Matsushita). Some of these companies offer products for the 800 to 1500-MHz operational range, as well as those with even lower insertion loss ($<\sim 1\,dB$) for the 1.8 to 2.4-GHz range. North American manufacturers of low-loss RF filters include Motorola and NORTEL (Canada) for cellular telephones and digital cordless phones and base stations (e.g., IS-95, IS-54) [13], [14]. Currently, almost all low-loss acoustic RF filters for mobile and wireless circuitry use appropriate leaky-SAW cuts of lithium niobate ($LiNbO_3$) and lithium tantalate ($LiTaO_3$), or lithium borate ($Li_2B_4O_7$) piezoelectric substrate materials in order to provide the required bandwidth for these applications. These filter types are increasingly being employed as replacements for the much bulkier microwave ceramic (MWC) duplexers employed in front-end RF stages, especially where size and portability of mobile/wireless equipment are becoming dominant system-design factors. Table 1.2 lists some worldwide suppliers of a variety of SAW components for different applications.

Figure 1.7 provides an example of a cellular phone employing SAW technology, as applied to a design for the North American Advanced Mobile Phone Service (AMPS) standards. Figure 1.8 graphs the performance of LSAW RF front-end filters for frequency selection of transmit and receive bands for AMPS cellular phone systems. Figure 1.9 illustrates a representative package for a SAW-based voltage-controlled oscillator (VCO). Rayleigh-wave oscillators are readily available at selected frequencies in the 300 to 800 MHz range, and are used in applications requiring correction of frequency drifts due to set accuracy, temperature stability, aging, and load pull. Maximum output powers are in the range of + 10 dBm, with noise floors as low as – 170 dBc/Hz. The tuning range is between 400–500 ppm, depending on the frequency. Surface transverse wave oscillators can also be employed for higher-frequencies or higher-power levels.

For many years IF filters using monolithic crystal filters (MCF) of the *bulk* acoustic wave type have been dominant in many analog cellular and paging systems. Their IF bandwidths can range from 10 to 30 kHz, centered mainly at frequencies in the 20 to 45 MHz range [13], [15]. Above 45 MHz, however, such crystal filters become increasingly fragile and expensive.

TABLE 1.2

SOME SUPPLIERS OF VARIOUS SAW COMPONENTS

North America	Andersen Labs, COM DEV, Motorola, NORTEL, Phonon, Raytheon, RF Monolithics, Sawtek, Texas Instruments, TRW, Vectron Technologies, Zenith.
Europe	C-MAC, GEC-Plessey Semiconductor, MICRONAS Semiconductor, Racal, Siemens, Siemens-Matsushita, TELEFILTER
Japan	Clarion, Epson, Fuji, Fujitsu, Hitachi, Hokuriko, Japan Energy, Japan Radio, JVC, Kinseki, Kyocera, Matsushita, Murata Manufacturing, NDK, NEC, Nikko Kyodo, OKI, Sanyo, Shoshin, Sumitomo Electric Industries, Taiyo Yuden, Toko, Tokyo Denpa, Toshiba, Toyocom
China	Beijing Chang Feng, Chongqing 26th Institute, Jiangxi Jinghua, Nanjing Electronic Devices Institute
Korea	Daewoo, Samsung
Russia	ONIIP, MRRI, Institute of Semiconductor Physics

FIG. 1.7. A cellular phone made by Fujitsu to AMPS specifications. (Courtesy of Fujitsu Ltd, Kawasaki, Japan.)

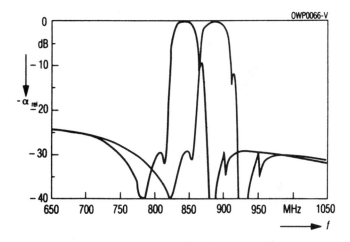

FIG. 1.8. LSAW RF front-end filter responses for cellular phones operating on AMPS standards. Tx: 824-849 MHz. Rx: 869–894 MHz. (Courtesy of Siemens-Matsushita Components GmbH & Co. KG, Munich, Germany.)

FIG. 1.9. Example of UHF SAW-based voltage-controlled oscillator (VCO) for precision tuning. Power levels of Rayleigh-wave oscillators are typically ~ +10 dBm, with noise floors as low as − 170 dBc/Hz. (Courtesy of Sawtek Inc., Orlando, Florida.)

Currently, however, the trend in mobile and wireless circuitry is to increase the IF frequency to aid suppression of spurious image responses in mixer stages. This has given increased impetus to the development of low-loss (<3 dB) and mid-loss SAW and pseudo-SAW filters and frequency control products. Manufacturers of such products and systems in the United States include Vectron Technologies Inc., Sawtek Inc., RF Monolithics, Phonon Corporation, and Andersen Labs. State-of-the-art products include, for example, timing extraction filters for SONET/SDH/ATM boards at varying frequencies up to 2488.32 MHz, as well as integrated voltage-controlled oscillators at standard SONET frequencies, employing an ASIC with an on-chip phase shifter [13].

1.6. Aims of This Text

Depending on the system and design concept, SAW-based mobile and wireless communications circuitry can employ many significantly different types of SAW/pseudo-SAW filter structures. For RF antenna duplexer, front-end, and interstage filter circuitry these can include selections from designs involving:

- Leaky-SAW longitudinally coupled resonator-filters
- Leaky-SAW interdigitated interdigital transducers (IIDTs)
- Floating-electrode unidirectional transducers (FEUDTs)

- Leaky-SAW ladder filters
- Leaky-SAW one-port impedance-element resonator structures

A still-greater variety of filters is currently available for IF filter-stage design in mobile and wireless communication circuitry. To date, these include:

- "Standard" SAW filters and delay lines
- Single-phase SAW unidirectional transducers (SPUDTs)
- SPUDT/reflector composite designs
- Transversely coupled (i.e., waveguide-coupled) SAW resonator-filters
- Nyquist filters for digital microwave radio
- Clock-recovery filters for timing in SONET compatible fiber-optic communications
- In-line coupled SAW resonator-filters
- Z-path SAW filters

In addition, SAW/pseudo-SAW devices are playing an ever-increasing part in phase-locked-loop and VCO circuitry for frequency synthesizers and in the development of ASICs for mobile and wireless applications. The common feature of these filters, is that they are *all* based on the operation of the elementary bidirectional SAW interdigital transducer, as devised by White and Voltmer in 1965 [1]!

This text examines the principles of operation and applications of some of the already listed RF and IF structures in mobile and wireless communications circuitry. As a necessary prelude to understanding their operating principles, however, it is necessary to examine some basic facets governing the operation of SAW IDTs and SAW filters, which involve the use of such bidirectional IDTs.

In addition, the text will consider aspects of SAW-based spread-spectrum techniques and systems, including those applied to short fixed codes, as well as SAW convolvers employing longer code sequences for both synchronous and asynchronous operation.

As an aid to the reader, Chapters 1 though 9 in Part 1 of this text deal with basic aspects relating to SAW and pseudo-SAW piezoelectrics, wave-propagation mechanisms, and devices. Also included in Part 1 is an examination of second-order effects that can corrupt a desired filter response.

In Part 2, Chapter 10 first reviews some current world-wide frequency- and bandwidth-allocations for analog and digital cordless and mobile phones, as well as for WAN and WLAN communications. Chapter 11 deals with SAW resonator structures and applications, including those for low-power unlicensed wireless. Chapter 12 covers some design techniques for implementation of low-loss filters using unidirectional transducers. Chapters 13 and 14 examine some current design approaches for implementation of LSAW RF front-end filters and antenna duplexers. Illustrative designs for frequently-used IF filters in mobile receivers and pagers are reviewed in Chapter 15.

Chapter 16 illustrates SAW-based techniques for spread-spectrum communications using fixed short-code structures. Chapter 17 expands on this to outline some SAW convolver designs that have been applied to packet-voice and packet-data spread-spectrum communications for both indoor and outdoor, communications.

Chapter 18 is concerned with a variety of surface-wave oscillator design techniques, including those applied to fixed-frequency oscillators, voltage-contralled oscillators (VCO), injection-locked oscillators for carrier recovery at gigahertz frequencies in binary phase-shift keying (BPSK)-modulated data systems, hopping oscillators for frequency-agile hybrid spread-spectrum designs—as well as a high-performance SAW-based frequency synthesizer technique for spread-spectrum, using SAW chirp-filters.

In separate sections, Chapter 19 first outlines Nyquist-filter concepts and then examines some illustrative SAW designs for Nyquist filters and slanted-IDT filters in digital microwave radio. This is followed by designs for clock recovery in SONET compatible fiber-optic links, including layered structures on diamond substrates. Chapter 19 concludes with highlights of on-board channelizing filters for upcoming low earth orbit (LEO) satellites, as well as a wideband (BW = 50%) SAW filter for a digital data terminal in a mobile earth station for INMARSAT-C satellite system. The postscript in Chapter 20 gives some of the author's views on trends in the application of SAW devices in mobile and wireless communications.

A Glossary is included to assist the reader with the deciphering of the many abbreviations used by the telecommunications industry, as well as those related to surface acoustic wave technology.

In preparing the coverage of both Parts 1 and 2, the author has attempted to emphasize the operation of the SAW/pseudo-SAW devices from a circuit viewpoint, rather than from a strictly physical one. In this way it is intended as an aid to circuit- and system-designers and engineers who need a working knowledge of such devices in the increasingly complex and sophisticated field of mobile, wireless, and personal communications. It is hoped that it can also serve as a reference and guide for courses on this subject, as well as for students and engineers who are contemplating entering this exciting field, where the demand for SAW-based systems is ever-increasing. Where appropriate, (and possible), systems- and device-oriented worked examples are given as an aid to familiarizing the reader with practical SAW design parameters and constraints.

1.7. REFERENCES

1. R. M. White and F. W. Voltmer, "Direct piezoelectric coupling to surface elastic waves," *Appl. Phys. Lett.*, vol. 17, pp. 314–316, 1965.
2. E. Dieulesaint and P. Hartmann, "Surface wave electro-acoustic wave filter," French patent no. 69 11765 16.04.69; USA patent no. 3 3633 132 4.01.72.
3. R. C. Williamson, "Case studies of successful surface-acoustic-wave devices," *Proc. 1977 IEEE Ultrasonics Symp.*, pp. 460–468, 1977.
4. C. S. Hartmann, "System impact of modern Rayleigh wave technology," in E. A. Ash and E. G. S. Paige (eds), *Rayleigh-Wave Theory and Applications*, Springer-Verlag, New York, pp. 238–253, 1985.
5. M. Lewis, "Surface skimming bulk waves, SSBW," *Proc. 1977 IEEE Ultrasonics Symposium*, pp. 744–752, 1977.
6. C. K. Campbell, "Applications of surface acoustic and shallow bulk acoustic wave devices," *Proc. IEEE*, vol. 77, pp. 1453–1484, Oct. 1989. (With 322 references).
7. W. R. Smith, "Basics of the SAW interdigital transducer," in J. H. Collins and L. Masotti (eds), *Computer-Aided Design of Surface-Acoustic Wave Devices*, Elsevier, New York, pp. 25–63, 1976.
8. K. Asai, I. Isobe, T. Tada and M. Hikita, "SAW timing-extraction filter and investigation of submicron process technology for Gb/s optical communications system," *Proc. 1995 IEEE Ultrasonics Symposium*, vol. 1, pp. 131–135, 1995.
9. Y. Shimizu and M. Tanaka, "A new cut of quartz for SAW devices with extremely small temperature coefficient by leaky surface waves," *Electronics and Communications in Japan*, Part 2, vol. 69, pp. 48–56, 1986.
10. See, for example, H. Yatsuda, T. Horishima, T. Eimura and T. Ooiwa, "Miniaturized SAW filters using a flip-chip technique," *Proc. 1994 IEEE Ultrasonics Symp.*, vol. 1, pp. 159–162, 1994.
11. See, for example, *Wireless Communications Products*, Fujitsu Limited, Tokyo, Data Book, Rev. 1.0, 1996.
12. Y. Shimizu, "Current status of piezoelectric substrate and propagation characteristics for SAW devices," *Japan J. Appl. Physics*, vol. 32, Part 1, No. 5B, pp. 2183–2187, May 1993.
13. C. S. Lam, D. S. Stevens and D. J. Lane, "BAW- & SAW-based frequency control products for modern telecommunications and their applications in existing and emerging wireless communications," *International Meeting on the Future trends of Mobile Communications Devices*, 22–23 Jan. 1996, Tokyo, Japan.
14. J. Saw, M. Suthers, J. Dai, Y. Xu, R. Leroux, J. Nisbet, G. Rabjohn and Z. Chen, SAW technology in RF multichip modules for cellular systems, *Proc. 1995 IEEE Ultrasonics Symp.*, vol. 1, pp. 171–175, 1995.
15. R. C. Smythe, "Piezoelectric and Electromechanical Filters," in E. A. Gerber and A. Ballato (eds), *Precision Frequency Control*, vol. 1, *Acoustic Resonators and Filters*, Academic Press, New York, pp. 185–228, 1985.
16. C. S. Lam, D-P Chen, B. Potter, V. Narayanan and A. Vishwanathan, "A review of the applications of SAW filters in wireless communications," *International Workshop on Ultrasonics Application*, Nanjing, China, Sept. 1996.

Basics of Piezoelectricity and Acoustic Waves

2.1. Introduction

This chapter highlights the basic properties of surface acoustic waves (SAW) and their generation (or detection) by an interdigital transducer (IDT) located on the free surface of a piezoelectric substrate. This coverage includes both the "true" SAW (or Rayleigh wave) and pseudo-SAW wave propagation, such as for leaky-SAW (LSAW), surface skimming bulk waves (SSBW), and surface transverse waves (STW). Further into this text an expanded discussion of RF filters on LSAW substrate cuts will be included; these filters now play a major role as low-loss high-performance components for front-end filtering and duplexing circuitry in mobile/wireless communications transceivers. Additionally, SSBW and STW oscillators are finding increased application due to their enhanced power-handling capabilities over Rayleigh-wave oscillators.

A detailed mathematical treatment of acoustic wave propagation in piezoelectrics is omitted in this coverage, as this may already be found in a number of texts [1]–[8]. Instead, emphasis is given to conveying an understanding of principles to readers interested in or involved with communications circuits and systems. However, an introductory mathematical outline of stress and strain relations in piezoelectric solids is given in order to relate the material aspects of SAW filter design.

This chapter will also examine the effects on surface-wave filter response by spurious bulk acoustic wave generation within the piezoelectric substrate. It is well known that the frequency response of Rayleigh-wave filters can be degraded by the generation of bulk acoustic waves by an excited IDT. Bulk waves that arrive at the output IDT will induce voltages that are in addition to those generated by the arriving Rayleigh wave. The resultant voltage due to both sources can degrade the in-band performance specifications on amplitude, phase or group-delay response and out-of-band rejection can be undesirably reduced. While little has been published on the effects of bulk-wave interference in leaky-SAW filters, it may be postulated that these degradations will be similar[1].

Such interference represents but one of a number of second-order effects that can occur in these devices. Other second-order effects will be examined in later chapters.

2.2. Surface Acoustic Waves

2.2.1. EXCITATION REQUIREMENTS

Surface acoustic waves (both Rayleigh and pseudo-SAW) can be generated at the free surface of an elastic solid. In the SAW devices considered in this text, the generation of such waves is achieved by application of a voltage to a metal-film IDT deposited on the surface of a piezoelectric solid. Two IDTs are required in the basic SAW or pseudo-SAW device configuration sketched in Fig. 2.1. One of these acts as the device input and converts signal voltage variations into mechanical acoustic waves. The other IDT is employed as an output receiver to convert mechanical SAW vibrations back into output voltages. Such energy conversions require the IDTs to be used in conjunction with elastic surfaces that are also piezoelectric ones. Note, however, that the surface outside the IDT regions need only be elastic, without being piezoelectric.

In that reciprocity applies to both systems, input and output IDTs may be likened to electromagnetic transmitting and receiving antennas. In principle, therefore, signal voltages can be applied to either IDT to give the same end result. (In practice, however, this will depend on the source and load impedances.) The aim of all this, of course, is to create some advantageous signal-processing function through the interaction of acoustic waves, rather than through electromagnetic ones, while enjoying the compact device dimensions attainable with such processing.

Detailed mathematical treatments of SAW propagation on the surface of an unbounded piezoelectric elastic surface are to be found in a number of texts [1]–[6]. Only the essentials of these will be considered here in order to highlight two of the most important practical properties relating to surface wave propagation on a piezoelectric substrate. These are wave velocity v and *electromechanical coupling coefficient* K^2 of the piezoelectric.

---

[1] Bulk-wave interference is not considered here for SSBW or STW structures, as these are inherently bulk-wave devices.

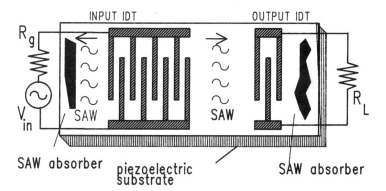

FIG. 2.1. Basic unapodized SAW/pseudo-SAW delay line on piezoelectric substrate, with metal thin-film input/output interdigital transducers. Note the absorbers that are sometimes used to absorb spurious SAW transmissions resulting from IDT bidirectionality. An alternative is to use nonrectangular substrates.

2.2.2. MECHANICAL MOTION OF SURFACE ACOUSTIC WAVES

Let us consider the propagation of a Rayleigh wave on an unbounded elastic surface. As pictured in the artistic representation (not to scale) of Fig. 2.2, the physical motion of this "true-SAW" wave type is associated mechanically with a time-dependent elliptical displacement of the surface structure. One component of this physical displacement is parallel to the SAW propagation axis, while the other is normal to the surface. Distance x relates to the SAW propagation axis, while y is a surface normal axis in a Cartesian coordinate system. (Cartesian coordinate notation x, y, z used here should not be confused with piezoelectric crystal axis notation X, Y, Z.) Surface particle motion is predominantly in the y-x-plane in Fig. 2.2. (This is only strictly true for certain crystal cuts, such as Y-cut Z-propagating lithium niobate.) The two wave motions are 90° out of phase with one another in the time domain, so that when one displacement component is maximum at a given instant the other will be zero. This has ramifications in SAW resonator design and will be considered later in the text. In addition, the amplitude of the surface displacement along the y-axis is larger than along the SAW propagation axis x. This can be appreciated intuitively, as it is "easier" for the crystal structure to vibrate in the unbounded direction than in the bounded one. The amplitudes of both of these SAW displacement components become negligible for penetration depths greater than a few acoustic wavelengths $\lambda\,(= v/f)$ into the body of the solid. (This phenomenon may be considered to be somewhat analogous to that of *skin depth*, relating electromagnetic wave penetration into a conductor.)

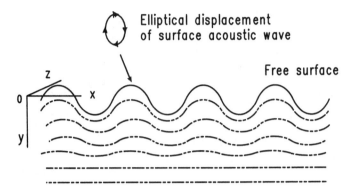

Interior of piezoelectric substrate

Fig. 2.2. Artistic representation (not to scale) of Rayleigh-wave motion on the surface of a piezoelectric substrate. While the illustration refers to a piezoelectric substrate, this is not a requirement for propagation but only for the excitation region.

Example 2.1 SAW and Electromagnetic Velocities and Wavelengths. A surface acoustic wave is generated on the surface of a piezoelectric *YZ*-lithium niobate substrate by means of an ac voltage applied to an IDT at a synchronous frequency of 1 GHz. (a) Given that the velocity of propagation of the SAW on this material is $v = 3488$ m/s, determine the acoustic wavelength λ. (b) Compare the value of this wavelength with that of an electromagnetic wave propagating in free space at the same signal frequency. (c) Determine the ratio between the SAW wavelength and the electromagnetic wavelength in this case. ■

Solution. (a) The SAW wavelength λ is given by $\lambda = v/f = 3488/(1 \times 10^9) = 3.488 \times 10^{-6}$ m $= 3.488\,\mu$m, where $1\,\mu$m $= 1$ micron $= 10^{-6}$ m. (b) The electromagnetic wavelength λ_e in this case is $\lambda_e = c/f$, where $c = 3 \times 10^8$ m/s is the velocity of light. Thus $\lambda_e = (3 \times 10^8)/(1 \times 10^9) = 0.3$ m. (c) The ratio of wavelengths is $\lambda/\lambda_e = (3.488 \times 10^{-6})/0.3 = 1.1 \times 10^{-5}$. ■

2.2.3. STRESS AND STRAIN IN A NONPIEZOELECTRIC ELASTIC SOLID[2]

First of all consider the relations between mechanical stress T and strain S for small static deformations of a nonpiezoelectric elastic solid. Stress is just the force F applied per unit area A of the solid. Moreover, stress, force

---

[2] An excellent coverage of tensor and matrix representations in crystallography is given in Reference [1], this chapter.

and area can all be represented as vector quantities (using bold face letter symbols) so that $T = F/A$. The units of T are N/m^2 when force F is expressed in newtons (N). In addition, the strain parameter S, which represents the fractional deformation due to force F, can be defined as $S = \Delta/L$ (dimensionless), where Δ is the fractional deformation of the solid of length L.

Stresses and strains exerted within an elastic solid can exist in compressional or shear form. With compressional stresses, for example, the applied force F is normal to the surface area A in Fig. 2.3. In either case they can be related proportionally by Hooke's Law for elastic deformation. For simple compressional stress and strain along the same axis, this can be written as

$$T = cS, \tag{2.1}$$

where c = elastic stiffness coefficient, also known as Young's modulus (N/m^2). In general, however, Eq. (2.1) must be formulated to accommodate all possible components of stress and deformation so that

$$(T) = (c):(S) \tag{2.2}$$

expressed as a *tensor equation*—with tensor terms identified by symbols (). Thus, for example, an expansion of tensor Eq. (2.2) for values of T_{xx} along the x-axis gives terms

$$T_{xx} = \begin{array}{l} c_{xxxx}S_{xx} + c_{xxxy}S_{xy} + c_{xxxz}S_{xz} \\ + c_{xxyx}S_{yx} + c_{xxyy}S_{yy} + c_{xxyz}S_{yz} \\ + c_{xxzx}S_{zx} + c_{xxzy}S_{zy} + c_{xxzz}S_{zz} \, . \end{array} \tag{2.3}$$

Because force and area vectors need not be aligned, the stress parameter T can be written as $T_{jk} = F_j/A_k$, where the first subscript j denotes the

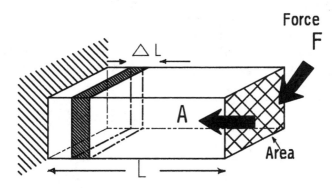

FIG. 2.3. Notation used here for parameters relating to the deformation of an elastic solid. Note that F and A parameters can be vector quantities.

direction of the force F while the second subscript k is the direction of the vector representing area A in Fig. 2.3. Likewise, tensor strain terms can be given as $S_{lm} = \Delta_l/L_m$ for deformation directions defined by the second two suffixes. Here, (c) is referred to as a *fourth-rank tensor*, as its components have four suffixes c_{jklm}. Similarly, (T) and (S) are classed as *second-rank tensors*. Where $j = k = l = m$ is specified along one axis (say, the x-axis, for example), $T_{xx} = c_{xxxx}S_{xx}$ relates compressional stress and strain along that axis.

It is not necessary to dwell long on the concept of tensors except to identify with the crystal classifications of various SAW piezoelectrics. Tensors are used to relate parameters that are dependent on more than one coordinate axis set (e.g., T_{xy}) to those measured in another (e.g., S_{yz}). Tensor equation (2.2) can be reduced to a *matrix* equation $[T] = [c][S]$ (denoted by symbols [ ]), by redimensioning $[T]$ and $[S]$ so that they each have only one suffix instead of two as in Eq. (2.3). To this end, tensor components of T and S can be reduced to matrix components given by

$$T_1 = T_{11}, \quad T_2 = T_{22}, \quad T_3 = T_{33}$$
$$T_4 = T_{32} = T_{23}, \quad T_5 = T_{31} = T_{13}, \quad T_6 = T_{12} = T_{21} \tag{2.4}$$

$$S_1 = S_{11}, \quad S_2 = S_{22}, \quad S_3 = S_{33}$$
$$S_4 = 2S_{32} = 2S_{23}, \quad S_5 = 2S_{31} = 2S_{13}, \quad S_6 = 2S_{12} = 2S_{21}. \tag{2.5}$$

In this way, the elastic stiffness constant is reduced to a 6×6 matrix $[c]$, with 36 possible independent values relating the six (reduced) components of stress to the six (reduced) components of strain. Moreover, from energy and symmetry considerations, these 36 possible independent terms can be reduced to a maximum of 21 for the most general crystal symmetry examples. A further reduction is made possible by an appropriate choice of reference coordinate axes in relation to crystal axes. For example, cubic crystals with coordinate reference axes x, y, z chosen parallel to crystal axes X, Y and Z have their number of independent elastic constant coefficient terms in $[c]$ reduced from 31 to just 3, as shown in Eq. (2.6).

$$CUBIC \begin{pmatrix} T_1 \\ T_2 \\ T_3 \\ T_4 \\ T_5 \\ T_6 \end{pmatrix} = \begin{pmatrix} c_{11} & c_{12} & c_{12} & 0 & 0 & 0 \\ c_{12} & c_{11} & c_{12} & 0 & 0 & 0 \\ c_{12} & c_{12} & c_{11} & 0 & 0 & 0 \\ 0 & 0 & 0 & c_{44} & 0 & 0 \\ 0 & 0 & 0 & 0 & c_{44} & 0 \\ 0 & 0 & 0 & 0 & 0 & c_{44} \end{pmatrix} \begin{pmatrix} S_1 \\ S_2 \\ S_3 \\ S_4 \\ S_5 \\ S_6 \end{pmatrix}. \tag{2.6}$$

Silicon (Si), which is not piezoelectric, falls into this class of cubically symmetric crystals, as does bismuth germanium oxide ($Bi_{12}GeO_{20}$).

On the other hand, the number of independent coefficients in the elastic constant matrix [c] reduces to five when the z-reference coordinate is chosen along the Z-axis of a hexagonal crystal. Piezoelectric zinc oxide (ZnO) falls into this hexagonal crystal class. It is used in thin-film or layered SAW circuits, sometimes in conjunction with silicon technology. Substrate materials such as lithium niobate and quartz are in the trigonal class of crystal structures, with a greater number of independent coefficients in the elastic coefficient matrix [c].

The elastic constant matrix for an isotropic or polycrystalline solid is the same as for a cubic crystal, except that any choice of reference axes can be employed. For isotropic crystals, the number of independent elastic coefficients reduces from three to two. The piezoelectric ceramic material lead-zirconium-titanate (PZT) is crystalline and, therefore, is isotropic elastically. Lithium niobate and quartz piezoelectrics come under the trigonal crystal classification with six independent elastic constants.

2.2.4. PIEZOELECTRIC INTERACTIONS

The stress-strain relations considered in the preceding have been tacitly applied to a nonpiezoelectric dielectric elastic solid. Application of an electric field to such a solid would have no effect on its mechanical stress-strain characteristics. To review the effect of applying an electric field of intensity E (V/m) to a simple nonpiezoelectric dielectric, consider the electrical relationships for the simple plane-parallel capacitor model of Fig. 2.4 containing a solid insulator. Here, the electric field E established by applied voltage V will cause a distortion of the otherwise neutral molecular charge distributions in the insulator; in turn, this will result in an accumulation of surface charge on the capacitor plates. The surface charge of density D (C/m^2) will be proportionally related to E by

$$D = \varepsilon_r \varepsilon_o E = \varepsilon E, \tag{2.7}$$

where ε_r = relative dielectric permittivity, or dielectric constant (dimensionless), and ε_o = permittivity of free space = 8.856×10^{-12} F/m.

The simple relation of Eq. (2.7) no longer holds for piezoelectric dielectrics. Because of coupling between electrical and mechanical parameters, the application of an electric field stimulus will give rise to mechanical deformations and vice versa. Mathematically, this interaction can be expressed in terms of a piezoelectric constant matrix [e] (with units of C/m^2) such that the electrical displacement density D is given by a matrix equation

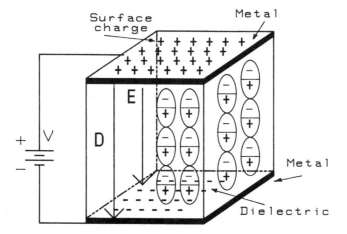

Fig. 2.4. Displacement density D and electric-field intensity E vectors in a simple plane-parallel capacitor containing a solid nonpiezoelectric dielectric.

$$D = [e][S] + [\varepsilon]E, \qquad (2.8)$$

where S = strain and E = electric field intensity as before. Here, permittivity ε is as measured at zero or constant strain. Equation (2.8) is a matrix equation employing the reduced coordinates for strain S in Eq. (2.5). The displacement density term is a three-dimensional one in x, y and z coordinates. Because the S term has six components from Eq. (2.5), the piezoelectric constant terms in $[e]$ will form a 3×6 matrix with 18 elements. The parameter $[\varepsilon]$ relating dielectric permittivity is a 3×3 matrix with 9 elements.

In addition, for piezoelectric materials the mechanical stress relationships are extended to

$$[T] = [c][S] - [e^t]E, \qquad (2.9)$$

where $[e^t]$ is now a 3×6 matrix and is the transpose of the piezoelectric constant $[c]$ (i.e., matrix rows and columns are interchanged.) Equations (2.8) and (2.9) are often referred to as constitutive equations.

The element values of $[e]$ will depend on the symmetry properties of the piezoelectric crystal. For lithium niobate and lithium tantalate piezoelectrics with trigonal crystal classification, these are

lithium niobate
and
lithium tantalate
(trigonal class)

$$[e] = \begin{pmatrix} 0 & 0 & 0 & 0 & e_{15} & -e_{22} \\ -e_{22} & e_{22} & 0 & e_{15} & 0 & 0 \\ e_{31} & e_{31} & e_{33} & 0 & 0 & 0 \end{pmatrix}. \quad (2.10)$$

Also in the trigonal class is quartz, with coefficients of $[e]$ given by

quartz
(trigonal class)

$$[e] = \begin{pmatrix} e_{11} & -e_{11} & 0 & e_{14} & 0 & 0 \\ 0 & 0 & 0 & 0 & -e_{14} & -e_{11} \\ 0 & 0 & 0 & 0 & 0 & 0 \end{pmatrix}. \quad (2.11)$$

Gallium arsenide, which is both a cubic-compound semiconductor and a piezoelectric, has piezoelectric constant coefficients

gallium arsenide
(cubic class)

$$[e] = \begin{pmatrix} 0 & 0 & 0 & e_{14} & 0 & 0 \\ 0 & 0 & 0 & 0 & e_{14} & 0 \\ 0 & 0 & 0 & 0 & 0 & e_{14} \end{pmatrix}. \quad (2.12)$$

For SAW wave propagation in piezoelectrics, it may be shown that the electromechanical coupling coefficient K^2 can be defined in terms of the piezoelectric coefficient e, elastic constant c and dielectric permittivity ε already considered herein, where

$$K^2 = \frac{e^2}{c\varepsilon} \quad (2.13)$$

and tensor subscripts have been dropped in Eq. (2.13). Appropriate constants depend on both the crystal cut and the propagation direction of the surface acoustic wave. Additionally, the parameter K^2 in Eq. (2.13) can be derived experimentally, as we shall later consider (see Example 2.3).

2.2.5. RAYLEIGH WAVE CONSIDERATIONS

Let us consider aspects of "true SAW" (i.e., Rayleigh-wave) propagation. Ordinarily, a description of the mechanism of elastic wave propagation in a piezoelectric medium would require an evaluation of both the mechanical equations of motion as well as the electromagnetic ones governed by Maxwell's equations. As the mechanical propagation of Rayleigh waves is at velocities in the order of 10^5 less than the velocity of light, however, it may be determined that the mechanical motion will dominate such wave transmission processes. The wave solutions for the mechanical wave propagation of the surface are indeed complex. It transpires, however, that the electrical potential Φ they induce at the surface of the piezoelectric may be modelled to a high degree of accuracy as a travelling wave of potential Φ (in volts) and displacement U (in metres) such that

$$\Phi = \Phi(x,t) \approx |\Phi| e^{j(\omega t - \beta x)} e^{-\beta|y|} \tag{2.14a}$$

$$U = U(x,t) \approx |U| e^{j(\omega t - \beta x)} e^{-\beta|y|}, \tag{2.14b}$$

with potential as sketched in Fig. 2.5. Here, $\omega = 2\pi f =$ angular frequency (rad/s) of the applied signal, while $\beta =$ phase constant (rad/m) such that $\beta\lambda = 2\pi$ and $\lambda = v/f$ is the acoustic wavelength at SAW velocity v. (The notation β, rather than k, is used here as perhaps being more familiar to electrical engineers than the interchangeable term k known as wave number or wave vector.) The first term $e^{j(\omega t - \beta x)}$ in eq. (2.14) relates the travelling wave distribution of potential (with magnitude $|\Phi|$) along the SAW propagation axis at the surface of the piezoelectric ($y = 0$). This potential is not just confined to the piezoelectric surface, or to the region of the travelling acoustic wave. It also extends above it by a distance on the order of one acoustic wavelength [6]. Because of the complex nature of the potential and field distribution along the y-axis, this variation contained in the term $e^{-\beta|y|}$ of eq. (2.14) is at best a very approximate one. The relative amplitude of potential Φ away from the free surface is reduced by a factor of $1/e$ when

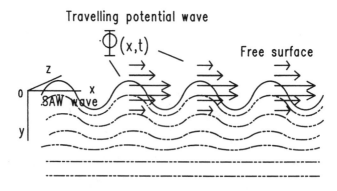

FIG. 2.5.   Simplistic artistic representation of Rayleigh wave on surface of piezoelectric substrate, accompanied by travelling wave of potential. Distribution is in x-y plane at the surface.

$\beta|y| = 1$, where distance y above the free surface is negative in the coordinate notation used.

The spatial variation of Φ at the piezoelectric surface will produce electric fields of intensity E (V/m). Along the SAW propagation axis, the longitudinal component E_x of the electric field will be given by $E_x = -\partial\Phi/\partial x$ for the coordinate axis notation of Fig. 2.5. There will also be a complex variation of the electric field intensity $E_y = \partial\Phi/\partial y$ extending above and below the free surface as depicted in Fig. 2.6.

It is often convenient to relate electric circuit processes in terms of an equivalent circuit commonly modelled in terms of either lumped or distributed inductance-capacitance (L-C) equivalent circuits. In such modelling, it is tacitly assumed that the propagating electromagnetic wave is in a purely transverse electromagnetic (TEM) mode.

As an aid to modelling some SAW transmission and reflection processes, it is also convenient to use equivalent transmission line circuit models and concepts. As with the electrical transmission line considered in the preceding, the accuracy of each model will be dictated by the assumed boundary and transmission conditions. Thus, for example, if the piezoelectric surface stresses associated with surface wave transmission are assumed to be predominantly compressional, the electric field associated with the potential Φ can be modelled as $E = E_x$ (and $E_y = 0$) at the surface. If, in this situation, it is further assumed that E_x is uniform within a defined penetration depth at the surface, then the SAW propagation can be modelled in terms of

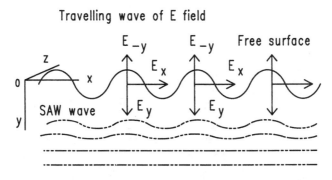

FIG. 2.6. A travelling wave of electric field accompanies the travelling wave of potential. The E-field can extend above the free surface of the piezoelectric by about one acoustic wavelength. Representation shown here is a simplistic artistic rendering of the complex field distributions.

a uniform travelling wave E_x with the SAW acoustic velocity v on an equivalent transmission line as sketched in Fig. 2.7, where

$$E = E_x \approx |E| e^{j(\omega t - \beta x)}. \qquad (2.15)$$

This particular simplification is used in the development of the *in-line Mason equivalent circuit* [7] for relating SAW propagation in terms of equivalent electrical parameters. Familiar electrical transmission line parameters such as characteristic impedance, given by $Z_o = \sqrt{L/C}$ in terms of distributed inductance L and capacitance C, can be reformulated in terms of piezoelectric parameters for an equivalent SAW transmission line model, within the constraints imposed by the assumed boundary conditions.

2.2.6. ELECTROMECHANICAL COUPLING COEFFICIENT K^2

The electromechanical coupling coefficient K^2 is a measure of the efficiency of a given piezoelectric in converting an applied electrical signal into mechanical energy associated with a surface acoustic wave. Here K^2 and SAW velocity v represent the two most important practical material parameters used in SAW filter design. As K^2 values are small, they are usually expressed as percentages, as shown in Table 2.1 for typical Rayleigh-wave piezoelectric substrates. Table 2.1 also lists values for IDT capacitance C_o (per finger pair per cm-length). Values shown for α_T, the temperature coefficient of delay (TCD), are first-order ones. While a variety of crystal cuts exist for SAW propagation on gallium arsenide (GaAs), the cut listed in Table 2.1 provides maximum K^2 and compatibility with other GaAs-

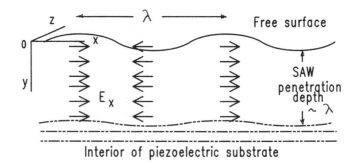

FIG. 2.7. This transmission line modelling of Rayleigh-wave propagation on a piezoelectric surface assumes that the travelling E-field is purely longitudinal in the direction of wave propagation, with a depth of about one acoustic wavelength below the free surface.

TABLE 2.1

DESIGN PARAMETERS OF COMMON PIEZOELECTRIC SUBSTRATES FOR RAYLEIGH-WAVE DEVICES

Material	Crystal Cut	SAW Axis	Velocity (m/s)	K^2 (%)	Temp Coeff (ppm/ °C)	Capacitance/ finger pair/ unit length C_o (pF/cm)	Applications
Quartz	ST	X	3158	0.11	~0 near 25°C	0.55	Precision oscillators, temperature-stable narrowband midloss IF filters, lowloss RF resonators
Lithium niobate $LiNbO_3$	Y	Z	3488	4.5	94	4.6	Wideband midloss IF filters
$LiNbO_3$	128°	X	3992	5.3	75	5.0	Wideband midloss IF filters
Bismuth germanium oxide $Bi_{12}GeO_{20}$	110	001	1681	1.4	120	—	Long delay lines
Lithium tantalate $LiTaO_3$	77.1° Rotated Y	Z'	3254	0.72	35	4.4	Oscillators Minimum diffraction cut
Gallium arsenide GaAs	(100)	⟨110⟩	<2841	<0.06	35	—	Semiconductor IC compatible

based integrated circuit technology. As well, ST-quartz (with a parabolic temperature dependence around an inversion point of 25°C and a frequency shift $\Delta f = 100$ ppm over -30°C to $+75$°C) has long been the standard for applications requiring minimum temperature sensitivity. Attention has been given recently to piezoelectric Langasite ($La_3Ga_5SiO_{14}$). This also has a low TCD, but enjoys a much higher coupling coefficient $K^2 = 0.3\%$ (compared with 0.11% for ST-quartz), with a slightly smaller SAW velocity $v = 2400$ m/s (compared with 3158 m/s for ST-quartz). The higher K^2 of Langasite enables greater filter fractional bandwidths to be achieved than with quartz.

In Eq. (2.13), K^2 was defined in terms of piezoelectric parameters as $K^2 = e^2/\varepsilon c$. This parameter may also be obtained experimentally, however, with recourse to the relation

$$K^2 = -\frac{2\Delta v}{v}, \tag{2.16}$$

where $|\Delta v|$ is the magnitude of the SAW velocity change that occurs when the free surface of the piezoelectric is shorted by a thin highly conducting metal film, and v is the unperturbed SAW velocity. The time-varying electric fields associated with SAW propagation will cause the metal surface film to accumulate charge as pictured in Fig. 2.8. If the SAW propagation is modelled as an equivalent electrical transmission line with distributed L-C, the additional charge accumulation will serve to increase C. Because electromagnetic wave velocity $v_e = 1/\sqrt{LC}$ on an electromagnetic transmission line, an increase in C will decrease v_e. From the SAW transmission line model, the SAW velocity v will likewise decrease when the highly conducting metal film is applied to the piezoelectric surface.

Example 2.2 Transmission Line Modelling and SAW Velocity Shift. Use transmission line modelling to show that $\Delta v/v = -0.5(dC/C)$ relates velocity change for SAW propagation on a piezoelectric, when a thin metal film is deposited on its surface. Here, dC/C is taken to represent the associate capacitance change in the equivalent SAW transmission line model. ■

Solution. Wave velocity v on a lossless distributed transmission line is given as $v = 1/\sqrt{LC}$. Differentiating this with respect to C gives $\Delta v/dC = -0.5L(LC)^{-3/2}$. Substitution of v yields $\Delta v/v = -0.5(dC/C)$. ■

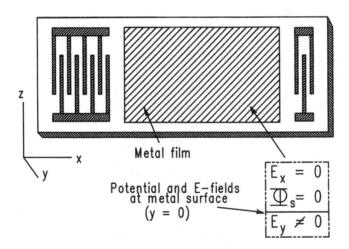

Fig. 2.8. The electromechanical coupling coefficient K^2 can be measured by observing the SAW propagation time between input/output IDTs, before and after a highly conducting metal film is deposited on the surface of the piezoelectric.

Example 2.3 Electromechanical Coupling Coefficient K^2. Electromechanical coupling coefficient K^2 is given in two different forms in Eqs. (2.13)
and (2.16). Use SAW transmission line modelling to show that these are
equivalent. ■

Solution. (a) First of all, when a thin perfectly conducting metal film is
deposited on top of the piezoelectric surface, the accumulation of charge
will be such that the *total* potential Φ_s and longitudinal electric field E_x are
zero. (A finite E_x would cause infinite current flow in the metal film!) Field
component E_y can still exist, however, so that field distribution can be
modelled as for a parallel-plate capacitor. (b) Next, assume that the metal
film does not degrade the SAW motion and relationships. From Eq. (2.8)
the increment in displacement density due to the charge on the metal is
$dD = eS$; however dD can be written as $dD = (d\varepsilon)E$ in terms of effective
permittivity change. Because the net potential Φ_s (due to SAW plus induced
charge) and E_x are zero at the metal surface, the force term T in Eq. (2.9)
must also be zero. (c) Now substitute for strain S from Eq. (2.9) and drop
superscripts and subscripts for convenience to get $(d\varepsilon)E = eS = e(eE/c) = e^2$
E/c. Therefore $(d\varepsilon)/\varepsilon = e^2/c\varepsilon$, but for the implied plane-parallel capacitor
$(d\varepsilon)/\varepsilon = (dC)/C$. From the result for Example 2.2, obtain $K^2 = e^2/c\varepsilon = (dC)/$
$C = -2(\Delta v)/v$ as required. ■

2.2.7. SAW Piezoelectric Crystal Substrates

Piezoelectric substrates typically employed in Rayleigh-wave filter design
are listed in Table 2.1. With reference to this table, YZ-lithium niobate
(LiNbO$_3$) substrates (i.e. Y-axis crystal cut, Z-axis propagation) with relatively high values of K^2 generally find application in wideband (BW ≈ 5 to
50%) SAW filters as well as in radar pulse compression filters with very
large time-bandwidth (TB) products (BW $\leq \approx 100\%$). As indicated, lithium
niobate crystals have a nonzero temperature coefficient of delay. The
propagation time τ between input and output IDTs separated by distance L
will be given by $\tau = L/v$. However, the SAW velocity v will be dependent on
the elasticity, density, and piezoelectric properties of the substrate used.
These parameters can all change with temperature. This will normally not
be critical in nonrecursive SAW filter applications at ambient temperatures.
It will be critical, however, if the filter is used in the feedback loop of an
oscillator. Temperature variations will result in phase shifts around the
loop, thereby reducing the stability of the oscillator. The 128°-rotated X-
propagating lithium niobate is a particular cut designed for reduced bulk
wave generation over its Y-Z counterpart.

ST-X quartz (i.e., stable temperature cut) has a value of K^2, which is
about forty times less than for lithium niobate. It finds application in

narrowband $(BW \leq \approx 5\%)$ filters and delay lines. Moreover, it is widely employed in SAW oscillator designs, because of its zero temperature coefficient of delay about room temperature. The stability of SAW oscillators using ST-X quartz may often be sufficiently high for use without oven stabilization in other than highly critical applications. Lithium tantalate $(LiTaO_3)$ with a higher K^2 than ST-X quartz, (but poorer temperature stability) has also found some use in oscillator designs.

As noted in the previous section, a relative newcomer to the SAW piezoelectric crystal field is that of Langasite $(La_3Ga_5SiO_{14})$. While it also has a low TCD suitable for temperature-stable device applications, it has a much higher electromechanical coupling coefficient $(K^2 = 0.3\%)$ than quartz, with a slightly smaller SAW velocity $(v = 2400 \text{m/s})$. Its advantage over quartz is that its larger K^2 value enables wider filter fractional bandwidths to be realized.

Of the remaining representative SAW substrates listed in Table 2.1, bismuth germanium oxide $(Bi_{12}GeO_{20})$ has found use in long delay line applications in view of compact structure size attainable with its comparatively low SAW velocity. The piezoelectric coupling of gallium arsenide (GaAs) is slightly less than that of quartz, while its attenuation is higher. As it is also a class III–V semiconductor of high repute in high-frequency electronic devices, much effort has been expended to develop SAW devices with this material for semiconductor integrated circuits [9].

In general, SAW piezoelectric substrates are anisotropic. That is, their SAW propagation characteristics are not constant in all directions. Unless avoided, this can lead to an undesirable second-order effect known as *beam steering*, which can degrade the response of a SAW bandpass filter. Avoidance of beam steering requires the use of piezoelectric substrates with crystal cuts such that the SAW velocity is either a maximum or a minimum along that particular cut. Improper alignment of an IDT with the required crystal cut can also result in beam steering, to a degree dependent on the anisotropic characteristics of the substrate material used.

2.2.8. TEMPERATURE COEFFICIENTS OF DELAY (TCD)

The effect of temperature on the operation of a SAW device is usually characterized in terms of a temperature coefficient of delay (TCD) given by α_T, where

$$\alpha_T = \frac{1}{\tau}\frac{d\tau}{T} = \frac{1}{L}\frac{dL}{dT} - \frac{1}{v_o}\frac{dv_0}{dT}, \qquad (2.17)$$

$\tau = L/v_o$ is the time delay between input and output IDTs separated by distance L, v_o is the SAW velocity, while T is the substrate measurement temperature. Coefficient α_T (with units of ppm/°C) is, of course, independent of the transducer separation L. The first term in Eq. (2.17) gives the change in α_T due to thermal expansion or contraction of the piezoelectric substrate, while the second term relates velocity shift as a function of temperature. Values of TCD for representative piezoelectric substrates are give in Table 2.1. With the exception of substrates such as ST-quartz, the slope of the TCD as a function of temperature for SAW, LSAW, and SSBW substrates is mainly negative, so that an increase in temperature will cause a downward shift in IDT center frequency and vice-versa.

For device stability as a function of temperature it is essential that α_T be as small as possible. This is especially the case in narrowband intermediate frequency (IF) stages of receivers for mobile communications. In this connection, it may be noted that a near-zero value of α_T is obtained for the ST-

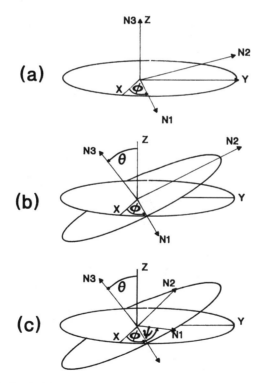

FIG. 2.9.   Right-handed Euler-angle sets. (a) Rotation angle ϕ in X-Y plane with SAW/pseudo-SAW propagation along $N1$-axis. (b) Cut-angle θ of the plate normal, away from the Z-axis. (c) Double-rotation cut angle ψ for propagation axis $N1$, when θ is nonzero. Single-rotation cut when $\psi=0$. (After Reference [14].)

X cut of quartz at an operational temperature $T_o \approx 20°C$, when the expansion and velocity components have equal values in Eq. (2.17) [10]. As a result, filters employing ST-X quartz or other substrates with very low TCD values are dominant in such IF circuit stages. With ST-quartz, however, a change of temperature in either direction around the operating point will cause the IDT center frequency to shift downwards.

2.2.9. EULER ANGLES AND CRYSTAL CUTS

Right-handed Euler-angle sets [11] and/or crystal-orientation axes are used to designate the piezoelectric crystal cut selected for SAW or pseudo-SAW propagation (i.e., leaky-SAW, SSBW and STW). Such Euler angle notation is illustrated in Fig. 2.9, in terms of angles ϕ, θ, and ψ, for "standard" cuts where propagation axis N1, N2 and N3 line up with crystalline axes X, Y, Z. In Fig. 2.9(a) the rotation angle ϕ is in the X-Y plane, while the desired SAW propagation axis is along the N1-axis. In Fig. 2.9(b) angle θ gives the cut-angle of the plate normal away from the Z-axis. In Fig. 2.9(c) rotation angle ψ further establishes the desired cut for propagation axis $N1$. Figure 2.10 shows a simple example for Y-Z lithium niobate, with Euler angles $\phi=0$, $\theta=90°$ and $\psi=90°$. The notation *double-rotation cut* is often used to

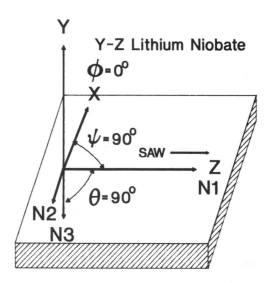

FIG. 2.10. Euler-angle example for Y-Z lithium niobate, with $\phi=0$, $\theta=90°$, and $\psi=90°$. (After Reference [12].)

describe crystal cuts, when angles θ and ψ are nonzero, as well as the use of *single-rotation cut* when $\psi = 0$.

2.3. General Equations for Surface Waves and Bulk Waves

The propagation of acoustic waves in an anisotropic piezoelectric crystal may be described in terms of a set of six linear equations, given in tensor notation [12]. These are:

Equations of motion:
$$\frac{\partial T_{ij}}{\partial x_i} = \rho \frac{\partial^2 u_j}{\partial t^2} \tag{2.18a}$$

From Maxwell's equations, (quasi-static):
$$\frac{\partial D_i}{\partial x_i} = 0 \tag{2.18b}$$

Electric field intensity:
$$E_i = -\frac{\partial \Phi}{\partial x_i} \tag{2.18c}$$

Piezoelectric mechanical stress from Eq. (2.9):
$$T_{ij} = c_{ijkl} S_{kl} - e_{nij} E_n \tag{2.18d}$$

Piezoelectric displacement density from Eq. (2.8):
$$D_m = e_{mkl} S_{kl} + \varepsilon_{mn} E_n \tag{2.18e}$$

Linear strain displacement:
$$S_{kl} = \frac{1}{2}\frac{\partial u_k}{\partial x_l} + \frac{\partial u_l}{\partial x_k} \tag{2.18f}$$

where $S =$ strain, $T =$ stress, $\rho =$ mass density, $u =$ mechanical displacement, $D =$ displacement density, $E =$ electric field intensity, and $\Phi =$ electric potential.

Solutions of Eqs. (2.18a) to (2.18f) that relate electric potential Φ and mechanical displacement u are assumed to be of the complex, travelling wave type in Eq. (2.14). Incorporation of such solution functions into Eq. (2.18) leads to four homogeneous equations in four variables ($n = 1, 2, 3, 4$). Three of these variable indices relate to the three Cartesian coordinates for displacement u, while the fourth index applies to potential Φ. The total solutions for u and Φ are taken to be linear combinations of the ensuing partial-wave equations applied to the piezoelectric. For the x, y, z spatial coordinate notation of Fig. 2.5, the travelling wave displacement and potential in the piezoelectric interior at reference time $t = 0$ are given by [13]

$$U_i = \sum_{n=1}^{4} u_i^{(n)} e^{j(\omega t - \beta x)} e^{-\alpha_n |y|} \quad (i = 1 \ to \ 3) \qquad (2.19a)$$

$$\Phi = \sum_{n=1}^{4} \Phi_4^{(n)} e^{j(\omega t - \beta x)} e^{-\alpha_n |y|} \qquad (2.19b)$$

where the (n) superscripts refer to the solution corresponding to the chosen root α_n, while $u_i^{(n)}$ (for $i = 1$ to 3, $n = 1$ to 4) are the mechanical amplitude coefficients, $\Phi_4^{(n)}$ (for $n = 1$ to 4) are the potential amplitude coefficients, and $\beta = \omega/v = 2\pi f/v = $ phase constant, where $v = $ wave velocity. Moreover, the root parameters α_n themselves are actually *propagation constants* (also known as decay constants), with units of metre^{-1}. As such, they may have a) pure real, b) pure imaginary or c) complex values as functions of frequency.

In describing surface wave and bulk wave propagation, it is often convenient to relate these as a function of inverse phase velocity $1/v_p$ (or *slowness*), rather than directly as a direct function of the phase velocity v_p parallel to the substrate surface. Moreover, the use of an *effective surface permittivity* ε_s parameter (which is usually plotted as a function of slowness s), represents another useful analytical tool. This effective surface permittivity has been derived in the form [13]

$$\varepsilon_s = \frac{\sigma}{\beta \Phi} \quad (at \ y = 0), \qquad (2.20)$$

where $\sigma = $ surface charge density (coulomb/metre^2), $\beta = $ phase constant, and $\Phi = $ surface potential (volts). Effective permittivity $\varepsilon_s = 0$ for an unmetallized surface with no free charges, while $\varepsilon_s = \infty$ for a metallized surface with zero potential Φ. The effective permittivity can be complex, so that $\varepsilon_s = \varepsilon_s' - j\varepsilon_s''$. Both the slowness *function* $s = 1/v_p$ and the effective permittivity ε_r are employed in the following sections.

Example 2.4 Complex Relative Permittivity $\varepsilon_r = \varepsilon_r' - j\varepsilon_r''$. A capacitor of plate area A and separation d contains a solid dielectric with relative permittivity $\varepsilon_r = \varepsilon_r' - j\varepsilon_r''$. Show that it is the ε_r'' component that is associated with ac-losses in the capacitor. ∎

Solution. The admittance of the capacitor is given by $Y = j\omega C = j\omega(\varepsilon_o \varepsilon_r A/ d) = j\omega\varepsilon_o(\varepsilon_r' - j\varepsilon_r'')A/d = \omega\varepsilon_o\varepsilon_r''A/d + j\omega\varepsilon_o\varepsilon_r'A/d$, where $\varepsilon_o = $ permittivity of free space. But the general admittance form is $Y = G + jB$, so that the loss conductance G is associated with ε_r'', while the ε_r' term is associated with phase shift of susceptance B.

2.4. Propagation Constants for Rayleigh Waves and Bulk Waves

The four surface wave propagation constants already described here relate to: a) the bulk longitudinal mode; b) the bulk vertically polarized shear mode; c) the electrostatic field; and d) the bulk horizontally polarized shear mode [14]. From consideration of the characteristics of a "true" SAW (i.e., the Rayleigh wave), and its confinement to the surface of the propagating medium, it can be deduced that all four propagation (decay) constants α_n must be positive for Rayleigh wave propagation to take place, so that

$$Re\big[\alpha_n\big] > 0 \quad (n = 1 \text{ to } 4). \tag{2.21}$$

In this situation the mechanical displacements given by U_i ($i = 1$ to 3) in Eq. (2.19a) all go to zero as penetration depth y becomes infinite. It has been demonstrated that all four modes ($n = 1$ to 4) are surface (Rayleigh) modes for high values of slowness function s (i.e., for low values of phase velocity), and positive real values of α_n [13], [15]. Under these conditions the transmission loss is ideally zero so that the effective surface permittivity can be approximated by $\varepsilon_s = \varepsilon'_s - j0$.

At the other velocity extreme, for zero slowness s, three of the modes are bulk wave modes corresponding to the *longitudinal* bulk wave, together with two *transverse shear* waves. The latter consist of a slow horizontally polarized shear wave and a fast vertically polarized shear wave. Such bulk wave propagation results in power loss. From Eqs. (2.19a) and (2.19b), however, it is seen that propagation into the piezoelectric interior is associated with *imaginary-axis* values of propagation constant α_n. As a result, under this condition the effective surface permittivity ε_s will be complex. In between these extremes of slowness function the bulk wave modes have cut-off values at which they degenerate into surface modes, as functions of velocity slowness function s. Note that the Rayleigh wave velocity is less than that for the bulk modes.

The above concepts are illustrated in the example of Fig. 2.11, relating the effective surface permittivity of Y-Z lithium niobate as a function of slowness number [13], [15]. Here, the pole and zero of ε_s at s_s and s_o correspond to surface wave propagation on a metallized and unmetallized surface, respectively. The transition at s_1 in the complex permittivity region corresponds to the longitudinal bulk wave, while that at s_2 corresponds to the transverse shear component of the bulk wave. The vertically polarized shear wave component is absent here because it does not couple with the electric fields and, therefore, does not affect the permittivity.

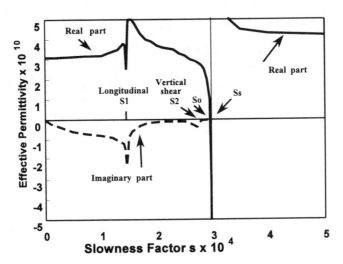

Fɪɢ. 2.11.   Effective surface permittivity of *Y-Z* lithium niobate, as a function of slowness parameter *s*. (After Reference [15].)

Example 2.5 Surface Wave Slowness on Y-Z Lithium Niobate. *Y-Z* lithium niobate has a free surface velocity $v_o = 3488$ m/s and coupling coefficient $K^2 = 0.045$. Determine the values of slowness function *s* for metallized and unmetallized surfaces. ■

Solution.   For the unmetallized surface the slowness function is $s_o = 1/v_o = 1/3488 = 2.86 \times 10^{-4}$ s/m. Also, $K^2 = |-2\Delta v/v_o|$ from Eq. (2.16), which gives $\Delta v = -K^2(v_o/2) = -0.045 \times 3488/2 = 78.48$ m/s. The corresponding slowness for the metallized surface is $s_s = 1/(3488 - 78.48) = 2.93 \times 10^{-4}$ s/m. Compare with the values shown in Fig. 2.11. ■

2.5.   Characteristics of Leaky-SAW and Shear Waves

2.5.1.   LEAKY-SAW FEATURES AND MERITS COMPARED WITH SAW DEVICES

Leaky-SAW and SAW devices both employ IDTs. In many instances they may be visually indistinguishable from one another. For example, operation in one or the other mode on quartz merely involves the choice of a different crystal axis for wave propagation. In operation, however, leaky-SAW devices possess a number of attractive features over their SAW counterpart, as listed in what follows.

1. Leaky-SAW velocities can be much higher than SAW ones, to the extent that LSAW devices can be designed for operation at frequen-

cies up to 1.6 times higher than for its SAW counterpart with the same lithographic geometry. This is particularly attractive for the fabrication of RF filters with fractional bandwidths of 4% or more, for mobile communications and cellular phones operating at gigahertz frequencies [16], [17].

2. LSAW piezoelectric crystal cuts can have much higher values of electromechanical coupling coefficient K^2, with corresponding increase in operating bandwidth. This is again particularly attractive for cellular or cordless communications applications.

3. Some LSAW piezoelectric crystal cuts of quartz, such as the LST-cut in Fig. 2.13, can have superior temperature stability coefficients of delay (TCD) over their SAW counterparts [18].

4. Because LSAW propagation is beneath the piezoelectric surface, such devices can be significantly less sensitive to surface contamination than SAW ones. The surface-fabricated IDT, however, is still susceptible to contamination. (One device-testing technique used by the author for observing whether LSAW or SAW mechanisms are involved requires dropping a small amount of a rapidly evaporating solution onto a nonmetallized portion of the wave-propagation surface. A temporary disappearance of the signal from the output IDT is indicative of SAW propagation, while no — or only partial — suppression of the signal indicates LSAW mechanisms are involved.)

5. As the LSAW penetrates further into the substrate than does the Rayleigh wave, corresponding acoustic power densities will be less. This means that LSAW devices, such as oscillators, are capable of handling larger powers before the onset of IDT degradation due to violent surface vibrations. They can also operate at higher powers before the onset of piezoelectric nonlinearities.

2.5.2. PROPAGATION AND DECAY CONSTANTS

In contrast with the conditions imposed on the four propagation (decay) constants for Rayleigh and bulk wave propagation, for leaky-SAW propagation the real part of one of these must be negative, so that the wave amplitude will increase with increasing penetration into the substrate. This gives

$$Re[\alpha_1] < 0$$
$$Re[\alpha_n] > 0 \quad (n = 2 \ to \ 4), \tag{2.22}$$

where α_1 is the decay coefficient for the slow shear wave. This combination of decay constants gives rise to a longitudinal wave, a slow shear wave, a fast shear wave, and a leaky-SAW. In this situation the velocity of the leaky-SAW is always faster than either the slow shear wave or the Rayleigh wave! [14]. The fact that the leaky-SAW wave has a higher velocity than the Rayleigh wave is particularly useful in the lithographic fabrication of surface wave devices in the gigahertz-frequency range because this allows larger separation of IDT fingers.

Figure 2.12 shows a simplified pictorial representation of leaky-SAW propagation, versus that for the "true" SAW (Rayleigh wave). As shown, the travelling wave of electric potential $\Phi(x,t)$ characterizing leaky-SAW propagation in the piezoelectric is of the form [19]

$$\Phi(x,t) \propto e^{-\gamma x} e^{j\omega t} \tag{2.23}$$

where $\gamma = \alpha + j\beta$ is the complex propagation constant per unit length, in terms of attenuation constant α and phase constant β of the leaky SAW.

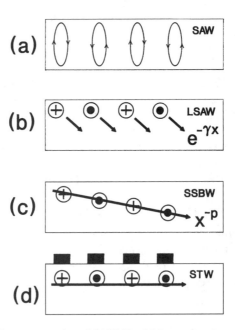

F_{IG}. 2.12. Artistic representation of SAW (Rayleigh wave) and pseudo-SAW wave motion in piezoelectric. (a) Rayleigh wave. (b) Leaky surface acoustic wave (LSAW). (c) Surface skimming bulk wave (SSBW). (d) Surface transverse wave (STW).

The attenuation constant α is often given in terms of nepers per acoustic wavelength, where 1 neper $= 20 \log_{10}(e) = 8.686$ dB [20].

2.5.3. LEAKY-SAW VELOCITY AND LOSS

As a leaky-SAW propagates into the interior of the piezoelectric substrate, rather than along its surface, it can be considered as a "complex" velocity, with components v_x and v_y, respectively, along the surface and normal coordinates of Fig. 2.5. The phase velocity v_p along the surface is $v_p = \text{Re}(v) = v_x$. Moreover, the transmission loss L_L (in decibels per acoustic wavelength) is given by [18]

$$L_L \approx 2\pi.20\log_{10}(e).\frac{Im(v)}{Re(v)} \approx 54.6\frac{Im(v)}{Re(v)} \quad (dB/\lambda). \qquad (2.24)$$

From Eq. (2.24) it is seen that, for minimum transmission loss, the leaky-SAW penetration should be kept to a minimum. For selected crystal cuts, as in Fig. 2.13, this propagation loss can be kept small enough to be acceptable for some device applications. A problem that can arise with leaky-SAW crystal cuts, however, relates to the difficulty of obtaining low values of TCD and of insertion loss with the same crystal cut [21], [22].

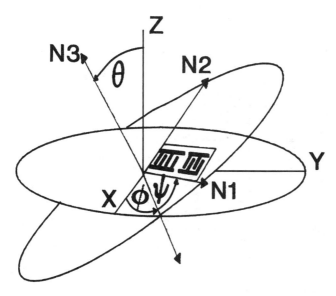

FIG. 2.13. First leaky-SAW crystal-cut on quartz, called an LST cut. (After Reference [18].)

2.5.4. FIRST AND SECOND LEAKY-SAW CHARACTERISTICS

Current commercial photolithographic camera resolutions with optimum resolutions of approximately 0.3–0.5 μm limit the upper frequency operational fundamental-mode capability of Rayleigh-wave devices to about the 2.0-GHz range. This limitation has, in part, prompted the study of alternative wave-propagation mechanisms for application to such lithography.

One such alternative relates to leaky-SAW wave propagation, as mentioned in the preceding. In this connection, however, two possible modes of leaky-SAW propagation can be obtained in selected crystal cuts, such as for lithium niobate (LiNbO$_3$) and lithium tantalate (LiTaO$_3$). One of these is the *first* leaky wave, with phase velocity between those of the slow and fast bulk shear waves. Figure 2.14 illustrates theoretical calculations of first leaky-SAW, slow and fast shear-wave, and Rayleigh-wave velocities on (ϕ, $\theta = 90°$, $\psi = 0°$)-cut LiNbO$_3$ [23]. Note that because the electromechanical coupling coefficient $K^2 = 2|\Delta v/v|$ is given by the difference between acoustic wave velocities on open- and metallized-surface, large separations between the corresponding leaky-SAW velocities correspond to large K^2-values.

Depending on the choice of piezoelectric crystal and cut, associated phase velocities and maximum frequencies of *first* leaky-SAW devices, (usually just called leaky-SAW), are about a factor of 1.3 higher than their

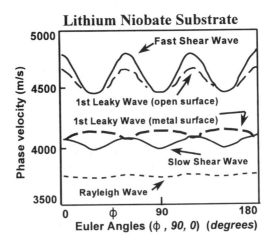

FIG. 2.14.   Theoretical computation for wave propagation in (ϕ, $\theta = 90°$, $\psi = 0°$)-cut lithium niobate, showing fast shear wave, slow shear wave, first leaky-SAW, and Rayleigh-wave velocities. (After Reference [23].)

Rayleigh-wave counterparts. At this time they find extensive application in RF-filter devices for mobile and cellular phone systems up into the 2-GHz regime. One current cut used for 800-MHz band RF-filter applications is that of 64° *Y-X* LiNbO$_3$ ($\phi=0°$, $\theta=64°$, $\psi=0°$), with $K^2=11\%$, and TCD $\approx$ 81 ppm/°C [17].

Ladder filters for antenna duplexers in RF front-end circuitry for mobile transceivers normally employ LSAW resonator elements fabricated on 36° *Y-X* LiTaO$_3$, with improved temperature stability over 64° *Y-X* LiNbO$_3$, as indicated in Table 2.2. As will be considered in a later chapter, the choice of the *film-thickness ratio* (h/λ) of the IDT aluminum (*Al*) metallization (where $h = Al$-film thickness, and λ = acoustic wavelength) on this substrate can influence considerably device insertion loss. In this respect, it has been demonstrated that the propagation loss arising from the leaky nature changes parabolically with both the IDT metal thickness and crystal-cut angle. Moreover, for (h/λ) ≈ 0.07 to 0.1 (i.e., (h/λ)% = 7 to 10%), a change of crystal cut from 36° *Y-X* LiTaO$_3$ to 40° ~ 42° *Y-X* LiTaO$_3$ offers minimum propagation loss without deteriorating other characteristics. At least one LSAW device manufacturer is already offering this differing cut for such applications, as shown in Fig. 2.17.

Also depending on the choice of crystal cut, the other wave propagation involves the so-called *second* leaky wave whose phase velocity is greater than that of the fast bulk shear wave. As they can have wave velocities about a factor of 2.0 higher than their Rayleigh-wave counterparts, devices employing *second* leaky-SAW propagation are currently under intensive study for application to mobile communications systems operating in the 2-GHz regime. Figure 2.15 illustrates theoretical calculations of *second* leaky-SAW, slow and fast shear-wave, longitudinal, and Rayleigh-wave velocities on ($\phi=90°$, $\theta=90°$, ψ)-cut LiNbO$_3$ [23]. Note, however,

TABLE 2.2

COMMON LEAKY SURFACE ACOUSTIC WAVE (LSAW) PIEZOELECTRIC SUBSTRATES
(After Reference [19])

Piezoelectric	Euler Angles ϕ θ ψ (degrees)			Velocity: metal surface (m/s)	K^2 (%)	Temp Coefficient (ppm/ °C)	Attenuation: metal surface (dB/λ)
LST-Quartz	0	15	0	3948	0.11	~0	0.00031
64° *YX*-LiNbO$_3$	0	64	0	4478	11.3	~−81	~0
41° *YX*-LiNbO$_3$	0	41	0	4379	17.2	~−80	0.0438
36° *YX*-LiTaO$_3$	0	36	0	4112	4.7	~−32	~0

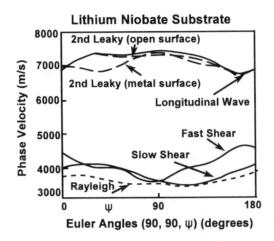

FIG. 2.15. Theoretical computation for wave propagation in (ϕ=90°, θ=90°, ψ)-cut lithium niobate, showing second leaky-SAW, slow shear wave, fast shear wave, and Rayleigh-wave velocities. (After Reference [23].)

that the corresponding K^2 values will be much less than for the first leaky-SAW.

2.5.5. ILLUSTRATIVE FIRST LEAKY-SAW CRYSTAL CUT ON QUARTZ

The preceding features are illustrated with the example of Fig. 2.13 to demonstrate the drastic change in propagation characteristics that can be obtained in wave propagation characteristics by the choice of Euler angles for a given piezoelectric. In the case of Rayleigh wave (i.e., true SAW) propagation on the temperature-stable ST-X cut of quartz discussed in the previous sections, the Euler angles involve a singly rotated Y-cut X-axis propagation cut with $\phi=\psi=0°$, and $\theta=132.75°$ with $K^2=0.11\%$, $v_o=3158$ m/s, and TCD ~ 0 in Table 2.1.

On the other hand, a singly rotated angle selection of $\theta=\psi=0°$, and $\theta=15.7°$ results in an extremely temperature-stable cut of quartz for *leaky*-SAW propagation, with a variation of TCD of less than ±10 ppm over −20°C ~ +80°C, $K^2=\sim 0.11\%$, and a leaky-SAW velocity $v=\sim 3940$ m/s. This has been termed an LST-cut [18].

2.5.6. OTHER LEAKY-SAW CRYSTAL CUTS

With reference to the Euler angle (ϕ, θ, ψ) notation of Fig. 2.9, other LSAW substrates currently in use for mobile communications RF-filter circuitry include 41° Y-X LiNbO$_3$ ($\phi=0°$, $\theta=41°$, $\psi=0°$), 64° Y-X LiNbO$_3$ ($\phi=0°$,

FIG. 2.16. Photograph of a lithium tantalate crystal boule, prior to wafer slicing. (Courtesy of Fujitsu Ltd., Kawasaki, Japan.)

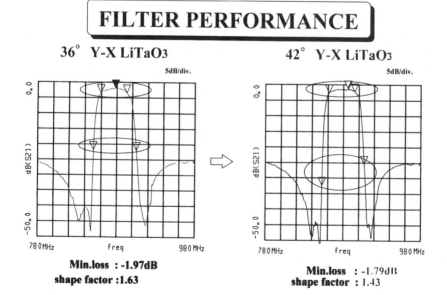

FIG. 2.17. Comparison of insertion loss (IL) and shape factor (SF) of 880-MHz LSAW RF filter design fabricated on both 36° *Y-X* LiTaO₃ and 42° *Y-X* LiTaO₃. The 36°-cut has IL=1.97 dB and SF=1.63, while improved 42°-cut has IL=1.79 dB and SF=1.43. (Courtesy of Fujitsu Ltd., Kawasaki, Japan.)

$\theta = 64°$, $\psi = 0°$), and 36° Y-X LiTaO$_3$ ($\phi = 0°$, $\theta = 36°$, $\psi = 0°$). Table 2.2 lists the significant design parameters for such substrates. Figure 2.16 is a photograph of a lithium tantalate crystal boule, prior to wafer slicing. As noted in Section 2.5.4, at least one SAW manufacturer is now supplying alternative cuts of Y-X LiTaO$_3$ (with cut angles between 40° and 42° as required), to provide both reduced insertion loss and shape factor, for some RF front-end filter designs. By way of illustration, Fig. 2.17 shows the relative performances of an 880-MHz LSAW RF front-end filter fabricated on 36° Y-X LiTaO$_3$ and on 42° Y-X LiTaO$_3$. As demonstrated, the sample filter on 36° Y-X LiTaO$_3$ had an insertion loss of 1.97 dB, and a shape factor of 1.63. On the other hand, the same design on 42° Y-X LiTaO$_3$ reduced the insertion loss to 1.79 dB, and the shape factor to 1.43. Reduction of insertion loss is of prime importance in RF front end filters for mobile/wireless receivers because front-end insertion loss roughly translates into an equivalent receiver noise factor.

2.6. Shallow Bulk Acoustic Waves (SBAW) and Piezoelectrics

2.6.1. SURFACE SKIMMING BULK WAVE AND SURFACE TRANSVERSE
WAVE PROPAGATION

Another type of acoustic wave propagation of application to surface wave devices involves SBAW propagation. Shallow bulk acoustic wave propagation involves longitudinal bulk waves with shear horizontal (SH) polarization, rather than the leaky-SAW or elliptically polarized Rayleigh waves considered so far in this text. These SH waves can be excited and detected by surface-deposited IDTs in similar fashion as for SAW or leaky-SAW devices. Whereas Rayleigh waves are generated by the electric field components parallel and normal to the excited IDT fingers, the SH waves are only excited by the parallel electric field. In general, there is no mathematical difference between SH-type SSBW and conventional bulk waves.

Because SBAW, Rayleigh-wave and leaky-SAW devices all use interdigital transducers, they are generally indistinguishable visually from each other. While they also have the same attractive features of small size, ruggedness and planar construction, the SBAW structures have essentially the same merits as for leaky-SAW, given in section 2.5.1, in that:

1. Bulk waves have higher velocities than SAW, so that SBAW devices can be designed for operation at fundamental frequencies of up to about 60% higher than with SAW IDTs of the same geometry.

2. Because the bulk waves in the SBAW devices propagate below the piezoelectric surface, the propagating wave is much less sensitive to

surface contamination than for SAW (although the IDTs can still be affected).

3. The SBAW have temperature coefficients superior to those of SAW in some cuts of quartz and lithium tantalate (LiTaO₃).

3. The SBAW filter resonators can operate at much higher powers than their SAW counterparts, (e.g., 25 dBm), before piezoelectric non-linearities set in.

5. Finally, SBAW devices can yield good suppression of spurious modes.

Shallow bulk acoustic wave (SBAW) devices may be grouped into two general categories, depending on whether or not energy-trapping structures are involved. Where there is no energy-trapping, the SH bulk waves are usually referred to as *surface skimming bulk waves* (SSBW) [24], [25]. As sketched in Fig. 2.19, when an energy-trapping grating structure—normally of a different period from the IDT electrodes, to avoid significant reflection at center frequency—is located between input and output IDTs to promote low-loss operation, the SBAW is termed a *surface transverse wave* (STW) [26], [27]. Figure 2.12 gives a simplistic depiction of the wave propagation directions and attenuation for SSBW and STW, as compared with SAW and leaky-SAW.

2.6.2. SURFACE SKIMMING BULK WAVE (SSBW) PIEZOELECTRIC CUTS

Desirable piezoelectric crystals and crystal cuts for SSBW devices are those that have: 1) large piezoelectric coupling to the SH bulk wave; 2) zero piezoelectric coupling to surface acoustic waves; 3) zero piezoelectric coupling to other bulk wave modes; and 4) a zero-, or low-temperature coefficient of delay. Most of these features are to be found in singly rotated Y-cuts of quartz. In terms of the Euler-angle notation of this chapter, Table 2.3 lists SSBW parameters for some singly rotated 90° propagation cuts on quartz [28] with zero coupling to SAW, as well as lithium niobate and lithium tantalate with negligible coupling to SAW [29]. Doubly rotated cuts of quartz have also been under study for SSBW [30], [31].

As seen from Table 2.3, the SSBW AT-cut of quartz is a most useful one, with its high-velocity SSBW and negligible temperature coefficient of delay. The SSBW cuts of lithium niobate and lithium tantalate, with propagation along the X-axis, have high values of electromechanical coupling coefficient K^2 but poorer temperature stabilities. For SSBW propagation at 90° to the X-axis, the SSBW velocity is approximately 60% higher than for SAW propagation along the X-axis. To demonstrate, a 1.766-GHz SSBW filter with an untuned insertion loss of 21 dB—and

TABLE 2.3

SOME PIEZOELECTRIC CRYSTALS SUITED TO SUNFECE SKIMMING BULK WAVE (SSBW)
PROPAGATION

Piezoelectric	Euler Angles ϕ θ ψ (degrees)			SSBW Velocity (m/s)	K^2 (%)	Temperature Coefficient (ppm/°C)
Rotated Y-cut ST-quartz	0	132.75	90	4990	1.89	−33
35.5° (AT) rotated Y-cut quartz	0	125.15	90	5100	1.44	±127 (−55 to +85°C)
36° rotated YX-LiTaO$_3$	0	36	0	4211	4.7	−45
37° rotated LiNbO$_3$	0	37.93	0	4802	16.7	−59

very smooth passband response—was obtained using a mask pattern for a
1.09-GHz SAW filter [25].

2.6.3.   SBAW VERSUS LEAKY-SAW PROPAGATION

The alert reader will note that the piezoelectric cut of 36° Y-X lithium
tantalate has been included in both the leaky-SAW Table 2.2 and the
SBAW Table 2.3. The reason for this is that, as the velocities of the two
wave types are close to one another, the wave type that is propagated is
strongly dependent on the surface boundary conditions and metallization
structure. Theoretical and experimental analyses have shown that leaky-
SAW is the dominant propagating mode on a metallized surface of this
piezoelectric, while SBAW dominates on the free surface. Also, for certain
boundary conditions both SBAW and leaky-SAW will contribute to signal
transfer [32], [33].

The surface potential of an SBAW wave is of the form [32], [33]

$$\Phi(x,t) \propto x^{-P} e^{j\omega t}: \quad \left(0.5 \leq P \leq 1.5\right) \tag{2.25}$$

where X = the distance between a sound source and an observation point.
From measurements of signal amplitude as a function of distance X, therefore, SBAW propagation can be distinguished from that of leaky-SAW,
with exponential amplitude decay, as given in Eq. (2.23). It may also be
noted that the decay rate for SBAW given in Eq. (2.25) is indicative of
large insertion losses in an SBAW device, relative to a comparable LSAW
structure.

2.7. Layered Structures for SAW and Pseudo-SAW Propagation

2.7.1. LAYERED STRUCTURES FOR SAW DEVICES

Thin-film layered SAW structures for diverse applications have been in production for many years. A notable example is in the use of ZnO thin-film video intermediate-frequency (IF) filters for domestic color TV sets. In these devices, and as sketched in Fig. 2.18, one pair of unapodized Al input/output IDTs is placed between a borosilicate glass substrate and a thin polycrystalline ZnO film deposited by RF sputtering. An additional Al counter thin-film counter electrode is deposited on top of the ZnO, to allow

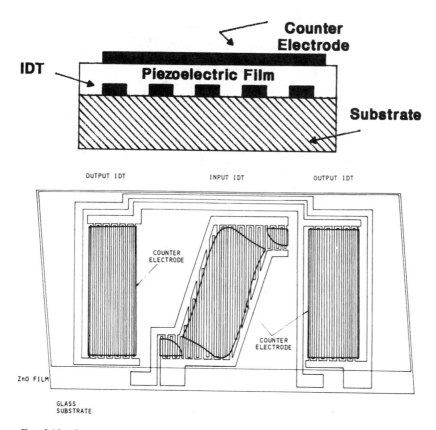

FIG. 2.18. Lower: geometry of structure for low-loss ZnO thin-film SAW video IF filter. Output IDTs are connected in parallel for low-loss operation. Upper: placement of counter electrodes. (Reprinted with permission from Yamazaki, Mitsuyu, and Wasa, [46], © IEEE, 1980.)

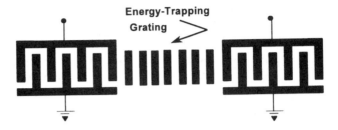

FIG. 2.19.   Energy-trapping grating in an STW filter. Grating period is normally different from IDT electrode period, to avoid significant reflections at center frequency.

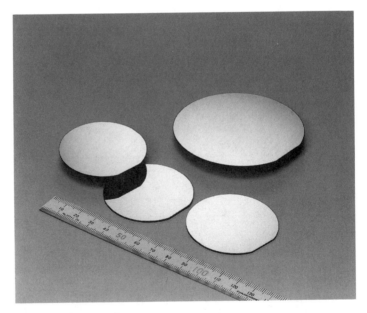

FIG. 2.20.   Photograph of Diamond/Si wafers for SAW device fabrication. (Courtesy of Sumitomo Electric Industries, Itami, Japan.)

for apodization weighting[3]. Such devices are used, for example, as low-loss IF filters around picture and sound carriers (Japan) of 58.75 MHz and 54.25 MHz, respectively, without a preamplifier.

Also among the layered structures employing thin- or thick-film piezoelectrics, the composite structure ZnO/AlN/glass can yield parameter values of $K^2 = 4.37\%$, TCD = 21 ppm/°C. and $v = 5840$ m/s. Sputtered piezo-

---

[3] The subject of apodization is introduced in Chapter 3.

electric zinc oxide (ZnO) films are not single crystal structures, but are polycrystalline ones, with c-axis orientation having a reported standard deviation of less than 2 percent [34]–[36].

Aluminum nitride (AlN) on glass is used as an inexpensive substitute for sapphire and ceramic substrates, while providing a large acoustic velocity. Wave propagation in this composite involves the Sezawa surface wave, which is the first higher-order higher-velocity mode of the Rayleigh wave. Surface acoustic wave delay lines operating above 1 GHz have also been reported using epitaxial AlN films on sapphire. (Sapphire is just aluminum oxide Al_2O_3.) One application of this latter structure relates to a SAW correlator in a one-chip RF integrated circuit for a spread spectrum transceiver [37].

2.7.2. LAYERED STRUCTURES FOR MOBILE COMMUNICATIONS APPLICATIONS

As fundamental operating frequencies for mobile and wireless communications extend into ever-higher gigahertz frequency bands, lithographic problems of fabricating IDTs with submicron line widths for SAW or

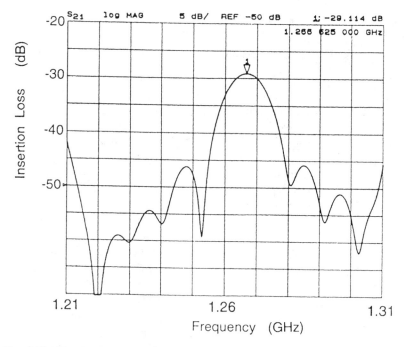

FIG. 2.21. Frequency response of a 1.3-GHz SAW filter fabricated on ZnO/Diamond/Si layered substrate. (Courtesy of Sumitomo Electric Industries, Itami, Japan.)

pseudo-SAW devices become increasingly severe. As one approach to overcoming such IDT fabrication-resolution limitations, intensive research and development activities have been applied to the creation of piezoelectric substrates with very high acoustic velocities. This is because, for a given fundamental operational frequency $f_o = v/\lambda_o$, the required IDT finger widths will increase with increasing acoustic velocity. The use of diamond as an acoustic substrate is particularly attractive because it has the highest sound velocity among all materials. Thus, when combined with a piezoelectric excitation layer (e.g., ZnO), it can be readily applied to the manufacture of SAW devices for gigahertz frequencies. For example, a 2.5-GHz SAW filter can be fabricated in this way with 1-μm electrodes, whereas submicron IDT lithography would be necessary with "conventional" SAW piezoelectric substrates. In this respect, it has been shown that for the first leaky-SAW mode in the ZnO/metal IDT/diamond/Si structure the acoustic velocity is $v_1 = 10,500 \, \text{m/s}$, with $K^2 = 1.5\%$. Moreover, it has been found that the second leaky-SAW mode in the IDT/ZnO/diamond/Si structure has an acoustic velocity $v_2 = 11,600 \, \text{m/s}$, with $K^2 = 1.1\%$ [38]. Figure 2.20 shows a photograph of Diamond/Si wafers for SAW device fabrication; Fig. 2.21 illustrates the frequency response of a sample 1.3-GHz SAW filter fabricated on a ZnO/Diamond/Si layered substrate.

2.8. Bleustein-Gulyaev-Shimizu (BGS) Waves On Ceramic Substrates

Ceramic bulk-wave resonators and SAW resonators are used in various consumer electronics products. Because of thickness limitations, the bulk-wave resonators are used primarily below 30 MHz and the SAW resonators are used primarily above 60 MHz because of device size and cost. Recently, however, the shear-horizontal (SH) waves described in the previous section have received increased attention because of their relatively large electromechanical coupling factors. These SH-type surface waves on piezoelectric ceramics such as $Pb(Zr, Ti)O_3$ (PZT) propagate as Bleustein-Gulyaev-Shimizu (BGS) waves, with velocity v ~ 2400 m/s. Their predominant displacement is parallel to the substrate surface and perpendicular to the propagation direction. It is important to note that BGS wave propagation requires the substrate to be polarized along the plane of the substrate (see Fig. 11.28). The BGS waves undergo complete reflection at the free edge of a substrate with high permittivity ($\varepsilon_r = 1200$ for PZT) [39]–[41]. This is in contrast to Rayleigh waves, in which shear vertical and horizontal displacements are converted again to vertical and longitudinal displacements at the free edge. As will be discussed in Chapter 11, very small BGS-wave resonators can be realized in this manner with reported frequencies of up to

160 MHz [39]–[41]. The PZT-1 ceramic substrates have large K^2 values ($K^2 = 22\%$) and a TCD of 9 ppm/°C, and are suitable for low-Q wideband resonators. The PZT-2 ceramics have small K^2 values ($K^2 = 4\%$) and TCD of 7 ppm/°C, and are suitable for high-Q narrow-band resonators. The low-Q resonators have been evaluated as delay elements in demodulators for DECT phones. The high-Q structures have been evaluated for operation as 75-MHz clocks in personal computers [39]–[41].

2.9. Effects of Acoustic Bulk Waves on Filter Performance

2.9.1. INTRODUCTION

The preceding sections have dealt with the launching and detection of surface acoustic waves on a piezoelectric solid surface by excitation of IDTs. In addition to the launching of surface waves, however, an excited IDT generating Rayleigh-waves or leaky-SAW can also generate *bulk* acoustic waves. While bulk waves are desirably employed in SSBW or STW designs, their presence in Rayleigh-wave or LSAW structures is highly undesirable, because they can cause serious degradation of the desired filter response. Bulk waves can propagate in any direction within the body of the substrate material. Those bulk wave components that arrive at the output IDT induce voltages in addition to those induced by the surface wave. Over the SAW filter bandwidth, interference between the two voltage components will cause distortion of the desired amplitude, phase and group delay characteristics for the SAW response alone. Out-of-band degradation will normally be evidenced by the reduced rejection levels in the amplitude response. In addition, because the input transducer is excited by a common source, power generation in the form of bulk waves will be at the expense of SAW power. This will affect input impedance levels and insertion loss.

The presence of bulk waves is also detrimental to the performance of other types of SAW devices, such as the *reflective array compressor* (RAC). The RAC is employed in SAW-based radar pulse compression systems requiring very large time-bandwidth (TB) products for signal processing gain. Typically the RAC employs several thousand periodically etched grooves on the surface of the piezoelectric substrate for reflection of SAW at various frequencies. For proper operation of these devices, it is essential that perturbations of the ideal phase response be minimal. These surface discontinuities can, however, cause bulk wave energy to be stored in reactive form, which can then affect the phase shift response to a serious degree—unless compensated for [42].

2.9.2. BULK ACOUSTIC WAVE MODES

Bulk acoustic waves can be characterized in terms of three modal types of mechanical excitation. One of these is a compressional wave, termed the *longitudinal bulk wave*, that is polarized in the direction of the acoustic wave propagation vector. The other two wave components, termed *transverse* or shear waves, have their vibrational modes (i.e., their polarizations) perpendicular to the acoustic propagation axis. In an isotropic solid these two transversely polarized bulk waves would propagate with the same velocity v_{bt}, which would be less than that of the longitudinal bulk-wave velocity v_{bl}.

In a finite anisotropic piezoelectric substrate, however, the two shear waves will have different velocities. Because the motion of the vertically polarized shear wave is less restricted, it will have a faster velocity than its horizontally polarized shear wave counterpart. These two shear waves are usually termed the *fast shear wave* and *slow shear wave*, respectively. In the anisotropic piezoelectric their velocities are again greater than that of the Rayleigh wave, and less than that of the longitudinal bulk wave component.

In typical Rayleigh-wave piezoelectric substrates, the velocity of the lowest transverse bulk wave is a factor of about 1.1 higher than the Rayleigh velocity v, while velocities of the longitudinal waves are typically a factor of two higher. As the finger spacing of the output IDT is fixed, these velocity differences imply that the bulk waves can (but may not necessarily) excite the IDT at or above the high-frequency end of the filter passband for which it was designed. The in-band degradation can be particularly troublesome as the bandwidth of the IDT is increased.

Bulk waves can be considered as radiating from each excited finger of the input IDT. As with a phased-array antenna, therefore, there will be preferential radiation angles ϕ at which the radiated bulk waves will be in-phase with maximum strength. With reference to Fig. 2.22, these in-phase maxima will occur at input signal frequencies f such that

$$fP = \frac{v_b}{\cos\phi}, \tag{2.26}$$

where P is the IDT finger spacing between electrodes of equal polarity, and v_b is the velocity associated with transverse or longitudinal bulk wave components, for propagation angles $0 \le \phi \le \pm 90°$. As a result there can be numerous undesirable contributions to the IDT output voltage as a function of frequency. This can be compounded by multiple reflections between substrate surfaces.

Figure 2.23 shows a classic illustration of bulk wave effects in a 30-MHz

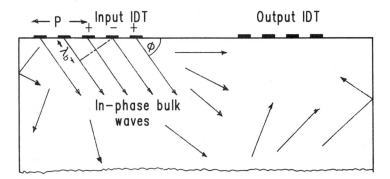

FIG. 2.22.   Excited IDT electrodes can cause bulk acoustic wave radiation, as well as surface waves. Bulk waves arriving at output IDT can seriously degrade the desired response. Analogous to an electromagnetic phased-array antenna, bulk waves can have maximum strength at specific propagation angles ϕ.

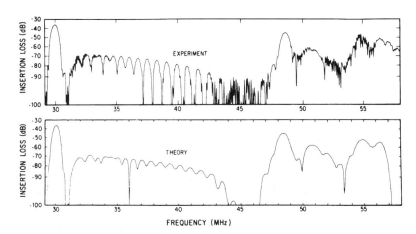

FIG. 2.23.   Upper plot gives experimental frequency response of 30-MHz SAW filter on ST-quartz, with out-of-band degradation due to bulk waves. Quartz thickness about 24 λ_o. Lower plot gives theoretical response. (Reprinted with permission from Wagers [43], © IEEE, 1976.)

narrowband Rayleigh-wave filter on a flat plate of ST-X quartz [43]. Here, the crystal was only $152/2\pi$ wavelengths thick and acted as an acoustic resonator with numerous transverse modes between top and bottom surfaces. Bulk wave reflections were associated in groups (also called *plate modes*), each of which satisfied a transverse resonance. As shown in both the experimental and theoretical responses, bulk-mode generation restricted the stopband rejection to only about 35 dB below the peak of the SAW response.

One common technique used to reduce the amount of bulk waves reaching the output transducer is to roughen the bottom surface of the piezoelectric substrate, and also coat it with a soft conductor such as silver epoxy. However, because bulk waves can travel in all directions within the substrate, some components will travel along the substrate surface together with the desired surface waves. Another technique for reducing the level of spurious bulk wave and spurious SAW reflections that arrive back at the IDTs is to use other than rectangular substrate geometries as sketched in Fig. 2.24.

One unique technique that can be used in some filter designs to circumvent bulk wave interference employs a *multistrip coupler* (MSC) (to be considered in detail at a later stage). Basically it consists of a series of parallel metal film strips placed between offset input and output IDTs. It serves to divert the surface waves so that they alone arrive at the output IDT. An MSC can only be employed on SAW substrates with large K^2 values, such as lithium niobate. It cannot be used in filter designs on ST-X quartz.

2.9.3. Acoustic Bulk Waves and IDT Bandwidths

In Rayleigh-wave or LSAW filter design, there is yet another problem due to acoustic bulk wave excitation. It relates to the amount of input power P converted into bulk wave energy through the input IDT. Thus $P = P_s + P_b$,

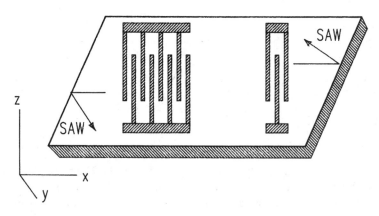

Fig. 2.24. Use of nonrectangular substrate geometry to reduce effect of spurious SAW reflections from edges of piezoelectric. Gives simpler lithographic fabrication process than with use of surface absorbers shown in Fig. 2.1.

where P_s represents the power in the excited SAW wave, while P_b relates that component radiated as bulk waves; P_b will be split in some way between transverse and longitudinal modes. The problem that arises is that the ratio P_s/P_b decreases drastically as the number of finger pairs in the exciting IDT is reduced. In Rayleigh-wave propagation on *Y-Z* lithium niobate, for example, the amount of input power converted into transverse bulk modes increases almost exponentially as the number of IDT finger pairs N_p is reduced below about $N_p = 5$, corresponding to a filter bandwidth BW% $\approx 100/N_p = 20\%$. Correspondingly, SAW power P_s decreases almost exponentially as bulk wave power increases. For *Y-Z*-lithium niobate filters with only 20% fractional bandwidth, as much as about 10% of the input signal power is undesirably expended into bulk waves generation [15].

Obviously, wideband SAW filter designs using the IDT structures of Fig. 2.1 usually will not meet typical performance specifications. Fortunately other IDT designs and geometries can be used to circumvent this bandwidth limitation, as will be considered at a further stage.

2.9.4. EXAMPLES OF HIGH-QUALITY SAW BANDPASS FILTER DESIGNS

The preceding sections on bulk wave interference have focused on some of the problems that can be encountered in SAW filter design. With good design techniques, as well as with the use of high-quality SAW substrates, these can be minimized or circumvented to obtain outstanding filter performance. A vital factor in attaining this is in the attention given to the selection and initial physical condition of the piezoelectric substrates, as well as in their treatment during the fabrication, testing and lead attachment stages of production. It must be emphasized that substrate surface displacements accompanying Rayleigh-wave propagation are extremely small. With typical signal inputs, the displacements may be on the order of only 10^{-5} λ (i.e., only a few angstroms of displacements!). It will then be appreciated that the propagating surface of the piezoelectric substrate must be extremely flat and unblemished if surface waves are to propagate without degradation. One arbitrary measure of a good-quality substrate surface is that it should appear to be scratch-free (and dust-free!) when examined under microscope magnification of 15,000 X. It is noted, however, that random surface scratches of less than about 0.1 μm do not substantially degrade device performance at frequencies below about 1 GHz [44]. Figure 2.25 demonstrates the level to which the design of Rayleigh-wave filters had risen by the early 1970s, in eliminating or circumventing numerous second-order effects such as those due to bulk waves (considered in preceding material). This particular Texas Instruments 287-MHz filter was designed

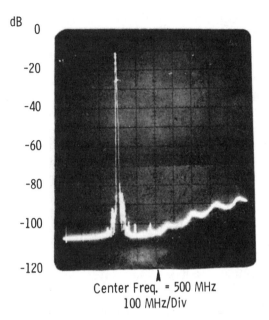

Center Freq. = 500 MHz
100 MHz/Div

Fɪɢ. 2.25.   Superior frequency response displayed by an early Texas Instruments 287-MHz
SAW filter on ST-X quartz. Sidelobe suppression greater than 70 dB up to 1 GHz. (Reprinted
with permission from Hays and Hartmann [45] © IEEE, 1976.)

on ST-X quartz and achieved sidelobe rejection levels of greater than 70 dB
from dc to 1 GHz [45].

Figure 2.26 shows an example of excellent Rayleigh-wave filter characteristics provided by commercially available devices. Here, the passband is flat
to within about 0.2 dB over the band from 33.9 MHz to 39.65 MHz. Adjacent channel rejection is better than 50 dB in a 50 Ω system. The near
rectangular amplitude response has a shape factor $SF \approx 1.2$, whereas shape
factor is *normally* defined as $SF = \Delta f_r / \Delta f$, where Δf_r is the maximum filter-
rejection bandwidth and Δf is the bandwidth measured to the 1-dB ampli-
tude points.

2.9.5.   A Cautionary Note on Shape Factor

The shape factor requirement for a given SAW filter design will depend on
the system application. For RF front-end filtering in mobile communications transceivers, the desired passband response is usually a highly rectangular one, with a flat passband amplitude response. This is not necessarily
the case for IF filter stages in mobile phone receivers! As will be considered

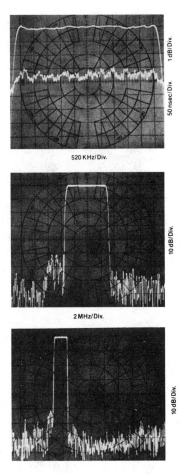

FIG. 2.26. Illustrative high-performance SAW IF filter (33.9 to 39.65 MHz) with low shape factor (<1.2), and small group-delay ripple (<±25 ns). Filter range 33.9 to 39.65 MHz. Passband flatness less than ±0.2 dB. Adjacent-channel rejection greater than 50 dB in 50-ohm system.

Top picture scales–Horizontal: 520 kHz/div. Vertical: 50 ns/div and 1 dB/div.
Middle picture scales–Horizontal: 2 MHz/div. Vertical: 10 dB/div.
Bottom picture scale: Vertical: 10 dB/div.
(Courtesy of Sawtek Inc., Orlando, Florida.)

in some detail in Part 2, the required shape factor for filters in IF stages will depend on the modulation scheme involved. For FM analog cellular systems the desired IF filter response will indeed be a highly rectangular one. However, other modulation schemes can have significantly different IF filter response requirements. One example of this relates to the Digital

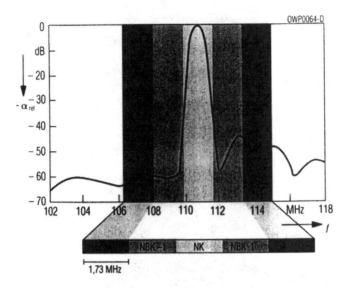

FIG. 2.27.   Required shape factor for SAW IF filter in desired and adjacent 1.728-MHz channels for Digital European Cordless Telephone (DECT) phone, with Gaussian Frequency Shift Keying (GFSK) and 0.5 Gaussian filter. (Courtesy of Siemens-Matsushita Components GmbH & Co. KG, Munich, Germany.)

European Cordless Telephone (DECT) architecture, which employs Gaussian frequency-shift keying (GFSK) with a 0.5 Gaussian filter. Not only will the IF filter in such a receiver be required to have a prescribed frequency response in the selected IF channel, it will also be required to have prescribed responses in *several adjacent channels*, to cater to interference suppression. An example of such an IF filter response for a DECT phone is shown in Fig. 2.27.

2.10.   Summary

This chapter has highlighted the basic characteristics of SAW and pseudo-SAW propagation on piezoelectric crystal substrates, as well as in layered structures and piezoelectric ceramics. The IDT was introduced as a means of launching or receiving surface waves. While the mathematical aspects of SAW propagation and piezoelectricity have not been treated to any great depth, sufficient detail has been included for a communications-oriented reader or designer to gain some insight into the materials aspects of SAW filter design.

The chapter also explored acoustic bulk wave excitation by an IDT as a highly undesirable effect in Rayleigh-wave or LSAW filter design that can cause severe degradation of filter performance as well as limit the fractional bandwidth obtainable with normal IDT structures.

2.11. REFERENCES

1. B. A. Auld, *Acoustic Fields and Waves in Solids,* Volume 1 and 2, John Wiley and Sons, New York, 1973.
2. D. P. Morgan, *Surface-Wave Devices For Signal Processing*, Elsevier, Amsterdam, 1985.
3. E. Dieulesaint and D. Royer, *Elastic Waves in Solids*, John Wiley and Sons, New York, 1980.
4. V. M. Ristic, *Principles of Acoustic Devices*, John Wiley and Sons, New York, 1983.
5. G. W. Farnell, "Elastic surface waves", in H. Matthews (ed.), *Surface Wave Filters*, John Wiley and Sons, New York, Chapter 1, 1977.
6. M. F. Lewis, "On Rayleigh waves and related propagating acoustic waves," in E. A. Ash and E. G. S. Paige (eds), *Rayleigh-Wave Theory and Application*, Springer-Verlag, Berlin, pp. 37–58, 1985.
7. G. S. Kino, *Acoustic Waves: Devices, Imaging, and Analog Signal Processing*, Prentice-Hall Inc., Englewood Cliffs, 1987.
8. See, for example, S. Datta, *Surface Acoustic Wave Devices*, Prentice-Hall, Englewood Cliffs, p. 54, 1986.
9. R. T. Webster and P. H. Carr, "Rayleigh waves on gallium arsenide," in E. A. Ash and E. G. S. Paige (eds), *Rayleigh-Wave Theory and Application, Springer-Verlag*, Berlin, pp. 122–130, 1985.
10. P. S. Cross and S. S. Elliott, "Surface acoustic wave resonators," *Hewlett-Packard Journal*, vol. 32, December 1981.
11. H. Goldstein, *Classical Mechanics*, Second Edition, Addison-Wesley Publishing Co., Reading, 1980.
12. A. J. Slobodnik, Jr., E. D. Conway and R. T. Delmonico (eds), *Microwave Acoustics Handbook, vol. 1A: Surface Wave Velocities*, Air Force Cambridge Research Laboratories, Bedford, Massachusetts, No. AFCRL-TR-73-0597, 1 October 1973.
13. R. F. Milsom, M. Redwood and N. H. C. Reilly, "The interdigital transducer," in H. Matthews (ed.), *Surface Wave Filters: Design Construction and Use*, John Wiley and Sons, New York, Chapter 2, pp. 59–60, 1977.
14. Y. Shimizu, "Leaky SAW propagation characteristics," *Proc. International Symposium on Surface Acoustic Wave Devices for Mobile Communication*, Sendai, Japan, 3–5 Dec. 1992, pp. 59–66, 1992.
15. R. F. Milsom, *"Bulk wave generation by the IDT,"* in J. H. Collins and L. Masotti (eds), *Computer-Aided Design of Surface Acoustic Wave Devices*, Elsevier, Amsterdam, pp. 64–81, 1976.
16. J. E. Padgett, C. G. Günther and T. Hattori, "Overview of wireless personal communications," *IEEE Communications Magazine*, pp. 29–41, January 1995.
17. C. K. Campbell, Longitudinal-mode leaky SAW resonator filters on 64° Y-X lithium niobate," *IEEE Trans. Ultrasonics, Ferroelectrics, and Frequency Control*, vol. 42, pp. 883–888, September 1995.
18. Y. Shimizu and M. Tanaka, "A new cut of quartz for SAW devices with extremely small temperature coefficient by leaky surface wave," *Electronics and Communications in Japan*, Part 2, vol. 69, pp. 48–56, 1986.

19. K. Yamanouchi and M. Takeuchi, "Applications for piezoelectric leaky surface waves," *Proc. 1990 IEEE Ultrasonics Symp*osium, vol. 1, pp. 11–18, 1990.

20. R. Chipman, *Theory and Problems of Transmission Lines*, Schaum's Outline Series, New York, pp. 35–37, 1968.

21. C. S. Lam, D. E. Holt and K. Hashimoto, "The temperature dependence of power leakage in LST-cut quartz surface acoustic wave filters," *Proc. 1989 IEEE Ultrasonics Symp*osium, vol. 1, pp. 275–279, 1989.

22. A. Isobe, M. Hikita and K. Asai, "Large K^2 and good temperature stability for SAW on new double-rotated cut of α-quartz with gold film," *Proc. 1993 IEEE Ultrasonics Sympo*sium, vol. 1, pp. 323–326, 1993.

23. S. Tonami, A. Nishikata, and Y. Shimizu, "Characteristics of leaky surface acoustic waves propagating on LiNbO$_3$ and LiTaO$_3$ substrates," *Japan Journal of Applied Physics*, vol. 34, Part 1, No. 5B, pp. 2664–2667, May 1995.

24. M. Lewis, "Surface skimming bulk waves, SSBW," *Proc. 1977 IEEE Ultrasonics Symp.*, pp. 744–752, 1977.

25. T. I. Browning, D. J. Gunton, M. F. Lewis and C. O. Newton, "Bandpass filters employing surface skimming bulk waves," *Proc. 1977 IEEE Ultrasonics Symp.*, pp. 753–756, 1977.

26. B. Auld. J. Gagnepain and M. Tan, "Horizontal shear surface waves on corrugated surfaces," *Electronics Letters,* vol. 12, pp. 650–652, 1976.

27. A. Renard, J. Henaff and B. A. Auld, "SH surface wave propagation on corrugated surfaces of rotated y-cut quartz and berlinite crystals," *Proc. 1981 IEEE Ultrasonics Symp.*, vol. 1, pp. 123–127.

28. K. F. Lau, K. H. Yen, R. S. Kagiwada and K. L. Wang, "Further investigation of shallow bulk acoustic waves generated by using interdigital transducers," *Proc. 1977 IEEE Ultrasonics Symp.*, pp. 996–1001, 1977.

29. K. H. Yen, K. F. Lau and R. S. Kagiwada, "Recent advances in shallow bulk acoustic wave device," *Proc. 1979 IEEE Ultrasonics Symp.*, pp. 776–785, 1979.

30. A. Ballato and T. J. Lukaszek, "Shallow bulk acoustic wave progress and prospects," *IEEE Trans. Microwave Theory & Tech.*, vol. MTT-27, pp. 1004–1012, 1979.

31. K. H. Yen, K. F. Lau and R. G. Kagiwada, "Shallow bulk acoustic wave filters," *Proc. 1978 IEEE Ultrasonics Symp.*, pp. 680–683, 1978.

32. M. Yamaguchi and K. Hashimoto, "Simple estimation for SSBW excitation strength," *Journal of Acoustical Society of Japan* (E) vol. 6, 1, pp. 51–54, 1985.

33. K. Y. Hashimoto, M. Yamaguchi and H. Kogo, "Experimental verification of SSBW and leaky SAW propagation on rotated Y-cuts of LiNbO$_3$ and LiTaO$_3$," *Proc. 1983 IEEE Ultrasonics Symp.*, vol. 1, pp. 345–349, 1983.

34. F. S. Hickernell, "Zinc oxide films for acoustoelectric device applications," *IEEE Trans. Sonics Ultrasonics*, vol, SU-32, pp. 621–629, Sept. 1985.

35. F. S. Hickernell, "The microstructural characteristics of thin-film zinc oxide for SAW transducers," *Proc. 1984 IEEE Ultrasonics Symp.*, vol. 1, pp. 239–242, 1984.

36. T. Shiosaki, Y. Mikamura, F. Takeda, and A. Kawabata, "High-coupling and high-velocity SAW using ZnO and AlN films on a glass substrate," *IEEE Trans. Ultrasonc., Ferroelec., Freq. Contr.*, vol. UFFC-33, pp. 324–330, May 1986.

37. K. Tsubouchi and N. Mikoshiba, "Zero-temperature coefficient SAW devices on AlN epitaxial films," *IEEE Trans. Sonics Ultrasonics*, vol. SU-32, pp. 634–644, Sept. 1985.

38. H. Nakahata *et al.*, "SAW devices on diamond," *Proc. 1995 IEEE Ultrasonics Symp.*, vol. 1, pp. 361–370, 1995.

39. M. Kadota, K. Morozumi, T. Ikeda, and T. Kasanami, "Ceramic resonators using BGS waves," *Japanese Journal of Applied Physics*, vol. 31, Supplement 31–1, pp. 219–221, 1992.

40. K. Morozumi, M. Kadota, and S. Hayashi, "Characteristics of Bleustein-Gulyaev-Shimizu wave resonators on ceramics at high frequencies," *Japanese Journal of Applied Physics*, vol. 35, Part 1, No. 5B, pp. 2991–2993, May 1996.
41. K. Morozumi, M. Kadota, and S. Hayashi, "Characteristics of BGS wave resonators using ceramic substrates and their applications," *Proc. 1996 IEEE Ultrasonics Symp.*, vol. 1, pp. 81–86, 1996.
42. R. C. Williamson, "Reflection grating filters," in H. Matthews (ed.), Surface Wave *Filters*, John Wiley and Sons, New York, Chapter 9, p. 404, 1977.
43. R. S. Wagers, "Spurious acoustic responses in SAW devices," *Proceedings of IEEE*, vol. 64, pp. 699–702, May 1976.
44. A. J. Slobodnik, Jr, "Materials and their influence on performance", in A. A. Oliner (ed.), *Acoustic Surface Waves*, (Topics in Applied Physics, Volume 24), Springer-Verlag, Berlin, Chapter 6, 1978.
45. R. M. Hays and C. S. Hartmann, "Surface-acoustic-wave devices for communications," *Proceedings of IEEE*, pp. 652–671, May 1976.
46. O. Yamazaki, T. Mitsuyu, and K. Wasa, "ZnO thin-film SAW devices," *IEEE Trans. Sonics Ultrasonics*, vol. SU-27, pp. 369–379, Nov. 1980.

—3—

Principles of Linear-Phase SAW Filter Design

3.1. Introduction

3.1.1. GENERAL CONCEPTS OF LINEAR-PHASE FILTERS

Filters with linear-phase response are employed in communications circuits, when distortion of the processed signal is to be avoided. A linear-phase filter is also termed a nondispersive one. In such a filter both phase velocity $v = \omega/\beta$ (where $\omega =$ angular frequency and $\beta = phase\ constant$) *and group velocity* $v = d\omega/d\beta$ *are ideally* constant and equal over the desired frequency band of the filter. All of the frequency components of an input signal will experience the same time delay and attenuation in passing through the filter, so that it emerges in undistorted form. To illustrate this point, consider a hypothetical example where a rectangular voltage pulse is applied to the input of a transmission line structure. The pulse may be represented mathematically in terms of its component frequencies as a Fourier series expansion of phase-related harmonics. Its transmission through the structure can be examined in terms of what happens to these harmonic components. If they are allowed to disperse through the filter with different velocities, the "reconstructed" output signal will not have the same shape as the input one. Additionally, the output signal will be a degraded replica of the input one if the harmonic components are attenuated by different amounts.

3.1.2. COMPARISON WITH L-C FILTERS

Conventional "linear-phase" passive L-C filters (also called Bessel filters) all have some inherent degree of phase nonlinearity. The degree to which their linearity of phase response is achieved over a prescribed frequency range increases with the order of the filter (i.e., with the number of reactive components). The resultant overall size of a passive L-C filter,

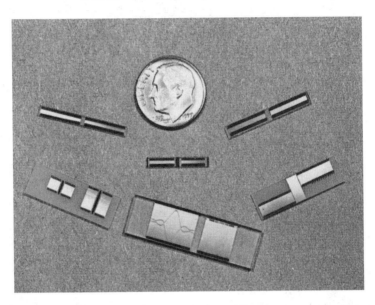

Fɪɢ. 3.1.   Some commercial Rayleigh-wave bandpass filters and resonators. (Courtesy of Sawtek Inc., Orlando, Florida.)

together with its complexity and cost, may render it unsuitable for many applications—especially mobile or airborne ones. By contrast, SAW filters with comparable performance are often tiny by comparison. Moreover, their sizes will decrease with increasing frequency. Figure 3.1 illustrates the size of some Rayleigh-wave filters for both bandpass-filter and resonator applications. Currently, ceramic-package sizes for commercial 2.4-GHz leaky-SAW RF front-end filters with 100-MHz bandwidth for Wireless Local Area Networks (WLAN), are typically 3.8 mm × 3.8 mm, with a height of 1.5 mm [1]. Figure 3.2 illustrates two examples of low-loss LSAW RF front-end filters for mobile communications. The larger package (3.8 mm) is a low-loss Tx-filter for a European GSM[1] digital cellular phone (Rx-band: 925–960 MHz; Tx-band: 880–915 MHz). The smaller package (3.0 mm) in Fig. 3.2 is a low-loss Rx-filter for Japan Personal Digital Cellular (PDC) radio (Rx-band: 1429–1453 MHz; Tx-band: 1477–1501 MHz). (For wireless cards which are required to fit into personal-computer (PC) slots, the *height* of the SAW-package can often be the most critical dimension!) The potential advantages of their small size are readily apparent, particu-

---

[1] GSM stands for Global System for Mobile Communications.

FIG. 3.2. Examples of low-loss RF front-end LSAW filter packages. Larger package (3.8 mm) is a Tx-filter for European GSM digital cellular phone (Rx-band: 925–960 MHz; Tx-band: 880–915 MHz). Smaller package (3.0 mm) is an Rx-filter for Japan Personal Digital Cellular (PDC) phone (Rx-band: 1429–1453 MHz; Tx-band: 1477–1501 MHz). (Courtesy of Fujitsu Ltd., Kawasaki, Japan.)

larly for mobile or wireless communications modems, because they can also be designed to meet exacting frequency response specifications that could otherwise require several hundred large reactive components if implemented in L-C form.

The alternating polarities of fingers in a SAW IDT restrict their signal processing capabilities to ac signals. They can not be implemented as lowpass filters, but only as bandpass (or bandstop) ones. In their bandpass operation, however, they have much more design versatility than *LC* filters. Moreover, and again in contrast to *LC* or other conventional analog filters, the amplitude response of a linear-phase SAW filter can be shaped to be nonsymmetric about center frequency. This type of filter finds application in equalizer circuitry in transmission channels. As these advanced design concepts require application of digital sampling techniques, their examination will be deferred to a later chapter.

In another significant contrast with passive *LC* circuitry, SAW filters can also be designed to operate very efficiently in harmonic modes up to about the ninth harmonic frequency. This can lead to a considerable simplification of the costly microelectronic lithographic processes required for gigahertz-device fabrication [2]–[4].

3.2. Scope of This Chapter

3.2.1. MODELLING SIMPLIFICATIONS USED HERE

At this point, the reader should note two important aspects of linear-phase SAW filter design, as they will be applied in this coverage. First, the linear-phase SAW-filter design concepts to be considered were originally developed for Rayleigh-wave filters, when such "true-SAW" designs were the only ones under development. To a good approximation, however, these concepts may also be applied to the design of IDTs for leaky-SAW filters (with somewhat different approximations being used for SSBW and STW designs). In this chapter, therefore, and for the remainder of Part 1 of this text, the designation "SAW" should be construed as applicable to both Rayleigh-wave and leaky-SAW filters.

Second, early SAW device-design techniques were applied to relatively low-frequency structures for intermediate-frequency (IF) stages. (Applications to RF front-end filters are relatively new developments.) For the most part, acoustic reflections from interdigital transducer (IDT) fingers were considered to be negligible in many such IF filter designs. However, and as will be demonstrated in Chapter 9, the frequency response of a surface-wave device can be strongly affected—either desirably or undesirably, depending on the application—by significant levels of acoustic reflection from the IDT fingers. The important design parameter that relates to this involves the *film-thickness ratio* (h/λ_o) where h = IDT metallization thickness[2] and λ_o = IDT center wavelength in the fundamental mode. (Film-thickness ratios (h/λ_o) are normally quoted for *aluminum* metallization, given in Angstrom (Å) units, where 1 Å = 10^{-8} cm. Moreover, because small-valued ratios are involved, the film thickness ratio is usually expressed as a percentage.) Until the end of Chapter 8, to illustrate SAW filter design considerations—without the added complication of IDT finger reflections—we will consider that IDT metallization thicknesses and operating wavelengths are such that $h/\lambda_o < \sim 1\%$, and that IDT finger-reflection effects may be reasonably neglected for such ratios. As will be introduced in Chapter 9 of this text, the effects of IDT finger reflections must be incorporated into high-frequency device designs where $h/\lambda_o > 1\%$; this is particularly the case for applications to low-loss RF front-end filters.

---

[2] Metallization-film thicknesses for IDT fabrication are normally in the range $h \approx 500$ Å to 2000 Å. (1 Å = 10^{-8} cm).Thicknesses much less than 500 Å can lead to poor electrical conductivity in the IDT. Moreover, the resultant series-resistance loss in input and output circuitry will contribute to the device insertion loss.

3.2.2. CHAPTER COVERAGE

Bearing the forementioned simplifications in mind, this chapter demonstrates the frequency-response capabilities of basic linear-phase SAW filters. To facilitate, the elementary delta-function model employed is based on digital sampling concepts. Apodization and window-function techniques are demonstrated as tools to prescribe a desired frequency response. These modelling techniques are applied here to the illustrative design of a wideband linear-phase SAW filter with a "flat" passband response, such as might be desired for IF filtering in a code-division multiple access (CDMA) spread-spectrum wireless receiver. Deviations from linear phase as a result of various second-order effects are also considered.

Example 3.1 IDT Film-Thickness Ratio. Determine the values of film thickness ratio (h/λ_o) for a leaky-SAW filter on 36° YX-LiTaO$_3$ for: (a) a 100-MHz design using an IDT with aluminum metallization film thickness $h = 1500$ Å; and (b) a 2.4-GHz design with the same metallization thickness h. For simplistic illustrative purposes here, assume that the acoustic velocity for this piezoelectric is not changed by the applied metallization.
Solution. (a) From Table 2.2, the acoustic velocity for 36° YX-LiTaO$_3$ is $v = 4112$ m/s. The acoustic wavelength λ_o at center frequency is $\lambda_o = v/f_o = 4112/(100 \times 10^6) \approx 4.1 \times 10^{-5}$ m $= 41\,\mu$m $(1\,\mu$m $= 10^{-6}$ m). Expressing h in metres, and deriving the film-thickness ratio as a percentage *gives* $h/\lambda_o = (1500 \times 10^{-10})/(41 \times 10^{-6}) = 3.6 \times 10^{-3} = 0.36$ %. (b) At 2.4 GHz the wavelength at center frequency is $\lambda_o = v/f_o = 4112/(2.4 \times 10^9) \approx 1.7 \times 10^{-6}$ m $= 1.7\,\mu$m. This gives $h/\lambda_o = (1500 \times 10^{-10})/(1.7 \times 10^{-6}) = 8.8 \times 10^{-2} = 8.8\%$

3.3. Deviations from Ideal Phase Response in SAW Filters

3.3.1. IDEAL LINEAR-PHASE RESPONSE

Consider the overall transfer function $H(f)$ of a SAW filter employing bidirectional IDTs, as sketched in Fig. 3.3. This can be evaluated with reference to the block diagram of Fig. 3.4, which shows that H(f) comprises three component terms. Two of these relate the individual responses $H_1(f)$ and $H_2(f)$ of input and output IDTs, respectively, while the third term is associated with the transmission path between the IDTs. The overall transfer function is:

$$H(f) = \frac{V_{out}}{V_{in}} = H_1(f)H_2^*(f)e^{-j\beta x(f)}, \qquad (3.1)$$

where $|H_2(f)| = |H_2^*(f)|$ here for conjugate response $H_2^*(f)$. In Eq. (3.1) the phase delay term between input and output IDTs is given as $e^{-j\beta x(f)}$ where

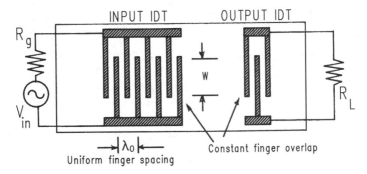

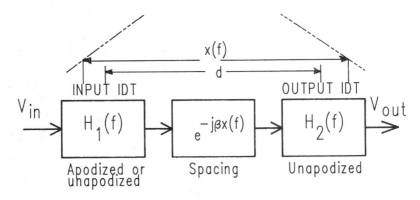

Fig. 3.3. Elementary structure comprising input and output IDTs with uniform finger spacing and constant finger overlap (apodization).

Fig. 3.4. Transfer-function components for the SAW filter of Fig. 3.2. In this specific example, with uniform IDTs and symmetric apodization patterns, the SAW emissions from each IDT are considered to emanate from phase centers distance d apart. Distance $x(f)$ is used instead, when this condition does not hold.

$\beta = 2\pi/\lambda = 2\pi f/v$ is the phase constant, and $x(f)$ is a frequency-dependent separation between those segments of input and output IDTs that are excited at signal frequency f. This relates to the general case where IDTs can have arbitrary finger separation. For the specific example of Fig. 3.3, where input and output IDTs have uniform finger separation as well as uniform finger apodization overlap, $x(f)$ reduces to $x(f) = d$, where d is the distance between the midpoints of the bidirectional input and output IDTs.

What does this signify? First, it is as if all the SAW emissions from the excited input IDT emanate from a common midsection, while those arriving at the output IDT are summing at that midsection. Moreover, at this axis the relative phase shifts of individual response functions $H_1(f)$ and $H_2(f)$ simply become $0°$ or $180°$, depending on which way the electrical leads are attached. As a result, the phase shift of this type of SAW filter is due entirely to the separation between IDT midsections, and given by $e^{-j\beta d}$. This will give a linear-phase response because phase angle $\angle \beta d$ varies linearly with frequency. The desired amount of filter time delay, or phase shift at a given frequency, is then simply achieved by selecting the spacing between IDTs. This type of SAW filter is usually known as a SAW *delay line*.

This linear relationship holds for the IDTs of Fig. 3.3, where the overlap of adjacent electrode fingers is constant (i.e., uniformly apodized). As will be demonstrated in what follows, it also holds true when either or both IDTs are apodized, provided that the apodization pattern is symmetric about the IDT central axis.

3.3.2. DEVIATIONS DUE TO SECOND-ORDER EFFECTS

The foregoing introductory presentation assumed that the SAW filter of Fig. 3.3 was operating under ideal conditions. As with all other filters, however, practical SAW filters are not ideal devices. Various second-order effects will perturb the ideal environment to some degree, and degrade the filter response. It is the task of the SAW designer to ensure that these perturbations are reduced to acceptable levels. This requires a knowledge of their sources, as well as circuit design techniques for dealing with them. Before examining the mathematics of SAW circuit models in this chapter and in Chapter 4, it is worthwhile to look ahead briefly and review the principal second-order effects that corrupt SAW filter responses, as listed in Table 3.1. These are:

TABLE 3.1

SOURCES OF SAW FILTER RESPONSE DEGRADATION

- 1. Electromagnetic feedthrough (crosstalk)
- 2. Triple-transit-interference (TTI)
- 3. (Electrode finger reflections)
- 4. Bulk wave interference
- 5. Circuit factor loading
- 6. Impedance mismatches
- 7. Diffraction
- 8. (Harmonic responses)
- 9. (Ground loops)

1. Electromagnetic feedthrough (crosstalk), which relates to the direct coupling of input signal from input to output IDTs, in the form of electromagnetic radiation. Essentially, the two IDTs act as capacitor plates, which may be separated by only a few acoustic wavelengths. The amount of electromagnetic feedthrough, which increases with frequency as the capacitive reactance decreases, can be significant. This feedthrough is coupled to the output IDT at the velocity of light, to interact with the SAW signal arriving there. This interaction gives rise to periodic ripples of amplitude and phase across the SAW filter passband at ripple frequency $f_r = 1/\tau$, where $\tau = d/v$ is the SAW propagation time between IDT phase centers separated by distance d. This can be one of the most troublesome sources of interference in SAW devices. (Others include parasitics due to packaging and bonding wires.)

2. Triple-transit-interference (TTI) is due to multiple SAW reflections between bidirectional input and output IDTs. This is normally associated with regenerative effects caused by current flow in the IDTs. Some of the SAW power received by the output IDT is reradiated into the piezoelectric. That portion from the output IDT that arrives back at the input can lead to further regeneration of a SAW wave. As a result, the main voltage signal induced in the output IDT is corrupted by additional voltages due to these multiple SAW reflections. The path differences between the main and doubly reflected SAW signals result in amplitude and phase ripple across the SAW passband, at a ripple frequency $f = 1/2\tau$.

3. The thin metal film IDT fingers deposited on the piezoelectric crystal surface introduce impedance and mass-loading discontinuities, so that a portion of the surface wave is reflected from both the front and back surface of each finger. This has two major effects. First, this will lead to a pronounced distortion of the radiation conductance of the IDT for film-thickness ratio $h/\lambda_o > \sim 1\%$, which can most beneficially be applied to the design of RF filters and resonator-filter structures, as well as some high-frequency IF filter designs (these will be neglected until introduced in Chapter 9). Further, multiple SAW reflections can occur between input and output IDTs to form an additional source of triple-transit interference (TTI).

4. Bulk-wave interference, which was already introduced in Chapter 2, will corrupt the passband amplitude and phase response, as well as reduce the out-of-band amplitude rejection levels. It will also lead to increased filter insertion loss.

5. Circuit factor loading results from the finite source and load impedances that are external to the SAW filter. Both the input and output impedances of a SAW filter are frequency-dependent parameters. The consequence of this is that an input voltage will be divided between source resistance and filter input impedance in a frequency-dependent way unless compensation is made. The same situation applies to the output transducer circuitry.

6. In the input IDT circuitry, signal power is divided between the generator resistance and a fictitious resistance associated with SAW power radiation into the acoustic substrate medium. Maximum power transmission by the SAW wave occurs only when the two resistances are matched (i.e., equal). The same situation arises with the load resistance connected to the output IDT circuit. Source or load changes result in insertion loss increases or changes due to mismatching.

7. Diffraction occurs in SAW IDTs in much the same way as it does in optical systems. Ideally, the SAW emissions induced by, and travelling under, excited IDT electrodes should have "flat" wavefronts, so that all parts of a wavefront launched by one IDT finger arrive at a receiving IDT finger after the same time delay. As in optics, however, the SAW wavefront will be spherical to a degree dependent on the aperture of the radiating source. This will corrupt the filter response. Anisotropy of the piezoelectric medium will also give rise to focusing or defocusing of the surface wave.

8. Excited IDT fingers can generate SAW waves at harmonic signal frequencies in addition to those generated at the fundamental signal frequency. (This may be a desirable or undesirable feature, depending on the application, and is listed in parentheses as a result.) The levels of such harmonic frequencies will be dependent on the relative width of the metal fingers with respect to their separations. They will also depend on the overall IDT geometry used. This can lead to undesirable levels of harmonic response outside the passband of the fundamental signal. In some applications, however, the SAW filter is designed to operate at a desired harmonic frequency, while suppressing the fundamental at the same time [5]–[7].

9. Ground loops can be particularly troublesome in the design of compact mobile and wireless communications circuitry, such as for wireless cards in personal-computer slots. With reference to Fig. 3.3, many SAW filter designs have one pad of both the input and output IDTs "grounded," for ease of fabrication. This can give rise to performance degradation, due to spurious coupling between circuit stages. One

method for reducing this in the receiver RF and mixer stages of such circuits is to employ RF resonator-filter designs with differential "ungrounded" output IDTs [8], which can then be connected directly to the dual-input stage of an active mixer.

3.4. Simple Modelling of an Ideal Linear-Phase SAW Filter

3.4.1. PROBLEMS WITH CONVENTIONAL MODELLING

While the modelling and synthesis of passive L-C filters is normally based on the use of pole-zero analytic techniques, these can not be applied to SAW filter design. The reason for this is that the concept of "poles" in conventional filter design relates to energy storage in the electric and magnetic fields of the reactive components. However, SAW filter processes do not relate to energy storage, but to energy transfer between input and output IDTs. As a result, the pole-zero transfer function of a SAW filter will contain only zeros, thus rendering it unsuitable for the design of this type of filter. Completely different modelling approaches must thus be employed for SAW filter design. An exception to this relates to SAW resonator operation (as considered in Chapter 11), where mechanical energy is indeed stored in the stress-strain fields within the elastic solid substrate. Despite this, it is usually better to use different modelling techniques for SAW resonators.

3.4.2. THE DELTA-FUNCTION MODEL

The *delta-function* model provides basic information on the transfer function response of a SAW filter [9]. It can only treat some of the second-order effects considered in the preceding. It cannot provide information on filter input-output impedance levels, circuit factor loading, harmonic operation, bulk wave interference or diffraction. Because it cannot cater to impedances, the transfer function determination yields only relative insertion loss as a function of frequency. Despite these modelling limitations, it can still provide excellent preliminary design information on the response of a SAW filter using bidirectional IDTs in input and output stages, and is introduced here for this reason. In the mathematical derivations given in what follows, the bidirectional nature of the IDTs will be ignored, as absolute values of insertion loss are not predicted by this model.

The delta-function model approximates the complex electric-field distribution between adjacent fingers of an excited IDT, as a discrete number of delta-function sources. As illustrated in Chapter 2, the propagation of a surface wave on a piezoelectric may be associated with travelling waves of

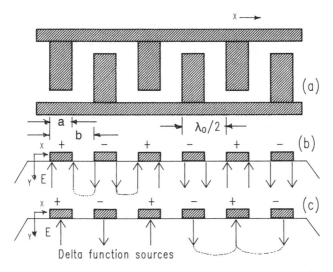

Fig. 3.5. (a) Bidirectional IDT with uniform IDT. (b) Delta-function modelling of E-field distribution in xy-plane under excited IDT, with sources at finger edges. (c) Simpler delta-function model employs only one source field under each electrode.

surface potential Φ and electric-field intensity E produced by and emanating from an excited IDT. While the distribution of the time-varying electric field under adjacent electrodes is complex, it may be approximated as normal to the piezoelectric surface. It is designated as E_{xy}, in the xy-plane, in the sketch of Fig. 3.5(a).

For illustrative purposes consider that Fig. 3.5(a) relates to the input IDT with response function $H_1(f)$. As depicted in this figure, electrode fingers alternate in voltage polarity, with centers spaced $\lambda_o/2$ at center frequency $f_o = v/\lambda_o$, so that SAW emissions under the IDT add constructively at this frequency. The metallization ratio η relating relative finger width and spacing is given by $\eta = a/b$. This determines the relative amount of harmonic frequency generation relative to the fundamental. It is shown in Fig. 3.5 with the usual value $\eta = 0.5$. The intensity of the electric-field distribution will be proportional to the instantaneous charge accumulation on adjacent electrode fingers due to input voltage V_{in} established by the time-dependent input signal. At any instant, adjacent electrodes have opposite voltage polarity and opposite charge accumulation. Because unlike charges attract, these migrate to the edges of the IDT fingers. The resultant charge distribution can be modelled as delta-function sources of electric-field intensity E_y at the finger edges. As depicted in Fig. 3.5(b), each electrode finger will have two delta sources of electric-field intensity associated with it, which amplitudes are proportional to the applied voltage. The directions and relative field polarities will alternate in pairs from finger to finger. A summation of these delta sources can be used to simulate the resultant electric-field intensity of one excited IDT in either the input or output stages. The overall transfer response can be obtained when this summation process is extended to include both IDTs.

While the delta-function model can be used as it stands, with two delta-function sources associated with each excited electrode, it can be further simplified for computational convenience. As the delta-function model cannot furnish information on harmonic performance, there is no special significance in employing a metallization ratio $\eta = 0.5$. And because any metallization ratio can be employed for this computation, there is no need to associate two delta-function excitation sources with each electrode. The two delta functions at finger edges can be replaced by one equivalent delta-function source at the center of each finger, sketched in Fig. 3.5(c).

The spatially distributed delta-function contributions may be summed at a convenient reference point along the x-axis. Consider that this reference point is taken as $x=0$ at the center of the IDT as shown in Fig. 3.6(a). Moreover, assume the IDT has an odd number N of electrodes,

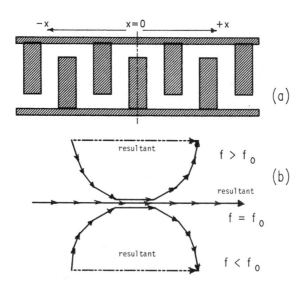

Fig. 3.6. (a) Reference axis $x=0$ at midpoint of uniform IDT. (b) When delta-function modelling is applied to excited IDT, the resultant relative phase angle of associated transfer function is always $0°$ or $180°$ at $x=0$.

so that $x = 0$ is at the center of an electrode. (The same end result will be obtained if N is taken to be an even number. The computation is slightly different, however, as the reference position $x = 0$ is at the midpoint between two electrodes in this latter instance.) Because only relative insertion loss can be obtained from this model, the amplitudes of the delta-function sources at each electrode can be normalized to a value $|E_y| = 1$. The summation of sources yields the frequency response $H_1(f)$.

Although the amplitudes of the delta sources are constant, their individual phase angles at the point of summation $x = 0$, will depend on the distance x from each source to this summation point. Individual phase-shift terms are $e^{-j\beta x_n}$ for electrodes located at discrete points $-x_N \leq x_n \leq +x_N$ over the length of the IDT. The resultant frequency response $H_1(\beta) = H_1(f)$ is

$$H_1\left(f\right) = \sum_{n = -(N-1)/2}^{(N-1)/2} \left(-1\right)^n A_n e^{-j\beta x_n}, \qquad (3.2)$$

where the term $(-1)^n$ relates the alternating electrode polarity, while A_n is an amplitude parameter proportional to the finger apodization overlap. This may be normalized to $A_n = 1$ for the uniformly apodized IDT.

Equation (3.2) may be reformulated by employing the trigonometric equivalent of the exponential term, so that

$$H_1\left(f\right) = \sum_{n = -(N-1)/2}^{(N-1)/2} \left(-1\right)^n A_n \left[\cos\left(\beta x_n\right) - j\sin\left(\beta x_n\right)\right]. \qquad (3.3)$$

The significance of choosing the reference axis $x = 0$ at the center of the IDT becomes apparent. From trigonometric identities $\cos(\theta) = \cos(-\theta)$ and $\sin(\theta) = -\sin(\theta)$, it is seen that the imaginary (i.e., $j \sin$) terms in Eq. (3.3) cancel out in pairs, so that the summation for $H_1(f)$ is entirely in terms of the purely real quantity $\cos(\beta x)$. When the reference axis is taken at the midpoint of the symmetric structure, the phasor relating $H_1(f)$ as a function of frequency will always lie along the real axis in the complex plane, as sketched in Fig. 3.6(b). Its relative phase at this point will always be just $0°$ or $180°$, depending on which way the electrical leads are connected. To obtain the bandpass response of the IDT about some center frequency f_o, now express frequency f in Eq. (3.3) as $f = \{(f - f_o) + f_o\}$. After some manipulation, an expansion of Eq. (3.3), with $A_n = 1$, yields a cosine series [10] given by:

$$H_1(f) = 1 + 2\cos\left(\pi \frac{f - f_o}{f_o}\right) + 2\cos\left(2\pi \frac{f - f_o}{f_o}\right)$$
$$+ \ .... \ + 2\cos\left(N_p \pi \frac{f - f_o}{f_o}\right), \tag{3.4}$$

where $N_p = (N-1)/2 \approx N/2$ for large odd N and $N_p = N/2$ for even N, and represents the number of electrode finger pairs in the IDT. Close to the center frequency f_o, Eq. (3.4) approximates a sinc function response given by

$$|H(f)| \propto \left|\frac{\sin X}{X}\right| \propto \left|\frac{\sin[N_p \pi (f - f_o)/f_o]}{N_p \pi (f - f_o)/f_o}\right|, \tag{3.5}$$

where the sinc function is defined as $\text{sinc}(X) = (\sin X)/X$.

An inspection of Eq. (3.5) indicates that the first nulls of the response about f_o occur when $\sin\{N_p \ \pi(f - f_o)/f_o\} = \sin(\pi)$, or $N_p = f_o/(f - f_o)$. This can be expressed as $N_p = f_o/(f - f_o) = 2/\text{BW}_{nn}$, where BW_{nn} is the fractional bandwidth between first nulls on either side of f_o. It may be shown that $\text{BW}_{nn} \approx 2.\text{BW}_4$, where BW_4 is the fractional bandwidth to the 4-dB amplitude points about the main response lobe. Expressed as a percentage, $\text{BW}_4\% \approx (100/N_p)\%$ where N_p is the number of electrode pairs in the IDT.

Figure 3.7 illustrates a computation of frequency response and relative insertion loss of one excited IDT with constant finger overlap, as derived from Eq. (3.2). Here, the IDT was designed for center frequency $f_o = 100\,\text{MHz}$, with $N_p = 20$ electrode pairs and $\text{BW}_4\% = (100/N_p) = 5\%$. Observe that the first sidelobes are $\approx 12\,\text{dB}$ below the main peak at center frequency f_o. This level is characteristic of sinc function responses. The preceding relations are applied to obtain $|H_1(f)|$ for the input IDT. The frequency response $|H_2(f)|$ of the output IDT is obtained in the same way. Its relative phase shift will also be just $0°$ or $180°$. The total amplitude response is $|H(f)| = |H_1(f)| \cdot |H_2(f)|$, while the overall phase shift will be due entirely to the phase delay $e^{-j\beta d}$ between centers of input and output IDTs. (In such uniform IDTs these midpoints are usually referred to as *phase centers.*) The phase shift $(-\beta d)$ will be perfectly linear as a function of frequency, in the absence of any second-order perturbations. This type of SAW filter with uniformly apodized IDTs is usually referred to as a delay line. It is employed at IF frequencies for implementing fixed time delays in a signal processing circuit. The purely real phase response obtained with a uniform IDT with constant apodization in Fig. 3.6 is not limited to this

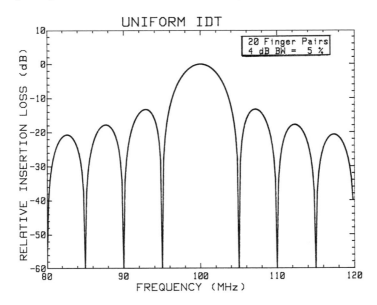

FIG. 3.7.   Delta-function modelling of magnitude response $|H_1(f)|$ of illustrative uniform IDT with $N_p = 20$ finger pairs, and center frequency $f_o = 100$ MHz.

specific geometric finger pattern. For nonuniform apodization, the amplitude parameter A_n in Eq. (3.3) will no longer be a constant, but will be proportional to individual finger overlaps. As long as the IDT finger apodization pattern is symmetric about the center axis, however, the sum of the imaginary terms in Eq. (3.3) will still be zero, so that the relative phase of the IDT transfer function will again be 0° or 180°. This pattern symmetry requirement is not just confined to apodization overlap. Symmetry also requires the finger spacings themselves to be mirror images about the central axis.

While the preceding discussions have specifically concerned linear-phase SAW filters, there are many signal processing uses for SAW filters with prescribed nonlinear-phase response. From the foregoing discussions on pattern symmetry, it can be seen that nonlinear-phase response in SAW filters can be obtained in one of two ways. One way is to use IDTs with uniform finger spacings and nonuniform apodization about the central axis. The other is to use uniform apodization and nonsymmetric finger spacing about the central axis. Certain types of SAW filters with quadratic phase response—known as *chirp* filters—are central to many SAW signal processing circuits.

Example 3.2 Linear-Phase SAW Filter with Slanted Finger IDTs. A
SAW filter employs input and output IDTs with slanted finger geometries
as shown in Fig. 8.1 of Chapter 8. From inspection of the IDT geometries,
determine whether or not the filter will have nominal linear-phase
response. ■

Solution. If a vertical line is drawn to bisect the input IDT, it is seen that
the two segments are symmetric about this axis. The same situation holds
for the output IDT. As a result, this SAW filter will have a nominal linear-
phase response. ■

Example 3.3 SAW Filter Sidelobe Suppression. A SAW filter with nominal linear-phase response employs identical uniformly apodized bidirectional IDTs in input and output stages. Each IDT has $N = 80$ electrodes.
Determine (a) the approximate 4-dB percentage fractional bandwidth of
each IDT and (b) their 3-dB fractional bandwidths; (c) indicate whether or
not the overall 4-dB filter bandwidth will be the same as in (a); and (d)
determine the approximate suppression level (in dB) of the first sidelobes
of the filter. ■

Solution. (a) The number of finger pairs in each IDT is $N_p = N/2 = 80/$
$2 = 40$. The 4-dB fractional bandwidth of each is $BW_4\% \approx 100/N_p = 100/$
$40 = 2.5\%$. (b) The 3-dB fractional bandwidth is about 0.9 of the 4-dB value.
Thus $BW_3\% \approx 0.9.BW_4\% \approx 0.9 \times 2.5 \approx 2.2\%$. (c) When the two IDT amplitude responses are multiplied, the resultant responses of each lobe will
become "sharper," although the nulls remain at the same frequencies. As a
result the overall 4-dB bandwidth will be less than for each IDT. (d) The
first sidelobe peaks of each IDT are suppressed by about 12 dB as shown in
Fig. 3.7. The first sidelobe suppression in the composite filter will be about
24 dB below the main peak. ■

3.5.   Fourier Transforms and IDT Finger Apodization

3.5.1.   FOURIER TRANSFORM PAIRS

Now consider the relationship between the frequency response and the
impulse response of a SAW filter. The impulse response is related to the
frequency response by the Fourier-transform pair

$$H(f) = \int_{-\infty}^{+\infty} h(t)e^{-j2\pi ft} dt \qquad (3.6a)$$

with a one-to-one correspondence between $h(t)$ and $H(f)$, where $h(t)$ is the
impulse

$$h(t) = \int_{-\infty}^{+\infty} H(f)e^{j2\pi ft} df, \qquad (3.6b)$$

response and $H(f)$ is the frequency response of the system. A knowledge of one response enables the other to be derived. Starting with the desired steady-state frequency response, the corresponding impulse response can be derived and this result can be used to synthesize the geometric pattern of the IDT. Its resultant spatial pattern represents a sampled version of this impulse response, converted to the spatial domain. Let the desired transducer impulse response $h(t)$ be expressed as

$$h(t) = a(t)e^{j\phi t}, \tag{3.7}$$

where $a(t)$ is the time-dependent amplitude response. The phase-weighting condition, which tells us that the IDT fingers can only be placed at physically *real* locations [11], leads to the requirement that the phase ϕ is

$$\phi(t_n) = 0 \quad or \quad \pi. \tag{3.8}$$

If we arbitrarily assume that the physical sampling is at the center of a finger gap, as shown in Fig. 3.8, Eq. (3.8) yields the required sampling times, and positions for each IDT finger. This is for the general case where the IDTs can have varying electrode spacing. For the uniform IDTs considered in this chapter the sampling period will, of course, be constant.

Particularly useful Fourier transform pairs to consider for SAW filter design are those relating to sinc-function responses in either the time domain or the frequency domain. Those sketched in Fig. 3.9 are as applicable at baseband (i.e., no carrier frequency). Those given in Fig. 3.10 are for the bandpass situations considered here. Observe that the time-domain amplitude response of Fig. 3.9 now becomes the modulation envelope in Fig. 3.10, while the zero frequency point translates to the carrier frequency.

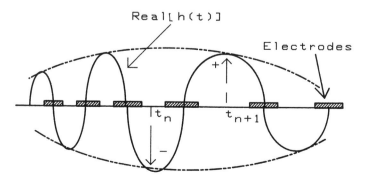

FIG. 3.8. Relationship between sampling times and electrode-finger placement.

BASEBAND TRANSFORMATIONS

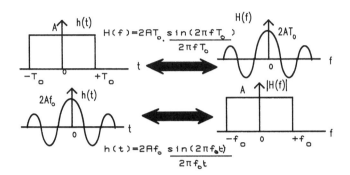

FIG. 3.9. Baseband transformation for Fourier-transform pairs involving the sinc function in the time and frequency domains.

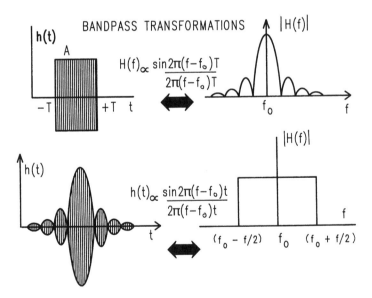

FIG. 3.10. Bandpass transformations for Fourier-transform pairs involving the sinc function in the time and frequency domains.

3.5.2. IMPULSE RESPONSE AND SAW FILTER APODIZATION GEOMETRY

The geometric pattern of an IDT is a unique feature of SAW filter design. It corresponds to a spatially sampled replica of the IDT impulse response. The rationale behind the introduction of Fourier transform concepts now becomes apparent. To appreciate this concept, the reader should reexamine the illustrative sinc-function frequency response of Fig. 3.7, which was computed for the uniform IDT with constant finger overlap in Fig. 3.6. Now compare these with the Fourier transform relations in Fig. 3.10. While the two responses have approximately the same shapes, there are two significant points of departure. With careful design, however, these can be accommodated to implement SAW filters with excellent agreement between theoretical and experimental responses. To elaborate on the preceding points, note first of all that the integration limits in the analog relationships of Eq. (3.6) extend to infinity. However, the impulse response of a SAW filter does not have infinite time duration. It lasts for only as long as the time taken for surface waves to propagate under the entire length of an IDT. In SAW filters, this will range from nanoseconds to microseconds, depending on the length of the IDTs. This is one source of approximation when designing SAW filters with recourse to the Fourier transform relations of Eq. (3.6). While the degree of correspondence will increase with the length of the IDT, some design trade-offs may be applied at this stage if the size substrate size and cost are to be kept within given constraints. On a second point of departure, the Fourier-transform relations in Eq. (3.6) relate to a continuously changing impulse response. Again, however, this is not the case for SAW filters, because the IDT fingers and their placement only give a spatially sampled approximation to the desired time domain impulse response of the analog SAW filter.

3.5.3. SINC FUNCTION APODIZATION OF THE IDT

Instead of using the uniform IDT of Fig. 3.6 with constant finger overlap apodization, consider what the effect will be on its frequency response if it is now given a sinc function apodization as sketched in Fig. 3.11. It can be anticipated that the frequency response of one such IDT should now approximate a rectangular (rather than a sinc function) bandpass response. Ideally, the frequency response would have the rectangular bandpass form of Fig. 3.10 if the impulse response of the SAW filter were of infinite time duration. This is of course not the case in practice due to truncation by the ends of the IDT. In the absence of any external feedback, a SAW filter is a *finite impulse response* (FIR) device.

In Fourier series expansions, truncation of the number of terms gives rise to Gibbs' phenomena. In SAW filter operation, this manifests as undesirable amplitude and phase ripples in the passband. The amplitude response also suffers an additional degradation in that the steepness of the transition at the band edges is reduced. This can lead to filter overlap with resultant interference in frequency division multiplexing

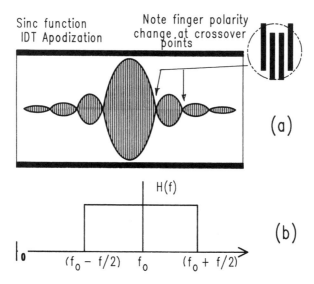

Sinc function
IDT Apodization

Note finger polarity
change at crossover
points

(a)

$H(f)$

$(f_0 - f/2)$ f_0 $(f_0 + f/2)$

(b)

FIG. 3.11.   (a) IDT with sinc-function apodization of electrode-finger overlap. Note finger polarity changes as sinc function goes through zero crossings! (b) Ideal rectangular bandpass frequency response for sinc-function apodization extending to infinity along substrate propagation axis.

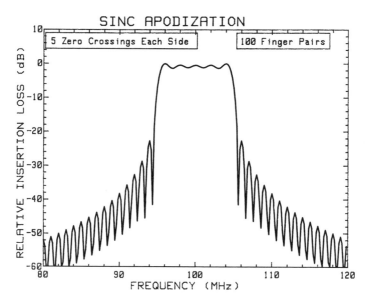

SINC APODIZATION

5 Zero Crossings Each Side   100 Finger Pairs

FIG. 3.12.   Example of amplitude of transfer-function response for IDT with sinc-function apodization extending to five zero-crossings on either side of main lobe of apodization pattern. The IDT has $N_p = 100$ finger pairs, and $f_o = 100$ MHz. Note Gibbs' ripple in passband.

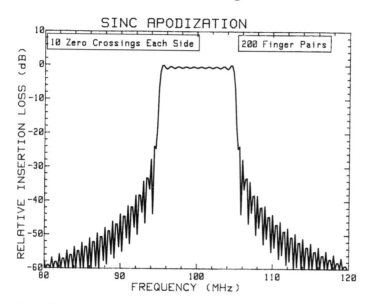

FIG. 3.13. Gibbs' ripple in the passband has same peak-to-peak amplitude but higher ripple frequency when sinc-function apodization of IDT pattern is extended to 10 zero-crossings, using longer IDT with $N_p = 200$ finger pairs.

channels. Both of these problems can be of serious concern to the SAW designer.

Gibbs' ripple effects are illustrated in Fig. 3.12, which shows the computed response of a sample IDT with uniform finger spacing and sinc function apodization. Here, the IDT was designed for operation about a center frequency $f_o = 100$ MHz, with $N_p = 100$ finger pairs. In this example, the sinc function apodization of the IDT pattern was truncated after only 5 zero crossings on either side of the main lobe. Three undesirable features to note here are: (1) the Gibbs' amplitude ripples in the passband; (2) the finite transition band; and (3) the relatively poor close-in stopband rejection of ≈ 24 dB. Figure 3.13 shows the computed bandpass response with the IDT length extended to $N_p = 200$ finger pairs, with 10 zero crossings of the sinc function about the main lobe. Observe that the peak-to-peak amplitude ripple is about the same as in Fig. 3.12, although the ripple frequency has increased. There is, however, a noticeable decrease in the transition band-width as well as improvement in close-in sidelobe rejection from ≈ 24 dB to ≈ 30 dB. In both cases, computations of the frequency response were obtained with the delta-function model.

Example 3.4 Filter Fractional Bandwidth Calculations.   The SAW IDT
response obtained in Fig. 3.12 was derived for a single IDT on YZ-lithium
niobate, with $N_p = 100$ finger pairs and sinc function finger apodization. The
calculation was based on a design with 5 zero crossings of the sinc function
apodization of the IDT pattern on either side of the main lobe. Calculate
the ideal fractional bandwidth for this design. ∎
Solution.   From Fig. 3.10 the impulse response is given as

$$h(t) = 2Af_o \frac{\sin\left(2\pi\left(f - f_o\right)t\right)}{2\pi\left(f - f_o\right)t}.$$

However, $|(f - f_o)| = \Delta f/2$ where Δf is the "rectangular" bandwidth. The sinc
function thus becomes zero when $\sin\{\pi\Delta ft\} = 0$ or $\Delta f.t = 1$. This gives $\Delta f =$
$1/t$ where $t = $ time for a SAW wave to traverse from the IDT center to the
first apodization zero crossing. Let $Z = $ number of zero crossings on either
side of the main lobe of the IDT pattern. This corresponds to $N_z = 0.5N/Z$
finger pairs to the first crossing. The associated distance is $d = N_z\lambda_o$ where
$\lambda_o = $ wavelength at center frequency. The time t for the SAW to traverse d
is $t = d/v$ where $v = $ SAW velocity $= \lambda_o f_o$. This gives $t = N_z\lambda_o/\lambda_o f_o = 0.5N_p/Zf_o$,
so that the bandwidth $\Delta f = 1/t = 2Zf_o/N_p$, or $\Delta f/f_o = (2Z/N_p)\times 100\%$ in
the general case. In this example $Z = 5$ and $N_p = 100$ so that $\Delta f/f_o = (2\times 5/$
$100)\times 100 = 10\%$. The reader should check this result with that illustrated
in Fig. 3.12. ∎

Example 3.5 Bandwidth Calculation Using a Longer IDT.   The IDT
length is now doubled over that given in Example 3.4, while the apodization
sinc function periodicity remains unchanged. Will this give a different fractional bandwidth over that derived for Example 3.4? ∎
Solution.   The number of IDT finger pairs is now $N_p = 200$. If the sinc
function periodicity remains unchanged there will be $Z = 10$ zero crossings
of the IDT apodization pattern on either side of the main response, instead
of 5 as in the previous example. The ratio of Z/N will remain unchanged,
however, so that the fractional bandwidth will be the same as that for
Example 3.4, although there should be a better approximation to a "rectangular" passband. Compare this result with that shown in Fig.3.12. ∎

3.6.   Use of Window Functions for Improved Bandpass Response

3.6.1.   THE NEED FOR WINDOW FUNCTION DESIGN TECHNIQUES

At first glance, the frequency response of the apodized IDT in Fig. 3.13
looks good. It may not be adequate, however, for IF filtering in applications
such as for CDMA spread-spectrum receivers. Typical design parameters

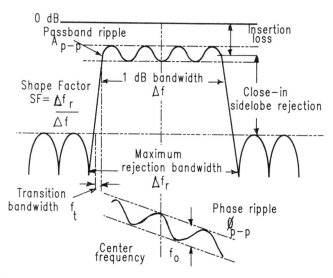

FIG. 3.14. Some design parameters required for specifying desired performance of a SAW bandpass filter. The shape factor (SF) for an ideal rectangular bandpass amplitude response would be SF = 1. (After Reference [12].)

for SAW bandpass filters must cater to a variety of specifications as shown in Fig. 3.14 for a modulation scheme where the desired bandpass response is a rectangular one with ideal shape factor (SF) = 1.0 [12].

The quality of a SAW linear-phase bandpass filter is determined mainly by three other parameters. These are: (a) the peak-to-peak amplitude ripple in the passband; (b) the close-in sidelobe rejection; and (c) the group delay $\tau = -d\phi/d\omega$, which gives the linearity of the phase response. Precision SAW bandpass filter specifications may call for amplitude ripples of less than 0.2 dB across the passband, close-in stopband rejections of greater than 50 dB and peak-to-peak group delay ripples of < 5 ns.

From Figs. 3.12 and 3.13 it was shown that increasing the length of an IDT with sinc function apodization improved its transfer function approximation to a rectangular bandpass response. Given this situation, and neglecting second-order effects, how could an extremely rectangular passband response be achieved without using excessively long apodized IDTs? The answer lies in the use of window function techniques, which modify the IDT apodization pattern so as to achieve the desired passband response with IDTs of modest length. Window functions are widely used in digital filter design to improve the shape of the passband response [12]. The impulse response of a digital FIR has but a finite length, due to the finite number of registers that store the impulse response coefficients. Finite impulse response length corresponds to the truncation of an infinite Fourier series. The abrupt truncation of a Fourier series gives rise to the Gibbs' ripple phenomena considered above. In the bandpass amplitude response, this is evidenced as an overshoot followed by a periodic ripple as illustrated in Figs. 3.12 and 3.13.

With the Gibbs' effect the largest ripple in the frequency response is theoretically about 9% of the size of the discontinuity. Most importantly, the ripple amplitude does not decrease as the length of the impulse is increased. It is confined instead to a smaller frequency range with increased ripple frequency, as evidenced from inspection of Figs. 3.12 and 3.13. Window (function) techniques are used to circumvent this problem and reduce this Gibbs' ripple response without resorting to excessive impulse response length. To achieve this, the impulse response of the digital filter under design is not abruptly truncated, but is instead tapered in some desired way so that the Fourier series converges. Desirable window functions are those that have (a) a main frequency response lobe which is small in width with (b) sidelobes that decrease rapidly with frequency. Many window functions exist for satisfying different design criteria. Examples of such include Kaiser, Kaiser-Bessel, Hamming, Cosine weighting, Taylor weighting and Dolph-Chebyshev types.

Indeed, the last-named was used at an early stage of SAW filter design [11]. While window function techniques are an extensive subject [13]–[16], their coverage in this text will be limited to demonstrating their applicability to SAW filter design, following some necessary mathematical review.

3.6.2. Use of Convolution in Window Function Techniques

Window function techniques involve convolution of two time-domain signals or two frequency response functions. To explain briefly the convolution process, consider operation in the time domain with two signals given by $f_1(t)$ and $f_2(t)$. The integral

$$f(t) = \int_{-\infty}^{+\infty} f_1(t) f_2(t - \tau) d\tau = f_1(t) \star f_2(t) \tag{3.9}$$

defines the convolution of $f_1(t)$ and $f_2(t)$. The right-hand side of Eq. (3.9) is a symbolic shorthand notation for the convolution integral, where symbol $\star$ signifies convolution.

While the convolution shown in Eq. (3.9) is carried out in the time domain, it can also be conducted in the frequency domain. Two important theorems result, namely

1. Time-Convolution Theorem: The time-domain convolution of functions $f_1(t)$ and $f_2(t)$ *corresponds to* multiplication of their respective frequency responses $F_1(\omega)$ and $F_2(\omega)$. That is, if

$$f_1(t) \leftrightarrow F_1(\omega) \quad and \quad f_2(t) \leftrightarrow F_2(\omega),$$

the Fourier transform of convolution integral $f(t)$ in Eq. (3.9) is

$$\mathscr{F}\left[f_1(t) \star f_2(t)\right] = F_1(\omega)F_2(\omega), \tag{3.10}$$

where $\mathscr{F}$ indicates the Fourier transform operation.

2. Frequency-Convolution Theorem: Convolution of the two responses $F_1(\omega)$ and $F_2(\omega)$ in the frequency domain corresponds to multiplication of their time domain responses $f_1(t)$ and $f_2(t)$.

In Eqs. (3.9) and (3.10) no restrictions have been placed on the forms of signals $f_1(t)$ and $f_2(t)$. Either or both can represent impulse responses, in

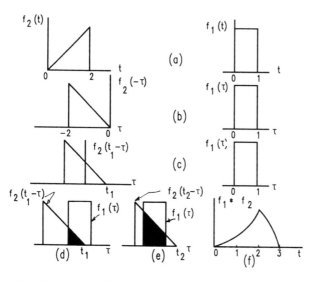

FIG. 3.15. Example of convolution in time domain. (a) Baseband signals $f_1(t)$ and $f_2(t)$. (b) Functions shifted to convolving time scale τ and f_2 is "flipped." (c) Sample time-shift $f_2(t_1 - \tau)$. (d) and (e) show two overlap times t_1 and t_2. (f) Resultant convolution result in Eq. (3.9).

which case they can be represented by $h_1(t)$ and $h_2(t)$, while the frequency responses $F(f)$, $F(\omega)$ can be replaced by $H(f)$, $H(\omega)$ in the notation of this text. In simple terms, the convolution process of Eq. (3.9) just represents a folding over of one of the responses (as given by the minus sign in the $-\tau$ term), together with a relative time displacement of one of the responses by amount τ, followed by multiplication to get the overlap areas (as given by the integration symbol). This convolution process is demonstrated in the illustrative sketch of Fig 3.15.

3.6.3. WINDOW FUNCTIONS FOR SAW IDTs

The IDTs of all SAW filters have a "built-in" window function. This is just the "rectangular" window function associated with the finite physical length of the IDT on the piezoelectric substrate as illustrated in Fig. 3.16. This shows time- and frequency-domain convolution equivalences with this rectangular window function. The convolution process follows the example of Fig. 3.15. The time-domain here corresponds to the spatial domain of the IDT. The resultant Gibbs' oscillation in the overall bandpass response appears in the bottom-left illustration. Now consider the example of Fig. 3.17, where a cosine window function has additionally been in-

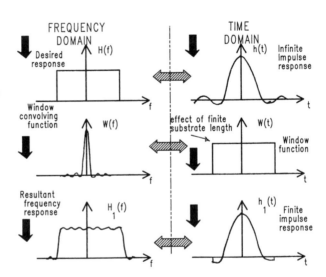

FIG. 3.16.   Window functions in both the time and frequency domains, and their influence on filter overall response. Note that a SAW filter has a "built-in" rectangular window in the time domain, associated with the finite length of IDT on piezoelectric substrate.

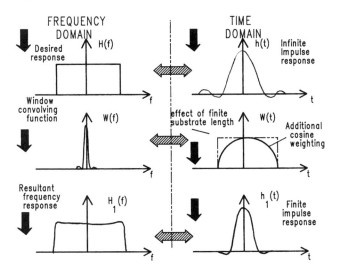

FIG. 3.17. A time-domain cosine-weighting function is used, in addition to "built-in" rectangular window of the SAW IDT. This is used here to reduce Gibbs' ripple in passband response of the SAW filter.

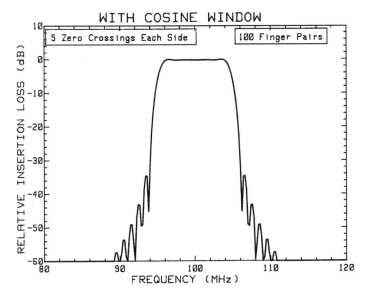

FIG. 3.18. Calculation of passband amplitude response for the IDT of Fig. 3.12, with additional cosine-weighting in the finger-overlap pattern, using weighting distribution of Fig. 3.16. Note that Gibbs' ripple is almost removed from passband response.

cluded to modify IDT apodization. Here, the "product" window function is (built-in rectangular window) × (cosine window) = (cosine window). At each finger location, the sinc function apodization value is multiplied with the corresponding cosine-window value, to give the resultant apodization overlap length for that finger. Figure 3.18 shows the computed response of the same IDT as in Fig. 3.12, with this additional cosine-weighting included. Note the drastic improvement in bandpass response.

3.7. Overall SAW Filter Response

Thus far only the response $H_1(f)$ of one IDT has been considered. The overall response $H(f) = H_1(f) \cdot H^*_2(f)$ may be readily determined by including the response $H_2(f)$ of the second IDT. (The notation $H^*_2(f)$ in the previous sentence just means the complex conjugate of $H_2(f)$.) This should simply be a wideband IDT with relatively few finger pairs, so that it does not degrade the desired bandpass response design incorporated into $H_1(f)$. Figure 3.19 shows the computed SAW filter response with a wideband output IDT used in conjunction with the cosine-weighted IDT of Fig. 3.18.

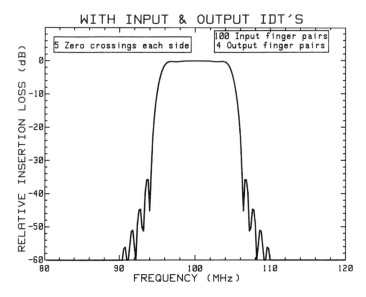

FIG. 3.19.   Delta-function model used to calculate overall amplitude response of 100-MHz SAW filter using cosine-weighted IDT of Fig. 3.17 as the input IDT, with a broadband IDT as receiver. Input IDT has $N_p = 100$ finger pairs with sinc function apodization and cosine-weighting. Uniform output IDT has $N_p = 4$.

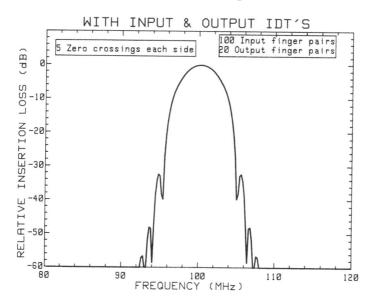

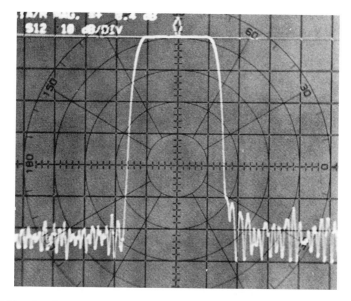

FIG. 3.20. Degradation of overall response of the 100-MHz filter when the output IDT is made too narrowband. Input IDT again has $N_p = 100$ finger pairs, but output IDT now has $N_p = 20$ finger pairs.

FIG. 3.21. Another example of a high-performance Rayleigh-wave bandpass filter, with shape factor SF = 1.33. Center frequency 192.25 MHz. Horizontal scale: 2 MHz/div. Vertical scale: 10 dB/div. (Courtesy of Sawtek Inc., Orlando, Florida.)

By contrast, Fig. 3.20 shows the reduced performance that results if the bandwidth of the output IDT is made too narrow for the desired response. Figure 3.21 illustrates the experimental response of a high-quality commercial SAW bandpass filter designed for operation at 192.25 MHz, with shape factor SF = 1.33. Note the close-in sidelobe rejection of about 60 dB.

To conclude this chapter, it should be observed that SAW filters are reciprocal devices in that it does not matter (theoretically) which IDT is used as the input one. In practice, however, there will generally be some observable difference in response, depending on the degree of mismatch between the impedance of each IDT and the load or source resistance to which it is attached.

3.8. Summary

Using delta-function modelling, this chapter has examined some basic design concepts for the realization of Rayleigh-wave and/or leaky-SAW filters, as well as the effects of various second-order perturbations on their performance. While the delta-function model is limited in its modelling scope, it can yield the frequency response of various bidirectional IDTs, including those with apodization patterns as well as window-function weighting.

3.9. REFERENCES

1. Fujitsu Limited, "*Wireless Communications Products,*" Data Book, 1996, Rev. 1.0, p. 1–14, Tokyo, Japan, 1996.
2. K. Asai, I. Isobe, T. Tada and M. Hikita, "SAW timing-extraction filter and investigation of submicron process technology for Gb/s optical communications system," *Proc. 1995 IEEE Ultrasonics Symposium*, vol. 1, pp. 131–135, 1995.
3. H. Engan, "Excitation of elastic surface waves by spatial harmonics of interdigital transducers," *IEEE Trans. Electron Devices*, vol. ED-16, p. 1014, 1969.
4. W. R. Smith, "Basics of the SAW Interdigital transducer," in *Computer-Aided Design of Surface Acoustic Wave Devices*, J. H. Collins and L. Masotti (eds), Elsevier, New York, pp. 25–63, 1976.
5. P. M. Naraine, C. K. Campbell and Y. Ye, "A SAW step-type delay line for efficient high-order harmonic mode excitation," *Proceedings of 1980 IEEE Ultrasonics Symposium*, pp. 322–325, 1980.
6. P. M. Naraine and C. K. Campbell, "Gigahertz SAW filters on YZ-lithium niobate without the use of sub-micron line widths," *Proc. 1984 IEEE Ultrasonics Symposium*, vol. 1. pp. 93–96, 1984.
7. C. K. Campbell, P. M. Smith, C. B. Saw and P. M. Naraine, "Design of a harmonic-mode SAW delay line with split-electrode and stepped-finger geometry," *Proc. 1986 IEEE Ultrasonics Symposium*, vol. 1, pp. 43–46, 1986.
8. See, for example, M. A. Sharif, M. A. Schwab, D. P. Chen and C. S. Hartmann, "Coupled resonators with differential input and/or differential output," *Proc. 1995 IEEE Ultrasonics Symp.*, vol. 1, pp. 67–70, 1995.

9. R. Tancrell and M. Holland, "Acoustic surface wave filters," *Proceedings of IEEE*, vol. 59. p. 393, 1971.
10. E. A. Guillemin, *The Mathematics of Circuit Analysis*, John Wiley and Sons, New York, Chapter VII, 1951.
11. C. S. Hartmann, D. T. Bell, Jr. and R. C. Rosenfeld, "Impulse model design of acoustic surface wave filters," *IEEE Trans. Microwave Theory and Techniques*, vol. MTT-21, pp. 162–165, April 1973.
12. "SAW Bandpass Filters", *Handbook of Acoustic Signal Processing*, vol. 1, Andersen Laboratories, Bloomfield, Connecticut, p. 4.
13. L. R. Rabiner and B. Gold, *Theory and Application of Digital Signal Processing*, Prentice-Hall, Englewood Cliffs, pp. 88–105, 1975.
14. F. J. Harris, "On the use of windows for harmonic analysis with the discrete Fourier transform," *Proceedings of IEEE*, vol. 66, pp. 51–83, Jan. 1978.
15. A. V. Oppenheim and R. W. Schafer, *Digital Signal Processing*, Prentice-Hall, Englewood Cliffs, 1975.
16. A. Antoniou, *Digital Filters: Analysis and Design*, McGraw-Hill, New York, 1979.

—4—
Equivalent Circuit and Analytic Models for a SAW Filter

4.1. Introduction

4.1.1. EARLY SAW FILTER CIRCUIT-DESIGN MODELS

This chapter begins by reviewing the mathematical relations for three circuit-response models, developed in the early days of SAW filter design. These are: 1) the *delta-function model* [1], [2]; 2) the *crossed-field model* [3]–[6]; and 3) the *impulse-response model* [7]. The reason for introducing them in this chapter is that each reveals an important aspect of SAW filter operation.

1) The delta-function model serves to demonstrate that an interdigital transducer can be modelled as a digital sampled-data structure. While it does not give any information on impedance or insertion-loss levels, it can often be used as a fast tool for obtaining an approximate IDT or overall frequency response. It also brings out the restrictions placed on apodizing both input and output IDTs in a SAW filter. Its significant limitations are that it does not give information on input/output impedance parameters or insertion loss. Moreover, it does not deal with the effects of acoustic reflections from IDT finger discontinuities. As will be developed in Chapter 9, such acoustic reflections from IDT finger discontinuities can be vital to the desired performance of many high-frequency filters for mobile/wireless communications circuitry, such as leaky-SAW RF antenna duplexers and RF inter-stage filters.

2) The SAW crossed-field model, (also known as Mason model [6]), does yield information on IDT frequency responses, as well as input and output impedance levels. Moreover, it can be used to obtain the radiation conductance parameters relating to the transmission or reception of a SAW by an excited IDT. The early crossed-field model does not incorporate the effects of acoustic reflections from IDTs,

99

which can significantly distort the frequency-response of the sinc-function (i.e., (sin X)/X) frequency response of the radiation conductance of an IDT. As highlighted in what follows and detailed in Chapter 9, the equivalent-circuit model [3], [4] is a lumped-parameter extension of the crossed-field model, which can incorporate segments to model the effects of acoustic reflections in IDT fingers due to mechanical or electrical loading.

3) The impulse-response model [7] is also reviewed in this chapter. While this model can incorporate circuit impedances and matching networks, it also does not cater to acoustic reflections from IDT electrode fingers. It is included here because it provides ready insight into the frequency scaling that must be employed in the determination of IDT apodization for broadband and chirp SAW filter design.

4.1.2. CURRENT ANALYTIC METHODS FOR SAW/LSAW FILTER DESIGN

This chapter briefly examines four current analytic methods used for SAW/LSAW filter design that will be detailed in later chapters. These are: 1) the *equivalent-circuit model*; 2) the *S-matrix model*; 3) the *P-matrix model*; and 4) the *coupling-of-modes* (COM) model. Formulas for the equivalent-circuit model, the *P*-matrix model, and the coupling-of-modes (COM) model can be converted into one another, and will give the same results if the internal modelling (i.e., IDT transmission, IDT reflection, and IDT transduction) is identical [8].

1) The equivalent-circuit model, (also known as the Smith equivalent circuit), is a lumped-element three-port model represented by two "acoustic" ports and one electric port[1] [3], [4] for determining the frequency response and impedance parameters for a single SAW IDT or a composite SAW filter. As extended, this equivalent-circuit model is based on the "crossed-field" Mason circuit, which was originally developed to describe the launching and detection of *bulk* acoustic waves. It can deal with IDTs with either split- or solid-electrodes [9]. Acoustic reflections from IDT fingers due to mechanical and electrical loading can be taken into account by using mismatched transmission line segments. The modelling of energy storage under IDT fingers, which can be important for LSAW filter applications, can also be included by introducing shunt susceptances. Recent modifications to

---

[1] The two "acoustic ports" of the IDT are usually designated as Port 1 and Port 2, while the electrical port is designated as Port 3.

the equivalent-circuit model also take the launching (and detection) of unidirectional surface acoustic waves into account [10].

2) The S-matrix model is a 3×3 scattering-parameter matrix[2] such as used in microwave circuit design, with incident and reflected voltages at each IDT port. This results in dimensionless S-matrix parameters, which can be converted into 3×3 transmission matrices or into input/ output impedance parameters [11]. Moreover, it can be applied to obtain the transmission and reflection coefficients of IDTs with acoustic reflections from IDT finger discontinuities.

3) The P-matrix model as developed by Tobolka [12] is a widely used variant of the S-matrix model. In contrast to the S-matrix, its matrix elements are in "mixed" form, where elements in the first two rows are dimensionless, while elements in the third row have the units of admittance.

4) Coupling-of-modes (COM) theory, originally developed by Pierce in 1954 [13], is an elegant method for modelling systems with time- or spatially varying parameters. Introduced into the SAW field in 1976 and 1977 [14]–[20], COM theory represents a widely used tool for the analysis of SAW/LSAW IDTs, filters, and reflection gratings. In some SAW/LSAW-related COM formulism procedures, an entire IDT or reflector may be determined in closed form, and represented by a P-matrix. As a result, there is sometimes no distinction made between COM and P-matrix determinations.

4.2.   The Delta-Function Model

4.2.1.   INTRODUCTION

The delta-function model [1], [2] is the simplest of the various models that describe the performance of a bidirectional SAW IDT. Although it is normally applied to IDTs with uniform finger spacings and constant or varying apodization overlap, it can also be applied to simple modelling of chirp filters with nonuniform IDTs. As illustrated in what follows, it can also be used to model relative degradation of passband response due to triple-transit-interference and/or electromagnetic feedthrough. Its major limitation is that it cannot be applied to obtain absolute insertion loss values because it has no provision for handling impedance levels. Despite this, it can be excellent for modelling frequency response in preliminary designs.

---

[2] In this S-matrix and P-matrix notation, the first suffix designates the destination port, while the second index designates the source port.

4.2.2. TRANSVERSAL FILTER EQUIVALENCE

The delta-function model is really a representation of an ideal transversal filter as sketched in Fig. 4.1. In the time domain, the output $y(t)$ of this filter is the superposition of the weighted outputs of a series of delay elements subjected to an input stimulus $x(t)$. For N delay elements, the output is

$$y(t) = \sum_{n=1}^{N} A_n x(t - T_n).$$ (4.1)

In terms of SAW IDT parameters, A_n corresponds to the finger overlap of the nth finger pair, while T_n is the accumulative delay time for the SAW to traverse to the nth finger pair. For a finite number N of such finger pairs, this filter has a finite impulse response (FIR). It will be a stable system (in the absence of feedback) because its impulse response $h(t)$ will be absolutely integrable (i.e., $\int h(t)dt < \infty$ when integrated over all time).

A shift in the time domain corresponds to a multiplication by $e^{-j\omega T_n}$ in translating to the frequency domain, so that the output frequency response $Y(\omega)$ is given by

$$Y(\omega) = \sum_{n=1}^{N} A_n \int_{-\infty}^{\infty} x(t - T_n) e^{-j\omega t} dt = \sum_{n=1}^{N} A_n e^{-j\omega T_n} X(\omega)$$ (4.2a)

$$Y(\omega) = H(\omega) X(\omega),$$ (4.2b)

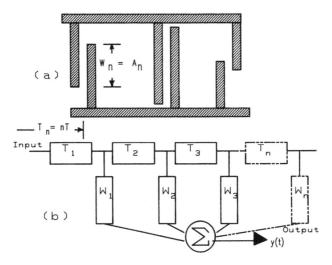

FIG. 4.1. (a) Finger placement and apodization of an IDT. (b) Transversal filter equivalent-circuit representation.

where $Y(\omega)$ is the Fourier transform of $y(t)$ and $X(\omega)$ is the Fourier transform of $x(t)$. From Eq. (4.2a), the frequency response $H(\omega)$ of the transversal filter is

$$H(\omega) = \sum_{n=1}^{N} A_n e^{-j\omega T_n},\qquad(4.3)$$

where $T_n = nT$ for uniform finger spacing, and $A_n = a_n e^{j\phi}$. Equation (4.3) corresponds to the delta-model formulation given in Eq. (3.2).

4.2.3. IDTs with Constant Finger Overlap

Now consider what restrictions, if any, are placed on the apodization of input and output IDTs in the SAW filter. To approach this, the SAW filter circuit of Fig. 4.2 is first examined, where the IDTs have arbitrary finger separations but constant finger overlap. In this instance, no special restrictions are placed on reference axis locations. A convenient reference axis to use is that at the left-hand edge of the input IDT. Likewise, the delta-function modelling of an excited IDT need not be restricted to one delta function of electric field per finger. If each A_n is proportional to its finger overlap, the complex potential P_n of the SAW at location x, due to excitation of an electrode at x, is

$$P_n = V_{in} A_n e^{-j\beta(x-x_n)},\qquad(4.4)$$

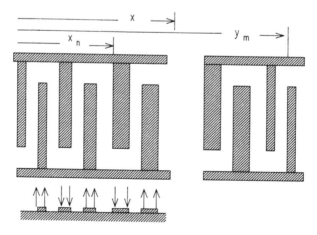

FIG. 4.2. SAW filter with unapodized input and output IDTs, and arbitrary finger widths. Electric-field distributions between excited electrodes are represented in this example by two delta-function sources per electrode.

where V_{in} = input voltage, which may be normalized to $V_{in} = e^{j\omega t}$. However, $\beta = 2\pi/\lambda = 2\pi f/v$ in terms of SAW velocity v, so that Eq. (4.4) becomes

$$P_n = A_n e^{-j\omega(x_n-x)/v} e^{j\omega t}, \qquad (4.5)$$

as the contribution from the delta sources at finger x. If location x is selected as $x = x_m$ at a point along the output IDT, the contribution from all of the input finger sources in the input IDT at this one location is

$$\sum_{n=1}^{N} P_n = \sum_{n=1}^{N} A_n e^{j\omega(x_n-x_m)/v} e^{j\omega t}, \qquad (4.6)$$

where N = total number of delta function sources in the input IDT. In this instance the number of electrodes $N_e = N/2$, as we are considering two delta functions per electrode.

Equation (4.6) gives the contributions of the input IDT at only one location along the output IDT. To get the total output voltage $V_{out}(t)$ it is necessary to sum over all of the output IDT fingers, to obtain

$$V_{out}(t) = \sum_{m=1}^{M} A_m \sum_{n=1}^{N} A_n e^{j\omega(x_n-x_m)/v} e^{j\omega t}, \qquad (4.7)$$

where finger overlap parameter A_m in the output IDT is similar to A_n in the input IDT, and M gives the total number of finger edges in the output IDT. Both A_m and A_n are proportional to the electric field gradient in the excited IDTs. In this particular delta-function example, they change their sign in pairs (i.e., if A_1 and A_2 are positive then A_3 and A_4 will be negative, etc.) Their amplitudes will also depend on the value of metallization ratio used. The overall transfer function response $H(f)$ is

$$H(f) = \left[\sum_{n=1}^{N} A_n e^{j\omega x_n/v}\right]\left[\sum_{m=1}^{M} A_m e^{j\omega x_m/v}\right]. \qquad (4.8)$$

The term inside the first bracket of Eq. (4.8) is just the Fourier transform of $\sum A_n \delta(x - x_n)$, where the delta function $\delta(x - x_n)$ represents the sources at finger edges. The same situation holds for the term inside the second bracket. Thus the total frequency response $H(f)$ is the product of the individual responses given by

$$H(f) = H_{in}(f) H_{out}^*(f) \qquad (4.9a)$$

or

$$|H(f)| = |H_{in}(f)||H_{out}(f)|. \qquad (4.9b)$$

In Eq. (4.9a), $H_{out}^*(f)$ is the complex conjugate of the transfer function of the output IDT frequency response $H_{out}(f)$. This is due to the sign change in the second exponential term in Eq. (4.8), where $|H_{out}(f)| = |H^*_{out}(f)|$ and $H^*_{out}(f)$ represents the complex-conjugate response (i.e., if $H = a + jb$ then $H^* = a - jb$). Note, most importantly, that Eq. (4.9) is valid only for the particular case where at least one of the IDTs has constant finger overlap.

4.2.4. PROBLEMS WITH APODIZING BOTH IDTs

If both IDTs are apodized as sketched in Fig. 4.3, with N fingers in the input and M finger pairs in the output IDT, the frequency response of a "shared" (N,M) electrode pair will take the form

$$\left(A_n A_m e^{j\omega(x_n - x_m)/v} \right) \times \left(a \text{ weighting factor } C_{nm} \right).$$

This weighting factor C_{nm} will be proportional to the acoustic beamwidth seen by the "active" part of the finger edge at x. In other words, if $W(x_n)$ represents the finger overlap at location x in the input IDT while $W(x_m)$ represents that at location x in the output IDT, then the overlap factor C_{nm} will either be $W(x_m)$ or $W(x_n)$, whichever is smaller, because that portion of the SAW that "misses" a finger overlap area does not contribute to its excitation. The filter transfer function $H(f)$ will be constrained as

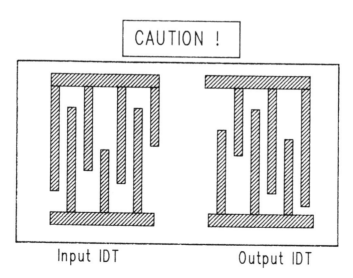

Input IDT   Output IDT

FIG. 4.3.   With the in-line structure, the frequency responses of input and output IDTs can not normally be designed independently, unless they are initially "pre-distorted" to compensate for inter-IDT spatial effects [21].

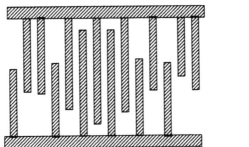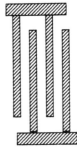

FIG. 4.4. A SAW filter can readily be designed with in-line IDTs, providing that only one is apodized. The unapodized IDT can be the input or output one, provided that input/output impedance matching is taken into consideration.

$$H(f) = \sum_{n=1}^{N} \sum_{m=1}^{M} C_{nm} A_n A_m e^{j\omega(x_n - x_m)/v}, \tag{4.10}$$

where binding term C_{nm} can **not** be split into two separate summations that represent the individual frequency responses of input and output IDTs. This can happen only if one of the IDTs is apodized as sketched in Fig. 4.4. As the filter is reciprocal, it does not matter which IDT has the apodization. In this event, the response $H(f)$ can be split into two summations relating input and output IDT responses, exactly as in Eq. (4.8). The uniform IDT will have a broadband response if it has only a few finger pairs, in which case it can be approximated as a constant, so that the overall SAW filter response reduces to

$$H(f) \approx \sum_{n=1}^{N} B_n e^{j\omega x_n/v}, \tag{4.11}$$

where $B_n = W(x_n)A_n$. Note that index N in Eq. (4.11) is the total number of delta-function excitation sources in the excited IDT. Two delta sources per finger were arbitrarily used in the preceding derivations. The simpler model with only one delta source per finger could just as readily have been employed; in which case N corresponds to the number of fingers in the IDT.

4.2.5. A Further Note on Apodization of Both IDTs

The apodization restrictions considered in the previous section apply to a SAW filter with two *in-line* components—an input IDT and an output IDT. As will be considered in Chapter 6, the overall frequency response may be readily designed as the product response of individually apodized input and

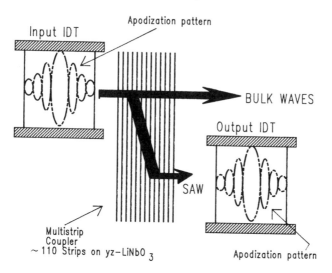

FIG. 4.5. The use of a multistrip coupler (MSC) permits apodization of both IDTs. It is restricted to use on piezoelectric substrates with large K^2 values, such as lithium niobate, because its use on quartz would require an excessive number of strips.

output IDTs (*if they are separated by a multistrip coupler*), *and displaced with respect to one another as sketched in Fig. 4.5.*

A disadvantage of using a multistrip coupler, however, is in the increased piezoelectric substrate area that is required for the device. In some instances the separate apodization of both IDTs without a multistrip coupler may be permitted, if the individual apodization designs are initially "predistorted" to compensate for the inter-IDT spatial effects discussed in the previous section [21].

A completely different way of designing the filter response of an IDT is to use the method of *withdrawal weighting* [22]. With this technique all of the fingers of the IDT have the same overlap. If all fingers were present, this would realize a "sinc"-function type of frequency response. In the withdrawal weighting designs, fingers are selectively removed from the IDT to achieve the desired nonsinc frequency response[3]. Withdrawal-weighted IDTs are used, for example, in filter designs for UHF mobile communications, where the IDT can not be apodized, in order to reduce the out-of-band filter sidelobe responses [23]. In some designs they are also incorporated with the length-apodized IDTs to realize out-of-band sidelobe suppression to the 60 to 70-dB level [22].

---

[3] The fingers must be symmetrically withdrawn if linear phase is to be preserved.

4.2.6. LINEAR-PHASE SAW AND DIGITAL FILTERS

The preceding mathematical derivations did not place any restriction on IDT finger spacings. These can be uniform or nonuniform. For the special case of a linear-phase SAW filter with uniform IDT finger spacing, we can set $x_n/v = nT$ in Eq. (4.11), where T = the time for the SAW wave to traverse between adjacent electrodes. In this way, Eq. (4.11) can be reduced to

$$H(f) = \sum_{n=0}^{N-1} B_n e^{j\omega nT}, \qquad (4.12)$$

which is exactly the relationship for the frequency response of a FIR digital linear phase filter. For withdrawal-weighting designs, selected coefficients B_n are simply set to zero in Eq. (4.12).

4.3. SAW Power Flow in Bidirectional IDTs

4.3.1. INHERENT INSERTION LOSS OF BIDIRECTIONAL IDTS

Before proceeding to the other models that yield information on the input or output impedances of a SAW IDT, it is appropriate to examine some basic power concepts applicable to bidirectional SAW IDTs. To this end, consider that the elementary bidirectional IDT of Fig. 4.6, with uniform finger spacing and constant finger overlap, is employed as the input IDT in a SAW filter. When it is excited by an ac stimulus within its frequency range, surface waves will flow outward in both directions along the propagation axis. If the incident signal power P_{in} is normalized to $P_{in} = 1$ W, power $P = 1/2$ W will be transported towards the output IDT, while $P = 1/2$ W will be lost for signal processing purposes (although this latter portion can contribute to interference if SAW waves are allowed to reflect from the

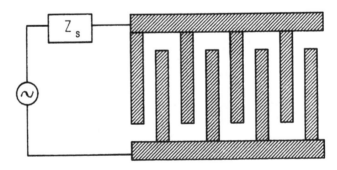

FIG. 4.6. Elementary bidirectional IDT with uniform finger-spacing, and constant acoustic aperture (finger overlap). Z_s is the source impedance.

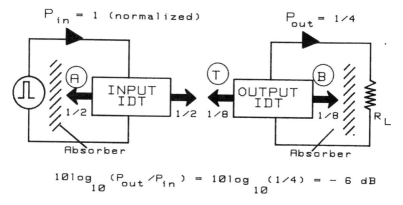

$$10\log_{10}(P_{out}/P_{in}) = 10\log_{10}(1/4) = -6 \text{ dB}$$

Fig. 4.7.   Power flow in bidirectional IDTs. Power flow at A and B is "lost" for signal-processing purposes. Power regenerated at T can contribute to triple-transit interference (TTI).

ends of the piezoelectric crystal). Thus the inherent loss of the input IDT will be 3 dB, as will be that for a bidirectional output IDT. The inherent overall loss of a SAW filter with such IDTs in input and output is then 6 dB, as illustrated in Fig. 4.7.

4.3.2. Triple-Transit-Interference

The forementioned power-loss problem is compounded by the fact that each IDT that absorbs incident SAW energy can also re-emit some of this on a regenerative basis. This is the same phenomenon encountered in an electromagnetic receiving antenna, in which part of the received energy is reradiated by the induced current in the antenna. In SAW filters employing bidirectional IDTs, this phenomenon gives rise to the *triple-transit-interference* (TTI) second-order effect introduced in Chapter 3. As sketched in Fig. 4.8, currents induced in the output IDT by the travelling surface waves can lead to their regeneration by that IDT. This regeneration will also be bidirectional. As well, that portion of the SAW energy arriving back at the input IDT can lead to further SAW emissions there. The regenerated wave arriving back at the output IDT after two traverses will be delayed by time 2τ from the main signal arriving there, where τ is the time for SAW propagation between input and output IDT phase centers.

Figure 4.8 shows relative power levels due to TTI in a SAW filter with matched source and load impedances. As shown, the output power level due to the main input signal is $P_{out} = 1/4$ W, relative to the input stimulus normalized to 1 W. This is because one-half of the SAW power absorbed by

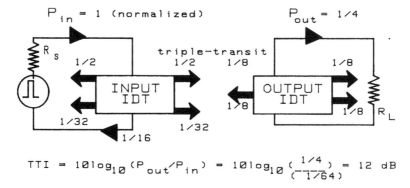

$$TTI = 10 \log_{10} (P_{out}/P_{in}) = 10 \log_{10} \left(\frac{1/4}{1/64}\right) = 12 \ dB$$

FIG. 4.8.   Power flow for triple-transit interference (TTI) in bidirectional IDTs.

the output IDT is reemitted as surface wave energy. As a result of multiple traverses, the additional output signal due to TTI has a power level $P_{TTI} = 1/64$ W. The ratio of the two power levels is $P_{TTI}/P_{out} = 1/16$ or $P_{TTI}/P_{out} = 10.\log (1/16) = -12\,dB$. This TTI level is significant compared to the main signal received at the output. The two voltage phasors will add and subtract as a function of frequency, giving rise to unacceptably large ripples in the output voltage signal. The ensuing ripple will be periodic at a frequency $f_r = 1/2\tau$. (This will show up in the crossed-field treatment considered in what follows.)

For the example considered here, which assumed matched source and load impedances, this 12-dB level of TTI suppression would yield a pass-band amplitude ripple ~ 4.4 dB peak-to-peak in the SAW filter response, together with a phase ripple ~ 28° peak-to-peak in a deviation from linear-phase response. This is clearly an unacceptable situation for linear-phase filter design! With bidirectional IDTs, the only way around this situation is to deliberately mismatch the source and/or load impedances. As an extreme mismatch example, there will obviously be no SAW regeneration and TTI if the output IDT is completely shorted out, because SAW emissions from the input would flow unperturbed under the output IDT. For this extreme condition, however, the insertion loss would be infinite. For less extreme mismatching, the result will be to increase the filter insertion loss by a finite amount above the inherent 6 dB value with bidirectional IDTs. Table 4.1 lists amplitude and phase ripple values commensurate with various levels of TTI suppression; achieved by mismatching source and/or load impedances by different amounts.

A rule of thumb for relating insertion loss and TTI when the former is greater than about 10 dB is.

TABLE 4.1

PEAK-TO-PEAK AMPLITUDE AND PHASE PASSBAND RIPPLE
VALUES FOR VARIOUS LEVELS OF TTI SUPPRESSION IN A SAW
FILTER WITH BIDIRECTIONAL IDTs AND CW INPUT EXCITATION

TTI Suppression (dB)	Amplitude Ripple (dB)	Phase Ripple (degrees)
12 (match)	4.46	28
16	2.78	18
20	1.74	11.4
24	1.09	7.2
28	0.69	4.6
32	0.43	2.9
36	0.27	1.8
40	0.17	1.1
44	0.11	0.72
48	0.07	0.46
52	0.04	0.29
56	0.02	0.18

$$TTI\ suppression \approx 6 + 2(insertion\ loss)\ \ (dB), \qquad (4.13)$$

where the factor of 6 in Eq. (4.13) relates to the minimum insertion loss of
the SAW filter with matched source and load impedances.

Example 4.1 TTI Suppression Level Calculations. A SAW filter employs bidirectional IDTs that are matched to source and load impedances.
The TTI suppression level is 12 dB referred to the main output signal. Show
that the corresponding peak-to-peak amplitude and phase ripple in the
passband are 4.4 dB and 28°, respectively. ∎

Solution. A TTI suppression of 12 dB corresponds to a power level ratio
of 1/16. To get voltage levels from power levels, normalize source and load
impedances to 1 ohm each for convenience. The power relation $V^2/R = P$,
yields $V_{TTI} = 1/4$ V relative to the received signal voltage normalized to 1 V.
The peak-to-peak amplitude ripple in the passband is obtained in dB as
$V_{p-p} = 20.\log_{10}[(1 + 0.25)/(1 - 0.25)] = 4.459$ dB. Next, phase angle deviations
from linearity will be a maximum when the TTI and the main signal voltage
phasors have a relative phase shift of $\pm 90°$. This gives the peak-to-peak
phase ripple as

$$\phi_{p-p} = 2\tan^{-1}(0.25/1) = 28°.$$

Check this result with that given in Table 4.1. ∎

Example 4.2 Calculating Insertion Loss for a Given Passband Ripple. A SAW filter employing bidirectional IDTs is required to have a maximum passband amplitude ripple of 0.07 dB. Determine the required level of filter insertion loss at center frequency. ∎

Solution. From Table 4.1, the TTI suppression is required to be 48 dB. Substitute this value in Eq. (4.13) to get $48 = 6 + 2 \times$ (Insertion Loss), so that the required insertion loss (IL) as IL = 21 dB. This can be arranged by appropriately mismatching source and/or load impedances with reference to their respective IDT impedances. ∎

4.3.3. ADDITIONAL DEGRADATION DUE TO ELECTROMAGNETIC FEEDTHROUGH

As noted in Chapter 3, electromagnetic feedthrough is one of the most serious of the second-order effects in its corruption of passband amplitude and phase response. It is mentioned here, because its manifestation as passband ripple is similar to, and sometimes confused with, that due to TTI.

The input and output IDTs may be considered to form two plates of a parasitic capacitor. Some of the input voltage signal will couple directly through to the output IDT in a frequency-dependent manner, at the veloc-

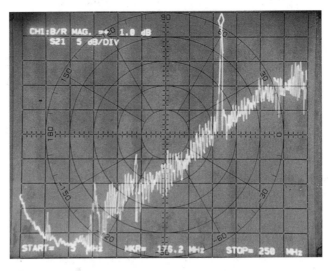

FIG. 4.9. Network analyzer measurement example of electromagnetic feedthrough in a 176-MHz SAW filter, due to bad design and/or bad packaging. Horizontal scale: 5 to 250 MHz. Vertical scale: 5 dB/div.

ity of light. As with TTI, the resultant output voltage signal will result from two voltage phasors associated with the main signal as well as electromagnetic feedthrough. Here, however, while the main signal is due to SAW waves of velocity v traversing from input to output IDTs, the electromagnetic feedthrough is at the velocity of light c and is instantaneous by comparison. The resultant output voltage will again have periodic ripples in the amplitude and phase response. In this case the ripple frequency will be $f_r = 1/\tau$.

In addition, while the effects of TTI will be manifest mainly in the filter passband, that of electromagnetic feedthrough will affect both the passband and stopband responses. Stopband rejection of electromagnetic feedthrough will normally worsen with increasing frequency, as illustrated in the experimental example of Fig. 4.9.

The delta-function model can be adapted to illustrate, on a relative basis, the degradation of filter response caused by TTI and/or electromagnetic feedthrough. To achieve this, it is merely necessary to include an additional voltage phasor component in the model, whose relative magnitude will depend on the amount of degradation, while its relative phase angle will relate to delay times of τ or 2τ appropriate to electromagnetic feedthrough or TTI, respectively.

4.4. The Crossed-Field Model

4.4.1. ELECTROACOUSTIC EQUIVALENCES

The SAW *crossed-field model* is derived from the Mason equivalent circuit employed for modelling acoustic *bulk* wave piezoelectric devices [3]–[5]. In this model the electric field distribution under the electrodes of an excited IDT is approximated as being normal to the piezoelectric surface, as if between the plates of a capacitor, as sketched in Fig. 4.10(a). An alternative model known as the *in-line model* is given in Fig. 4.10(b) for reference. Experimentally it was found that the crossed-field model yielded better agreement with experiment than the in-line one, for IDTs fabricated on high-K^2 piezoelectric substrates such as lithium niobate. (The in-line model is applicable, however, to IDT designs on quartz substrates.)

In the adaptation of the Mason equivalent circuit to SAW filter design, each IDT is represented by a three-port network, shown in Fig. 4.11. In the terminology followed here, Ports 1 and 2 represent electrical equivalents of "acoustic" ports, while Port 3 is a true electrical port. The electrical equivalents in Ports 1 and 2 are those for an acoustic, and passive, SAW transmission line. Port 3 is where the actual signal voltages are applied or detected.

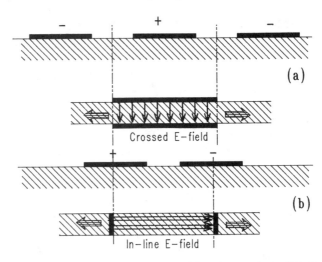

FIG. 4.10. Simplistic representations depicting: (a) an instantaneous *E*-field direction in the crossed-field model; and (b) an instantaneous *E*-field distribution for the "in-line" model.

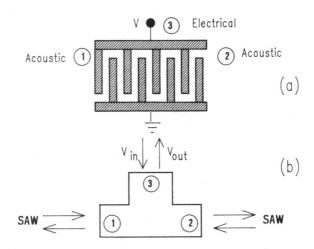

FIG. 4.11. (a) Representation of the SAW IDT as a three-port network, Ports 1 and 2 are normally assigned to the "acoustic" ports, while Port 3 is the electrical port. (b) In the crossed-field model, acoustic signals at Ports 1 and 2 are converted to equivalent electrical transmission-line parameters.

At this point it is most important for the reader to note that all of the subsequent derivations in this chapter apply to an IDT where acoustic reflections from IDT finger discontinuities are neglected! This approximation can be considered to hold for electrodes with low values of film-thickness ratio ($h/\lambda \ll \sim 1\%$). As will be seen, the result of making this approximation is that the radiation conductance parameter $G_a(f)$ to be derived will have a sinc-function (symmetric) amplitude response about center frequency. This will **not** be the case for the RF or IF filters considered in Part 2 of this book, where frequencies and/or IDT metallization thicknesses can be such that $h/\lambda > \sim 1\%$!

For all three ports to be treated in equivalent electrical terms, the acoustic parameters at Ports 1 and 2 must be converted to electrical equivalents. At these ports acoustic forces F (in newtons) are transformed to electrical equivalent voltages V, while mechanical SAW velocities v are transformed to equivalent currents I. In terms of a common proportionality constant ϕ these transforms are

$$V = F/\phi \tag{4.14a}$$

$$I = v\phi,$$

where parameter ϕ is interpreted as the turns-ratio of an equivalent acoustic-to-electric transformer. In turn, these definitions allow the mechanical characteristic impedance $Z_m = F/v$ of the piezoelectric substrate to be expressed as an equivalent transmission line characteristic impedance Z_o.

For a uniform acoustic wave propagating in a substrate of density ρ (kg/m³) and large cross-sectional area A (m²), the mechanical impedance can be written as

$$Z_m = \rho v A \quad (kg/s), \tag{4.14b}$$

while the equivalent electrical characteristic impedance is

$$Z_o = \frac{Z_m}{\phi^2} \quad (ohm). \tag{4.15}$$

These definitions enable the electrical characteristic admittance $G_o = 1/Z_o$ of the equivalent SAW transmission line to be derived as

$$G_o = K^2 C_s f_o \quad (mho), \tag{4.16}$$

where K^2 = electromechanical coupling constant, f_o = IDT center frequency and C_s = static capacitance of one periodic section; C_s may also be expressed as $C_s = C_o W$, where C_o = capacitance/finger pair/unit length (pF/cm) and W = finger apodization overlap (cm).

4.4.2. THREE-PORT ADMITTANCE MATRIX [Y] FOR AN IDT

In terms of a three-port admittance applicable to Fig. 4.12, the equivalent current-voltage relations for a single IDT are given as

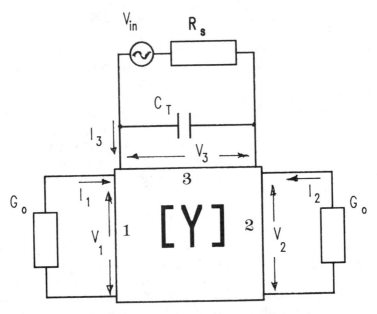

Fig. 4.12. Three-port equivalent admittance network representation for an IDT in the crossed-field model. Here, G_o is the equivalent electrical characteristic conductance of a SAW transmission line. Note that the total IDT capacitance C_T can be considered to be "outside" the remaining admittance configuration.

$$\begin{pmatrix} I_1 \\ I_2 \\ I_3 \end{pmatrix} = \begin{bmatrix} Y \end{bmatrix} \begin{pmatrix} V_1 \\ V_2 \\ V_3 \end{pmatrix} \tag{4.17}$$

where the 3×3 admittance matrix $[Y]$ may be expanded as

$$\begin{bmatrix} Y \end{bmatrix} = \begin{pmatrix} Y_{11} & Y_{12} & Y_{13} \\ Y_{21} & Y_{22} & Y_{23} \\ Y_{31} & Y_{32} & Y_{33} \end{pmatrix}. \tag{4.18}$$

Under the notation used here, matrix subset elements Y_{11}, Y_{12}, Y_{21}, Y_{22} are just those relating to an equivalent SAW transmission line.

A first stage of simplification can now be applied. From transmission-line symmetry and reciprocity, $Y_{11} = Y_{22}$ and $Y_{21} = Y_{12}$. An additional simplification can be applied to the remaining matrix elements involving electrical port 3, by setting $Y_{32} = -Y_{13}$. The negative sign is required here, because a voltage applied to port 3 causes a SAW to emanate from both sides of the

IDT with the same potential Φ. With these simplifications, Eq. (4.18) reduces to

$$[Y] = \begin{pmatrix} Y_{11} & Y_{12} & Y_{13} \\ Y_{12} & Y_{11} & -Y_{13} \\ Y_{13} & -Y_{13} & Y_{33} \end{pmatrix}. \qquad (4.19)$$

These matrix elements for the crossed-field model may be derived to be [3]

$$
\begin{aligned}
Y_{11} &= -jG_o \cot(N\theta) \\
Y_{12} &= jG_o \csc(N\theta) \\
Y_{13} &= -jG_o \tan(\theta/4) \\
Y_{33} &= j\omega C_T + j4NG_o \tan(\theta/4),
\end{aligned}
\qquad (4.20)
$$

where $G_o = 1/Z_o = $ characteristic admittance (mho), $C_T = NC_s = $ total IDT capacitance (F), $N = $ number of electrode pairs (periods), $C_s = $ capacitance of one finger pair (F) and $\theta = 2\pi(f/f_o) = $ electrical transit angle, in radians, through one electrode pair (i.e., one period).

Unfortunately, the matrix elements in Eq. (4.20) "blow up" at center frequency, when $\theta = 2\pi$. The impedance and transfer functions remain finite, however, and may be calculated by expanding the matrix for frequencies near center frequency. When this is done, the input admittance Y_3 at center frequency f_o may be expressed as

$$Y_3(f_o) = \left.\frac{I_3}{V_3}\right|_{f_o} = G_a(f_o) + j2\pi f_o C_T, \qquad (4.21)$$

where $G_a(f_o) = $ the radiation conductance at center frequency f_o such that

$$G_a(f_o) = 8K^2 f_o C_s N^2 \quad (mho). \qquad (4.22)$$

However, from Eq. (4.16) the equivalent characteristic admittance G_o of the SAW transmission line is given as $G_o = K^2 C_s f_o$. This can be substituted in Eq. (4.22), to give the input radiation conductance at center frequency as

$$G_a(f_o) = 8N^2 G_o \quad (mho). \qquad (4.23)$$

For frequencies near center frequency Eq. (4.23) may be generalized as

$$
\begin{aligned}
G_a(f) &\approx G_a(f_o)\left|\frac{\sin x}{x}\right|^2 \\
&\approx 8N^2 G_o \left|\frac{\sin x}{x}\right|^2 : (\text{neglecting IDT finger reflections!})
\end{aligned}
\qquad (4.24)
$$

where $x = N\pi(f - f_o)/f_o$ in the **unperturbed** sinc-function relation $\mathrm{sinc}(X) = (\sin X)/X$, which does not incorporate effects of acoustic reflections at IDT finger discontinuities!

Equation (4.21) may also be generalized as

$$Y_3(f) = G_a(f) + j2\pi f C_T \quad (mho) \tag{4.25}$$

in approximating the input admittance.

The equivalent circuit input impedance of Fig. 4.12 is sketched in Fig. 4.13. The same principles can be applied to model the output IDT. In this case, the equivalent excitation source for output IDT is a current

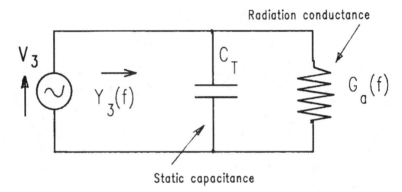

FIG. 4.13. Crossed-field equivalent admittance at electrical Port 3, for an IDT at center frequency. Fictitious conductance $G_a(f)$ relates power generation from an excited IDT. (Analogous to radiation resistance in an electromagnetic antenna.) C_T is the total static capacitance of IDT. Radiation susceptance $B_a(f) = 0$.

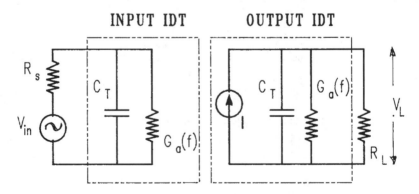

FIG. 4.14. Input/output equivalent circuit for a SAW filter in the crossed-field model. Output IDT is considered to be excited by a high-impedance current source proportional to the SAW amplitude. Here, R_s and R_L are the filter source and load resistances.

source. (The use of a current source here can be deduced intuitively, since the output IDT can be visualized as "looking back" into a high impedance source.) The overall equivalent circuit of the SAW filter, including source and load impedances is then as sketched as in Fig. 4.14.

Example 4.3 IDT Radiation Conductance Calculation. A SAW filter is fabricated on YZ-lithium niobate. Its input and output IDTs have constant finger overlap. The input IDT has $N_p = 50$ finger pairs and an apodization width $W = 100$ acoustic wavelengths at center frequency $f_o = 400\,\text{MHz}$. Consider that the capacitance/finger pair/cm is $C_o = 4.6\,\text{pF/cm}$. Determine the numerical value of unperturbed radiation conductance G_a at f_o. ∎

Solution. From Table 2.1, the electromechanical coupling coefficient for YZ-lithium niobate is $K^2 = 4.6\%$, while the SAW velocity is 3488 m/s. The acoustic wavelength at center frequency is $\lambda = v/f_o = 3488/(400 \times 10^6) = 8.72 \times 10^{-6}\,\text{m}$. The apodization overlap of 100 λ_o corresponds to $8.72 \times 10^{-2}\,\text{cm}$. The capacitance C_s per finger pair is $C_s = C_o \times 8.72 \times 10^{-2} = 4.6 \times 8.72 \times 10^{-2} \approx 0.40\,\text{pF}$. From Eq. (4.22) the unperturbed radiation conductance at 400 MHz is $G_a = 8K^2 f_o C_s N^2 = 8 \times 0.046 \times (400 \times 10^6) \times (0.4 \times 10^{-12}) \times 50^2 = 0.14\,\text{mho}$. ∎

Example 4.4 IDT Impedance Matching of Source Resistance. The input IDT in Example 4.3 is to be employed with a matched source resistance R_s. What is the numeric value of R_s for match? Assume that the input capacitive reactance component has been tuned out by a parallel-connected inductance. ∎

Solution. For match under this condition, $R_s = 1/G_a = 1/0.14 \approx 7\,\Omega$. ∎

Example 4.5 Characteristic Admittance/Impedance of SAW Transmission Line. Determine the numerical values of (a) the unperturbed characteristic admittance (conductance) G_o and (b) characteristic impedance Z_o of the equivalent SAW transmission line in Example 4.3. ∎

Solution. (a) From Eq. (4.23) the unperturbed radiation conductance G_a (f_o) at center frequency is $G_a(f_o) = 8N^2 G_o$, where $N = N_p = $ number of electrode pairs in the IDT. This gives $G_o = G_a/8N^2 = 0.14/(8 \times 50^2) \approx 7 \times 10^{-6}\,\text{mho}$. (b) $Z_o = 1/G_o \approx 143\,\text{k}\Omega$. ∎

4.5. Application to Overall SAW Filter Response

4.5.1. Use of Overall Two-Port Electrical Network

In obtaining the overall SAW filter response, the only concern is with the ratio of voltages at electrical Ports 3 in both input and output IDTs, as depicted in the block diagram of Fig. 4.15. From standard circuit theory for obtaining admittance parameters in Fig. 4.15, current-voltage relations are used at input and output

$$I_a = Y_{aa}V_a + y_{ab}V_b \tag{4.26a}$$

and

$$I_b = y_{ba}V_a + y_{bb}V_b. \tag{4.26b}$$

These yield driving-point admittances,

$$\left(at\ input\right)\quad y_{aa} = \left.\frac{I_a}{V_a}\right|_{V_b=0} \tag{4.27a}$$

$$\left(at\ output\right)\quad y_{bb} = \left.\frac{I_b}{V_b}\right|_{V_a=0}. \tag{4.27b}$$

Likewise, short-circuit transfer admittances are

$$\left(input\text{-}to\text{-}output\right)\quad y_{ba} = \left.\frac{I_b}{V_a}\right|_{V_b=0} \tag{4.28a}$$

$$\left(output\text{-}to\text{-}input\right)\quad y_{ab} = \left.\frac{I_a}{V_b}\right|_{V_a=0}. \tag{4.28b}$$

Input signal voltage V_{in} is related to V_a by $V_{in} = V_a + I_a R_s$, where R_s = source resistance, while load voltage across load resistance R_L is $V_L = V_b$. From Eqs. (4.26), (4.27), and (4.28) the input-output voltage transfer function $H(f)$ is

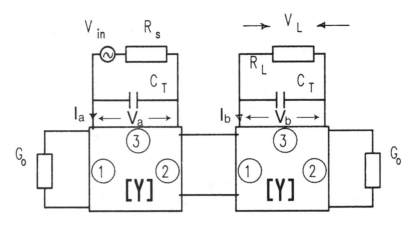

FIG. 4.15. SAW filter frequency response can be computed using three-port electrical admittances applied to the input and output IDTs.

$$H(f) = \frac{V_L}{V_\in} = \frac{y_{ab}R_L}{\left(1 + y_{aa}R_s\right)\left(1 + y_{bb}R_L\right) - y_{ab}^2 R_s R_L}. \qquad (4.29)$$

In Eq. (4.29) the transfer admittance term $y_{ab}^2 R_s R_L$ is just that associated with triple-transit-interference as considered earlier in the chapter. The term y_{ab}^2 implies two additional transits due to TTI.

4.5.2. SUBSTITUTION OF THE THREE-PORT PARAMETERS INTO THE TWO-PORT NETWORK

Equation (4.29) may now be translated into admittance terms associated with the three-port IDT formalism of the previous section [24]. Here, y_{aa} is the short-circuit input admittance measured with the output IDT short-circuited. However, input admittance y_{aa} in Eq. (4.29) just corresponds to the IDT input three-port admittance in Eq. (4.25), so that

$$y_{aa} = Y_3(f) = G_a(f) + j2\pi f C_T \quad (input\ IDT), \qquad (4.30)$$

where

$$G_a(f) = 8N^2 G_o \left| \frac{\sin\left[N\pi(f - f_o)/f_o\right]}{\left[N\pi(f - f_o)/f_o\right]} \right|^2 \quad (input\ IDT) \qquad (4.31)$$

is again an unperturbed radiation conductance (i.e., finger reflections neglected) and $N = N_p =$ number of input IDT finger pairs. In the same way the output driving point admittance y_{bb} is

$$y_{bb} = G_a^0(f) + j2\pi f C_T^0 \quad (output\ IDT), \qquad (4.32)$$

where the superscripts on the right-hand side of Eq. (4.32) indicate G_a and C_T values for the output IDT. At the output IDT, therefore, the unperturbed radiation conductance is

$$G_a^0(f) = 8M^2 G_o \left| \frac{\sin\left[M\pi(f - f_o)/f_o\right]}{\left[M\pi(f - f_o)/f_o\right]} \right|^2 \quad (output\ IDT), \qquad (4.33)$$

where $M = M_p =$ number of finger pairs in output IDT. Note that the unperturbed sinc-function terms in Eqs. (4.31) and (4.33) have to be "squared," as conductance is a pure real parameter in a passive device!

The remaining term y_{ab} in Eq. (4.29) is a transfer admittance, involving both input and output parameters. These may again be related to three-port admittance parameters, to give

$$y_{ab} = 8NMG_o \frac{\sin\left(N\pi\Delta f/f_o\right)}{N\pi\Delta f/f_o} \cdot \frac{\sin\left(M\pi\Delta f/f_o\right)}{M\pi\Delta f/f_o} e^{j\left[\pi\left(1-\left(N+M\right)\Delta f/f_o\right)-\phi\right]}; \quad (4.34)$$

where $M = M_p$ = number of finger pairs in the output IDT, $\Delta f = (f - f_o)$ and ϕ is a phase parameter relating to the separation between IDT phase centers. Note, too, that the sinc functions are *not* squared for y_{ab} in Eq. (4.34) because phase information is already included. Indeed, the phase angle variations in the exponential term with frequency give rise to the periodic passband ripple due to TTI.

The tacit assumption in Eq. (4.34) is that there are no acoustic reflections from IDT finger discontinuities to Y_{ab}. Strictly speaking, this equation should be applied to IDTs with split-electrode geometry, which minimize spurious finger reflections at center frequency. These have yet to be considered in this text.

4.5.3. INSERTION LOSS AND EFFECTIVE TRANSMISSION LOSS

In the SAW filter literature, the term *Insertion Loss* is normally applied to situations, such as in many network analyzer measurement techniques, involving equal source and load impedances. Where differing source and load impedances are involved it is better to use an *effective transmission loss* (ETL) definition [24]. This is the ratio of power received at the load to the maximum available power available from the generator. For source and load resistances R_s and R_L, which may be unequal, the effective transmission loss (ETL) is defined as

$$ETL = 20\log_{10}\left| \frac{V_{in}\sqrt{\left(R_L/R_s\right)}}{2V_L} \right| \quad (dB). \quad (4.35)$$

Substitution of Eq. (4.29) into Eq. (4.35) gives the ETL in decibels as

$$ETL = 20\log_{10}\left| \frac{\left[\left(1 + y_{aa}R_s\right)\left(1 + y_{bb}R_L\right) - y_{ab}^2 R_s R_L\right]\sqrt{R_L/R_s}}{2R_L y_{ab}} \right| \quad (dB). \quad (4.36)$$

Example 4.6 Effective Transmission Loss (ETL) of SAW Bandpass Filter. A SAW bandpass filter has equal input and output IDTs with $N = M = 40$ uniformly spaced finger pairs in each. The input and output impedances are matched to source and load resistances $R_s = R_L$ in this instance. Tuning circuits are incorporated in input and output stages to remove the capacitive reactance terms in y_{aa} and y_{bb}. From the effective transmission loss equation for a SAW filter, determine: (a) the mean value of the ETL in the passband; (b) the minimum and maximum values of ETL across the passband due to TTI; and (c) the peak-to-peak amplitude ripple due to TTI. ■

Solution. Use Eq. (4.36) to obtain the ETL values. Because $R_s = R_L$ in this example, the ETL equation is the same as for insertion loss. For the input IDT $y_{aa} = 1/R_s$ under the matched condition so that $y_{aa}R_s = 1$ in Eq. (4.36). Likewise, $y_{bb}R_L = 1$ for the output IDT. From Eq. (4.34) at center frequency $|y_{ab}| = 8NMG_o = 8N^2 G_o = G_a(f_o)$ in Eq. (4.31). However, $G_a(f_o) = 1/R_L$, so that $|y_{ab}|R_L = 1$ and $y_{ab}^2 R_s R_L = 1$ in Eq. (4.29). (a) To get the mean value of ETL neglect the TTI ripple term $y_{ab}^2 R_s R_L$ in Eq. (4.29) and Eq. (4.36). Substitution of the remaining parameters in Eq. (4.36) gives $\text{ETL} = 20\log_{10}|4/2| = 6\,\text{dB}$ for the matched SAW filter. (b) Minimum ETL will occur when the TTI is additive to the main signal. Including the TTI effect in Eq. (4.36) gives $\text{ETL} = 20\log_{10}|(4-1)/2|\,\text{dB} = 3.52\,\text{dB}$. Maximum ETL with TTI effects occur when $\text{ETL} = 20\log_{10}|(4+1)/2| = 7.95\,\text{dB}$. (c) From (b) the peak-to-peak TTI ripple in the passband is then $(7.95 - 3.52) = 4.43\,\text{dB}$, which is the result given in Table 4.1. ■

4.6. Impulse-Response Model

4.6.1. IMPLEMENTATION OF THE MODEL

The impulse-response model [7] yields additional information on the SAW transducer response over that given by the simple delta-function model, in that circuit impedances and matching networks can be included. It also provides ready insight into frequency scaling that must be employed in the determination of apodization quantities for broadband or chirp SAW filters. In essence, the impulse-response model uses Fourier transform pair relations to determine the impulse response $h(t)$ of a SAW filter, given its desired frequency response $H(f)$.

In the design procedure, one-half cycle of a sine wave in the impulse response $h(t)$ is placed between electrodes of opposite polarity, as depicted in the IDT of Fig. 4.16, for arbitrary finger spacing, uniform finger overlap, and constant metallization ratio η. The location of the zero-crossing of each $h(t)$ sample is an entirely arbitrary choice. In Fig. 4.16 it is sketched as occurring at the midpoint of each electrode. Next, the amplitude of each $h(t)$ half-cycle sample is multiplied by a frequency-scaling parameter $f_i^{3/2}(t)$, where f_i is the instantaneous frequency corresponding to the finger spacing at that point in the IDT. This 3/2 scaling factor is required from conservation of energy principles, as will be outlined in what follows [7]. Finally, the overall $h(t)$ response is multiplied by $F\sqrt{W}$ where constant F is a measure of

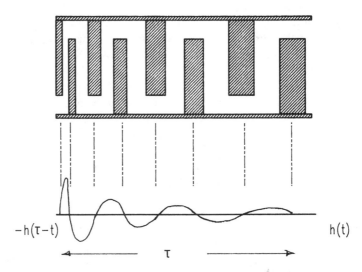

FIG. 4.16. Construction of an impulse response for an excited IDT with constant finger overlap and variable finger spacing.

the efficiency of electromechanical coupling between IDT voltage excitation and SAW generation in a given piezoelectric substrate. Parameter W is the overlap width (or acoustic beamwidth) of an electrode finger pair.

Constant F may be evaluated by solving the impulse response for an N finger pair IDT and comparing the result with that obtained from the crossed-field model. In this way it is found that

$$F\sqrt{W} = 4\sqrt{\left(K^2 C_s\right)},\qquad(4.37)$$

where K^2 = electromechanical coupling coefficient and C_s = capacitance of an electrode finger pair, which varies linearly with the acoustic beamwidth W. The resultant impulse response $h(t)$ obtained from these steps is

$$h(t) = 4\sqrt{\left(K^2 C_s\right)}f^{3/2}(t)\sin\theta(t),\qquad(4.38)$$

where

$$\theta(t) = 2\pi\int_0^t f_i(\tau)d\tau.\qquad(4.39)$$

In Eq. (4.39), τ is the time taken for the SAW to traverse the entire length of the IDT as sketched in Fig. 4.16, while $t = x/v$ is the time for the SAW to traverse to a given coordinate point x at velocity v along reference x-axis. For an IDT with N finger pairs, the impulse response of Eq. (4.38) is

$$h(t) \propto 4\sqrt{\left(K^2 C_x\right)} f_o^{3/2}(t) \sin\left(2\pi f_o t\right) \quad for\ 0 \le t \le N/f_o$$
$$= 0 \quad otherwise,$$

(4.40)

where f_o is the synchronous frequency appropriate to the spacing of each electrode pair and is constant for uniform IDT finger spacing. The Fourier transform of $h(t)$ in Eq. (4.40) is approximated by

$$\left|H(f)\right| \approx 2\sqrt{\left(K^2 C_s f_o\right)} \cdot N \cdot \frac{\sin\left[N\pi\left(f - f_o\right)/f_o\right]}{\left[N\pi\left(f - f_o\right)/f_o\right]}.$$

(4.41)

Equation (4.41) has the same unperturbed sinc-function dependence as the other models.

In the frequency domain, the total SAW energy $E(f)$ radiated by the IDT in both directions, when excited by a unit impulse voltage $V_{in}(f) = 1$, will be

$$E(f) = 2\left|H(f)\right|^2,$$

(4.42)

where the factor of two relates the bidirectionality of the IDT. Now, although the SAW is emitted in both directions, the equivalent radiation conductance $G_a(f)$ corresponds to the total $E(f)$, although half of this is lost in each IDT for signal processing purposes. Thus,

$$G_a(f) = E(f) = 2\left|H(f)\right|^2.$$

(4.43)

From Eqs. (4.41) and (4.43) the *unperturbed* radiation conductance is

$$G_a(f) = 8K^2 C_s f_o N^2 \cdot \left|\frac{\sin\left[N\pi\left(f - f_o\right)/f_o\right]}{\left[N\pi\left(f - f_o\right)/f_o\right]}\right|^2,$$

(4.44)

in accord with the result of Eq. (4.24) derived from the crossed-field model. Moreover, the *unperturbed* radiation susceptance $B_a(f)$ is obtained by taking the Hilbert transform of Eq. (4.44), so that

$$B_a(f) = \frac{8N^2 G_o \left[\sin\left(2X\right) - 2X\right]}{2X^2},$$

(4.45)

where $X = N\pi(f - f_o)/f_o$, $G_o =$ characteristic admittance of the equivalent SAW transmission line in the crossed-field model and $N =$ number of finger pairs. Radiation susceptance $B_a(f)$ can be regarded as relating to energy storage in the stress-strain fields associated with SAW excitation. It is a reactive electrical parameter (i.e., $\pm jB$) which goes to zero at center fre-

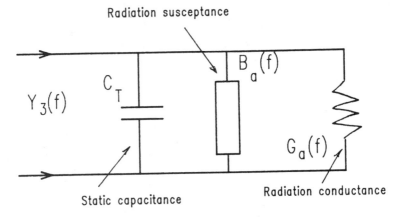

FIG. 4.17. Equivalent electrical admittance of an IDT in the impulse-response model. The radiation susceptance is zero at center frequency. Around center frequency it is also often small compared with the susceptance of IDT finger-capacitance C_T.

quency. Near center frequency it is normally quite small when compared to the IDT capacitance C_T, and is often omitted in calculations as a result. The equivalent circuit of the IDT with G_a, B_a and C_T is shown in Fig. 4.17. The $f_o^{3/2}$ frequency scaling factor is normally not applied unless the SAW filter is a wideband or chirp one. In such cases, however, it is necessary to scale the IDT finger apodization overlap by $f^{-3/2}$ in the event that a constant time-domain impulse response and/or a flat frequency domain passband amplitude is required.

4.6.2. ENERGY CONSERVATION AND THE $F^{3/2}$ FACTOR

Suppose that we consider two IDTs with the same number of finger pairs and the same amount of finger overlap (i.e., the same acoustic beamwidth), but with different finger spacing. The input admittance of the higher frequency IDT will then be equal to the admittance of the lower frequency IDT scaled to the higher frequency and multiplied by a scaling factor α:

$$Y_s(f) = \alpha Y_o \frac{f}{\alpha}. \tag{4.46}$$

Likewise, the impulse responses for the two will be similar except for an amplitude scaling factor β, so that

$$h_s(t) = \beta h_o(\alpha t). \tag{4.47}$$

Because the energy in both impulse responses is obtained by integrating over the same number of half-cycles in each instance,

$$\frac{E_s}{E_o} = \frac{\beta^2}{\alpha}. \tag{4.48}$$

The energy absorbed by each input admittance is obtained by a similar integration in the frequency domain, giving the frequency scaling result

$$\frac{E_s}{E_o} = \alpha^2. \tag{4.49}$$

From Eqs. (4.48) and (4.49) we obtain the 3/2 scaling factor in

$$\beta = \alpha^{3/2}. \tag{4.50}$$

4.7. Highlights of Current Circuit Models for SAW Filter Design

4.7.1. EQUIVALENT CIRCUIT MODEL

As will be considered in more detail in Chapter 9, Figure 4.18 illustrates a lumped equivalent circuit model for one IDT finger-section [3]. It is often referred to as either the Smith equivalent circuit or the Mason equivalent circuit. This is based on the Mason crossed-field model [6] discussed earlier in this Chapter. The particular illustration in Figure 4.18 gives a lumped equivalent circuit for a single solid-electrode IDT with a metallization ratio $\eta = 0.5$. The three-port circuit for an entire IDT with N such sections is obtained by connecting the acoustic ports in cascade and the electrical ports in parallel [25]. It incorporates differing characteristic impedances for metallized and unmetallized regions sections in one IDT over a distance of half-period of $\lambda_o/2$. The acoustic velocity under a metallization layer will decrease with increasing film-thickness ratio h/λ, with a commensurate decrease in the center frequency of the IDT[4]. Impedance parameter relationships are discussed in Chapter 9.

The transformer polarities will be dictated by the excitation polarity of the electrode. For convenience, a 1:1 normalized transformer ratio is shown in Fig. 4.18. In practice, this transformer ratio is a function of both the acoustic aperture and the electromechanical coupling coefficient K^2 of the piezoelectric substrate.

Figure 4.18 also includes parallel-connected shunt susceptances at electrode discontinuities. These are incorporated when it is necessary to consider energy storage at discontinuities. While energy storage is normally not a problem for SAW filters operating in the *fundamental* mode at low frequencies, it can be of significance for harmonic operation of SAW IDTs, as well as for leaky-SAW RF filter designs where film-thickness ratio $h/\lambda > \sim 1\%$.

---

[4] Determinations of SAW/LSAW velocity shifts due to IDT metallization loading are made with reference to a normalized *self-coupling coefficient* κ'_{11}, as introduced in Chapter 9.

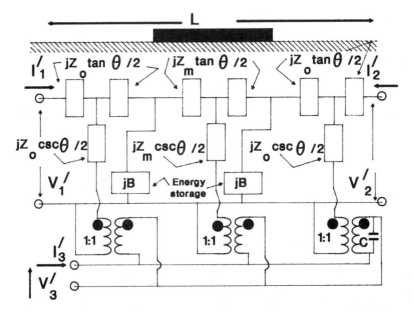

<small>Fɪɢ. 4.18. An equivalent circuit for an IDT finger, with impedance discontinuities between metallized and unmetallized regions. Example is for metallization ratio $\eta = 0.5$ and $L = \lambda_o/2$. Energy storage at discontinuities, (for LSAW and/or harmonic SAW designs) is represented here by shunt susceptances B.</small>

Although beyond the scope of this book, It may be noted that the forementioned lumped-equivalent circuit model may be extended to a more generalized form, by using electric-field distributions between electrodes—rather than using the bulk-wave electric-field distributions associated with the crossed-field model. This generalized circuit is capable of giving an accurate characterization of an IDT, for both fundamental- and harmonic-mode operation. In this way, the modelling of an IDT may be applied to solid- or split-electrodes, with arbitrary polarity sequences, as well as arbitrary metallization ratios [26].

4.7.2. S-Matrix Parameter Modelling

The S-matrix parameter modelling of an IDT utilizes the same 3×3 scattering-matrix concepts as applied to microwave networks, as formulated in Fig. 4.19(a). The dimensionless matrix elements relate to ratios between incident and output voltages at the acoustic and electrical ports of the IDT. For matched acoustic loads, the S_{11} and S_{22} parameters correspond to the input and output reflection coefficients, respectively. For power conservation in a lossless three-port system, we have $|S_{11}^2| + |S_{12}^2| + |S_{13}^2| = 1$. If IDT finger reflections are neglected, the reflection coefficients S_{11} and S_{22} parameters will only involve the load (or generator) reflection coefficient, as shifted clockwise on the Smith Chart (i.e., towards the generator).

In evaluating scattering parameter relationships at input/output acoustic ports of an IDT, it is essential to specify the reference axes for these determinations. As sketched in Fig. 4.19(b) the reference axes are usually (but not always) located $\lambda_o/8$ distant from the edges of an electrode. (This places the reference axes midway between electrodes in a uniform IDT with solid-electrode fingers, and metallization ratio $\eta = 0.5$.) From transmission-line theory, input and output impedances Z_{in} and Z_{out} at these reference axes are

$$Z_{in} = Z_o \frac{\left(1 + S_{11}\right)}{\left(1 - S_{11}\right)} \tag{4.51a}$$

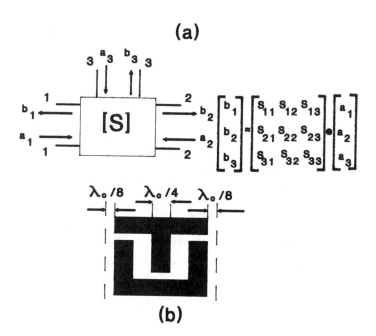

(a)

(b)

FIG. 4.19. (a) *S*-matrix terms applied to a three port SAW/LSAW IDT. (b) When using *S*- or *P*-matrices, it is essential to specify the reference-axis locations for acoustic Ports 1 and 2. As shown, these are normally taken as $\lambda_o/8$ from a finger edge.

$$Z_{out} = Z_o \frac{(1+S_{22})}{(1-S_{22})}, \tag{4.51b}$$

where $Z_o = 1/G_o$ in a lossless substrate, and G_o = characteristic admittance (conductance) of the piezoelectric substrate given in Eq. (4.16).

Example 4.7 S11 Reflection Coefficient For IDT With Negligible Finger Reflections. An IDT has N_t alternating and uniform solid-electrode fingers spaced $\lambda_o/2$ apart at center frequency $f_o = v/\lambda_o$. It is connected to a load impedance $Z_L = 1/Y_L$. The IDT film-thickness ratio h/λ_o is small enough for acoustic reflections from electrode fingers to be neglected. The radiation conductance at center frequency is $G_a(f_o)$ mho, while the IDT capacitance is C_T farads. Derive a relationship for the input S_{11} reflection coefficient at center frequency, as measured at a reference axis $\lambda_o/8$ in front of the edge of the end finger. Assume that the acoustic propagation is lossless under the IDT. ■

Solution. It may be shown that, for uniform unapodized IDTs, the load reflection coefficient ρ_L is given by [27],

$$\rho_L = -\frac{G_a(f)}{Y_t(f)}$$

where

$$Y_t(f) = Y_L + Y_{idt} = Y_L + \left[G_a(f) + j\left(\omega C_T + B_a(f) \right) \right],$$

and $B_a(f)$ = radiation susceptance (i.e., Hilbert transform of $G_a(f)$). If we consider that the load is connected at the midpoint of the IDT, the physical length d from midpoint to reference axis will be $d = 0.5 \times N_t \times \Lambda$, where $\Lambda = \lambda_o/2$. This gives the electrical length $\beta_o d$ at center frequency, where $\beta_o = 2\pi/\lambda_o$. For lossless acoustic propagation, transmission line theory gives

$$S_{11}(f_o) = \rho_L \exp(-j\beta N_t \Lambda)$$

at center frequency, where the minus sign in the previous equation corresponds to a clockwise rotation on a Smith Chart from load to generator, and a factor of 2 has been included, because the reflection coefficient repeats every half-wavelength. ■

4.7.3. P-MATRIX PARAMETER MODELLING

The 3×3 *P*-matrix is formulated as shown in Fig. 4.20. Matrix elements for the first two rows are dimensionless, as for the *S*-matrix, so that $P_{11} = S_{11}$, $P_{12} = S_{12}$, $P_{21} = S_{21}$, $P_{22} = S_{22}$, $P_{13} = S_{13}$, and $P_{23} = S_{23}$. The remaining transfer parameters P_{31}, P_{32}, and P_{33} have the dimensions of admittance.

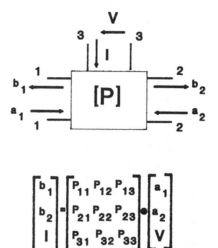

FIG. 4.20. In contrast to the 3×3 S-matrix terms, which are dimensionless, the P-matrix has "mixed" units.

In the P-matrix modelling of a SAW filter structure, the individual IDTs are broken down into individual segments for modelling: a) delays for metallized and unmetallized regions; b) discontinuities at IDT finger edges; and c) excitation regions under charged segments of an excited IDT finger. P-matrices are assigned to each cell, and finally combined to obtain an overall IDT response function. For calculating the overall response of a SAW filter with input and output IDTs, the 3×3 P-matrix is extended into a 4×4 P-matrix, in order to obtain the overall admittance matrix [8].

Example 4.8 Power Relations in an IDT for a Matched-Load Condition. A receiving IDT with negligible finger reflections and losses is matched to an external load at centre frequency. The IDT capacitance is "tuned out" by an inductance at center frequency f_o. a) Determine the power-transfer distribution under these conditions. b) Compare this result with that for the case of a short-circuit load. ∎

Solution. a) From Example 4.6, at center frequency $Y_{L} = G_a(f_o)$. From Eq. (4.45) the unperturbed radiation susceptance B_a has a value $B_a = 0$ at center frequency. With the shunt effect of C_T removed by external tuning, from Example 4.7 the load reflection coefficient reduces to

$$\rho_L = -\frac{G_a(f)}{Y_t(f)} = -\frac{1}{2}$$

at center frequency. Because S-parameters involve voltage ratios, the squares of these must be considered for power ratios. This gives $|S_{11}^2| = 1/4$ and $|S_{12}^2| = |(1 + \rho_L)^2| = (1 + (-0.5))^2 = 1/4$. In addition, $|S_{13}^2| = 1 - |S_{11}^2| - |S_{12}^2| = 1/2$. Under matched conditions, therefore, 1/2 of the incident power is absorbed by the load, 1/2 is regenerated in the reverse direction, and 1/2 is transmitted in the forward direction.

b) From Example 4.6, $\rho_L = 0$ for a short-circuit load. This gives $|S_{11}^2| = 0$, $|S_{12}^2| = 1$, and $|S_{13}^2| = 0$. Thus the acoustic wave propagates through the IDT as if it were "invisible" to the wave. This last result is, however, an extremely simplistic one. In practice, all metallization discontinuities will cause surface acoustic wave reflections. ∎

4.7.4. COUPLING-OF-MODES (COM) MODELLING

The Coupling-of-Modes (COM) analytical approach to the modelling of SAW/LSAW transducers and filters is a sophisticated method for modelling systems with time-varying or spatially varying parameters [13]–[20]. This model deals with acoustic waves propagating in the forward and reverse directions in a distributed structure (such as an IDT or reflection grating), and incorporates their coupling interactions as they pass through it. For two counter-propagating waves $a(x)$ and $b(x)$ in a uniform IDT, the wave distributions can be simplified to the form

$$a(x) = R(x)e^{-j\beta_g x/2} \qquad (4.52a)$$

$$b(x) = S(x)e^{+j\beta_g x/2}, \qquad (4.52b)$$

where $R(x)$ and $S(x)$ are slowly varying acoustic-wave amplitudes, and $\beta_g = 2\pi/p$, where p is the pitch of the uniform IDT. This allows the three differential COM equations for the two acoustic ports and the electrical port to be expressed in matrix form as

$$\begin{pmatrix} \dfrac{dR}{dx} \\[2mm] \dfrac{dS}{dx} \\[2mm] \dfrac{dI}{dx} \end{pmatrix} = \begin{pmatrix} -j\delta & jk_{12} & j\zeta \\ -jk_{12}^* & j\delta & -j\zeta^* \\ -j2\zeta^* & j2\zeta & j\omega C_s \end{pmatrix} \cdot \begin{pmatrix} R \\ S \\ V \end{pmatrix}, \qquad (4.53)$$

where δ is a "detuning" parameter, for frequencies off the centre frequency, κ_{12} is a mutual-coupling coefficient, as will be considered in Chapter 9, ζ is an excitation coefficient, and $2C_s p$ is the static capacitance of one electrode-pair. For uniform IDTs these equations can be integrated, so that all parameters of the P-matrix or S-matrix of the IDT elements can be described analytically.

4.8. Summary

This chapter first highlighted the distinguishing features of three early circuit models for obtaining frequency responses of SAW IDTs and filters. It was emphasized that all of these early models applied to IDT structures in which acoustic reflections from IDT finger discontinuities were neglected. (This was a reasonable approximation for designing filters in the 100-MHz range, for example, where film-thickness ratios h/λ were minimal.) From the delta-function model we saw that the frequency responses of input and output IDTs could not be separated when both IDTs were apodized and in-line. The crossed-field model enabled impedances to be included, so that absolute insertion loss (or effective transmission loss) could be obtained, together with their frequency dependences. From the crossed-field model a radiation conductance was obtained with a sinc-function frequency response as a result of the neglecting finger reflections. Finally, the impulse-response model could be used to provide additional information on IDT radiation susceptance as well as radiation conductance, while also providing some ready insight into the 3/2 frequency scaling factor for IDTs.

The chapter concluded with the highlights of four current analytical models for obtaining the frequency response and impedance characteristics of SAW/LSAW structures. It was noted that all of these models could incorporate effects of finger reflections, and resultant velocity shifts, as related to self-coupling coefficient κ_{11}. These current models have included: a) Smith's equivalent-circuit model; b) an S-Matrix model; c) a P-matrix modelling approach; and d) relationships for coupling-of-modes (COM) modelling, incorporating a mutual-coupling coefficient κ_{12}.

4.9. REFERENCES

1. R. H. Tancrell and M. G. Holland, "Acoustic surface wave filters," *Proc. IEEE*, vol. 59, p. 393, 1971.
2. R. H. Tancrell, "Principles of Surface Wave Filter Design," in H. Matthews (ed.), *Surface Wave Filters*, John Wiley and Sons, New York, Chapter 3, 1977.
3. W. R. Smith, Jr., "Studies of Microwave Acoustic Transducers and Dispersive Delay Lines", Ph.D. Thesis in Department of Applied Physics, Stanford University, pp. 1–147, 1969.

4. W. R. Smith, H. M. Gerard, J. H. Collins, T. W. Reeder and H. J. Shaw, "Analysis of interdigital surface wave transducers by use of an equivalent circuit model," *IEEE Trans. Microwave Theory and Techniques*, vol. MTT-17, pp. 856–864, November 1969.

5. W. R. Smith, H. M. Gerard, J. H. Collins, T. W. Reeder and H. J. Shaw, "Design of surface wave delay lines with interdigital transducers," *IEEE Trans. Microwave Theory and Techniques*, vol. MTT-17, pp. 865–873, November 1969.

6. W. P. Mason, "*Electromechanical Transducers and Wave Filters*," van Nostrand-Reinhold, 2nd Edition, Princeton, New Jersey, pp. 201–209, 399–409, 1948.

7. C. S. Hartmann, D. T. Bell, Jr. and R. C. Rosenfeld, "Impulse response model design of acoustic surface-wave filters," *IEEE Trans. on Microwave Theory and Techniques*, vol. MTT-21, pp. 162–175, April 1973.

8. C. C. W. Ruppel, W. Ruile, G. Scholl, K. C. Wagner and O. Männer, "Review of models for low-loss filter design and application," *Proc. 1994 IEEE Ultrasonics Symposium*, vol. 1, pp. 313–324, 1994. (With 108 references).

9. K. Inagawa and M. Koshiba, "Equivalent networks for SAW interdigital transducers," *Proc. IEEE Trans. Ultrason., Ferroelec., Freq. Contr.*, vol. 41, pp. 402–411, May 1994.

10. K. Nakamura and K. Hirota, "Equivalent circuits for directional SAW-IDT's based on the coupling-of-modes theory," *Proc. 1993 IEEE Ultrasonics Symposium*, vol. 1, pp. 215–218, 1993.

11. See for example: R. G. Brown, R. A. Sharpe, W. L. Hughes, R. E. Post, *Lines, Waves, and Antennas*, 2nd edition, The Ronald Press, New York, ch. 5, 1973.

12. G. Tobolka, "Mixed matrix representation of SAW transducers," *IEEE Trans. Sonics Ultrason.*, vol. 26, No. 3, 1979.

13. J. R. Pierce, "Coupling-of-modes of propagation," *J. App. Phys.*, vol. 25, pp. 179–183, 1954.

14. Y. Suzuki, H. Shimizu, M. Takeuchi, K. Nakamura and Y. Tamada, "Some studies of SAW resonators and multiple-mode filters," *Proc. 1996 IEEE Ultrasonics Symposium*, pp. 297–302, 1976.

15. H. A. Haus, "Modes in SAW grating resonators," *J. App. Phys.*, vol. 49, pp. 4955–4961, 1977.

16. P. S. Cross and R. V. Schmidt, Coupled surface-acoustic-wave resonators,' *Bell Syst. Tech. j.*, vol. 56, pp. 1447–1482, October 1977.

17. P. V. Wright, "A coupling-of-modes analysis of SAW grating structures," *Ph.D. Thesis in Electrical Engineering*, Massachusetts Institute of Technology, Cambridge, Massachusetts, April 1981.

18. D-P Chen and H. A. Haus, "Analysis of metal strip SAW gratings and transducers," *IEEE Trans. Sonics Ultrason.*, vol. SU-32, pp. 395–408, May 1985.

19. C. S. Hartmann, P. V. Wright, R. J. Kansy and E. M. Garber, "An analysis of SAW interdigital transducers with internal reflections and the application to the design of single-phase unidirectional transducers," *Proc. 1982 IEEE Ultrasonics Symposium*, pp. 40–45, 1982.

20. H. A. Haus, *Waves and Fields in Optoelectronics*, Prentice-Hall, New Jersey, ch. 7, 1984.

21. T. Motsoela and P. M. Smith, "Doubly apodized SAW filters," *Proc. 1995 IEEE Ultrasonics Symposium*, vol. 1, pp. 95–98, 1995.

22. A. J. Slobdonik Jr., K. R. Laker, T. L. Szabo, W. J. Kearns and G. A. Roberts, "Low sidelobe SAW filters using overlap and withdrawal weighted transducers," *Proc 1977 IEEE Ultrasonics Symposium*, pp. 757–762, 1977.

23. T. Morita, Y. Watanabe, M. Tanaka and Y. Nakazawa, "Wideband low loss double mode SAW filters," *Proc. 1992 IEEE Ultrasonics Symposium*, vol. 1, pp. 95–104, 1992.

24. A. J. deVries, "Surface Wave Bandpass Filters," in H. Matthews (ed.), *Surface Wave Filters*, John Wiley and Sons, New York, Chapter 6, 1977.

25. W. R. Smith, "Basics of the SAW interdigital transducer,", in J. H. Collins and L. Masotti (eds), *Computer-Aided Design of Surface Acoustic Wave Devices*, Elsevier, New York, pp. 25–63, 1976.
26. W. R. Smith and W. F. Pedler, "Fundamental- and harmonic-frequency circuit-model analysis of interdigital transducers with arbitrary metallization ratios and polarity sequences," *IEEE Trans. Microwave Theory Tech.*, vol. MTT-23, pp. 853–864, Nov. 1975.
27. C. B. Saw, "Single-phase unidirectional transducers for low-loss surface acoustic wave filters,' *Ph.D. Thesis in Electrical Engineering*, McMaster University, Hamilton, Ontario, Canada, July 1988.

—5—

Some Matching and Trade-Off Concepts for SAW Filter Design

5.1. Introduction

The basics of linear-phase SAW filter design were introduced in Chapter 3, together with an introductory review of various second-order effects that cause degradation of passband and/or stopband frequency responses. Some of these second-order effects are considered here in more detail, as they affect trade-offs in the design of linear-phase SAW/pseudo-SAW filters with bidirectional transducers. Although these second-order effects can never be totally eliminated, they can often be reduced to levels where exceptionally good bandpass- or resonator-filter performance can be realized.

Principal second-order effects that detract from SAW/pseudo-SAW filter performance can be due predominantly to:

1) Mismatching of external circuitry at filter input and output
2) Bulk waves
3) Diffraction
4) Triple-transit-interference (TTI)
5) Electromagnetic feedthrough
6) Spurious harmonic responses
7) Temperature-dependent degradation
8) [Acoustic reflections from IDT fingers, depending on the application]

In the foregoing listing, acoustic reflections from IDT fingers are shown in brackets. Such reflections may, or may not, contribute to a degradation of the filter response. It depends on the application. For example, the design of typical leaky-SAW RF front-end ladder- or resonator-filters for mobile transceivers desirably incorporates IDT finger reflections. This chapter will consider matching and mismatching of Rayleigh-wave filters, (with the same principles considered applicable to pseudo-SAW structures) and Chapter 6 will be concerned with the remaining second-order effects.

Before proceeding with matching considerations, however, it is instructive to consider some general specifications and requirements for RF filters in mobile and wireless communications circuitry. Current RF frequencies for mobile telephone systems vary from about 450 MHz up into the 2-GHz band. RF receiver front-end filters for those bands typically have insertion losses on the order of 3 dB or less over a required environmental temperature range of from -30 to $+70°$C. Filter fractional bandwidths can be up to about 5%. In-band ripple specifications are typically on the order of 0.6 dB. Typical designs operate in 50-ohm circuitry.

As insertion loss in a front-end filter of a receiver will degrade its signal-to-noise (S/N) performance, it is essential that this be kept to a minimum. This can place demanding specifications on the filter design, as well as on its matching networks, because even an increase of 0.5 dB in front-end insertion loss can be undesirable in many such systems. Moreover, input/output impedance matching of the filter must be effective over a range of frequencies around the midband point.

As measured on a network analyzer in the 50-ohm Smith Chart mode, Fig. 5.1 shows the input/output impedance characteristics of a sample leaky-SAW, 880-MHz RF filter on 64° Y-X LiNbO$_3$, as fabricated with interdigital interdigitated transducers (IIDTs)[1] [1]. Figure 5.1(a) shows the input/output impedance variations (for S_{11} and S_{22}, respectively) over the frequency range from 678.5 to 1078.5 MHz, with no external matching. (The 50-ohm matching point is at the center of the Smith chart in each instance.) Figure 5.1(b) shows the improved input/output impedance matching responses when one-port leaky-SAW resonator impedance elements were added[2] while Fig. 5.1(c) shows the still-improved matching response with the further addition of tuning inductances. This structure realized a passband insertion loss of 2.5 dB and a sidelobe attenuation of 45 dB in the specified stopband.

The preceding second-order listing does not represent all of the factors that can cause response degradation. For example, attainable levels of sidelobe suppression can depend critically on the dimensional accuracies attained in fabricating the IDT, as well as in its alignment with the SAW propagation axis of the piezoelectric crystal. This becomes particularly crucial in current mobile circuits operating up into the 2-GHz band, requir-

---

[1] This was for a receiver filter for a portable telephone transceiver in Japan, requiring a filter passband from 870 to 887 MHz, and a stopband from 925 to 842 MHz.

[2] One-port leaky-SAW impedance-element resonator structures, with frequency responses equivalent to lumped LCR resonators, are introduced in Chapter 13. Such impedance-element resonators are also used as "building blocks" in the design of leaky-SAW ladder filters for antenna duplexers.

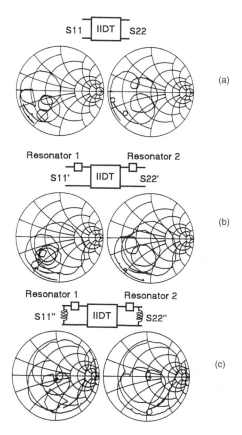

Fɪɢ. 5.1. Network analyzer input/output impedance display for 50-Ω, 880-MHz, leaky-SAW-IIDT RF filter on 64° *Y-X* LiNbO$_3$, over frequency range 678.5 to 1078.5 MHz. (a) IIDT filter. (b) IIDT filter and resonators. (c) IIDT filter, resonators and tuning inductors. Reprinted from Yatsuda, Takeuchi and Horishima, [1]. (Courtesy of Japanese Journal of Applied Physics.)

ing the fabrication of IDTs with finger widths of less than 1 μm (1 μm = 10^{-6} m), in filter packages as small as 3.2×2.5×0.9 mm^3 [2].

A complete SAW filter network will include external circuit elements connected to input and output SAW transducers. The external elements may be entirely passive. Some designs may also incorporate transistor circuitry for changing impedance and/or signal voltage levels. Regardless of what is used, however, the overall filter performance should be dominated by the response of the SAW filter and not by that of the external circuitry, otherwise there would be no point in having the SAW filter in the circuit.

External circuit influences are usually referred to as *circuit factor loading*. Obviously, there must be some frequency breakpoint above which the circuit factor loading can corrupt the desired filter response. This breakpoint is a function of the type of piezoelectric substrate used. It is usually referred to as the *intrinsic fractional bandwidth* $(\Delta f/f_o)_I\%$, above which the insertion loss (theoretically) increases at 40 dB/decade. Substrates with a large electromechanical coupling coefficient K^2, such as YZ-lithium niobate, $(\Delta f/f_o)_I\% \approx 24\%$, are suitable for the design of filters with wider bandwidths. On the other hand, with $(\Delta f/f_o)_I\%$ at only about 4.4% for ST-quartz, this primarily limits this substrate to narrowband IF filter or oscillator applications in mobile transceivers. Currently, ST-quartz is the preferred substrate in mobile-transceiver IF circuitry, because its first-order temperature coefficient of delay (TCD) is approximately zero at 25°C.[3]

5.2. Bandwidth Limitations in Linear-Phase SAW Filter Design

5.2.1. INTRODUCTION

Source and load circuit parameters affect the overall frequency response of a SAW filter network. The external circuitry also places a limitation on the maximum fractional bandwidth that a SAW filter can have before the network response becomes adversely dominated by the frequency response characteristics of the input and output circuitry.

Consider the equivalent circuit of the SAW filter network sketched in Fig. 5.2, with fixed resistances as source and load impedances. The signal voltage will be shared between the source resistance and the parallel combination of IDT susceptance $j\omega C_T$ and unperturbed radiation conductance $G_a(f)$ considered here. As the unperturbed radiation susceptance $B_a(f)$ is zero at center frequency [see Eq. (4.45)], we will assume in what follows that the overall IDT susceptance is due entirely to the IDT electrode capacitance C_T. The combination of source resistance and shunt capacitance then forms a first-order RC lowpass filter. Unless its response is compensated for, the frequency response of the input IDT will be degraded by the 20 dB/decade roll-off due to the RC filter. A similar situation holds for the output circuit. The simplest compensation involves incorporation of a series or shunt inductance L to tune out C_T at center frequency (i.e., setting $2\pi f_o L = 1/2\pi f_o C_T$). Because tuned circuits have only finite bandwidths, design trade-offs will be required—especially for wideband filters.

---

[3] The temperature coefficient of delay of ST-quartz is parabolic around the zero-point at $\sim 25°C$. Thus, the center frequency of an IDT on this substrate will shift *downwards*, when the operating temperature goes *above or below* $\sim 25°C$!

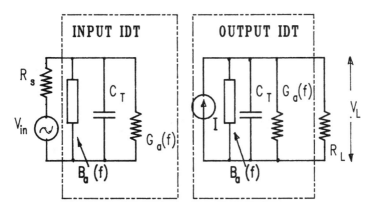

Fig. 5.2. Equivalent circuit for SAW filter.

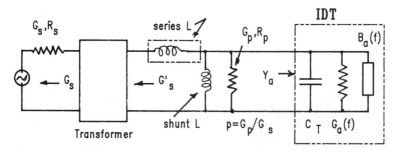

Fig. 5.3. Matching circuit shown for input IDT can include shunt-loading resistance R_p, a series- or shunt-connected inductance L, as well as a transformer, depending on specifications. A similar arrangement can be employed at the output IDT.

This leads us to investigate the relationships between external and internal components that dictate the overall SAW filter response.

5.2.2. MATCHING NETWORKS

At the outset in this presentation, the term "matching" as applied to linear-phase SAW filter design should be viewed with caution. In some designs, the goal is to *mismatch* the input and/or output circuits in attaining design trade-offs. As a result, the term "matching" as used in this chapter should be interpreted to mean the *degree* of matching or mismatching required.

Figure 5.3 illustrates various components that might be employed in the input circuitry. A similar one can be applied to the output IDT. Some designs may, of course, allow operation without matching components. Matching circuitry may simply include either a series- or shunt-connected lumped inductance in input and/or output, to tune out the IDT capacitance C_T at center frequency f_o, such that $2\pi f_o L = 1/2\pi f_o C$. Many high-frequency communications circuits operate with fixed impedance levels of 50 or 72 Ω. As a result, design trade-offs can also require the use of shunt-loading resistors and/or transformers, as shown. While the transformers are normally wire-wound ones, they can be replaced by microstrip structures if either the operating frequencies are sufficiently high or packaging sizes permit. The photographs in Figs. 5.4 and 5.5 illustrate examples of matching structures (utilizing thin-film spiral inductors) in some commercial SAW filter designs.

5.2.3. Radiation Q and External Q in SAW Filter Design

The *maximum intrinsic fractional bandwidth* of a SAW filter with bidirectional IDTs is the maximum bandwidth up to which its insertion loss can (theoretically) be held to 6 dB before external circuit responses prevail. This is governed by the value of K^2 of the piezoelectric substrate. The expression for intrinsic bandwidth in terms of K^2 may be readily determined in terms of the Qs of its component circuitry.

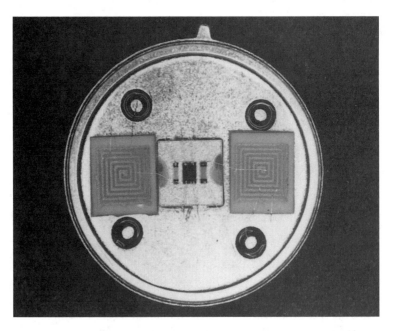

Fig. 5.4.   Example of commercial 750-MHz SAW bandpass filter with printed inductor tuning elements in input and output IDTs. (Courtesy of Crystal Technology, Inc., Palo Alto, California.)

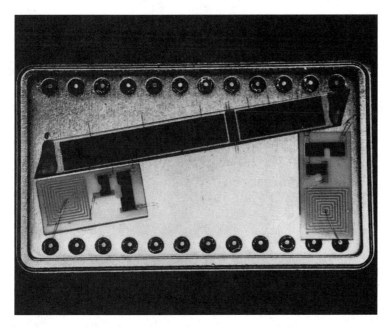

Fig. 5.5.   Another example of a commercial 150-MHz SAW bandpass filter with hybrid tuning elements in input and output stages. (Courtesy of Crystal Technology, Inc., Palo Alto, California.)

Two separate Q parameters are applied to the derivation of intrinsic bandwidth. These are: 1) Q_r, the acoustic radiation Q of the SAW IDT; and 2) Q_e, the external circuit Q. By way of comparison with microwave terminology [3], consider Q_r as corresponding to the "unloaded Q" (usually given as Q_o) of a one-port resonant cavity, while Q_e relates the "external Q." For such a cavity, the overall loaded Q (normally designated as Q_L) is given by $1/Q_L = 1/Q_r + 1/Q_e$, as shown in Fig. 5.6. When the cavity is perfectly matched at one frequency, the Q relations are $Q_e = Q_r$ and $Q_L = Q_r/2$. These relations must be applied to both IDTs in the SAW filter. Input and output circuitry will have only the same Qs in the event that these stages are identical.

The acoustic radiation Q of each SAW IDT is defined as

$$Q_r = \frac{f_o}{\Delta f_r} = \frac{2\pi f_o C_T}{G_a(f_o)}, \qquad (5.1)$$

where $\Delta f_r = $ 3-dB bandwidth of Q about center frequency f_o, $C_T = $ total IDT capacitance and $G_a(f_o)$ is the radiation conductance of the IDT at center frequency [4]. Likewise, the external Q associated with each SAW IDT is

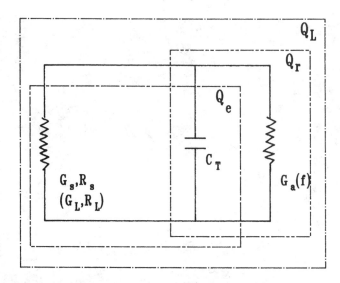

FIG. 5.6. Notation used here for acoustic radiation Q_r, external Q_e, and loaded Q_L as applied to IDT equivalent circuit. IDT finger capacitance C_T value is also assumed to include small radiation susceptance $B_a(f)$.

$$Q_e = \frac{f_o}{\Delta f} = \frac{2\pi f_o C_T}{G_s}, \tag{5.2}$$

where $\Delta f = $ 3-dB bandwidth of the external circuit. Here, $G_s = 1/R_s = $ source conductance. For the load circuit, G_s is replaced by $G_L = 1/R_L$ where R_L is the load resistance. From Eqs. (4.22) and (4.24) the radiation conductance G_a near center frequency is

$$G_a \approx G_a(f_o) \approx 8K^2 f_o C_s N^2 \quad (mho) \tag{5.3a}$$

$$\approx 8K^2 f_o C_s W N^2 \quad (mho), \tag{5.3b}$$

where $C_s = $ capacitance/finger pair (F), $C_o = $ capacitance/finger pair/unit finger length, $W = $ acoustic aperture of the IDT fingers (m) and $N = $ number of finger pairs. Equation (5.3b) is useful as it indicates that the radiation conductance is proportional to the acoustic aperture of the IDT, while $f_o C_o/G_a = $ constant, is proportional to $1/K^2$. Equation (5.3b) may also be expressed as

$$G_a(f) \approx 8K^2 f_o C_T N \quad (mho), \tag{5.4}$$

at center frequency, where $C_T = C_s N = $ total IDT capacitance (F). Now substitute Eq. (5.4) into Eq. (5.1) to obtain

$$Q_r = \frac{f_o}{\Delta f} = \frac{\pi}{4K^2 N}. \tag{5.5}$$

The frequency response of the overall SAW filter network will be dominated by that portion having the higher Q. If the SAW filter is to serve its purpose, this requires $Q_r \geq Q_e$ to hold. In terms of circuit operation, this means that when the SAW filter response dominates, its overall insertion loss can be kept to a low of 6 dB, commensurate with the use of matched bidirectional IDTs. The transition region, where the SAW filter starts to lose its dominance, occurs for $Q_r = Q_e$. At frequencies above those corresponding to this breakpoint, the overall insertion loss cannot be held to 6 dB. Its value at higher frequencies is dictated by the frequency response characteristics of the external circuitry.

5.2.4. MAXIMUM INTRINSIC FRACTIONAL BANDWIDTH

To obtain an approximate estimate of the maximum fractional bandwidth from the preceding relationships, consider that the input IDT is unapodized and matched to the input source conductance with $G_s = G_a$ at center frequency. From Eqs. (5.2) and (5.5) the external $Q_e|_t$ at the response transition threshold may be approximated as

$$Q_{e|t} = \left.\frac{f_o}{\Delta f}\right|_t = \frac{\pi}{4K^2 N}. \tag{5.6}$$

From the discussion following Eq. (3.5), the 4-dB fractional bandwidth of an unapodized IDT is $BW_4 \approx 1/N$, where $N = N_p = $ number of finger pairs. This is related to the 3-dB fractional bandwidth by $BW_3 \approx 0.9 BW_4 \approx 0.9(1/N)$. As an approximation here, simply consider that these two bandwidths are equal. This gives the radiation Q_r as

$$Q_r = \frac{f_o}{\Delta f} \approx N. \tag{5.7}$$

From Eqs. (5.6) and (5.7) the maximum fractional bandwidth for which the insertion loss of the SAW filter with bidirectional IDTs can be held to 6 dB is

$$\left.\frac{\Delta f}{f_o}\right|_t \approx \frac{2\left(K^2\right)^{1/2}}{\sqrt{\pi}} = \sqrt{\frac{4K^2}{\pi}}. \tag{5.8}$$

Equation (5.8) was derived under a number of assumptions. While one of these was that the IDTs were unapodized, it is found in practice that the result is also approximately valid for apodized IDTs [5]. Note that mass-loading and finger reflection effects were omitted and the IDT fingers were considered to have zero resistance. Additionally, the derivation ignored the parasitic resistance of electrode fingers and shunt capacitance of the pads. The bandwidth result of Eq. (5.8) is often referred to as the maximum *intrinsic* bandwidth of the SAW filter [4]. Some SAW filter matching designs also include the shunt loading resistor shown in Fig. 5.3, which additionally serves to lower the loaded Q and increase the permissible maximum bandwidth, at the expense of parasitic loss [6]. [To include this, replace $G_a(f_o)$ by $(G_a(f_o) + G_p)$ in the denominator of Eq. (5.1), where $G_p = 1/R_p$ and R_p is the shunt loading resistor across the IDT.]

Table 5.1 lists the maximum intrinsic bandwidths for representative SAW piezoelectric substrates. Values given are for bidirectional transducers with conjugate matched inputs and outputs. (The term *conjugate matching* means that if the input IDT impedance is $(R_1 - jX_1)$ at center frequency, then the source resistance is required to be $(R_1 + jX_1)$.)

5.2.5. ABOVE THE MAXIMUM FRACTIONAL BANDWIDTH

The bandwidths given in Table 5.1 for bidirectional IDTs are those values for which it is theoretically possible to hold overall the insertion loss of the SAW filter to 6 dB by conjugate matching. Above these bandwidth limits, circuit response is dominated by external circuitry. With resistive source and load impedances and in the absence of any matching, the external resistors and IDT capacitances will simply act as RC lowpass filters. The attenuation of first-order filters has the form $1/(1 + jX(f))$, so that those associated with each IDT will roll off at 20 dB/decade above cut off. In this situation the insertion loss of the SAW filter network will start to increase at 40 dB/decade as shown for ST-quartz and YZ-lithium niobate in the logarithmic plot of Fig. 5.7, for which

TABLE 5.1

MAXIMUM FRACTIONAL BANDWIDTHS OF
RAYLEIGH-WAVE SUBSTRATES

Substrate	K^2	$(\Delta f/f_o)_1 \%$
ST-quartz	0.0016	4.5
Lithium tantalate	0.0072	9.5
YZ-lithium niobate	0.045	23.9
128° YX-lithium niobate	0.053	25.9

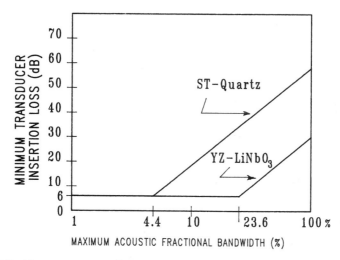

FIG. 5.7. Theoretical logarithmic plot of maximum acoustic fractional bandwidth (%) versus insertion loss, for SAW filters with bidirectional IDTs on ST-quartz and YZ-lithium niobate.

$$IL = 10\log_{10}\frac{\pi^2}{\left(2K^2\right)^2}\left(\frac{\Delta f}{f_o}\right)^4. \tag{5.9}$$

The foregoing relations indicate the maximum bandwidths limits that can be employed with a given substrate. They are quite limited in their applicability, however, because such filters would not normally be designed for 6-dB insertion loss due to excessive TTI in passband amplitude and phase responses—as given in Table 4.1. This calls for trade-offs between bandwidth, insertion loss, triple-transit interference, and voltage standing-wave ratio (VSWR).

5.3. Design Trade-Offs

5.3.1. ACOUSTOELECTRIC TRANSFER FUNCTION T_{13}

Acoustoelectric transfer functions and acoustic reflection relations for the two-port SAW filter can be obtained from the three-port equivalent circuit for an IDT as introduced in Chapter 4. There, the notation designated suffix "3" as the electrical port, while suffixes "1" and "2" designated "acoustic" ports. With this notation, the electroacoustic transfer function T_{13} is defined for one IDT as [6],[7]

$$T_{13}(f) = \sqrt{\left(2Q_e/Q_r\right)}\,\frac{A(f)}{C(f)}. \tag{5.10}$$

In Eq. (5.10), Q_e and Q_r are as defined in Eqs. (5.1) and (5.2). Here, Q_r is the 3-dB bandwidth of $A(f)$. In this new notation, the frequency response of the IDT and SAW filter is given in terms of $A(f)$ and $C(f)$, where $A(f)$ is termed the *array factor* and $C(f)$ is the *circuit factor* [6],[7]. Parameter $A(f)$ gives the "bare" response of the IDT, such as might be calculated with the delta-function model. It is interpreted as corresponding to a voltage ratio; $C(f)$ is also a voltage ratio, and accounts for actual internal and external impedances. This gives the overall frequency response of the SAW filter as $H(f) = A(f)/C(f)$ where circuit loading is included; $C(f)$ is set to $C(f) = 1$ for the "bare" IDT response as considered in previous chapters. Equation (5.10) may be reexpressed for electroacoustic *power* transfer as

$$\left|T_{13}(f)\right|^2 = \frac{2Q_e}{Q_r}\frac{A(f)^2}{C(f)^2}. \tag{5.11}$$

The factor of 2 accounts for the inherent 3-dB loss associated with a bidirectional IDT. The leading term $(2Q_e/Q_r)$ in Eq. (5.11) yields an estimate of the insertion loss of *one* IDT at center frequency, so that $IL \approx 10\log_{10}(2Q_e/Q_r)$ dB per IDT. The overall power transfer from input to output of the SAW

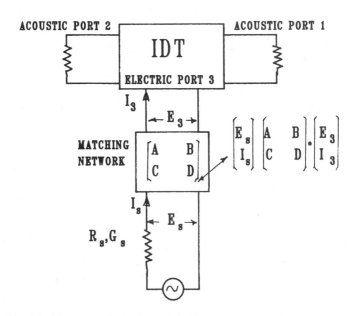

F<small>IG</small>. 5.8.   Matching network may be modelled in terms of an $ABCD$ matrix, with element values given in Table 5.2.

filter is the product response $|T_{13}|^2 \cdot |T'_{31}|^2$, where $|T'_{31}|^2$ is the acousto-electric power transfer at the output IDT.

Figure 5.8 shows the block diagram representation of an input IDT, where the matching network is an arbitrary one, symbolized in terms of $ABCD$ matrix parameters such that

$$\begin{pmatrix} E_s \\ I_s \end{pmatrix} = \begin{pmatrix} A & B \\ C & D \end{pmatrix} \cdot \begin{pmatrix} E_3 \\ I_3 \end{pmatrix},$$ (5.12)

where E_s, I_s, E_3 and I_3 are as depicted in this figure. (For the load circuit, replace E_s, I_s by E_L, I_L and R_s by R_L.) Using the $ABCD$ notation, the general form of the circuit factor $C(f)$ is

$$C(f) = A + BY_a + R_s(C + DY_a),$$ (5.13)

where Y_a = IDT input admittance or $Y_a = G_a(f) + B_a(f) + j\omega C_T$. Calculations of IDT transfer functions are simplified when mass loading and finger reflections are ignored. This is easily handled for situations where IDT capacitance dominates in Y_a so that $Y_a \approx j\omega C_T$. Values of $ABCD$ matrix elements without series matching, as well as series and shunt matching impedances Z_{ser} and Y_{sh}, respectively, are given in Table 5.2.

5.3.2. THE ACOUSTIC REFLECTION COEFFICIENT FUNCTION T_{11}

One other key trade-off parameter in linear-phase SAW filter design is the acoustic reflection coefficient $T_{11}(f)$ or $T_{22}(f)$ at the acoustic ports of the bidirectional IDT. This ties in with triple-transit-interference levels so it must be determined for both input and output IDTs. For the input circuit of Fig. 5.8 it is defined as [6]

$$T_{11}(f) = \frac{Q_e}{Q_r} \frac{|A(f)|^2}{N(f)},$$ (5.14)

TABLE 5.2

CIRCUIT FACTORS $C(f)$ AND $N(f)$ FOR MATCHING OF INPUT IDT [After Reference 6]

Matching Network	$ABCD$ Matrix		Circuit Factors $C(f)$, $N(f)$
No matching (plain wires!)	1 0		$C(f) = N(f) = 1 + R_s Y_a$
Shunt admittance Y_{sh} (L and/or resistor R_p)	1 0		$C(f) = N(f) = 1 + R_s(Y_a + Y_{sh})$
Series impedance Z_{ser} (L and/or resistor)	1 Z_{ser}		$C(f) = 1 + Y_a(R_s + Z_{ser})$
	0 1		$N(f) = R_s(Y_a + 1/(R_s + Z_{ser}))$

where $N(f)$ is a second circuit factor given by

$$N(f) = R_s Y_a + \frac{R_s(A + CR_s)}{(B + DR_s)}.$$ (5.15)

Representative values of $N(f)$ are given in Table 5.2 along with those of $C(f)$. As shown, $C(f) = N(f)$ for no matching, or matching with a shunt admittance Y_{sh}, in which case T_{11} and T_{13} are related by

$$T_{11}(f) = \frac{|T_{13}(f)|^2}{2} C(f) \quad (for\ B = 0, D = 1).$$ (5.16)

5.3.3. Mismatch Parameters

The relationship for maximum intrinsic fractional bandwidth derived in Section 5.2 applied to matched conditions in input and output IDT stages. In SAW bandpass filter designs employing bidirectional IDTs, however, operation under matched source and/or load conditions is generally undesirable because the TTI suppression is then only 12 dB. This leads to excessive amplitude and phase ripple across the passband, as listed in Table 4.1 of Chapter 4. To examine operation and trade-offs under mismatch conditions, we may define a mismatch parameter q applicable to either input or output stages such that

$$q = q(f_o) = \frac{G_a}{G_s}$$ (5.17)

at center frequency f_o. [Equation (5.17) applies to the input IDT. For the output IDT simply replace source conductance $G_s = 1/R_s$ by load conductance $G_L = 1/R_L$. In either instance $G_a = G_a(f_o) = $ radiation conductance of the corresponding IDT at center frequency.]

Further, consider a second mismatch parameter p to relate the effect of including a shunt load resistance across the IDT in Fig. 5.3. This loading mismatch parameter is defined at center frequency as [6]

$$p = p(f_o) = \frac{G_p}{G_s},$$ (5.18)

where $G_p = 1/R_p = $ shunt loading conductance. In terms of these mismatch parameters q and p, the insertion loss for one IDT at center frequency $[(H(f_o) = C(f_o) = 1]$ from Eq. (5.11) becomes [6]

$$|T_{13}(f_o)|^2 = \frac{2q}{(1 + q + p)^2},$$ (5.19)

while the acoustic power reflection loss is now

$$\left|T_{11}(f_o)\right|^2 = \frac{q^2}{(1+q+p)^2}.$$ (5.20)

For $q \ll 1$ in Eqs. (5.19) and (5.20), a 2-dB decrease in T_{11} due to a change of q is at the expense of a 1-dB increase in insertion loss. On the other hand, with $|T_{11}(f_o)/T_{13}(f_o)|^2$ independent of p, a variation of pad resistance R_p does not gain much in the trade-off between insertion loss and reflection suppression.

5.3.4. TRADE-OFF ON VSWR

Because any mismatching of the input and/or output circuits will result in values of voltage standing wave ratios (VSWR) greater than unity, some trade-off between mismatch and VSWR may be required if the latter proves to be critical. The form of the VSWR equation employed here is a general one as can be applied to lumped element networks [8], and is defined in terms of mismatch parameters q and p at center frequency as [7]

$$VSWR = (q+p) \quad for \ (q+p) \geq 1$$ (5.21a)

or

$$VSWR = \frac{1}{(q+p)} \quad for \ (q+p) \leq 1.$$ (5.21b)

5.3.5. EVALUATING THE UNPERTURBED RADIATION CONDUCTANCE $G_a(f_o)$

For a given value of source or load resistance, a desired value of transmission loss for each IDT at center frequency f_o, due to mismatching, is achieved by scaling the unperturbed radiation conductance $G_a(f_o)$ of each. Given the desired frequency response $H(f)$, this is done by scaling the acoustic aperture. From Eq. (5.4) recall that $G_a(f_o)$ is proportional to the total IDT capacitance C_T. For a uniform IDT, C_T may be expressed as

$$C_T = C_o NW = C_s N \quad (F),$$ (5.22)

where C_o = capacitance/finger pair/unit length, $N = N_p$ = number of finger pairs and W = acoustic aperture. Values of C_o and electromechanical coupling coefficient K^2 for representative Rayleigh-wave substrates, as given in Table 2.1, are repeated in Table 5.3 for convenience.

TABLE 5.3

RAYLEIGH-WAVE SUBSTRATE PARAMETERS USED IN DETERMINATION OF
RADIATION CONDUCTANCE

Substrate	SAW Propagation Axis	K^2 %	C_o (pF/cm)
ST-quartz	X	0.11	0.55
YZ-lithium niobate (LiNbO$_3$)	Z	4.5	4.6
128° YX-lithium niobate	X	5.3	5.0
Lithium tantalate (LiTaO$_3$)	Z	0.72	4.4

Example 5.1 Insertion Loss and Power Transmission Loss of SAW Filter. A SAW delay line is fabricated on YZ-lithium niobate for operation at $f_o = 100$ MHz in a circuit with 50-Ω source and load impedances. The uniform input IDT has $N_{pi} = 40$ finger pairs, while the broadband output IDT has $N_{po} = 10$ finger pairs. Both transducers have apodization widths of 80 λ. Determine: (a) the power transmission loss of the input IDT; (b) the power transmission loss of the output IDT; and (c) the overall insertion loss associated with impedance mismatches alone. Assume that shunt tuning inductances have been employed in input and output stages to tune out the capacitances of the IDTs. ∎

Solution. a) From Table 5.3, $K^2 = 4.5\% = 0.045$ and $C_o = 4.6$ pF/cm. From Table 2.1, the SAW velocity is =3488 m/s. The acoustic wavelength at center frequency is $\lambda_o = v/f = 3488/(100 \times 10^6) \approx 34.8 \times 10^{-6}$ m. The apodization width $W = 80\lambda_o = 80 \times 34.8 \times 10^{-4} = 2.8 \times 10^{-1}$ cm, so that $C_s = WC_o = (2.8 \times 10^{-1}) \times (4.6 \times 10^{-12}) = 1.3 \times 10^{-12} = 1.3$ pF/finger pair. Total IDT capacitance C_T is $C_T = C_s N_{pi} = 1.3 \times 10^{-12} \times 40 = 52$ pF. In Eq. (5.4) $G_a(f_o) = 8 \times 0.045 \times (100 \times 10^6) \times (52 \times 10^{-12}) \times 40 \approx 2 \times 10^{-3}$ mho = 2 millimho. Input mismatch parameter $q = G_a/G_s = G_a R_s \approx 0.1$. From Eq. (5.16) for p=0, the input power transmission loss is $|T_{13}|^2 = 2q/(1+q)^2 = (2 \times 0.1)/(1.1)^2 = 0.16$, with insertion loss $-10\log_{10}(0.16) = 8.0$ dB for the input IDT.

(b) By scaling the number of finger pairs in (a), the radiation conductance of the output IDT is $G'_a \approx (2 \times 10^{-3}) \times (10/40)^2 \approx 1.2 \times 10^{-4}$ mho. The output mismatch parameter $q' = G'_a/G_L = G'_a.R_L \approx 6 \times 10^{-3}$, so that $|T'_{31}|^2 \approx 2q' = (2 \times 6 \times 10^{-3})$, for an insertion loss of ≈ 19 dB.

(c) The overall insertion loss due to mismatching is IL = 8 + 19 = 27 dB. Note that Eq. (5.4) for G_a is really only strictly valid for unapodized IDTs with negligible finger reflections. To determine G_a for an apodized IDT, it can be modelled as a number of parallel channels of uniform IDTs [4]. The constituent radiation conductances for each channel can be summed to obtain the total for the apodized IDT. ∎

5.3.6. TRIPLE TRANSIT SUPPRESSION

If the input and output IDTs have the same loading (i.e., $R_s = R_L$), the (two-transducer) triple-transit-suppression (TTS) is approximated as [6]

$$TTS \approx IL - 20\log_{10} q + 6 \quad (dB), \tag{5.23}$$

with TTS = 12 dB for the matched condition with $q = 1$ at source and load, as identified in Table 4.1 of Chapter 4.

Example 5.2 Input Voltage Standing Wave Ratio (VSWR) of SAW Filter.   Determine the input VSWR at center frequency for the circuit of the previous example. ■
Solution.   In the input circuit of Example 5.1, $p = 0$ and $q \approx 0.1$ at center frequency. From the definition for VSWR in Eq. (5.21b), VSWR = $1/q = 10$. ■

Example 5.3 Reducing the VSWR with a Shunt Loading Resistor.   The value of VSWR = 10 in the previous example is considered to be too high. It is to be reduced to VSWR = 5 by the use of a shunt loading resistor R_p across the input IDT. Determine the required value of R_p. ■
Solution.   From Eq. (5.21b), VSWR = $1/(q+p)$, where $q = 0.1$ and VSWR = 5, which yields the required value $p = (1 - 5q)/5 = 0.1$. With source conductance $G_s = 1/R_s = 1/50 = 20$ mmho, we obtain $G_p = G_s p = (20 \times 10^{-3}) \times 0.1 = 2$ mmho. The shunt loading resistance is $R_p = 1/G_p = 1/(2 \times 10^{-3}) = 50 \, \Omega$ in Fig. 5.3. ■

5.3.7. BANDWIDTH AND CIRCUIT FACTOR $C(f)$ TRADE-OFF

For illustration purposes, we examine a SAW filter with equal input and output IDTs so that they can each be described by the same array factor $A(f)$, giving $|A(f)|^2$ as the "bare" SAW filter response in the absence of circuit factor loading. Because the actual response of the SAW filter network will be a composite of $|A(f)|^2$ and $|C(f)|^{-2}$, some design trade-offs will have to be applied in order to achieve the desired final response. To expand on this, consider the case where shunt tuning is employed. From Eq. (5.13) and Table 5.2, the frequency dependence of the circuit factor may be approximated near center frequency as

$$C(f) \approx 1 + j2Q_e \frac{(f - f_o)}{f_o}. \tag{5.24}$$

For a given bandpass design the frequency response of $|A(f)|^2$ and $|C(f)|^{-2}$ might look as sketched in Fig 5.9, together with the composite

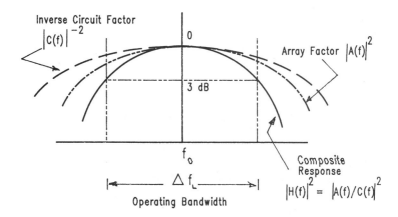

FIG. 5.9.   Array Factor $A(f)$, Circuit Factor $C(f)$, and Composite Response $H(f)$ for SAW filter example. Illustration assumes identical input and output IDTs and circuits. (After Reference [6].)

response $|H(f)|^2 = |A(f)/C(f)|^2$. As shown in Fig. 5.9, the 3-dB bandwidths of both $|A(f)|^2$ and $|C(f)|^{-2}$ must be greater than that of the composite response. Some trade-off between the two will be required depending on the limitations imposed on frequency responses attainable with $A(f)$ and $C(f)$.

Example 5.4 Circuit Factor $C(f)$ and Array Factor $A(f)$ of SAW Filter.   A 100-MHz Rayleigh-wave filter network is to employ equal input and output IDTs, together with source and load resistances of 50 Ω. The 3-dB bandwidth of $|H(f)|^2$ is to be 20 MHz. Shunt tuning is to be incorporated into the circuit factor response $C(f)$ in input and output stages. The response of the SAW filter is broadened so that the array factor $A(f)$ of each IDT rolls off at 0.7 dB at the band edges of 90 MHz and 110 MHz. What is the required insertion loss of each circuit factor $C(f)$ in input and output at these band edges? ∎

Solution.   As given, the array factor $A(f)$ for each IDT rolls off by 0.7 dB at the band edges of 90 MHz and 110 MHz, for a total roll-off of 1.4 dB. The total circuit factor response $|C(f)|^{-2}$ must provide the additional amount of $(3 - 1.4) = 1.6$ dB, so that the insertion loss of each $C(f)$ is $1.6/2 = 0.8$ dB at the band edges. ∎

Example 5.5 Required Turns Ratio For Mismatch Transformer.   The input and output IDTs of Example 5.4 have radiation conductances $G_a(f_o) = 1.24$ millimho at center frequency. Shunt inductance tuning is employed in input and output IDTs to tune out electrode capacitance C_T at center frequency, at which the TTI specifications require a VSWR = 1.6. If the SAW filter is to operate with 50-Ω source and load resistances, determine the turns ratio for the mismatch transformer required. Assume that no shunt loading resistance is included. ■

Solution. With no shunt resistor included, $p = 0$ in Eq. (5.21b), which yields $q = G_a/G_s' = 1/\mathrm{VSWR} = 0.62$ for the input stage in Fig. 5.3 (as well as for the output one). Where $G_a = 1.24$ mmho, $G_s' = 1.24/0.62 = 2$ mmho. Note that G_s' is the conductance looking into the mismatch transformer from the IDT. With an actual source conductance of $G_s = 1/R_s = 1/50 = 20$ mmho, a transformer with turns ratio $\mathrm{TR} = \sqrt{(G_s/G_s')} = \sqrt{(20/2)} \approx 3$ is required to transform the 50-Ω source (and load) to $G_s' = 2$ mmho. ■

5.3.8. ILLUSTRATIVE EXAMPLES

Figure 5.10 shows the measured frequency response of an 880-MHz RF filter incorporating tuning elements in input and output circuitry. This example is for a low-loss interdigitated interdigital (IIDT) leaky-SAW RF filter on 64° *Y-X* LiNbO$_3$, as employed for a mobile telephone in Japan, with a passband of 870 to 887 MHz and an upper stopband of 925 to

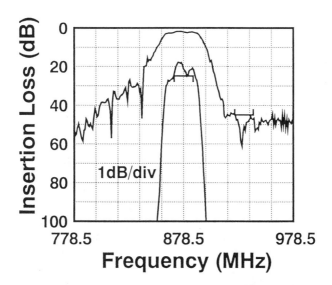

FIG. 5.10. Measured frequency response of 880-MHz IIDT RF filter for mobile transceiver in Japan, matched for 50-Ω using inductors and one-port resonators in input and output. Insertion loss is 2.5 dB, with 45 dB attenuation in upper stopband. Reprinted from Yatsuda, Takeuchi and Horishima, [1]. (Courtesy of Japanese Journal of Applied Physics.)

Fɪɢ. 5.11. Tuning and matching layout for a 2488-MHz SAW delay-line VCO for a SONET application. (Courtesy of Andersen Laboratories, Bloomfield, Connecticut.)

942 MHz. Here, the passband insertion loss is less than 2.5 dB. The upper stopband suppression is greater than 45 dB, and is an increase of about 15 dB over the unmatched structure [1].

Figure 5.11 shows the tuning and matching layout of a 2488-MHz SAW delay-line VCO for a SONET application, while Figure 5.12 shows the tuning and matching layout for a 500-MHz SAW delay-line VCO.

5.4. Matching of Filters Employing Single-Phase Unidirectional Transducers

While low-loss surface wave filters employing single-phase unidirectional transducers (SPUDTs) are not considered until Part 2, it is to be noted that these do not obey the same TTI rules as for filters with bidirectional IDTs. In contrast to filters with bidirectional IDTs, the passband insertion loss of SPUDT-based filters decreases with increased matching of the electrical ports. In addition, however, the passband TTI ripple will also decrease as acoustic reflection and piezoelectric generation begin to cancel out [9]. This is illustrated in Fig. 5.13 for the measured frequency response of a 211-MHz SPUDT filter before and after application of matching.

FIG. 5.12. Tuning and matching for a 500-MHz SAW delay-line VCO. (Courtesy of Andersen Laboratories, Bloomfield, Connecticut.)

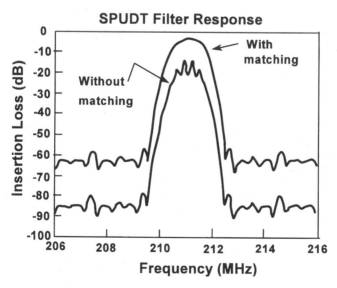

FIG. 5.13. Frequency response of an illustrated SPUDT-based filter showing characteristic reduction of both insertion loss and TTI as matching is increased. Horizontal scale: 206 to 216 MHz. Vertical scale: 0 to – 100 dB. (After Reference [9].)

5.5. Summary

The matching of RF front-end filters can be of extreme importance in mobile/wireless receivers, in order to minimize insertion loss and maximize receiver S/N performance. Matching levels can also be critical in IF filter designs for attaining passband ripple specifications with minimum TTI induced ripple. This chapter has examined some matching techniques and principles as applied to the frequency response of linear-phase SAW filter networks. Relationships were derived for the maximum intrinsic fractional bandwidth of a Rayleigh-wave filter with bidirectional IDTs, as constrained by such external circuit loading. Several working examples were also presented to illustrate matching trade-offs for meeting specifications on TTI and VSWR. While the material has been applied to Rayleigh-wave filter design, the basics concepts are also generally applicable to the matching of pseudo-SAW filters. As also noted, SPUDT-based filters do not obey the same rules for insertion loss and TTI levels as for SAW filters employing bidirectional IDTs. In these SPUDT structures both passband insertion loss and TTI ripple reduce as the degree of matching is increased.

5.6. REFERENCES

1. H. Yatsuda, Y. Takeuchi and T. Horishima, "Surface acoustic wave filter composed on interdigitated interdigital transducers and one port resonators," *Japan J. Appl. Phys.*, vol. 33, pp. 2979–2983, May 1984.
2. H. Yatsuda, T. Horishima, T. Eimura and T. Ooiwa, "Miniaturized SAW filters using a flip-chip technique," *Proc. IEEE Trans. Ultrason., Ferroelec., Freq. Contr.*, vol. 43, pp. 125–130, Jan. 1996.
3. See, for example, R. G. Brown, R. A. Sharpe, W. L. Hughes and R. E. Post, *Lines, Waves and Antennas*, John Wiley and Sons, New York, Third Edition, 1987.
4. R. H. Tancrell, "Principles of Surface Wave Filter Design," in H. Matthews (ed.), *Surface Wave Filters*, John Wiley and Sons, New York, Ch. 3, 1977.
5. P. B. Snow, "Matching Networks and Packaging Structures," in H. Matthews (ed.), *Surface Wave Filters*, John Wiley and Sons, New York, Ch. 5, 1977.
6. W. R. Smith, "Key trade-offs in SAW transducer design and component specification," *Proc. 1976 IEEE Ultrasonics Symposium*, pp. 547–552, 1976.
7. W. R. Smith, "Basics of the SAW Interdigital Transducer," in J. H. Collins and L. Masotti (eds), *Computer-Aided Design of Surface Acoustic Wave Devices*, Elsevier, New York, pp. 25–63, 1976.
8. P. R. Geffe, *Simplified Modern Filter Design*, John F. Rider Publisher, New York, p. 14, 1963.
9. C. S. Lam, D-P Chen, B. Potter, V. Narayanan and A. Vishwanathan, "A review of the applications of SAW filters in wireless communications," *International Workshop on Ultrasonics Application*, Nanjing, China, Sept. 1996.

—6—

Compensation for Second-Order Effects in SAW Filters

6.1. Introduction

6.1.1. SOME IF FREQUENCIES AND BANDWIDTHS FOR MOBILE PHONE SYSTEMS

Rayleigh-wave filters employing only bidirectional IDTs have insertion losses that are typically in the 10- to 35-dB range. Insertion loss degrades signal-to-noise (S/N) ratios in receiver stages so this limits such SAW filters to intermediate frequency (IF) signal processing stages, where signal levels are desirably in the millivolt range. Because of the excellent frequency responses that can be obtained with such "standard" filters, they are still routinely employed in some wideband IF signal-processing stages of mobile communications circuitry. Currently, temperature-stable ST-quartz is the preferred substrate for IF signal processing, so that acoustic processing involves Rayleigh-wave propagation[1]. As will be examined in what follows, ST-quartz is an anisotropic piezoelectric, with associated second-order effects that must be taken into consideration when designing surface wave filters on this material.

In recent years, these standard IF filters have been supplemented by low-loss SAW IF filters which have been introduced, in part, to cater to the increased sensitivity required for low-power cordless phones, as well as to permit reductions in their size and weight. As will be introduced in Part 2, these low- to mid-loss IF filters include single-phase unidirectional transducers, Z-path filters, in-line resonator-filters and waveguide-type transversely coupled resonator filters [2].

A listing of operating frequencies and channel specifications for some

---

[1] As noted in Table 2.2 of Chapter 2, a competitor to ST-quartz is the leaky-SAW LST-cut, with a TCD variation of less than ± 10 ppm over − 20°C ~ +80°C, with K^2 = ~0.11% and acoustic velocity v = ~3940 m/s [1].

major world-wide analog and digital cellular and cordless phone networks as well as for spread-spectrum channels involving Code-Division Multiple Access (CDMA) is provided in Chapter 10.

As tabulated in Chapter 10, channel spacings for mobile telephone systems are currently in the range of 10 to 30 kHz for narrowband analog cellular systems. Those for spacings for digital mobile phone are 200 kHz for the Global System for Mobile Communications (GSM), and Personal Communications Networks (PCN), while 1.728 MHz is the channel spacing for the Digital European Cordless Telephone (DECT). Spread-spectrum CDMA signalling in licensed CDMA bands involves much larger IF bandwidths (e.g., 1.23 MHz for North American IS-95 digital cellular).

To date, the IF-filter center frequencies for mobile phone systems do not follow any common guideline—they tend to depend on the design concept. As a result, IF frequencies can range from about 45 to 460 MHz. Some designs use the higher IF frequencies to reduce image-channel interference. Fabrication tolerances and temperature-induced frequency drift in the higher-frequency filters will govern typical bandwidth variations and obtainable selectivity of the Rayleigh filters on ST-quartz. Moreover, In digital cellular and cordless phone systems employing differential quadrature phase-shift keying (DQPSK), the upper usable IF center frequency will also be limited by the speed of the follow-up A/D converters.

Wideband CDMA-IF filters require a "rectangular" passband response with low values of amplitude and group-delay ripple in the passband. These can be designed using "standard" bidirectional-IDT designs, catering to reduced second-order degradation due to triple-transit interference, electromagnetic feedthrough, and bulk wave interference.

Intermediate Frequency (IF) filters for narrowband analog phone transceivers, with fractional bandwidths of up to about 0.2%, employ mainly low-loss two-pole in-line, or waveguide-coupled, resonator-filters with insertion losses of about 2 to 3 dB. Four-pole resonator-filters may be employed when out-of-band rejection levels of more than 60 dB are required. These four-pole designs can be formed by cascading two-pole resonator-filters to yield insertion losses of < ~4 dB.

Intermediate Frequency (IF) filters for GSM and PCN mobile telephone systems typically employ low-loss SPUDT filters and Z-path filters, as will be introduced in later chapters. At lower IF frequencies, however, SPUDT filter chip lengths can be more than 30 mm. Such sizes are undesirably large for many mobile-phone systems. Moreover, their performance can be corrupted by second-order diffraction and beam steering in the anisotropic ST-quartz piezoelectric.

6.1.2. Scope of This Chapter

This chapter examines some second-order effects that can degrade Rayleigh-wave filter design, together with some design trade-offs affecting their performance. (While concepts invoked here may also be considered as generally applicable to leaky-SAW filters, surface skimming bulk wave (SSBW) and surface transverse wave (STW) are excluded from such generalities because they involve bulk wave propagation.) As pointed out in Chapter 4, it depends as well on the application as to whether a second-order effect is to be considered a degradation. The second-order effects examined in this chapter include: a) IDT finger reflections; b) bulk wave interference in Rayleigh-wave filters; c) diffraction; d) beam-steering; e) acoustic attenuation; f) TTI; and g) electromagnetic feedthrough.

Temperature-stable ST-quartz is currently the dominant piezoelectric substrate for such applications. While some of the general diffraction concepts given here are applicable to IF filter designs on leaky-SAW substrates, the "SAW" coverage and examples in this chapter are related strictly to Rayleigh-wave filters.

6.2. Bulk Waves and the Multistrip Coupler

6.2.1. Preamble

As reviewed in Chapter 2, all excited IDTs that generate surface waves will also generate bulk waves to some extent. Because bulk waves (with transverse- and longitudinally polarized components), have higher velocities than the surface waves, interference between such bulk waves and SAW at the receiving IDT will be most pronounced at the high-frequency end of the filter passband. Under these conditions, the half-wavelength IDT finger spacing for efficient SAW reception also yields efficient bulk wave transduction, if these are coupled to the electric fields under the IDT. The result can lead to: (a) passband distortion; (b) reduction of out-of-band suppression; and (c) increased insertion loss, as depicted in Fig. 6.1.

Various fabrication techniques can be employed to reduce the severity of the bulk wave interference. One technique involves the use of very-thin piezoelectric substrates (e.g., 0.05 cm thick), to reduce the total volume available to bulk waves. In some instances the bottom surface of the substrate is roughened so as to scatter the emitted bulk waves away from the receiver. Adhesive materials can also be applied to the bottom surface to absorb impinging bulk waves. Yet again, some substrates have a slot milled into the bottom surface between input and output IDTs. However, as bulk waves can radiate in all directions, none of these techniques

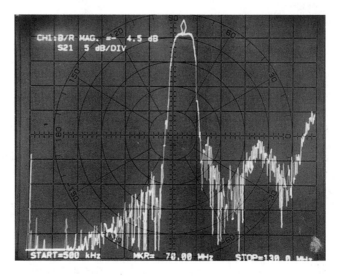

FIG. 6.1. Network analyzer measurement on 70-MHz linear-phase SAW filter, with 15 percent fractional bandwidth, showing significant bulk-wave interference in upper stopband. Horizontal scale: 500 kHz to 130 MHz. Vertical scale: 5 dB/div.

will serve to combat those components that travel along the surface with the SAW.

Not all piezoelectric substrates have the same susceptibility to bulk wave radiation by the IDT; for example, ST-quartz is rated as being much more prone to bulk wave excitation than YZ-lithium niobate [3]. Some Rayleigh-wave designers employ 128°-rotated YX-lithium niobate as having still less bulk wave excitation than the YZ-type. No hard and fast rule exists as to which substrate to use. The final decision will depend on the application and on permissible operating specifications, such as operating temperature variations.

As an added complication in some wideband IDT designs, it is found that with YZ-lithium niobate the amount of radiated bulk wave power relative to SAW power increases drastically as the number of finger pairs is reduced below about $N_p = 5$, corresponding to $BW_4\% = 20\%$ in unapodized IDTs for SAW delay line applications [4]. While no comparable data are readily available for 128°-rotated YX-lithium niobate, or for the leaky-SAW cuts on this substrate, experimental experience indicates that the same situation holds.

One technique for circumventing (rather than eliminating) the bulk wave problem involves the use of a multistrip coupler (MSC). The MSC does not decrease the amount of bulk wave generation. Instead, it acts to steer the

SAW and bulk waves in different directions so that only the SAW arrives at the receiving IDT. A bonus to its use in linear-phase filter design is that it enables both input and output IDTs to be apodized. This technique can be used, for example, in the design of wideband CDMA IF filters for mobile phones. The trade-off here, of course, is in the increased substrate size required to accommodate the MSC, which will limit the number of devices that can be lithographed on to a piezoelectric crystal wafer.

6.2.2. USE OF THE MULTISTRIP COUPLER FOR 100% SAW ENERGY TRANSFER

A multistrip coupler (MSC) can be employed in linear-phase Rayleigh-wave filter design as a means of physically separating the paths travelled by bulk and surface waves radiated from the input IDT. (The author is not aware of any designs involving the use of an MSC on leaky-SAW substrates.) The receiving IDT is placed so as to capture only the surface waves. In this way bulk wave effects are circumvented rather than eliminated. While the MSC technique can theoretically be applied to all SAW substrate piezoelectrics, in practice it is used mainly for high K^2 substrates (such as lithium niobate), due to the very large dimensions that could be involved with low K^2 substrates, as well as severe second-order reflection effects.

The invention of the MSC represented a major advance in the design of Rayleigh-wave filters [5], [6]. It is a deceptively simple structure, consisting merely of a number of thin metal strips placed between offset input and output IDTs, as shown in Fig. 6.2. The transverse length of the MSC strips is twice that of the individual IDTs of equal acoustic aperture. The metal used in the MSC is the same as that for input and output IDTs, thus enabling the entire structure to be fabricated in one process step.

The MSC may be considered as consisting of two physical tracks 1 and 2. A surface wave launched by the input IDT into track 1 with complex amplitude A_1 sets up a potential difference between adjacent strips. In turn, this potential difference generates a surface wave into track 2, which emerges with complex amplitude B_2, to impinge on the output IDT. Bulk waves that are generated by the input IDT into track 1 do not have their paths modified by the MSC and stay in track 1, thereby missing the output IDT.

In a simplistic analysis of the MSC, consider that the surface wave generated by the input IDT has a uniform spatial distribution across track 1. (The fact that the input IDT may be apodized does not significantly affect the end result.) Further consider that the surface wave motion causes each metal strip to have two mechanical modes of vibration. As pictured in Fig. 6.3, one vibrational mode corresponds to each end of the strip being clamped, while

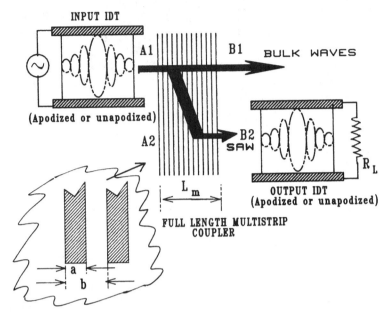

FIG. 6.2. SAW (Rayleigh wave) filter with full-length multistrip coupler (MSC) for 100 percent SAW energy transfer is used to circumvent bulk-wave interference. This MSC use also allows both IDTs to be apodized if desired. Insert shows MSC geometry.

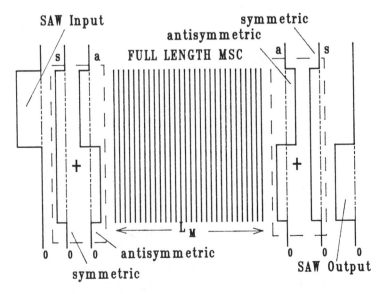

FIG. 6.3. Symmetric and antisymmetric mechanical vibration modes of a full-length multistrip coupler, for 100 percent SAW energy transfer.

the in-between portion vibrates symmetrically. This is termed the symmetric mode. The other modal vibration is where the center of each strip is assumed to be physically clamped while the "free" ends vibrate up and down in opposite directions. This latter condition is termed the antisymmetric mode. The interaction between the vibrating strips and the surface wave travelling under the MSC causes the SAW to propagate as two components with slightly different velocities. This difference in velocity through the MSC will cause a phase difference ϕ_d between the two SAW components, given by [7]

$$\phi_d = \left(\phi_s - \phi_a\right) = 0.5RK^2\beta L_M, \tag{6.1}$$

where ϕ_s = SAW phase shift in the symmetric mode
  ϕ_a = SAW phase shift in antisymmetric mode
  R = a filling factor
  K^2 = electromechanical coupling coefficient
  $\beta = 2\pi/\lambda$ = phase constant at acoustic wavelength $\lambda = v/f$
  L_M = length of the MSC

A_1 and A_2 represent the complex amplitudes of incident surface waves on tracks 1 and 2; B_1 and B_2 represent those for surface waves leaving these tracks. Their input-output amplitudes can be related by a matrix equation [8]

$$\begin{pmatrix} B_1 \\ B_2 \end{pmatrix} = \begin{pmatrix} \cos\left(\dfrac{\phi_d}{2}\right) & -j\sin\left(\dfrac{\phi_d}{2}\right) \\ -j\sin\left(\dfrac{\phi_d}{2}\right) & \cos\left(\dfrac{\phi_d}{2}\right) \end{pmatrix} \cdot \begin{pmatrix} A_1 \\ A_2 \end{pmatrix}. \tag{6.2}$$

From Eq. (6.2) it is seen that for 100% SAW energy transfer from track 1 to track 2 the phase difference must be $\phi_d = \pi = 180°$. With the phase difference proportional to the number of strips in the MSC, its corresponding length $L_{M\Pi}$ for 100% energy transfer is

$$L_{M\Pi} = \frac{\lambda}{RK^2}. \tag{6.3}$$

To minimize the overall size of the SAW filter structure, it is best to use an MSC with the smallest permissible length. With K^2 fixed for a given substrate, while λ is also fixed for a given operating frequency, this means that the value of filling factor R in Eq. (6.3) must be set to a maximum. Representative values of R are plotted for three values of metallization ratio $\eta = 0.1$, $\eta = 0.5$, and $\eta = 0.9$ in Fig. 6.4(a), as a function of parameter $\gamma = b/\lambda$ where b = strip separation shown in the inset. As shown in Fig. 6.4(a), the

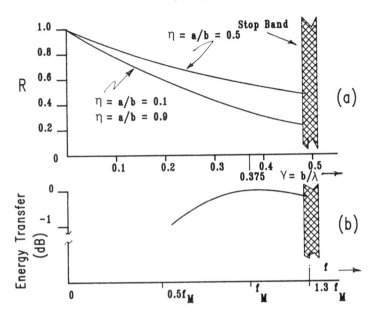

Fig. 6.4.   (a) MSC filling factor R, as a function of parameter $\gamma = b/\gamma$. (b) Ideal frequency response of a full-length MSC, for 100 percent SAW energy transfer at λ. [After Reference [7].)

highest values of R occur for $\eta = 0.5$ (i.e., equal strip widths and spacings). The choice of $\gamma = 0.375$ corresponds to 100% SAW energy transfer from track 1 to track 2. This is shown in Fig. 6.4(b), which plots SAW energy transfer between tracks as a function of frequency.

It is most important to realize that the separation between MSC strips must **not** correspond to $\lambda_o/2$ at the filter center frequency $f_o = v/\lambda_o$. If this were the case, the MSC would reflect the incident SAW waves at this frequency, rather than transmit them[2]. To operate well away from this stop band, the MSC strip separation typically places it about 1.3 times above the midband frequency, as indicated in Fig. 6.4.

Example 6.1 Multistrip Coupler on YZ-Lithium Niobate.   A multistrip coupler design is included in the design of a SAW bandpass filter on *YZ*-lithium niobate, with bandwidth $BW_3\% = 10\%$ and center frequency

---

[2] The same situation prevails for the placement of waveguiding metal strips between IDTs in a surface transverse wave (STW) filter.

$f_o = 100$ MHz. The MSC strips have metallization ratio $\eta = 0.5$, with strip separations given by $\gamma = b/\lambda = 0.375$. Determine the number of strips N_M in the MSC for full energy transfer between MSC tracks at 100 MHz. ■

Solution. From Fig. 6.4(a) the required filling factor is $R \approx 0.6$ for $\eta = 0.5$ and $\gamma = 0.375$. (Note that γ represents the reciprocal of the number of MSC strips per wavelength.) Use this value of R in Eq. (6.3), together with $K^2 = 0.045$ for YZ-lithium niobate. This gives the MSC length in wavelengths as $L_{M\Pi} \approx 37 \; \lambda$, where $\lambda = \lambda_o = v/f_o = 3488/(100 \times 10^6) = 34.88 \mu$m. The number of strips in the MSC is $N_M = L_{M\Pi}/\gamma = 37/0.375 \approx 98$ strips. From Fig. 6.4(b) the stopband edge of the MSC will be at $f_s = 1.3$ $f_M = 1.3 \times (100$ MHz$) = 130$ MHz. This is well outside the passband of the SAW filter. ■

Example 6.2 Multistrip Coupler on ST-Quartz. If the design of Example 6.1 were to be implemented on ST-quartz rather than YZ-lithium niobate, what would be the required length and number of metal strips in the MSC? ■

Solution. The required MSC length would be increased in inverse proportion to the respective values of electromechanical coupling coefficient K^2. Using $K^2 = 0.0011$ for ST-quartz we obtain $L_{M\Pi} = 37 \times (0.045/0.0011) = 1513 \; \lambda$, while the required number of strips is $N_M = 1513/0.375 \approx 4036$. Because of second-order reflections, such an MSC would be impractical on ST-quartz. ■

6.2.3. USE OF MULTISTRIP COUPLER WITH APODIZED IDTs

In Chapter 4 it was noted that the overall transfer function of a SAW filter with in-line apodized IDTs can not readily be decomposed into two separate identifiable responses for each IDT because the fingers of the output IDT would "see" SAW wavefronts of various widths emanating from IDT fingers of different apertures. A different situation exists, however, with the use of the multistrip coupler. In this case, the acoustic wavefronts of different widths that enter track 1 are converted to a uniform SAW wavefront on leaving track 2. The fact that the output IDT "sees" an incoming uniform SAW wavefront also allows it to be apodized. As a crude analogy, the effect of the MSC may be likened to a frosted glass screen placed in front of a group of separate light sources to give a diffused illumination over the entire screen. Figure 6.5 is a photograph of the actual lithographic pattern for a SAW filter with doubly apodized IDTs offset about an MSC. Observe that this represents a SAW filter with nonlinear phase response because the input IDT apodization is not a symmetric one.

168   *Fundamentals of Surface Acoustic Waves and Devices*

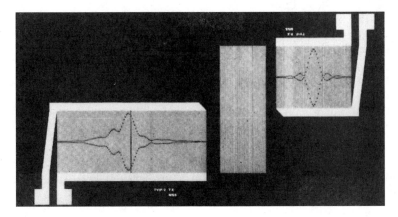

Fig. 6.5. Photolithographic mask pattern for a prototype 45-MHz SAW filter for a TVIF application, with nonsymmetric magnitude and nonlinear phase response built into the apodization of the input IDT.

6.2.4. Location of the MSC

Whereas the MSC is intended for use with travelling surface waves, its exact location between offset IDTs is *theoretically* not critical. In practice, however, its position and exact composition can indeed be critical in precision SAW filter designs that call for 50 dB or more out-of-band rejection levels. In the time domain this means that triple-transit and spurious reflections from all sources must be about 60 dB below the main impulse response peak of the SAW filter. In this event, both the placement and performance of the MSC can be critical to the attainment of such specifications.

6.3. Diffraction and Diffraction Compensation

6.3.1. Introduction

The diffraction experienced by SAW waves radiating from narrow IDT apertures is the same type of phenomenon experienced in optical structures, when light is passed through a narrow slot. Its effect may be more troublesome with apodized IDTs than with unapodized ones. It can give rise to increased insertion loss, passband distortion, and a reduction of out-of-band rejection. In some apodized IDT designs, diffraction compensation is realized by scaling the individual finger lengths in a prescribed manner.

Tied in with diffraction is the problem of SAW energy loss due to acoustic beam-steering. This relates to the accuracy with which the IDTs can be aligned with the desired propagation axis (i.e., the *pure-mode axis*) on the piezoelectric substrate. While acoustic attenuation of the surface wave is usually not significant below about 200 MHz, the loss increases at f^2 and can be a significant factor in the design of SAW filters and delay lines for gigahertz frequencies.

As sketched in Fig. 6.6, the surface-wave beam radiated by an IDT will be subject to some angular spreading, due to the finite width of the IDT aperture. If this angular spreading is sufficiently large, a significant fraction of it will not be intercepted by the receiving IDT. This will result in increased insertion loss. In addition, the received signal will suffer amplitude and phase distortion because each finger of the IDT intercepts a curved wavefront rather than the desired "flat" one. Passband distortion and reduced out-of-band rejection will ensue. As diffraction is a "two-way" process, its effects cannot be overcome by changing the relative sizes of input and output IDTs, or in reversing their role as input/output transducers.

This beam-spreading constitutes diffraction of the surface wave. It is analogous to optical diffraction from a slit or aperture. The diffraction patterns would indeed be equivalent to those in classical optics if the substrate were isotropic. Because piezoelectric single crystals are anisotropic (i.e., surface wave propagation characteristics vary with direction), the associated diffraction problem can be extremely complex. For instance, the

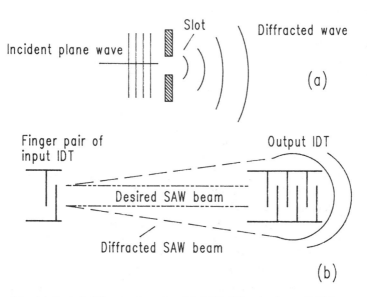

FIG. 6.6. (a) Optical diffraction by a slot. (b) SAW diffraction by an IDT finger pair.

variation of surface wave velocity with propagation direction can cause the beam spreading to increase or decrease. A decrease in beam-spreading is referred to as *autocollimation*. This is usually a desirable phenomenon in surface wave devices, in that a small fabrication misalignment of IDT with the desired pure-mode axis may not corrupt the filter response. Lithium niobate has such autocollimation properties in certain "focussing" directions. The other category includes ST-quartz, where IDT fabrication alignment with the pure mode axis can be exceedingly critical to the design of long Rayleigh-wave delay lines for avoidance of a multipath in indoor mobile phone relay networks, or in SPUDT IF filters for mobile phones.

As with classical optics, the diffraction can be decreased by employing wider acoustic apertures. In some instances an improvement of up to 30 dB in out-of-band rejection can be realized just by increasing the IDT finger apodization overlap [9]. Typical IDT apertures can range from 10 λ to 100 λ. Aperture widths above about 100 λ may often be limited by constraints on crystal size and apodization ratios, as well as by limitations on radiation conductance and circuit-factor $C(f)$.

6.3.2. FRESNEL AND FRAUNHOFER REGIONS IN SAW DEVICES

Two approaches can be used to relate the surface wave diffraction pattern of a radiating IDT. One is to consider that each aperture corresponds to the overlap region of a finger pair [9]. The other is to identify diffraction apertures with individual finger lengths [10].

As a brief introduction to this vast subject, consider first of all that the piezoelectric substrate can be approximated as an isotropic one. The spreading of the wave by passage through an acoustic aperture is then related to the diffraction equation of classical optics as

$$F = \frac{\lambda D}{W^2}, \tag{6.4}$$

where F is a dimensionless (Fresnel) parameter, $2W$ is the aperture width and D = distance from the IDT aperture to the point of evaluation. With $\lambda = v/f$, diffraction losses are inversely proportional to frequency. Values of $F < 1$ in Eq. (6.4) correspond to the so-called *Fresnel* or *near-field* region, in which the surface wave radiation pattern is essentially in the form of a parallel beam with amplitude and phase ripples across its profile, as illustrated in Fig. 6.7. In this situation, most of the acoustic energy is contained within a strip equal to that of the aperture. Values of $F > 1$ correspond to the Fraunhofer or far-field region, in which the angular pattern remains constant as the distance from the aperture is increased, while the phase distri-

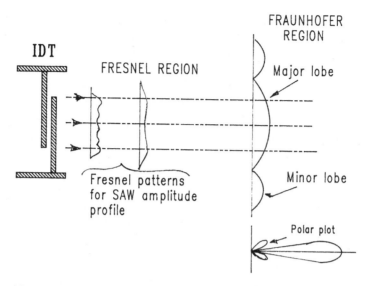

FIG. 6.7. Fresnel and Fraunhofer SAW transverse amplitude profiles. Input and output IDTs should be located within Fresnel regions for minimum diffraction degradation.

bution is commensurate with that from a circular aperture. There is, of course, no abrupt transition between the two regions.

As it is usually desirable (or even essential) that all or most of the acoustic energy radiated by the input IDT be intercepted by the output IDT, this means that input and output IDTs should be located within the Fresnel regions of each other. The diffraction phenomenon is a reciprocal one, so it does not matter which IDT is made to be the transmitting one as far as diffraction interference is concerned. Fresnel and Fraunhofer regions are depicted in the simplified illustration of Fig. 6.7.

Example 6.3 Diffraction Equation and Parameters. A Rayleigh wave is radiated from and diffracted by an IDT aperture $2W = 2\,\text{mm}$ on a piezoelectric substrate that is considered to be reasonably isotropic. The acoustic wavelength at $f = 300\,\text{MHz}$ is $\lambda = 10\,\text{m}$. For such a substrate, the separation D between midpoints of input and output IDTs in the Fresnel zone is to correspond to $F \leq 0.5$. Determine the value of D. ∎

Solution. From Eq. (6.4), $D = FW^2/\lambda = 0.5 \times (10^{-1})^2 /(10 \times 10^{-4}) = 5\,\text{cm}$. ∎

Because piezoelectric crystals are anisotropic, precise diffraction calculations for surface wave propagation can be exceedingly complicated and outside the scope of this text. For purposes of approximation, however, the diffraction relation for isotropic materials in Eq. (6.4) may be amended to include anisotropic parameters. For Rayleigh waves, the diffraction parameter F takes the more general form

$$F = \frac{\lambda D \left(1 - 2 A_d \right)}{W^2}, \qquad (6.5)$$

where anisotropy parameter A_d can take on either positive or negative values. A positive value of A_d yields a larger allowable separation between input and output IDTs over the isotropic case for a given Fresnel number $F < 1$. A negative value of A_d results in decreased allowable separation. Values of A_d for some Rayleigh-wave substrates are given in Table 6.1. (Comparable parameters for leaky-SAW piezoelectrics are not available.)

The author has employed this to advantage in the design of linear-phase SAW filters with 50% bandwidth on lithium niobate, using *slanted* IDT finger geometries, which were offset as much as ±6 degrees [11].

As indicated in Table 6.1, the use of GaAs with nominal ⟨001⟩ orientation and (110) wave propagation angle is most useful for integrated-circuit (IC) compatible SAW devices, and consistent with that used for GaAs IC development [12]. Lithium niobate substrates have positive values of A_d, allowing greater separation between IDTs than with a nonisotropic substrate. This means that although the SAW beam might be launched slightly off the propagation (i.e., pure-mode) axis, autocollimation causes it to return (within limits) to that axis.

6.3.3. SAW Diffraction and the Slowness Surface

If a surface acoustic wave were to be propagated on an isotropic plane surface, its velocity $v(\phi)$ would be the same in all propagation directions. This is not the case for SAW propagation on the plane surface of an anisotropic surface such as a piezoelectric substrate. In this situation, and as

TABLE 6.1

ANISOTROPY PARAMETER A_D FOR POWER FLOW ON
SAW SUBSTRATES

Substrate	A_d
YZ-lithium niobate	0.54
ST-quartz	−0.18
YZ-lithium tantalate	0.10
⟨001⟩, (110) Gallium arsenide (i.e. ⟨001⟩ orientation with SAW propagation along (110) direction)	≈0: most useful cut [12]

sketched in Fig. 6.8, the "beam" of elastic energy produced by an excited IDT may not necessarily be along the propagation wavefront [13]. This is obviously an undesirable situation in a SAW filter design, as it would lead to a reduction in coupling to the output IDT even if this were in the near field of the launching IDT [14]. In order to have the SAW elastic energy beam emanate parallel to the propagation wave vector, both the SAW propagation direction and crystal cut are chosen to ensure a high degree of crystalline symmetry along this axis. So-called *pure-mode* axes, such as the Z-propagation axis on YZ-lithium niobate, are those that satisfy this symmetry criterion.

As introduced in Chapter 2, a useful polar plot of surface velocity as a function of propagation angle ϕ is given in terms of the inverse velocity $1/v(\phi)$, and depicted as a *slowness curve*. The rationale for using this is that, along a given propagation axis, the direction of the SAW energy beam is normal to curvature of the slowness surface about that axis. Figure 6.8 illustrates such a hypothetical curve for a planar (x,y)-surface. Here, k_x and k_y represent the propagation vector components along x- and y-axes, respectively, where the IDT is placed normal to the representative x-axis. The resultant wave vector is thus $k(\phi)^2 = k_x^2 + k_y^2$. (Recall that wave vector $k = 2\pi/\lambda = 2\pi f/v$ just corresponds to phase constant $\beta = 2\pi/\lambda = 2\pi f/v$ in electrical engineering notation.) Angle ψ is the angle of the SAW energy beam, relative to an axis normal to the IDT fingers. In most SAW filter designs the piezoelectric crystal orientation angle θ is chosen for pure-mode propagation, so that the SAW phase velocity $v(\phi)$ is symmetrical about angle $\phi = 0$ for small angular displacements, with beam angle $\psi = 0$.

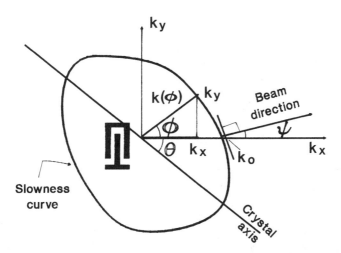

FIG. 6.8. Example of a "slowness" curve, demonstrating that the "beam" of elastic energy from an excited IDT may not necessarily be along the propagation wavefront.

One method used to approximate the resultant angular distribution of the diffracted wave in the (x,y) plane from a knowledge of the slowness curve characteristics employs a *parabolic approximation* to the angular distribution. In this approximation wave vector k_x is assumed to be a quadratic function of k_y. In this method a *parabolic anisotropy parameter* B_p is approximated as [14]

$$B_p \approx 1 + \frac{d\gamma}{d\theta}, \tag{6.6}$$

in terms of the angle parameters of Fig. 6.8.

Equation (6.6) is only valid for large acoustic apertures of the IDT. Moreover, its validity also depends on how well the slowness curve can be approximated by a quadratic in the vicinity of $\phi = 0$. This will only be the case for certain piezoelectrics such as ST-quartz, which are not autocollimating ones (see Section 6.3.2). Equation (6.6) will not be valid for autocollimating piezoelectric substrates such as YZ-lithium niobate. Table 6.2 lists values of anisotropy parameter B_p for ST-quartz and YZ-lithium tantalate.

The diffraction pattern of an excited uniform SAW IDT of acoustic aperture W on a substrate such as ST-quartz may be computed using the forementioned parabolic approximation, as applied to standard Fresnel integrals. These standard integrals are given by [15]

$$C(\xi) = \int_0^\xi \cos\left(\frac{\pi t^2}{2}\right) dt \tag{6.7a}$$

$$S(\xi) = \int_0^\xi \sin\left(\frac{\pi t^2}{2}\right) dt, \tag{6.7b}$$

where ξ is a dimensional parameter, and $C(\infty) = S(\infty) = 0.5$.

TABLE 6.2

PARABOLIC ANISOTROPY PARAMETER $B_p = 1 + (d\gamma)/(d\theta)$ FOR ILLUSTRATIVE SAW SUBSTRATES

Piezoelectric substrate	Pure-mode velocity v_0 m/s	Crystal cut	SAW axis	$B_p = 1 + (d\gamma)/(d\theta)$
ST-Quartz	3158	ST	X	1.38
YZ-Lithium Tantalate	3230	Y	Z	0.789

From the preceding equations the magnitude $|\psi(x,y)|$ of the SAW diffraction field in the (x,y) plane of Fig. 6.8 is obtained as [15]

$$\left|\Psi(x,y)\right| = \left|C(\xi_2) - C(\xi_1) + j\left[S(\xi_2) - S(\xi_1)\right]\right|. \tag{6.8}$$

For the case of SAW diffraction from a uniform IDT, the limits for the integrals in Eq. (6.8), in terms of parabolic anisotropy parameter B_p are [14]

$$\xi_1 = \left(\hat{y} - \frac{\hat{W}}{2}\right)\left[\frac{2}{\left|B_p\right|\hat{x}}\right]^{1/2} \tag{6.9a}$$

$$\xi_2 = \left(\hat{y} + \frac{\hat{W}}{2}\right)\left[\frac{2}{\left|B_p\right|\hat{x}}\right]^{1/2}, \tag{6.9b}$$

where the "hat" sign in Eqs. (6.9a) and (6.9b) indicates distances in acoustic wavelengths, W = acoustic aperture of the IDT, x = distance along propagation axis from end of IDT, y = transverse distance measured from the midpoint of the acoustic aperture, and $|B_p|$ = parabolic anisotropy parameter. Example 6.4 illustrates the application of these relationships to a diffraction calculation.

Example 6.4 Diffraction Computation Using Parabolic Anisotropy Parameter B_p.   Uniform SAW IDTs with center frequency $f_o = 192.6$ MHz are to be employed in a two-pole waveguide-coupled IF resonator filter on ST-quartz [16]. The IDTs have an acoustic aperture $W = 25\lambda_o$. Compute and plot the SAW diffraction pattern over 40 acoustic wavelengths at a distance $x = 10\lambda_o$ from the end of the IDT, along the propagation axis. Compute the diffraction pattern over 4 points per acoustic wavelength along the transverse y-axis, at $x = 10\lambda_o$. ∎

Solution.   From Table 6.2, the pure mode SAW velocity for ST-quartz is $v_o = 3158$ m/s, while the parabolic anisotropy parameter is $B_p = 1.38$ for the X-propagating axis. The Fresnel integrals in Eqs. (6.7a) and (6.7b) are given by

$$C(\xi_2) = \int_0^{\xi_2} \cos\left(\frac{\pi t^2}{2}\right) dt; \quad C(\xi_1) = \int_0^{\xi_1} \cos\left(\frac{\pi t^2}{2}\right) dt$$

$$S(\xi_2) = \int_0^{\xi_2} \sin\left(\frac{\pi t^2}{2}\right) dt; \quad S(\xi_1) = \int_0^{\xi_1} \sin\left(\frac{\pi t^2}{2}\right) dt,$$

where the integral limits for computation over 100 points along a cross section $40\lambda_o$ along the y-axis in Eqs. (6.9a) and (6.9b) are

$$\xi_1 = \left(\hat{y} + \frac{\hat{W}}{2}\right)\sqrt{\frac{2}{|B_p|\hat{x}}} = \left[\frac{(y-50)}{2.5} + \frac{25}{2}\right]\sqrt{\frac{2}{1.38 \times 10}},$$

$$\xi_2 = \left(\hat{y} - \frac{\hat{W}}{2}\right)\sqrt{\frac{2}{|B_p|\hat{x}}} = \left[\frac{(y-50)}{2.5} - \frac{25}{2}\right]\sqrt{\frac{2}{1.38 \times 10}}.$$

Apply these limits in Eq. (6.8) to obtain the diffraction plot as given here:

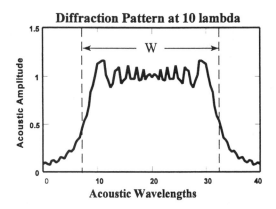

Diffraction Pattern at 10 lambda

EXAMPLE 6.4. Computation of SAW Fresnel diffraction on ST-quartz at distance $x=10\lambda_o$ from uniform IDT of aperture $W=25\lambda_o$. Parabolic anisotropy parameter $B_p=1.38$. ∎

6.3.4. BEAM-STEERING LOSSES

Beam steering is associated largely with fabricational alignment inaccuracies. If the IDTs are misaligned with respect to the desired pure-mode axis, the receiving IDT does not intercept all of the incident SAW beam, thereby giving rise to additional insertion loss. This can be particularly troublesome with substrates such as ST-quartz with negative values of anisotropy parameter A_d. Unlike diffraction losses which are frequency-dependent, beam-steering losses are independent of frequency but proportional to IDT width for a given substrate and degree of IDT misalignment. Table 6.3 illustrates additional contributions to insertion losses for YZ-lithium niobate and ST-quartz due to diffraction as well as beam steering [7]. The microsecond time delays in row A correspond to a 3-dB diffraction loss at 1 GHz for equal

TABLE 6.3

TIME DELAYS BETWEEN IDTs FOR 3-dB DIFFRACTION AND
BEAM STEERING LOSSES IN *YZ*-LITHIUM NIOBATE AND
ST-QUARTZ (After Reference [7])

	Substrate	
Loss Mechanism	*YZ*-lithium niobate (μs)	ST-quartz (μs)
A. Diffraction (at 1 GHz, with IDTs of 40λ width)	29	0.07
B. Beam steering (with equal $200\,\mu$m IDTs offset $0.2°$)	7.4	3.8

unapodized input and output IDTs with a finger overlap of 40λ at 1 GHz, as calculated from Eq. (6.5). Those in row B correspond to 3-dB insertion loss due to beam steering with IDTs of finger widths of $200\,\mu$m and misalignments of $0.2°$ [7].

6.3.5. DIFFRACTION COMPENSATION

The SAW diffraction problem is most troublesome with apodized IDTs, with each IDT finger contributing a different diffraction profile. Some IDT designs employ *diffraction compensation* by suitably correcting the length of each IDT finger in the apodized IDT to compensate for the diffraction degradation that would otherwise ensue. This compensation requires the adjustment of both the amplitude and phase of the SAW radiated by each IDT finger and is achieved by using IDTs with split-electrode geometries. In this way the lengths of both segments of each split-electrode are adjusted first in unison to give the required mean value of amplitude compensation for that finger pair. Then the relative lengths of elements of each split-electrode pair are adjusted to provide phase compensation. This procedure can best be understood with reference to the delta-function model. Instead of using one delta-excitation source for each finger in a single electrode IDT, we now use two delta sources for each split electrode. By varying the overall and relative lengths of the split-electrode segments we can control the amplitude and phase of the resultant phasor representing the delta-function excitation of each finger pair, as sketched in Fig. 6.9. As shown in Fig. 6.10, the use of dummy fingers is also routinely applied to many IDT

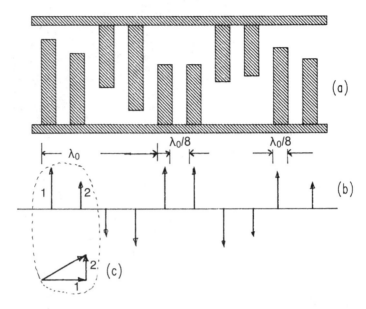

Fɪɢ. 6.9. (a) Overall and relative lengths of split-electrodes are adjusted for diffraction compensation. (b) Delta-function modelling of IDT finger excitation. (c) Phasor summation of delta functions for one split-electrode pair can have any desired resultant phase. Adjacent phasor components have 90° phase shift.

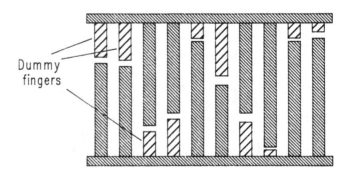

Fɪɢ. 6.10. "Dummy" electrodes are routinely used to improve the SAW amplitude profile and reduce diffraction in IDTs with solid- and split-electrode geometries.

designs as a means of realizing a more uniform SAW profile and velocity over the entire aperture, thereby reducing diffraction. Figure 6.11 illustrates a portion of a SAW-IDT pattern, with 0.5 μm line widths and gaps, for an on-board bandpass filter in an L-band (1600-MHz) satellite.

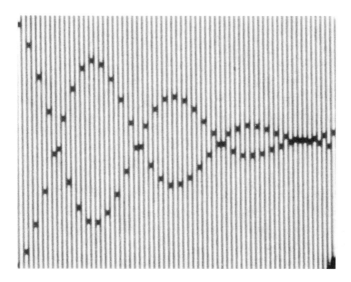

FIG. 6.11. Portion of a SAW-IDT pattern with $0.45\,\mu m$ line widths and gaps, used for an L-band (1600-MHz) on-board bandpass filter in satellite. (Courtesy of COM DEV, Cambridge, Ontario, Canada.)

6.3.6. ATTAINABLE SAW DELAY LINE LENGTHS

With careful control of diffraction and beam steering factors, exceedingly long SAW delay lines can be realized. For Rayleigh-wave structures these delays are on the order of $3\,\mu s$ per cm of substrate path. High-quality ST-quartz and lithium niobate crystals, commercially available up to 25 cm in length, can provide up to about $75\,\mu s$ of delay without folding the acoustic path. Cascaded SAW delay line modules have been reported with sufficiently low insertion loss as to provide a maximum delay of $400\,\mu s$ [17]. And 1.9-GHz Rayleigh-wave delay lines have been applied to add delays $> 2\,\mu s$ between diversity antennas employed for direct-sequence spread-spectrum in unlicensed PCS spread-spectrum CDMA indoor communications [18].

6.4. Acoustic Attenuation

The propagating coherent surface acoustic wave can lose energy in a number of ways, including: (1) viscous damping in the crystal lattice of the piezoelectric substrate, through which the SAW energy is dissipated as a minuscule amount of heat; (2) acoustoelectric losses (e.g., due to resistive films on $LiNbO_3$ or to special dopings of semiconductors); and (3) crystal imperfections. In *vacuum*, SAW energy loss is approximately proportional

TABLE 6.4

ACOUSTIC ATTENUATION OF SOME SAW SUBSTRATES
(with Frequency F in gigahertz) [19]

Substrate	Acoustic attenuation α (in dB/μs)
YZ-lithium niobate	$0.88F^{1.9} + 0.19F$
YX-quartz	$2.15F^2 + 0.45F$
001, 110 bismuth germanium oxide	$1.45F^{1.9} + 0.19F$

to the *square* of the operating frequency at, or near, room temperature. There will also be another small ultrasonic loss contribution when the SAW surface is surrounded by air or some inert atmosphere, due to the coupling of ultrasonic energy to the surrounding atmosphere. As this atmosphere loading loss is only proportional to frequency, it can often be ignored by comparison with the f^2 dependence due to SAW scattering.

The insertion loss contribution due to acoustic attenuation losses will be in addition to those due to mismatching, diffraction, beam steering, and IDT ohmic loss. Below about 200 MHz, such acoustic attenuation is usually small compared with the other insertion loss contributions and can often be disregarded. Quantitative results [19] for acoustic attenuation (in dB/μs), on YZ-lithium niobate, YX-quartz and bismuth germanium oxide are given in Table 6.4, where frequency F is in gigahertz. The first term gives the SAW attenuation in vacuum, while the second one gives additional attenuation due to air loading.

***Example 6.5 Acoustic Attenuation of SAW Delay Line on* YX-Quartz.** A 1.96-GHz SAW delay line is to be employed in a diversity antenna system for spread-spectrum CDMA indoor communications. It is to be fabricated on YX-quartz to give a delay of 2 μs. Determine the acoustic attenuation at this frequency. ∎

Solution. From Row 2 of Table 6.4, the acoustic attenuation α in dB/μs is $\alpha = (2.15 \times 1.96^2) + (.0.45 \times 1.96) = 9.14$ dB/μs. The total acoustic loss is $(9.14 \times 2) = 18.28$ dB. This will be in addition to the "normal" insertion loss, without such attenuation. ∎

6.5. Triple-Transit Effects and Unidirectional IDTs

In bidirectional IDTs, TTI can be due to regenerative and/or non-regenerative reflections in IDTs. These give rise to a periodic ripple in the amplitude and phase response across the passband, of frequency $f_r = 1/2\tau$

where τ = time separation between phase centers of the IDTs. The regenerative effect is due to the re-excitation of a surface wave by an IDT in which a voltage has been induced by the passing surface wave. This can be decreased by mismatching the IDT, at the expense of increased insertion loss. (As an extreme example, there would be no regenerative TTI if the output IDT were to be shorted out, but the insertion loss would be infinite.) As already discussed, nonregenerative TTI due to finger reflections can be reduced by the use of split-electrode IDT geometries.

Triple-transit-interference can be circumvented in some narrowband filter designs involving unidirectional transducers, where fractional bandwidths are limited to a maximum of about 10%. Several techniques of differing complexity have been developed over the years for obtaining unidirectionality of surface wave propagation [20]. As outlined in Fig. 6.12, two principle methods (with variants) in current use relate to the single-phase unidirectional transducer (SPUDT) [21]–[23], and to the floating-electrode unidirectional transducers (FEUDTs) [24]. The SPUDT is normally used for IF filter fabrications on ST-quartz, while FEUDTs have been successfully developed for leaky-SAW RF filter design [24]. Both of these structures will be examined in later chapters.

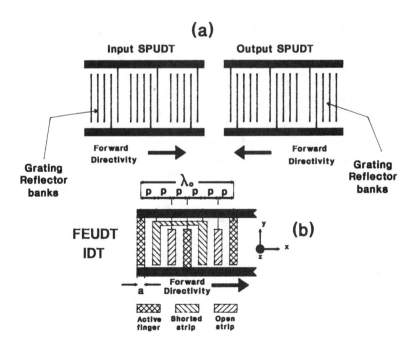

FIG. 6.12. (a) The basic SPUDT filter shown here can have several variants. (b) The basic FEUDT filter also has several variants.

There is one significant difference between the frequency responses of the particular SPUDT and FEUDT structures of Fig. 6.12. Whereas the SPUDT structure of Fig. 6.12(a) is a periodically segmented one, it can be configured to act either as a comb filter with a multiplicity of bandpass responses or as a single bandpass filter. On the other hand, the FEUDT structure of Fig. 6.12(b) has only a single bandpass response.

The insertion loss capability of these unidirectional structures is typically less than 3 dB. All of these designs (and their variants) invoke some trade-offs between insertion loss, fractional bandwidth, and close-in sidelobe suppression. The particular SPUDT design of Fig. 6.12(a), which is examined in Chapter 12, can be applied to the design of frequency-hopping oscillators, such as might be used in hybrid spread-spectrum modems for indoor communications. An alternative SPUDT design, widely used in IF filtering for mobile-phone channels, is highlighted in Chapter 15.

6.6. Electromagnetic Feedthrough

Electromagnetic feedthrough (also referred to as "crosstalk") is one of the villains of SAW filter design, particularly at high frequencies. Feedthrough is due to the fact that input and output IDTs effectively form a capacitor, so that electromagnetic energy can be coupled directly from input to output IDT. The resultant signal induced into the output IDT is a composite of that due to the desired SAW signal and to the undesired feedthrough. This causes amplitude and phase ripple across the passband at ripple frequency $f_r = 1/\tau$ where τ is the time separation between phase centers of the IDTs. Because the susceptance of the equivalent series capacitance between IDTs increases with frequency, so will the severity of the feedthrough. Minimization of feedthrough can often be accomplished only after laborious trial and error designs (which information tends to become proprietary). These can involve: a) very careful choice of metal packaging structure and dimensions; b) the judicious placement of ground connections to avoid ground loop effects; c) the placement of thin metal film ground strips between input and output IDTs; and d) the use of balanced transformers in input and output structures (with trade-off penalties on fabrication complexity, cost, and size). As shown in Fig. 6.13, the balanced transformer converts the normally unbalanced input leads so as to a balanced arrangement so as to suppress feedthrough. Such transformers can be undesirably large or expensive for mobile/wireless applications. As an alternative technique, therefore, some SAW RF filter designs with differential outputs are also under development for use with balanced-mixers [25].

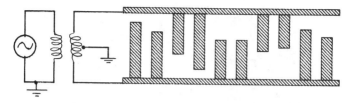

FIG. 6.13. Some SAW-IF filter designs incorporate center-tapped transformers in input IDT circuitry, to convert the structure from an unbalanced to a balanced one.

6.7. Undesirable IDT Finger Reflections

Depending on the application, finger reflections may have a desirable or undesirable effect on the performance of a SAW filter. In IF filter designs employing only bidirectional IDTs, unless compensated for, finger reflections usually degrade the amplitude and phase response. In some types of low-loss RF filters and duplexers for mobile communications receivers on the other hand, finger reflections are used to enhance (or even drastically change) the filter characteristics. This chapter deals with the undesirable effects of IDT finger reflections in SAW-IF filters, while Chapter 9 considers enhancing features of finger refections, as applied to the design of SAW and leaky-SAW RF filters and resonator-filters.

A standard technique employed to reduce undesirable finger reflection effects involves the use of IDTs with split-electrode geometries, as shown in Fig. 6.14. As illustrated in Fig. 6.15, for electrode widths and spacings of $\lambda_o/8$ at center frequency, differential path lengths are such that the SAW reflections from each split-electrode pair cancel out at center frequency, rather than add on as for single-electrode IDTs [26].

One disadvantage of the split-electrode geometry is in the increased lithographic resolution required for fabricating IDTs. For SAW filters operating in the fundamental mode, this means that if the maximum attainable device frequency with single-electrode IDTs is 1 GHz, using a sophisticated photolithograhic camera, the same photolithography can yield only split-electrode IDTs operating at maximum fundamental frequencies of 500 MHz. In some instances this problem may be circumvented by operating the IDTs in a harmonic mode [24].

Two other design IDT design parameters may change with the use of split-electrodes structures. First of all, the effective electromechanical coupling coefficient K_e^2 is not the same as for single electrodes. For YZ-lithium niobate, for example, $K_e^2 \approx 0.88 \times K^2 \approx 0.04$ for split-electrode IDTs [27]. In addition, the capacitance C_T of a split-electrode IDT is 1.4 times larger than for an equivalent single-electrode one [28].

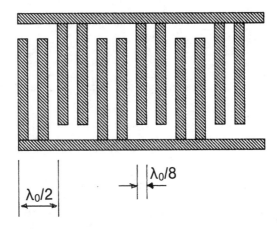

FIG. 6.14. IDT with split-electrode geometry and metallization ratio $\eta = 0.5$.

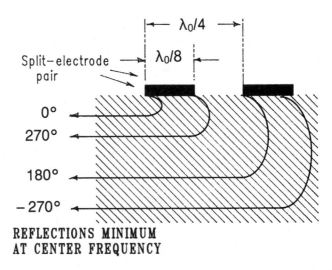

FIG. 6.15. SAW reflections from edges of split electrodes give a resultant minimum at center frequency. Used in filter designs where IDT finger reflections are undesirable.

6.8. Desirable Harmonic Operation of SAW Devices

6.8.1. IDT LITHOGRAPHIC RESOLUTION CONSTRAINTS IN GIGAHERTZ DEVICES

As SAW technology extends into higher gigahertz-frequency regimes, such as the 2.4-GHz filters now being employed for WLAN spread-spectrum systems [29], as well as for timing-extraction (i.e., clock) filters in optical

TABLE 6.5

TOLERANCES FOR λ/4-FINGER IDTs IN ST-CUT
RAYLEIGH-WAVE AND SSBW (SH) TIMING FILTER WITH
PERMISSIBLE FREQUENCY RANGE 2.48832 ± 0.00024 GHz
(After Reference [30])

Rayleigh-Wave on ST-quartz	$L = 0.3\,\mu\text{m} \pm 0.06\,\mu\text{m}$
	$h = 300\ \text{Å} \pm 30\ \text{Å}$
SSBW (SH)-wave on quartz	$L = 0.5\,\mu\text{m} \pm 0.16\,\mu\text{m}$
	$h = 500\ \text{Å} \pm 8\ \text{Å}$

communications systems[3] [30], the fabrication of IDTs and reflection gratings becomes increasingly exacting—due to the lithographic resolution limitations of current commercial mass-production microelectronic fabrication systems (currently ~ 0.1–0.3 μm).

To illustrate this, Table 6.5 lists the IDT dimensional tolerances required for a 2.48832-GHz timing-extraction filter in an optical communications link, where long-term phase stability of 20° is required over a 20-year period. The parameter tolerances (L = finger width, h = metallization thickness) in Table 6.5 are for fundamental-frequency Rayleigh-wave (ST-cut) propagation on quartz and SSBW propagation on quartz (as sketched in Fig. 6.22), with a large coupling between the shear-horizontal (SH) wave and the IDT [2].

One technique for reducing fabrication tolerance restrictions is to employ piezoelectric substrates and surface-wave modes such SSBW, STW and LSAW, with higher velocities. A drastically different technique is to operate the IDTs in harmonic modes.

An advantage of operating Rayleigh-wave devices in harmonic modes relates to the suppression of bulk-wave interference. From Chapter 2, it will be recalled that bulk waves propagate with higher velocities than surface waves. This can result in considerable interference in, and above, the passband of a SAW filter. By operating well above the fundamental frequency, however, we can escape from bulk wave interference. This is because bulk waves travel, for the most part, in the interior of the propagating piezoelectric. They do not participate in the periodic data-sampling[4] imposed by the interdigital transducer (IDT), and thus do not appear in harmonic passbands.

---

[3] Timing-extraction filters for fiber optics are discussed in more detail in Chapter 19.
[4] See Chapter 7.

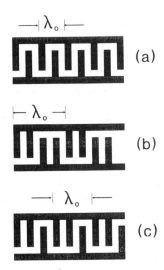

FIG. 6.16. Some SAW linear-transducer structures for fundamental and harmonic operation. (a) Solid-electrode ("two-finger") IDT for odd harmonics. (b) Split-electrode ("four finger") IDT for odd harmonics. (c) "Three-finger" IDT for even- and odd-harmonic operation.

6.8.2. IDT ELEMENT FACTOR[5]

An IDT can operate in harmonic modes, depending on its electrode geometry, (two-finger, three-finger, four finger etc.) as well as on the value for metallization ratio η. For example, the solid-electrode SAW IDT of Fig. 6.16(a) with $\eta = 0.5$ will yield only odd-harmonic frequency responses of differing amplitude, (including some with differing phase), with zero response at the third harmonic.

The *elemental* field distributions between, and below, an individual finger of an excited IDT will have a harmonic frequency response given by (analog) element factor $\rho(f, \eta)$. When this is included, the overall fundamental and harmonic frequency response of the IDT is given by the product $H(f) = \rho(f, \eta) A(f)$, where $A(f)$ is the *array factor* giving the harmonic-independent response of the IDT.

Spatial-harmonic components of the element factor for a given excited IDT may be derived from boundary-solution techniques. One method involves obtaining a solution for the voltage distribution between two oppositely polarized electrodes, by expanding it in terms of *Legendre polynomials* [31]–[34]. Another method is an iterative one involving the

---

[5] The term *element factor* is also referred to as *elemental charge density*.

Fourier spectrum of the voltage distribution throughout the entire IDT [35]. It may also be noted that an exact analytic solution to the electrostatic problem has been given by Peach [36].

As derived from such boundary-solution methods, for: a) the solid electrode ("two-finger"); b) the split-electrode ("four-finger") IDT; and c) the "three-finger" IDT in Figs. 6.16, 6.17, 6.18, and 6.19 illustrate the respective relative amplitudes of element-factor magnitude for these three structures. Note that the solid-electrode and the split-electrode IDTs have only odd-harmonic response capability, while the three-finger IDT can have both even and odd harmonic responses.

Figure 6.20 shows the experimental frequency response of a fifth-harmonic Rayleigh-wave delay line operating at 2 GHz. This particular example employed a stepped-finger IDT structure with metallization ratio $\eta = 0.5$. Note that the 2-GHz response is well outside bulk wave interference around the 400-MHz fundamental, which is almost completely suppressed here [37], [38].

6.8.3. MEANDER-LINE IDT FOR HARMONIC OPERATION

Figure 6.21 illustrates a section of a meander-line IDT designed for third-harmonic operation at frequencies in the 2.5-GHz range [30]. This design was aimed at relaxing the very severe tolerance limitations on "standard"

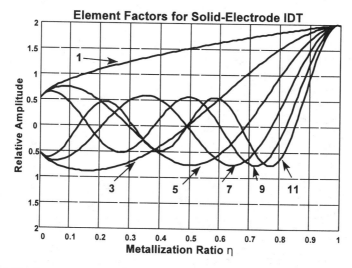

FIG. 6.17. Odd-harmonic element-factor responses as a function of metallization ratio η, for the solid-electrode IDT of Fig. 6.16(a).

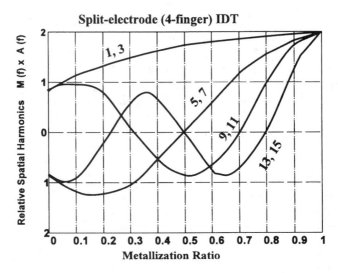

FIG. 6.18. Odd-harmonic element-factor responses for the split-electrode IDT of Fig. 6.16(b).

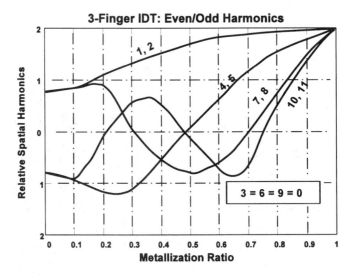

FIG. 6.19. Even/odd element-factor responses for three-finger IDT of Fig. 6.16(c). Harmonic multiples of 3 have a value of zero here.

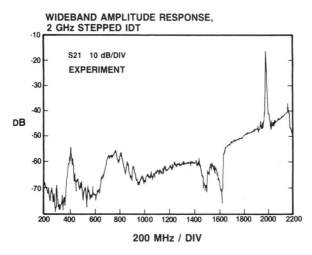

FIG. 6.20. Measured response of 2-GHz SAW delay line, using scaled version of 1-GHz design. Fundamental response at 400 MHz is almost completely suppressed. Horizontal scale: 200 to 2200 MHz. Vertical scale: 10 dB/div. (Reprinted with permission from Naraine and Campbell, [38], © IEEE, 1984.)

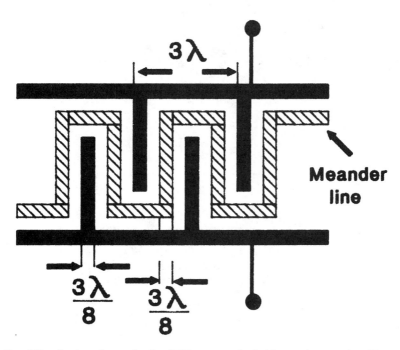

FIG. 6.21. Section of meander-line IDT structure for 3rd harmonic operation. Electrode and meander line widths are $3\lambda_o/8$. Period is $3\lambda_o$. (After Reference [30].)

solid-electrode IDTs with $\lambda/4$ electrode widths. In this design the width of both the electrodes and the meander line is $3\lambda_o/8$, so that the IDT period is $3\lambda_o$. With the inclusion of the meander line, however, this IDT essentially behaves as a split-electrode IDT with metallization ratio $\eta = 0.5$. As shown in Fig. 6.18, this has a strong response at the third harmonic. The meander-line structure of Fig. 6.21 will also have increased input impedance, due to the electrical series path between alternately excited electrodes. This permits a repetitive interlacing of input and output IDTs to decrease the insertion loss. As already mentioned, Fig. 6.22 illustrates the IDT orientation for the SSBW device fabrication on quartz.

Figure 6.23 illustrates the experimental response of a third-harmonic

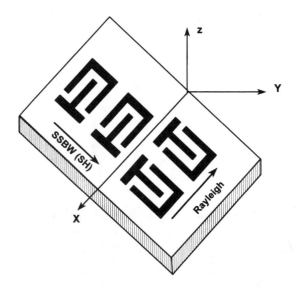

Fig. 6.22. Showing the orientation for Rayleigh-wave propagation on ST-X cut on quartz, and the 90° rotation cut for SSBW (SH) wave propagation at a velocity about 1.6 times higher.

TABLE 6.6

TOLERANCES FOR $3\lambda/8$-FINGER IDTs IN ST-CUT
RAYLEIGH-WAVE AND SSBW (SH) TIMING FILTER WITH
PERMISSIBLE FREQUENCY RANGE 2.48832 ± 0.00024 GHz
(After Reference [30])

Rayleigh-Wave on ST-quartz	$L = 0.45\,\mu\text{m} \pm 0.09\,\mu\text{m}$
	$h = 300\,\text{Å} \pm 30\,\text{Å}$
SSBW (SH)-wave on quartz	$L = 0.75\,\mu\text{m} \pm 0.24\,\mu\text{m}$
	$h = 500\,\text{Å} \pm 8\,\text{Å}$

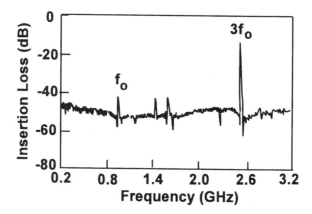

FIG. 6.23.   Experimental response of 2.5-GHz third-harmonic Rayleigh-wave filter on ST-quartz, using meander-line IDT structure of Fig. 6.21. Horizontal scale: 0.2 to 3.2 GHz. Vertical scale: − 80 to 0 dB. (After Reference [30].)

($3f_o$) 2.5-GHz filter of this $3\lambda_o/8$ construction, as fabricated on ST-quartz. The insertion loss at 2.5 GHz was about 14 dB, with a Q of about 700. Note the significantly reduced fundamental at f_o. Table 6.6 illustrates the relaxed lithographic tolerances with this construction, for both a Rayleigh-wave and SSBW filter of quartz, as compared with the conventional $\lambda_o/4$ IDT geometry of Table 6.5.

6.9.   Summary

This chapter first reviewed some of the frequency parameters for IF filtering in various analog and digital phone systems. This was followed by an examination of several second-order effects that can corrupt the responses of such filters, as well as techniques for reducing such corruption. Reduction of second-order influences is generally accompanied by trade-offs between insertion loss, fractional bandwidth, and close-in sidelobe suppression. Temperature-stable ST-quartz is currently the dominant piezoelectric substrate for such applications.

The chapter concluded by illustrating the use of harmonic IDT techniques as a means of reducing second-order effects associated with bulk wave and/or lithographic resolution degradation in gigahertz devices.

6.10.   REFERENCES

1. Y. Shimizu and M. Tanaka, "A new cut of quartz for SAW devices with extremely small temperature coefficient by leaky surface wave," *Electronics and Communications in Japan*, Part 2, vol. 69, pp. 48–56, 1986.

2. C. C. W. Ruppel et al., "SAW devices for consumer communications applications," *IEEE Transactions on Ultrason., Ferroelect., and Freq. Control, vol. 40, pp. 438–452, Sept. 1993.*

3. R. H. Tancrell, "Principles of surface wave filter design," in H. Matthews (ed.), *Surface Wave Filters*, John Wiley and Sons, New York, Chapter 3, 1977.

4. R. F. Milsom, "Bulk Wave Generation by the IDT," in J. H. Collins and L. Masotti (eds), *Computer-Aided Design of Surface Acoustic Wave Devices*, Elsevier, Amsterdam, pp. 64–81, 1976.

5. F. G. Marshall, C. O. Newton and E. G. S. Paige, "Theory and design of the multistrip coupler," IEEE Trans. Microwave Theory and Techniques, vol. MTT-21, pp. 206–215, April 1973.

6. F. G. Marshall, C. O. Newton and E. G. S. Paige, "Surface acoustic wave multistrip components and their applications," *IEEE Trans. Microwave Theory and Techniques*, vol. MTT-21 pp. 216–225, 1973.

7. G. W. Farnell and E. A. Adler, "An overview of acoustic surface wave technology," Final Report to Communications Research Center, Ottawa, Canada, on DSS Contract 36001-3-4406, 12 August 1974.

8. E. A. Ash, "Fundamentals of Signal Processing Devices", in A. A. Oliner (ed.), *Acoustic Surface Waves*, Topics in Applied Physics, vol. 24, Springer-Verlag, New York, Chapter 4, 1978.

9. M. R. T. Tan and C. A. Flory, "Minimization of diffraction effects in SAW devices using a wide aperture," *Proc. 1986 IEEE Ultrasonics Symp.*, pp. 13–17, 1986.

10. I. Streibl, "SAW Diffraction Compensation for LiNbO3," Technical Report No: CERL-84-02, Department of Electronics, Carleton University, K1S 5B6, Canada, March 1984.

11. C. K. Campbell, Y. Ye and J. J. Sferrazza Papa, "Wide-band linear phase SAW filter design using slanted transducer fingers," *IEEE Transactions on Sonics and Ultrasonics*, vol. SU-29, pp. 224–228, July 1982.

12. T. W. Grudkowski, G. K. Montress, M. Gilden and J. F. Black, "Integrated circuit compatible surface acoustic wave devices on gallium arsenide," *IEEE Transactions on Microwave Theory and Techniques*, vol. MTT-29, pp. 1348–1356, December 1981.

13. G. W. Farnell, "Elastic surface waves," in H. Matthews (ed.), *Surface Wave Filters*, John Wiley and Sons, New York, Chapter 1, 1977.

14. D. P. Morgan, *"Surface-Wave Devices For Signal Processing,"*, Elsevier, New York, Chapter 6, 1985.

15. J. M. Stone, *Radiation and Optics*, McGraw-Hill, New York, Chapter 10, 1953.

16. C. K. Campbell, P. M. Smith and P. J. Edmonson, "Aspects of modelling the frequency response of a two-port waveguide-coupled SAW resonator-filter," *IEEE Trans. on Ultrason., Ferroelect., and Freq. Control*, vol. 39, pp. 768–773, Nov. 1992.

17. M. F. Lewis, C. L. West, J. M. Deacon and R. F. Humphryes, "Recent developments in SAW devices," *IEE Proceedings (Great Britain)*, vol. 131, Part A, pp. 186–215, June 1984.

18. S. G. Gopani, J. H. Thompson and R. Dean, "GHz SAW delay line for direct sequence spread spectrum, CDMA in-door communication system," *Proc. 1933 IEEE Ultrasonics Symp.*, vol. 1, pp. 89–93, 1993.

19. A. J. Slobodnik, Jr.,"Materials and Their Influence on Performance," in A. A. Oliner (ed.), *Acoustic Surface Waves*, Topics in Applied Physics, vol. 24, Springer-Verlag, New York, Chapter 6, 1978.

20. R. C. Rosenfeld, R. B. Brown and C. S. Hartmann, "Unidirectional acoustic surface wave filters with 2 dB insertion loss,' *Proc. 1974 IEEE Ultrasonics Symp.*, pp. 425–428, 1974.

21. C. S. Hartmann, P. V. Wright, R. J. Kansy and E. M. Gerber, "An analysis of SAW interdigital transducers with internal reflections and the application to the design of single-

phase unidirectional transducers," *Proc. 1982 IEEE Ultrasonics Symposium*, pp. 40–45, 1982.

22. M. F. Lewis, "Low loss SAW devices employing single stage fabrication," *Proc. 1983 IEEE Ultrasonics Symp.*, pp. 104–108, 1983.

23. C. K. Campbell and C. B. Saw, "Analysis and design of low-loss SAW filters using single-phase unidirectional transducers," *IEEE Trans. Ultrason., Ferroelec., and Freq. Control*, vol. UFFC-34, pp. 357–367, May 1987.

24. M. Takeuchi, K. Yamanouchi, K. Murata and K. Doi, "Floating-electrode-type SAW unidirectional transducers using leaky surface waves and their application to low-loss filters," *Electronics and Communications in Japan*, Part 3, vol. 76, pp. 99–110, 1993.

25. M. A. Sharif, M. A. Schwab, D. P. Chen and C. S. Hartmann, "Coupled resonator filters with differential input and/or differential output," *Proc. 1995 IEEE Ultrasonics Symp.*, vol. 1, pp. 67–70, 1995.

26. R. F. Mitchell and D. W. Parker, "Synthesis of acoustic-surface-wave filters using double electrodes," *Electronics Letters*, vol. 10, p. 512, 1974.

27. W. R. Smith and W. F. Pedler, "Fundamental- and harmonic-frequency circuit model analysis of interdigital transducers with arbitrary metallization ratios and polarity sequences," *IEEE Trans. Microwave Theory and Techniques*, vol. MTT-23, pp. 853–864, Nov. 1975.

28. S. Datta, *Surface Acoustic Wave Devices*, Prentice-Hall, Englewood Cliffs, p. 178, 1986.

29. *Wireless Communications Products; Data Book*, Rev. 1, Fujitsu Limited., Tokyo, 1996.

30. K. Asai, I. Isobe, T. Tada and M. Hikita, "SAW timing-extraction filter and investigation of submicron process technology for Gb/s optical communications system," *Proc. 1995 IEEE Ultrasonics Symposium*, vol. 1, pp. 131–135, 1995.

31. H. Engan, "Excitation of elastic surface waves by spatial harmonics of interdigital transducers," *IEEE Trans. on Electron Devices*, vol. ED-16, pp. 1014–1017, 1969.

32. H. Engan, "High-frequency operation of surface-acoustic-wave multielectrode transducers," *Electronics Letters*, vol. 10, pp. 395–396, 1974.

33. H. Engan, "Surface acoustic wave multielectrode transducers," *IEEE Trans. Sonics and Ultrasonics*, vol. SU-22, pp. 395–401, 1974.

34. M. Abramowitz and I. A. Stegun (eds), *Handbook of Mathematical Functions*, Dover, New York, 1965.

35. C. S. Hartmann and B. G. Secrest, "End effects in interdigital surface wave transducers," *Proc. 1972 IEEE Ultrasonics Symp.*, pp. 413–416, 1972.

36. R. C. Peach, "A general approach to the electrostatic problem of the SAW interdigital transducer," *IEEE Trans. Sonics and Ultrasonics*, vol. SU-28, pp. 96–105, March 1981.

37. P. Naraine, C. K. Campbell and Y. Ye, "A SAW step-type delay line for efficient high order harmonic mode excitation," *Proc. 1980 IEEE Ultrasonics Symposium*, vol. 1, pp. 322–325, 1980.

38. P. M. Naraine and C. K. Campbell, "Gigahertz SAW filters on YZ-lithium niobate without the use of sub-micron line widths," *Proc. 1984 IEEE Ultrasonics Symp.*, vol. 1, pp. 93–96, 1984.

—7—

Designing SAW Filters for Arbitrary Amplitude/Phase Response

7.1. Introduction

7.1.1. THE IDT AS A SAMPLED-DATA STRUCTURE

This chapter examines the modelling-representation and operation of a surface acoustic wave filter as a sampled-data structure. The sampled-data structure is imposed by the IDT pattern and is independent of the acoustic wave-type propagating in the underlying piezoelectric substrate. As a result, the "SAW" filter discussions given here will apply to both SAW and pseudo-SAW filter structures.

The uniform and apodized SAW filters considered so far have tacitly applied to those having passband frequency responses with symmetric amplitude and linear phase around the center frequency. There are, however, many signal-processing applications that call for bandpass filters with some prescribed degree of nonlinear phase and/or nonsymmetric magnitude response around center frequency. A simple example is that for a TVIF filter with nonsymmetric amplitude and nonlinear phase response. Figure 7.1 illustrates the amplitude/phase response of a 43.5-MHz TVIF filter stage that originally required a three-stage LC filter for implementation, and is now readily obtained by a single SAW-IF filter [1].

Such response-shaping can be readily implemented by appropriate apodization of one IDT in the SAW filter. This capability arises because the geometric pattern of an IDT represents a (spatial) sampled-data approximation of its impulse response. Knowing the velocity of the underlying acoustic wave, the equivalent time response is easily modeled. The frequency response of the IDT may be computed by applying a discrete Fourier transform of the impulse-response function. Thus, by judicious apodization and finger placement of either the input or output IDT (but not both), a SAW filter with prescribed amplitude and phase response can be obtained.

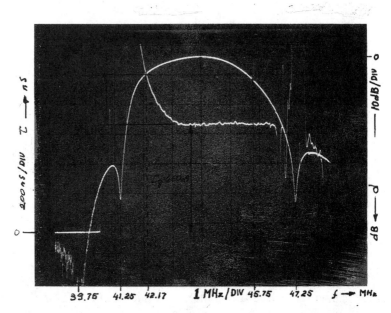

FIG. 7.1.   Example of nonsymmetric amplitude and nonlinear group delay of a 43.5-MHz TVIF filter. Magnitude scale: 10 dB/div. Group delay scale: 200 ns/div with 600 ns delay at midband. (This oscilloscope trace photograph is taken directly from a laboratory notebook.)

In modelling an IDT as a representation of a sampled-data function, finger periodicity must be considered. So far in the text, IDT fingers have been considered to be uniformly spaced, and $\lambda_o/2$ apart, at filter center frequency $f_o = v/\lambda_o$. In terms of sampled-data equivalents, this gives the sampling frequency f_s as $f_s = 2f_o$, for sampling time $T = 1/2f_o$. This condition also results in maximum acoustic finger reflections at center frequency f_o. For filter designs where IDT finger reflections are undesirable, however, it would be best to move the sampling frequency out of the passband altogether. This sampling shift can be achieved using appropriate IDT finger geometry. This ability to have differing sampling and center frequencies can be important for the fabrication of gigahertz-band SAW filters for mobile/wireless systems, otherwise limited by lithographic resolution.

This sampling technique can also be used to increase the phase slope (i.e., effective Q) of a SAW delay line, without increasing its overall length, in devices for mobile systems where component size is critical. When used as a feedback element in a delay-line SAW oscillator, for example, the increased oscillator Q would (theoretically) result in a decrease of phase noise interference in adjacent channels.

7.1.2. ANALOG AND DIGITAL CONSIDERATIONS

SAW devices are inherently *analog ones*. They can, however, be considered to be analog/digital hybrids, as they find ready application in *digital* systems, such as for digital cellular, digital microwave radio [2], [3] and spread-spectrum communications [4], [5]. Moreover, some SAW filter designs require a comprehensive knowledge of digital sampling techniques. It is the unique structure of an IDT that gives the SAW device its versatility in both analog and digital signal processing circuits and systems. As noted in the preceding, its geometry corresponds to a spatially sampled version of the desired finite impulse response (FIR) of the SAW filter. Additionally, the fact that it is spatially sampled also opens up the gateway to digital sampling considerations and interpretations. As a result, many sampling and design techniques for digital signal processing of FIR filters can be applied to linear-phase SAW filters.

For illustrative purposes, only bidirectional IDT geometries with uniform finger spacing are examined in this chapter. Thus for IDTs with uniform single-electrodes[1] the finger spacing is $\lambda_o/2$ at filter center frequency f_o. The sampling frequency is $2f_o$, while the sampling time is $T = 1/(2f_o)$. In many IF Rayleigh-wave filter designs it is undesirable to have a finger-sampling frequency of $2f_o$ in IDTs with single electrodes due to the level of ensuing finger reflections.[2] As shown, the split electrode IDT can be used to combat this problem, with a trade-off in fabrication resolution. In terms of sampling concepts, what happens is that the finger sampling frequency changes from $2f_o$ to $4f_o$. These higher sampling rates can be applied to the realization of a *prescribed* nonlinear phase and/or a nonsymmetric passband magnitude response in a single SAW filter employing bidirectional IDTs with uniform finger spacing. The prescription is built into only one IDT. This is usually the input one, while the unapodized uniform output IDT is configured as a broadband receiver.

Let us review first some fundamental concepts relating to the use of both positive and negative frequency regions in filter frequency response modelling, as can be applied to the synthesis of realizable SAW filters [6], [7].

---

[1] Single electrodes in IDTs are also termed *solid* electrodes.
[2] As considered in Part 2 of this book, finger reflections are often an essential feature in the design of many leaky-SAW filter designs!

7.2. Negative and Positive Frequency Concepts in IDT Design

7.2.1. Realization of Real Functions and Real Responses

The frequency and impulse response of a linear electrical system are inter-related by the Fourier transform pair

$$H(f) = \int_{-\infty}^{\infty} h(t) e^{-j2\pi f t} dt \qquad (7.1a)$$

$$h(t) = \int_{-\infty}^{\infty} H(f) e^{j2\pi f t} df, \qquad (7.1b)$$

where the evaluation of impulse response $h(t)$ is over the frequency range $-\infty \le f \le +\infty$. This is required for $h(t)$ to be obtained as a physically realizable electrical response, as illustrated by the following example, relating the *real* function:

$$\cos(\omega t + \theta) = \frac{e^{+j(\omega t + \theta)} + e^{-j(\omega t + \theta)}}{2}, \qquad (7.2)$$

where phase offset term θ is included for generality. In the expansion of Eq. (7.2) the exponentials can be regarded as electrical phasor components of equal amplitude that rotate at equal rates in opposite directions in the complex plane, as depicted in Fig. 7.2. Rotations of the first phasor term in Eq. (7.2) can be arbitrarily considered as applying to the positive frequency domain, while the second component relates to the negative frequency one.

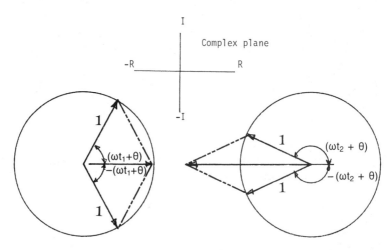

Fig. 7.2. $2.\cos(\omega t + \theta)$ represented as two oppositely rotating phasors at times t_1 and t_2.

As illustrated, the two phasors will always add up to give a purely real resultant along the real axis, with positive or negative amplitude values. If the "negative" frequency component is absent, however, the result will no longer yield a *pure real* function. While this frequency concept was applied to one simple example, it holds for the derivation of all realizable impulse responses $h(t)$. The same principle also applies to the derivation of the spatial impulse pattern of the SAW IDT. What it means is that the inverse Fourier transform of Eq. (7.1b) must be evaluated over the entire frequency range $-\infty \leq f \leq +\infty$ in order to obtain physically realizable IDT geometries.

7.2.2. Hermitian Conjugate Responses

The term "Hermitian conjugate response" is often applied to describe the performance of a realizable filter[6]. This terminology just means that the filter has symmetric magnitude response $|H(\omega)| = |H(-\omega)|$ and antisymmetric phase response $\phi(\omega) = -\phi(-\omega)$ between positive and negative frequency domains, as depicted in Fig. 7.3. This is easily appreciated in term of the cosine function example of Fig. 7.2, where the magnitudes of the exponential terms are equal (i.e., symmetric), while their phase angles are equal and opposite (i.e., antisymmetric). With the forementioned in mind, consider a desired SAW filter bandpass response of the general form

$$H(\omega) = |H(\omega)| e^{j\phi(\omega)} \qquad (7.3)$$

over angular frequency range ω_1 to ω_2, with arbitrary phase response $\phi(\omega)$. From Eq. (7.1b), the physically real impulse response $h(t)$ can be derived and applied to the determination of the SAW-IDT geometry. Instead of

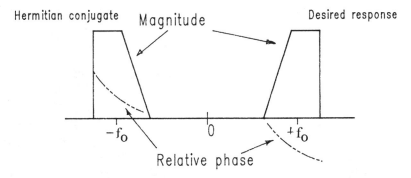

Fig. 7.3.   Filter response function $H(f)$ in positive frequency domain and its Hermitian conjugate in negative frequency domain, with symmetric magnitude and antisymmetric phase.

$h(t)$, the spatial impulse response $h(x)$ can be employed because IDT finger spacing x is linearly related to time for an IDT with uniformly spaced fingers. Selecting the spatial formulation for illustration yields

$$h(x) = C \int_{-\omega_2}^{-\omega_1} |H(\omega)| e^{j[\phi(\omega)-\beta x]} d\omega + C \int_{+\omega_1}^{+\omega_2} |H(\omega)| e^{j[\phi(\omega)-\beta x]} d\omega, \qquad (7.4)$$

where C = a geometric constant and $\beta = 2\pi/\lambda$ = phase constant; and $h(x)$ reduces to the required real function form

$$h(t) = 2C \int_{+\omega_1}^{+\omega_2} |H(\omega)| \cos[\phi(\omega) - \beta x] d\omega. \qquad (7.5)$$

Equation (7.5) can be solved using computer-aided design techniques such as with the inverse discrete Fourier transform (IDFT), where the IDT apodization pattern is obtained by sampling impulse response function $h(x)$ at uniform intervals, and suitably scaling geometric parameter C. Sampling intervals are normally taken to coincide with points of maximum spatial amplitude in Eq. (7.5). In addition, suitable window functions for truncating (spatial) impulse response $h(x)$ can be incorporated.

7.3.   The IDT as a Sampled-Data Structure

7.3.1.   Effects of Sampling

The focus here is on SAW bandpass filters. Generally, the sampling of the impulse response for this type of filter will be straightforward. It may be shown that the impulse response $h(t)$ of an arbitrary bandpass filter can be expressed as

$$h(t) = A(t)\cos(\omega_o t) + B(t)\sin(\omega_o t), \qquad (7.6)$$

where $A(t)$ and $B(t)$ are pure real functions of time. Thus, from an informational point of view, it is necessary merely to sample $h(t)$ fast enough to be able to reconstruct $A(t)$ and $B(t)$ from these samples. This sampling time will be $1/(2\Delta\omega)$, where $\Delta\omega$ is the bandwidth of the bandpass filter. Observe that two samples are required per time period—one for $A(t)$ and the other for $B(t)$. This sampling-rate (or frequency) is readily achieved for most SAW filter designs, as it is much faster than that dictated by the uniform sampling theorem. In the case of the single-electrode and split-electrode IDT geometries considered so far, the sampling frequency is $2f_o$ or $4f_o$, respectively, which is well above the rate required for digital filter applications.

To further understand the significance of these sampling rates as they affect SAW filter design, consider the bandpass response characteristic sketched in Fig. 7.4(a), with the Hermitian conjugate response included for realizability. If the IDT in question is a bidirectional one with uniform spacing of single-electrode fingers, its sampling frequency will be $2f_o$. Because the geometric nature of the IDT establishes it as a sampled-data structure, this will cause the responses in both positive and negative frequency domains to repeat at frequency intervals of $2f_o$ [7]. In this instance the shifted responses overlap as illustrated in Fig. 7.4(b), so that the resultant response will have only a symmetric magnitude response. This poses no problems as long as this is what is desired. If the magnitude response is to be nonsymmetric about f_o, however, a higher sampling frequency must be employed to separate the responses. Figure 7.4(c) shows the result when the sampling frequency is $4f_o$, as for split-electrode IDTs. Here, the responses in positive and negative frequency domains repeat every $4f_o$ without overlap. This makes it possible to implement arbitrary amplitude and/or phase response in the SAW-IDT design, even when the fingers of the bidirectional IDT are uniformly spaced, as considered in this chapter. The realization of arbitrary amplitude and/or phase response with bidirectional IDTs using split electrode geometry can be attained using exactly the same finger-weighting technique described in Chapter 6 for diffraction compensation [8]. The required lengths of individual finger segments in the split-electrode

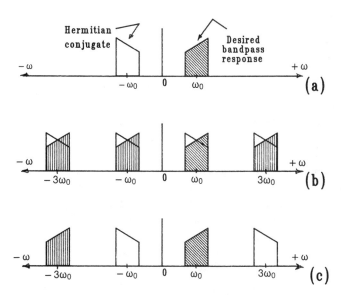

FIG. 7.4.   (a) An illustrative desired bandpass magnitude response. (b) Effect of sampling the response at $2f_o$ with solid electrodes. (c) Sampling at $4f_o$ with split electrodes removes aliasing ambiguity from (b).

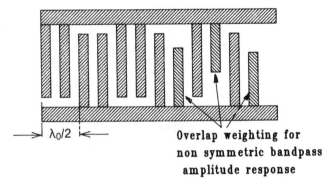

Overlap weighting for
non symmetric bandpass
amplitude response

Fig. 7.5.   Overlap-weighting of the split-electrode IDT can be applied to nonsymmetric bandpass designs, as well as for diffraction compensation.

IDT of Fig. 7.5 can again be derived from the delta-function model, to cause the desired magnitude and phase response. Diffraction compensation can be additionally included, using the finger weighting technique outlined in Chapter 6.

7.3.2.   A Design Example

Figure 7.6(a) illustrates a design example where a 27-MHz SAW filter was required to have a passband with symmetric magnitude response about center frequency f_o, as well as a phase shift of $+90°$ at center frequency in addition to the inherent linear phase shift and group delay due to the separation between input and output IDTs. This design evaluation must include the Hermitian conjugate response as considered here. In this example, the impulse response $h(t)$ was obtained by applying an inverse discrete Fourier transform (IDFT) to the total frequency response. Because a computer deals only with numbers, the negative-frequency scale can be dispensed with in the computations, so that the desired and conjugate frequency responses may be arranged as in Fig. 7.6(b). Here, the design was derived from a baseband prototype with sampling time of 1 s and sampling frequency of 1 Hz and over the frequency range from −0.5 to +0.5 Hz. In this particular case, the desired bandpass response was centered at 0.25 Hz, so that the Hermitian conjugate was centered at −0.25 Hz. The shifted Hermitian conjugate after sampling at $T = 1$ s thus appears at 0.75 Hz as shown. A 1024-point FFT algorithm was employed in the IDFT computations, as indicated in Fig. 7.6(b). The points were distributed as 512 pairs of samples to store both the real- and imaginary-components of the

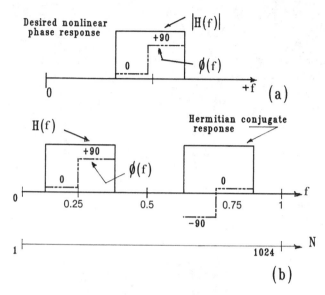

FIG. 7.6. (a) Design example requiring relative phase shift of +90° at center frequency. (b) Result normalized to 1/4 Hz for inverse discrete Fourier transform with 1024-point array, including Hermitian conjugate response.

overall frequency response function. Figure 7.7 is a photograph of the IDT pattern obtained in this manner, after suitable scaling of the normalized impulse response values [7], [9]. Split-electrode IDT geometry was employed.

This illustration was chosen to demonstrate a most important feature of SAW-IDT design: that the spatial impulse-response pattern $h(x)$ need not look exactly like that for $h(t)$ on an oscilloscope! As long as the sequential segments of $h(x)$ preserve the same evolutionary order as for $h(t)$, they can be separated and displaced as shown. This can be employed to promote a more uniform SAW aperture and thereby reduce diffraction degradation. Excluding their vertical shifts, observe that the segments of the IDT apodization pattern in Fig. 7.7 are nonsymmetric about the center axis. Figure 7.8 shows the experimental response for a 27-MHz SAW filter incorporating the IDT of Fig. 7.7 in conjunction with a broadband output one [7]. To highlight the 90° phase shift at center frequency, the additional linear-phase term was subtracted out with the network analyzer instrumentation employed. Note that the dip in the magnitude response of Fig. 7.8(a) is due to the truncation of the impulse response in this example, in agreement with the predicted response.

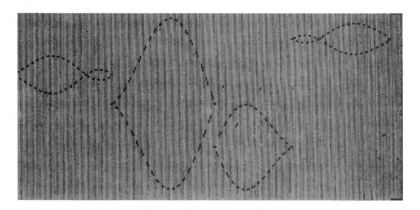

Fig. 7.7. Split-electrode IDT for example of Fig. 7.6. Individual segments of the apodization pattern are displaced to promote uniformity of the SAW beam and reduce diffraction. (Reprinted with permission from Reilly, Campbell, and Suthers [9], © IEEE, 1977.)

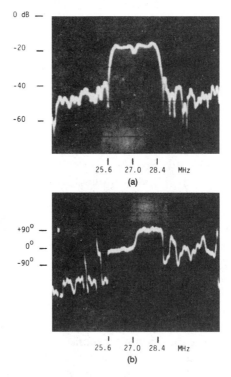

Fig. 7.8. Response of example of Fig. 7.6 fabricated on $LiNbO_3$, with wideband output IDT. (a) Magnitude of dip at 27-MHz center frequency results from truncation of IDT. (b) Relative phase shift of $+90°$. The residual linear phase delay was subtracted out by the network analyzer. (Reprinted with permission from Reilly, Campbell, and Suthers [9], © IEEE, 1977.)

Example 7.1

∠ Two positions give conjugate differential phase ∠

Example 7.1 Simple Technique for Changing Sign of Filter-Phase Response. After designing the forementioned SAW linear-phase filter with relative phase response of +90° at center frequency, the engineer responsible suddenly realized that the design was supposed to have been for a −90° phase shift. Could this mistake be rectified in a hurry with minimum effort and calculation without resorting to a completely new design calculation? ∎

Solution. Yes. All that had to be done was for the relative positions of input and output IDT to be spatially reversed as sketched in Example 7.1 here. This ordering does not concern linear-phase IDTs, because their IDT patterns are symmetric. With the nonlinear phase designs, however, it is essential that the two IDTs be laid down correctly with respect to one another.

7.3.3. APPLICATION TO GROUP DELAY PERFORMANCE

In linear-phase SAW filter design, the maximum attainable group delay $\tau = -d\phi/d\omega$ is proportional to the separation between IDTs, and is limited by substrate length as a result. However, even if the SAW designer is limited to a given substrate length, the maximum group delay attainable can be increased by using a nonlinear phase design, with *negative* values of relative phase shift added at center frequency [10]. Figure 7.9 illustrates the group delay of such a SAW filter, with −90° relative phase shift added at center frequency, to create a 50% increase in group delay. Maximum attainable group delays can increase even further by using a relative phase shift of −180°. This technique can be used to increase the phase slope (i.e., effective Q) of a SAW delay line without increasing its overall length. When used as a feedback element in a delay-line SAW oscillator in a mobile or

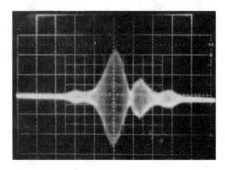

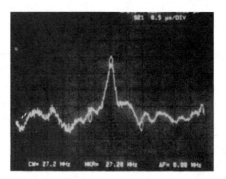

FIG. 7.9.   (a) Impulse response of SAW filter with −90° relative phase shift. Horizontal scale: 0.5 μs per large division. (b) Measured group delay over range ±0.88 MHz about center frequency f_o = 27.3 MHz. Vertical scale: 0.5 μs/div. (Reprinted with permission from Campbell [10], © IEEE, 1982.)

wireless link, the increased oscillator Q would (theoretically) result in a decrease of phase-noise interference in adjacent channels.

7.4.   Sampling the IDT Fingers at Other Rates

So far, the considerations of Hermitian conjugate and sampling relations for bidirectional IDTs with uniform finger spacing here have been limited to those with finger sampling rates of $2f_o$ and $4f_o$. In the particular design technique examined, split-electrode geometry was applied to attain nonsymmetric bandpass responses. It is possible to extend these concepts, however, so as to yield nonsymmetric bandpass response while using single-electrode geometries. This is done by shifting the sampling frequency (or its multiples) directly from the passband, and requires a double application of Hermitian conjugate parameters. As shown in Fig. 7.10, this is achieved by

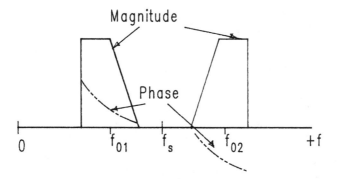

Fig. 7.10. Conjugate response in positive and negative frequency domains is used to move sampling frequency f_s out of the passband. Allows use of solid-electrode IDT geometry in design of SAW filter with arbitrary magnitude/phase response.

employing Hermitian conjugate bandpass responses in the positive-frequency domain as well as the negative one [11]. As shown, the bandpass responses about center frequencies f_{o1} and f_{o2} need not be symmetric as long as they are Hermitian conjugates about midfrequency f_o. The IDT finger sampling frequency (i.e., the synchronism frequency) can then be set to $2f_s$ using single-electrode IDT geometry. In this way any SAW reflections from the single-electrode IDT structure pose no problems because they will all be out-of-band ones.

Either the upper or lower Hermitian conjugate response can be selected as the desired one. All that has to be done is to design the single-electrode wideband output IDT in the same conjugate way, except that it is given a different midfrequency and sampling rate $2f'_s$. In this way, only one input bandpass function will overlap with one output one in obtaining the overall response. One fabrication advantage of this technique is that if the upper bandpass response is chosen as the desired one, then the sampling rate $2f_s$ is smaller than it would be ordinarily. In this way, the upper frequency limit for SAW filter fabrication with a given lithographic system can be increased over that attainable with a "normal" IDT design.

7.5. Summary

Some techniques have been illustrated for designing SAW filters with arbitrary amplitude and/or phase response. The approach used considered the bidirectional IDT with uniform finger spacing as a periodic sampled-data structure. Additionally, a technique was demonstrated for employing a reduced sampling rate $2f_s$ as one means of increasing the upper-frequency fabrication limit for SAW filter in a given lithographic system.

7.6. REFERENCES

1. A. J. deVries and R. Adler, "Case history of a surface-wave TV IF filter for color television receivers," *Proc. of IEEE*, vol. 64, pp. 671–676, 1976.
2. E. Ehrmann, H. R. Stocker, C. Ruppel and W. R. Mader, "SAW-filters for spectral shaping in a 140 Mbit/s digital radio system using 16 QAM," *Proc. 1983 IEEE Ultrasonics Symp*, pp. 17–22, 1983.
3. J. C. B. Saw, T. P. Cameron, and M. S. Suthers, "Impact of SAW technology on the system performance of high capacity digital microwave radio," *Proc. 1993 IEEE Ultrasonics Symp.*, vol. 1, pp. 59–63, 1993.
4. D. Malocha, J. H. Goll and M. A. Heard, "Design of a compensated SAW filter used in wide spread spectrum signals," *Proc. 1979 IEEE Ultrasonics Symp.*, pp. 518–529, 1979.
5. K. Hoshkawa, K. Komine, N. Araki, and H. Suzuki, "Surface acoustic wave devices for pulse position modulated spread spectrum communications systems," *Proc. 1995 IEEE Ultrasonics Symp.*, vol. 1, pp. 151–154, 1995.
6. T. J. Boege, G. Chao and W. J. Drummond, "Design of arbitrary phase and amplitude characteristics in SAW filters," *Proc. 1976 IEEE Ultrasonics Symp.*, p. 313–316, 1976.
7. R. F. Mitchell, "Basics of the SAW Filter: A Review," in J. H. Collins and L. Masotti (eds), *Computer-Aided Design of Surface Acoustic Wave Devices*, Elsevier, Amsterdam, pp. 111–132, 1976.
8. R. F. Mitchell and D. W. Parker, "Synthesis of acoustic-surface-wave filters using double electrodes," *Electronics Letters*, vol. 10, p. 512, 1974.
9. J. P. Reilly, C. K. Campbell and M. S. Suthers, "The design of SAW bandpass filters exhibiting arbitrary phase and amplitude response," *IEEE Trans. Sonics and Ultrasonics*, vol. SU-24, pp. 301–305, 1977.
10. C. K. Campbell, "Application of the inverse discrete Fourier transform to the design of SAW filters with nonlinear phase shift," *Proc. 1982 IEEE Ultrasonics Symp.*, pp. 46–49, 1982.
11. M. S. Suthers, C. K. Campbell and J. P. Reilly, "SAW bandpass filter design using Hermitian function techniques," *IEEE Trans. Sonics and Ultrasonics*, vol. SU-27, pp. 90–93, 1980.

—8—

Interdigital Transducers With Chirped or Slanted Fingers

8.1. Chapter Coverage

8.1.1. IDTs With Slanted Fingers

Up to this point, the IDT coverage in this text has dealt with those that have uniformly spaced electrodes placed normally to the acoustic propagation axis. The two differing IDT structures considered in this chapter digress from this geometry. In outlining their operation the assumption is made that finger reflections from IDT finger discontinuities are minimal, and that Rayleigh-wave propagation is involved.

The structure relates to filters and resonators employing IDTs with slanted-finger[1] geometries. The significant features of midloss SAW filters employing slanted-electrode IDTs include: a) wideband performance capability; b) flat passband responses; and c) small group-delay ripple [1]–[6]. Moreover, excellent out-of-band rejection can be attained by appropriate weighting of the electrode fingers [5]. These combined features make them suitable for code-division multiple-access (CDMA) IF filters in mobile phone circuits. Resonator applications of slanted IDTs have also been reported as a means of increasing the tuning-bandwidth of a voltage-controlled oscillator (VCO) in mobile circuitry over that attainable with a standard one-port SAW resonator [6]. As illustrated in Chapter 19, 70-MHz slanted-finger IDT-SAW IF filters with 10% bandwidth have also been employed in a nonregenerative digital microwave radio system with message signals at 6 Mb/s, with a requirement for very-low passband ripples in the IF filter response; 70-MHz slanted-finger IDT-SAW IF filters with 50% bandwidth are also incorporated in satellite communications systems such

---

[1] The slanted-finger IDTs are also known as slanted-electrode IDTs, as well as fan-shaped IDTs.

as for Inmarsat-C, in digital data terminals for Mobile Earth Stations (MES).

8.1.2. THE LINEAR FM CHIRP FILTER

The other type of IDT considered in this chapter concerns the linear FM SAW chirp filter that has been employed—together with its variants—in radar systems since the early days of SAW development [7]–[11]. In recent years, SAW chirp-filter techniques have garnered increased attention for their use in satellite on-board high-selectivity frequency demultiplexers. These can be implemented using a SAW chirp Fourier transform (CFT) preprocessor for a follow-up DSP processor to reduce on-board power consumption. This power-saving trend is increasing as ground-terminal antennas for personal communications are becoming smaller, with increasing emphasis placed on the satellite antenna and its multiple "spot-beam" capability [12], [13]. For mobile indoor spread-spectrum communications employing SAW convolvers (see Chapter 17), focussing chirp filter techniques have been applied as a means of increasing convolver efficiency, while maintaining the wide bandwidths required [14], [15].

By themselves, linear FM SAW chirp filters are inherently dispersive structures with nonlinear (quadratic) phase response. As will be demonstrated in this chapter, however, the judicious use of two linear FM filters can realize a variable delay line without dispersion. Such capability can find application in direct-sequence spread-spectrum techniques for CDMA indoor communications systems, where it is necessary to add time delays between diversity antennas [16].

8.2. Interdigital Transducers with Slanted-Finger Geometries

8.2.1. LINEAR-PHASE FILTERS USING SLANTED-FINGER IDTs

Figure 8.1 illustrates the configuration of a linear-phase SAW filter employing unweighted slanted-finger IDTs at input and output ports. Very wide fractional bandwidths can be attained with these filter types. In early studies of such filters the author readily attained bandwidths of up to 50%. Figure 8.2 shows the measured frequency response of one such 70-MHz filter, with 25% bandwidth and passband ripple of less than 0.6 dB. For 25% bandwidth designs it is necessary to use finger-slant angles of as high as ±7° [1]. Such severe slant-angle deviations from the normal propagation axis require device fabrication on an autocollimating substrate such as Y-Z LiNbO$_3$, to avoid beam-steering losses. (See Table 6.2 of Chapter 6.) However, IF filters with much smaller bandwidth requirements can be

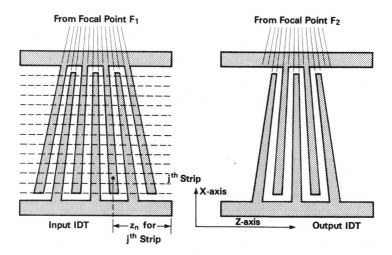

FIG. 8.1. Geometry of SAW filter using unweighted slanted-finger IDTs in input and output stages. Maximum finger-slant angle (exaggerated here) limited to $\pm 7°$ on LiNbO$_3$, with smaller maximum angles for ST-quartz. (Reprinted with permission from Campbell, Ye, and Sferrazza Papa, [1], © IEEE, 1982.)

fabricated on temperature-stable ST-quartz substrates. For example, the fractional bandwidth needed for a USA 100-MHz CDMA filter would only be 1.23%.

As demonstrated in Example 2.2 of Chapter 3, symmetry of the slanted IDT structure ensures nominal linear phase response. In Fig. 8.1, the synchronous finger spacing at the upper extremities of each IDT finger overlap dictates the high-frequency edge of the passband, while that at the bottom extremities conforms to the lower edge of the passband.

The transfer response of each slanted-finger IDT may be calculated by modelling it as a parallel assembly of a large number S of "standard" linear-phase IDTs. Using the delta-function model and large S (e.g., S = 60), the overall transfer function $H(f)$ of one IDT can be derived as a summation of sinc-function terms

$$H(f) = \sum_{j=1}^{S} A_j(f) f_j^{1/2} \frac{\sin\left(N_p \pi \left(f - f_j\right)/f_j\right)}{N_p \pi \left(f - f_j\right)/f_j}, \tag{8.1}$$

where N_p = number of finger pairs, f_j = center frequency of the jth modelling strip and $A_j(f)$ = an amplitude-weighting term for the jth strip. As illustrated, a differing number of finger pairs can be used in input and output IDTs as an aid to smoothing the overall passband response. This is in

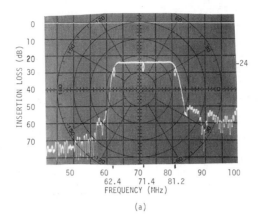

(a)

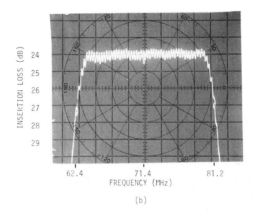

(b)

FIG. 8.2. (a) Amplitude response of 70-MHz SAW filter with BW ≈ 25% and insertion loss 24 dB, using slanted finger IDTs on LiNbO₃. Horizontal scale: 40–100 MHz. Vertical scale: 10 dB/div. (b) Expanded vertical scale of 1 dB/div, showing ripple less than 0.6 dB. (Reprinted with permission from Campbell, Ye, and Sferrazza Papa, [1], © IEEE, 1982.)

contrast to the design of wideband linear-phase filters using chirp filters in a nondispersive configuration [9]. With this latter technique, the input/output chirp filters are ideally required to be identical, with the result that any passband distortions due to a single chirp are potentially magnified in the overall transfer response.

To demonstrate the attainable bandwidths with the slanted-finger technique, Fig. 8.2 shows the experimental response of a 70-MHz linear phase filter with a bandwidth of 25% [1]. In this 50-Ω design, $N_p = 30$ finger pairs

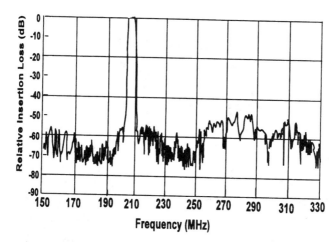

Fig. 8.3.   Experimental response of 200-MHz SAW filter on quartz with BW = 2%, using weighted slanted-finger IDTs. Passband ripple is 0.1 dB. Out-of-band rejection >50 dB. Horizontal scale: 150–330 MHz. Normalized vertical scale: 10 dB/div. Actual unmatched insertion loss ~ 28 dB. (After Reference [5].)

was used in both input and output IDTs, using aluminum thin-film electrodes of ≈ 500 Å thickness and a maximum tilt angle of ±6.7° about a normal-to-the-SAW propagation axis. No input/output matching was employed, and the peak-to-peak amplitude ripple across the band was less than about 0.6 dB.

With the elementary design of Fig. 8.2, it is very difficult to achieve out-of-band rejection levels of better than 40 dB. Improved out-of-band rejection car be achieved, however, by application of suitable weighting functions to the slanted fingers. Figure 8.3 illustrates the experimental response of an illustrative 200-MHz SAW filter on ST-quartz, with such weighting applied to the slanted-fingers of each IDT. This particular design example yielded a 2% bandwidth, in-band ripple of only 0.1 dB and out-of-band rejection greater than 50 dB [5].

8.2.2.   One-Port SAW Resonators Using Slanted-Finger IDTs

While resonator structures are not considered until Chapter 11, it may be noted that one-port SAW resonators employing slanted-finger IDTs have been developed for voltage-controlled oscillator (VCO) applications in mobile systems. The advantage of this type of resonator is that it enables a wider tuning range to be attained than with a "conventional" one-port

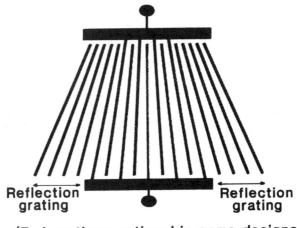

(End gratings optional in some designs)

FIG. 8.4.   A one-port SAW resonator configuration using slanted-finger IDT and slanted reflection gratings, for use in voltage-controlled oscillator. (Reflection gratings may be dispensed with in some designs [6].)

SAW resonator [6]. Figure 8.4 illustrates the basic structure, which will be examined in Chapter 11.

8.3.   The IDT for a SAW Linear FM Chirp Filter

8.3.1.   GENERAL CONSIDERATIONS

Figure 8.5 illustrates the geometry of an unapodized metal finger IDT for a linear FM SAW chirp filter. Although sketched here with single-finger (i.e., solid-finger) geometry for ease of illustration, these transducers normally employ split-electrodes and dummy fingers to reduce response degradation due to spurious SAW reflections and diffraction.

Let us examine first the operation of this chirp filter in nonmathematical terms. From an inspection of the chirp IDT pattern in Fig. 8.5, and from previous considerations of the correlation between apodization geometry and impulse response, it is apparent that this IDT will have a nonlinear phase response. The mathematical outline given in what follows will serve to demonstrate that the linear FM chirp filter has a quadratic phase response.

Now consider the chirp IDT as a receiver of incoming surface acoustic waves. This will have maximum efficiency in generating an IDT voltage in the region where the finger period is also that of the surface wave. The same

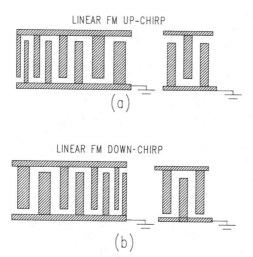

FIG. 8.5. (a) Elementary linear FM up-chirp SAW filter. (b) Corresponding down-chirp configuration. Filters are shown unapodized, with broadband output IDTs for singly dispersive operation. Broadband IDTs can alternatively be used at inputs, due to reciprocal nature of the devices.

situation will hold when the IDT acts as a transmitter. Surface waves will be excited principally by regions of the IDT whose finger periods are in synchronism with the excitation frequency. For a linear FM chirp filter with center frequency f_o and bandwidth B, the excitation frequency at the "low end" of the IDT (i.e., maximum finger spacings) will be $(f_o - B/2)$, while that at the "high end", (i.e., minimum finger spacing) will be $(f_o + B/2)$.

While the bandwidth B of the chirp filter is set by the finger spacing pattern of the IDT, its impulse response $h(t)$ will be a finite one of duration T, corresponding to the time for the surface waves to traverse the entire length of the chirp IDT. For a linear FM chirp, the frequency components of $h(t)$ will increase or decrease linearly over time T, depending on whether the design is an up- or a down-chirp one.

In the absence of second-order effects, both SAW chirp filters in Fig. 8.5 would ideally have the same overall frequency response *magnitude* $|H(f)| = |H_c(f)| \, |H_2(f)|$, where $|H_c(f)|$ is the frequency response of the chirp IDT and $|H_2(f)|$ is that of an assumed uniform broadband IDT. As one is an up-chirp structure and the other is a down-chirp one, however, their individual dispersive phase shifts will be of opposite sign, as will be the frequency-time slopes of the impulse response $h(t)$.

The nonlinear phase shift gives rise to dispersion (i.e., nonconstant group delay as a function of frequency), as illustrated in the experimental example

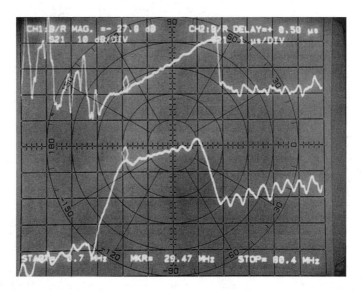

FIG. 8.6.   Network-analyzer responses of unapodized linear FM up-chirp SAW filter. Chirp bandwidth $B = 25\,\text{MHz}$ with $TB \approx 62.5$. Horizontal scale: 8.7–80.4 MHz. Upper trace gives group delay on vertical scale of 1 μs/div. Lower trace gives amplitude response on vertical scale of 10 dB/div.

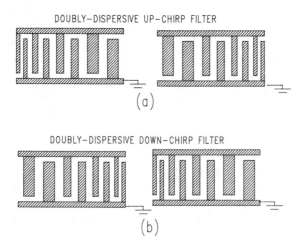

FIG. 8.7.   (a) Unapodized in-line doubly dispersive up-chirp SAW filter. (b) Doubly dispersive down-chirp counterpart.

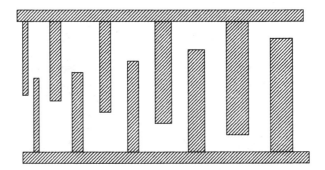

F<small>IG</small>. 8.8.   Chirp IDT with $f^{-3/2}$ apodization correction.

of Fig. 8.6 for a simple 25- to 50-MHz linear FM up-chirp SAW filter with $TB = 62.5$. This particular example employed the up-chirp IDT geometry of Fig. 8.5(a), so that the group delay increases with time over the chirp response. The SAW filter types of Fig. 8.5 are referred to as a *singly dispersive, in-line* FM chirp filters. Those of Fig. 8.7 that employ two chirp IDTs in mirror-image positions are termed as being *doubly dispersive in-line* chirps [11]. For the same individual chirp IDT geometry in Figs. 8.5 and 8.7, the latter structures will have twice the dispersion and twice the TB product of the singly dispersive counterpart.

If a constant-amplitude impulse response is required, it is necessary to apodize the IDT with an $f^{-3/2}$ overlap weighting, as shown in Fig. 8.8, to compensate for the $f^{+3/2}$ amplitude-weighting inherent in the impulse response of an IDT with constant finger overlap, as given in Eq. (4.40) of Chapter 4. In the mathematical derivations in the following section it will be assumed that this has been done.

8.3.2.   I<small>MPULSE</small> R<small>ESPONSE</small> R<small>ELATIONSHIP</small>

Now consider the phase response of the linear FM chirp filter. To obtain this, first express the impulse response $h(t)$ in the general form

$$h(t) = A(t)e^{-j\phi(t)}, \tag{8.2}$$

where $\phi(t)$ relates the differential phase shift for SAW waves that emanate from different finger regions of the chirp IDT and $A(t)$ is a voltage amplitude term. If it is assumed that $f^{-3/2}$ apodization-weighting has been applied, we obtain the impulse response amplitude as $|H(t)| = |A(t)|$ in Eq. (4.40), where

$$A(t) = \begin{pmatrix} 1 \\ 0 \end{pmatrix} \quad \begin{array}{l} for \ -T/2 \leq t \leq +T/2 \\ otherwise \end{array}, \tag{8.3}$$

and T is the dispersion of the chirp IDT. Next, because angular frequency ω is simply the rate of change of phase ($\omega = d\phi/dt$), the phase angle $\phi(t)$ in Eq. (8.2) is

$$\phi(t) = 2\pi \int f(t) dt, \tag{8.4}$$

where $f(t)$ is the instantaneous frequency. For the linear up-chirp IDT

$$f(t) = f_o + \frac{B}{T} t = f_o + \mu t \tag{8.5a}$$

or

$$\omega(t) = \omega_o + \frac{2\pi B}{T} t = \omega_o + 2\pi \mu t, \tag{8.5b}$$

where dispersion time $- T/2 \leq t \leq + T/2$ for convenience, so that $f(t) = f_o$ when $t = 0$. The FM chirp slope $\mu = B/T$ (in MHz/μs) is positive for an up-chirp filter and negative for a down-chirp are. From Eqs. (8.4) and (8.5) $\phi(t)$ becomes

$$\phi(t) = 2\pi \int \left(f_o + \frac{B}{T} t \right) dt = 2\pi \left(f_o t + \frac{B}{2T} t^2 \right) \quad for \ -T/2 \leq t \leq +T/2, \tag{8.6}$$

with a linear term in t and a quadratic term in t^2. The impulse response can be expressed as

$$h(t) = e^{-j2\pi \left[f_o t + (B/2T) t^2 \right]} \quad for \ -T/2 \leq t \leq +T/2, \tag{8.7}$$

with $A(t) = 1$.

From Eq. (8.6) the group delay time $T = - (d\phi/d\omega)$, which may be expressed as $T = - [(d\phi/dt)/(d\omega/dt)]$, is just a linear term $T = \{$a constant $\pm t\}$ for $0 \leq t \leq T$ over the chirp range, where the "+" sign applies to an up-chirp- and the "−" sign to a down-chirp configuration. This result may be compared with the illustrative response of Fig. 8.6.

In the chirp design, the IDT fingers must be positioned in accordance with the prescription for $h(t)$ [11], [17]. To this end, a convenient sampling of $h(t)$ is at its zero-crossings at times t_n, with finger positions $x_n = v t_n$ in terms of SAW velocity v. In doing so, the phase characteristics of $\phi(t)$ in Eq. (8.6) must be retained, giving

$$\phi(t_n) = n\pi \quad for \ integer \quad n = 0, \ \pm 1, \ \pm 2, \ \dots. \tag{8.8}$$

From Eqs. (8.6) and (8.8)

$$2\pi\left(f_o t_n + \frac{B}{2T}t_n^2\right) = n\pi,\tag{8.9}$$

is obtained from which

$$t_n = f_o\frac{T}{B}\left(-1+\left(1+\frac{nB}{Tf_o^2}\right)^{1/2}\right),\tag{8.10}$$

with IDT finger locations at

$$x_n = vt_n.\tag{8.11}$$

Finally, the limits on integer n are $T(f_o - B/4)$ on the low-frequency side and $T(f_o + B/4)$ on the high-frequency side of the chirp response, giving a total number of fingers N (not pairs) in the linear FM chirp IDT as

$$N = 2Tf_o + 1.\tag{8.12}$$

Example 8.1 Linear FM Chirp Filter Parameters. A linear FM chirp IDT is to be fabricated on YZ-lithium niobate for operation at a center frequency $f_o = 20$ MHz. Required time-bandwidth parameters are $T = 10\,\mu s$ and $B = 10$ MHz, together with a metallization ratio $\eta = 0.5$. Determine: (a) the time-bandwidth product; (b) the total number of fingers in the IDT; (c) the finger positions x_n; (d) the IDT finger width at the low-frequency end of the band; and (e) the IDT finger width at the high-frequency end. ■
Solution. (a) The time-bandwidth product is $TB = (10 \times 10^{-6}) \times (10 \times 10^6) = 100$.
(b) From Eq. (8.12) $N = 2Tf_o + 1 = [\{2 \times (10 \times 10^{-6}) \times (10 \times 10^2)\} + 1] = 401$.
(c) From Eqs. (8.10) and (8.11), $x_n = vt_n = 3488t_n$ gives

$$x_n = 0.069\left[-1+\left(1+0.0025n\right)^{1/2}\right].$$

The values of n are obtained from the limits at $-T(f_o - B/4)$ and $+T(f_o + B/4)$ so that $-175 \le n \le 225$. (d) At the low-frequency end the synchronous frequency is $f_{sL} = f_o - B/2 = 15$ MHz, giving the finger width (= finger gap) at that end as $\lambda/4 = v/f_{sL} = 3488/(4 \times 15 \times 10^6) = 58\,\mu m$. (e) Similarly, at the high-frequency end the finger width (= finger gap) $= \lambda/4 = 3488/(4 \times 25 \times 10^6) = 34\,\mu m$, ($1\,\mu m = 10^{-4}$ cm). ■

8.3.3. FREQUENCY RESPONSE OF THE CHIRP IDT

Having determined the fingers positions of the chirp IDT from Eqs. (8.10) and (8.11) it is a straightforward matter to calculate the frequency response of the chirp filter. In terms of the delta-function relationship and notation of Eq. (3.2) in Chapter 3, this can be expressed as either

$$H_c(f) = \sum_{n=-(N-1)/2}^{+(N-1)/2} (-1)^n A_n f^{1/2} e^{-j\beta x_n} \qquad (8.13\text{a})$$

for constant apodization, or

$$H_c(f) = \sum_{n=-(N-1)/2}^{+(N-1)/2} (-1)^n A_n e^{-j\beta x_n}, \qquad (8.13\text{b})$$

for 3/2 weighting, where $H_c(f)$ is the frequency response of the chirp IDT, in the absence of any circuit-factor loading (corresponding to the array factor $A(f)$), β = phase constant and A_n is normalized to $A_n = \pm 1$ for delta-function modelling, with one delta function per electrode. In Eq. (8.13a) the $f^{1/2}$ factor is a consequence of energy conservation. This term is removed if an $f^{-3/2}$ finger weighting is applied as indicated in Eq. (8.13b). The response magnitude $|H_c(f)|$ can be readily determined in either instance, using delta-function modelling with one delta function per single-electrode. The plot shown in Fig. 8.9 is for the apodized weighting response of Eq. (8.13b). This is for a linear FM chirp IDT with centre frequency $f_o = 100$ MHz, dispersion $T = 4\,\mu$s, bandwidth $B = 20$ MHz and $N = 801$ electrode fingers. Note the slight increase in the amplitude response with increasing frequency. Math-

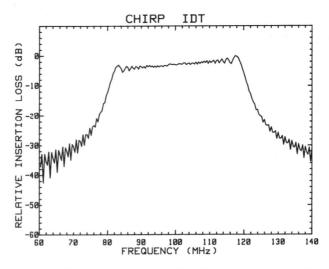

Fig. 8.9. Computed frequency response of 100-MHz linear FM chirp IDT, using delta-function modelling. The IDT has design parameters $B = 40$ MHz, $T = 4\,\mu$s, and $N = 801$ fingers.

ematically, this can be attributed to the increase in the delta-function density with increasing frequency.

The overall filter response must include that of the output IDT. In addition, circuit-factor loading and other second-order effects will degrade the passband response. If desired, the slope of the magnitude response can be adjusted by varying the chirp IDT finger weighting around the nominal $f^{-3/2}$ scaling choice.

8.3.4. THE SAW CHIRP FILTER AS A PLURALITY OF BANDPASS FILTERS

The linear FM SAW chirp filter can be regarded as a plurality of linear-phase SAW filters, of differing center frequency that are connected in parallel. When the structure is excited at a frequency f_i, the SAW energy is radiated principally from that region of the IDT where the interdigital period corresponds to (v/f_i). The "effective number" of finger pairs N_e excited by this signal can be estimated empirically as those for which the total phase error over the "active" region must not exceed 180°. In this manner, N_e may be approximated around midband as [18]

$$\frac{N_e}{N_p} = \frac{f_i}{f_o\sqrt{(TB)}}, \tag{8.14}$$

where N_p = total number of electrode pairs and f_o = center frequency. In actuality there will be fewer participating finger pairs at the band edges. This explains the relatively broad transition bands illustrated in Figs. 8.6 and 8.9.

8.3.5. INPUT ADMITTANCE OF A CHIRP FILTER

In many cases it is necessary to know, or specify, the input admittance $Y_c(f)$ of the SAW chirp filter. As the equivalent input impedance circuit in Chapter 4 is quite general, it can include the linear FM SAW chirp filter so that

$$Y_c(f) = G_{ac}(f) + j2\pi f C_{Tc} + jB_{ac}(f) \tag{8.15a}$$

$$\approx G_{ac}(f) + j2\pi f C_{Tc}, \tag{8.15b}$$

where $G_{ac}(f)$ is the radiation conductance of the linear FM chirp filter and C_{Tc} is its capacitance. In Eq. (8.15b) as well as in Chapter 4, it is assumed that the radiation susceptance $B_{ac}(f)$ is either much smaller than C_{Tc} or incorporated into that term.

If the chirp filter is regarded as a plurality of bandpass filters, then Eqs. (8.15a) and (8.15b) may be used to approximate $G_{ac}(f)$ at midband $f = f_o$. To do this, recall that the unperturbed radiation conductance $G_{al}(f)$ for a linear-phase IDT at center frequency is

$$G_{al}(f_o) \approx 8K^2 f_o C_s N_p^2 \quad (mho), \tag{8.16}$$

where C_s = capacitance/finger pair and N = total number of uniform finger pairs. In this way Eq. (8.14) is substituted into Eq. (8.16) to estimate $G_{ac}(f_o)$ for the chirp filter as

$$G_{ac}(f_o) \approx 8K^2 f_o C_s N_e^2 \approx 8K^2 f_o C_s \frac{N_p^2}{(TB)}, \tag{8.17}$$

or

$$G_{ac}(f_o) \approx \frac{G_{al}(f_o)}{(TB)} \quad (same\ IDT\ finger\ numbers!). \tag{8.18}$$

Equation (8.18) relates the unperturbed radiation conductances of a linear FM chirp IDT and a linear-phase one when they have: a) the same apodization width; b) the same number of finger pairs; and c) the same centre frequency. If they have the same fingers numbers, however, they will not have the same fractional bandwidth.

Instead of using Eq. (8.18) to estimate the radiation conductance, an alternative method is to consider that the two IDTs have the same fractional bandwidth ($\Delta f/f_o$): in which case they will have differing numbers of IDT electrodes as sketched in Fig. 8.10. Given the fractional bandwidth of the chirp IDT we can determine the number of finger pairs in the hypothetical linear-phase IDT. (For example, $N_p = 2$ in a linear-phase delay line filter with BW% = 50%). Next, obtain the radiation conductance $G_{al}(f_o)$ for the hypothetical linear-phase filter. Finally, approximate the radiation conductance of an unapodized chirp IDT with the same fractional bandwidth as

$$G_{ac}(f_o) \approx G_{al}(f_o)(TB) \quad (same\ fractional\ bandwidth!). \tag{8.19}$$

It may also be shown [17] that the radiation Q_r values of unapodized chirp and linear-phase IDTs are the same if they have the same acoustic bandwidth and acoustic aperture. With $Q_r = 2\pi f_o C_T / G_a(f_o)$ where $C_T = C_s N$ = total IDT capacitance, this means that the total capacitance C_{TC} of the chirp filter is greater than that of a corresponding linear-phase IDT by the time bandwidth product, or

$$\begin{matrix} (chirp\ IDT) \Rightarrow \\ (linear\ IDT\ with\ same\ \Delta f/f) \Rightarrow \end{matrix} \frac{C_{TC}}{C_T} = (TB). \tag{8.20}$$

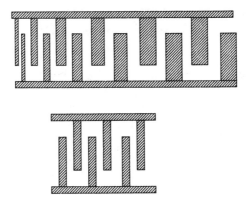

FIG. 8.10.   Linear FM chirp IDT and linear phase uniform IDT, requiring differing numbers of fingers for the same fractional bandwidth.

8.4.   Variable Delay Lines Using a SAW Chirp Filter

8.4.1.   REQUIREMENTS FOR NON-DISPERSIVE PHASE RESPONSE

Delay lines with programmable control of differential delay time τ_d can find many useful applications in radar and communications circuit instrumentation. The circuit now examined employs two linear FM chirp filters to attain nondispersive variable delay of a time-varying signal [18]. This type of circuit can find ready application as an adaptive filter for cancelling multipath effects in satellite-to-ground communications circuits [19].

First, consider what would happen if an attempt was made to implement variable delay with a single linear FM SAW chirp filter, as sketched in Fig. 8.11(a). Here, the input signal is mixed with a local oscillator, so that the difference frequency is in the range accommodated by the chirp filter. The delay can be varied over a total differential range $\Delta\tau_d$ by varying the frequency of the local oscillator. As shown, a second mixer is required to restore the original signal frequency at the output. While this simple circuit does indeed provide differential delay, the problem with it is that it is dispersive. However, this does not pose any problems if the input signal is a CW one. Its dispersion renders it unsuitable for variable delay of a digitally modulated waveform because different frequency components of the emerging waveform would be delayed by differing amounts, causing it to become distorted.

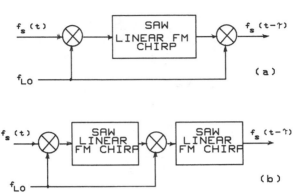

FIG. 8.11.　(a) Circuit with single linear FM SAW chirp filter provides variable delay, but with undesired dispersion. (b) Use of two linear FM SAW chirp filters, combined with intermediate spectral inversion, can provide variable delay without dispersion.

8.4.2. VARIABLE DELAY WITHOUT DISPERSION

To provide distortion-free pulse transmission it is necessary to use two series-connected linear FM SAW chirp filters as shown in Fig. 8.11(b), in conjunction with a common voltage-controlled local oscillator (VCO). In this arrangement, the input signal is up-converted in the first mixer stage and down-converted in the second mixer stage. This second conversion spectrally inverts the output from the first chirp filter. In this way, and although both chirp filters have the same chirp slope, the second chirp filter serves to remove the residual dispersion introduced by the first [19].

While the two chirp filters in Fig. 8.11(b) have the same chirp slope and chirp direction, they may not have the same time-bandwidth products. This is due to the fact that the input chirp filter design has to accommodate the bandwidth of the input signal as well as the frequency range of the VCO. As a result, the time bandwidth product of the input chirp filter will normally be larger than the second one. If a large variable delay range is required it may also be necessary to employ reflective array compressors (RACs)[2] as the chirp components in the circuit [20]. This point is illustrated in Example 8.3 which directly follows Example 8.2.

Example 8.2 Local Oscillator Frequency for Linear FM SAW Chirp Filter. A linear FM SAW chirp filter is to be used as a variable delay line

---

[2] The reflective array compressor (RAC) is a sophisticated version of the SAW chirp filters considered here, with *TB*-products as high as 10,000, as used in high-resolution radar systems [9].

in a satellite-to-ground communications link, with center frequency $f = 250\,\text{MHz}$, bandwidth $B = 40\,\text{MHz}$ and dispersion time $T = 10\,\mu s$. It is used in the variable delay line circuit of Fig. 8.11(a). The signal to be delayed is a CW one centered at frequency $f_s = 2\,\text{GHz}$. Determine the frequency requirements for the local oscillator in this circuit. Assume that the inherent linear delay between phase centers of input and output IDTs is $7\,\mu s$. ■

Solution. We can select the local oscillator frequency f_{LO} to be above or below that of the signal frequency f_s. The latter choice in this instance yields the local oscillator frequency $f_{LO} = 2\,\text{GHz}-250\,\text{MHz} = 1.75\,\text{GHz}$. In this quiescent condition, the midportion of the chirp filter will be "activated," so that the inherent delay through the filter will be $7\,\mu s$ in this state. To obtain a total differential delay of $10\,\mu s$, the local oscillator frequency will have to be tuned over the range given by $f_{LO} \pm \Delta f/2 = 1.75\,\text{GHz} \pm 20\,\text{MHz}$, so that the output signal delay will range from 2 to $12\,\mu s$ over the oscillator tuning range. ■

Example 8.3 SAW Delay Line for Adaptive Cancellation of Multipath Signals Using Two-Stage SAW Down-Chirp Filters.

A variable SAW delay line employs two linear FM RAC chirp filters in the cascaded arrangement of Fig. 8.11(b). It is included in the 70-MHz IF stages of a satellite-to-ground communications system, to provide for adaptive cancellation of multipath signals arriving at the receiver up to $30\,\mu s$ after the direct signal. The two chirp structures are in the form of down-chirp SAW reflective array compressors (RACs) because of the large time-bandwidth product required. If the input IF signal range is to be $f_s = 70 \pm 5\,\text{MHz}$, determine: (a) the minimum bandwidth required for input RAC-1; (b) its minimum dispersion; as well as (c) its minimum time-bandwidth (TB) product. ■

Solution. (b) The minimum bandwidth $(B_1)_{min}$ of RAC-1 will be required to accommodate the input signal range Δf_s plus the frequency range Δf_{VCO} of the VCO, so that $(B_1)_{min} = \Delta f_s + \Delta f_{VCO}$. For minimum TB product, $\Delta f_s = \Delta f_{VCO}$, giving $(B_1)_{min} = (10 + 10) \times 10^6 = 20\,\text{MHz}$ for RAC-1.

(b) To determine the minimum dispersion time $(T_1)_{min}$ for RAC-1, a total differential delay $\Delta T_d = 30\,\mu s$ is required for a frequency excursion $\Delta f_{VCO} = 10\,\text{MHz}$. However, allowance most be made for opposite frequency excursions of f_s and f_{VCO}. As a result, the required minimum dispersion of RAC-1 will be given by $(T_1)_{min} = 2 \times \Delta T_d = 2 \times 30 \times 10^{-6} = 60\,\mu s$. c) The minimum time-bandwidth product for input RAC-1 is $(T_1 B_1)_{min} = (60 \times 10^{-6}) \times (20 \times 10^6) = 1200$. ■

Example 8.4 SAW Delay Line for Adaptive Cancellation of Multipath Signals Using Two-Stage SAW Down-Chirp Filters—Time Bandwidth Calculations for Output RAC.

For the variable delay line adaptive

cancelling of Example 8.3, determine the minimum time-bandwidth parameters for the design of output RAC-2. ■

Solution. For the design of RAC-2 the only concern is with the frequency excursions Δf_{VCO} of the local oscillator. The minimum bandwidth requirement for RAC-2 is $(B_2)_{min} = \Delta f_{VCO} = 10\,MHz$. Because the chirp slope μ of RAC-2 must be the same as for RAC-1 for nondispersive delay, the dispersion of RAC-2 will be just one-half that for RAC-1, giving $(T_1)_{min} = 30\,\mu s$. The time bandwidth product is $(T_2 B_2)_{min} = (30 \times 10^{-6}) \times (10 \times 10^6) = 300$.

The midband frequencies of RAC-1 and RAC-2 will not be equal. That for RAC-2 will be just the same as the IF frequency, namely, 70 MHz in this example. For RAC-1, however, the midband frequency will be the sum or difference between the local oscillator and IF frequencies. If the midpoint of the VCO is 165 MHz, the mid-band of RAC-1 will be 165–70 = 95 MHz, if the difference frequency is selected from the input mixer. ■

8.5. Summary

This chapter has introduced the slanted-finger SAW IDT and the linear FM chirp IDT, together with some aspects of the application of these structures to mobile communications systems. To date, these structures have been designed for Rayleigh-wave propagation. The slanted-finger IDT structure may be employed in the design of wideband IF filters, with low levels of passband ripple, such as for CDMA applications. As will be examined in more detail in Part 2, it has also been applied to the design of one-port SAW resonators for increasing the tuning range of a voltage-controlled oscillator, over that attainable with a conventional one-port SAW resonator design.

Circuit features and operation of the basic linear FM SAW chirp filter were then examined. It was noted that these types of filters can be employed in satellite on-board SAW chirp Fourier transform (CFT) preprocessors, for high-selectivity frequency demultiplexers. As examined in Part 2, other applications include chirp focussing transducers for convolvers in spread-spectrum modules for indoor mobile communications. As illustrated in actual examples, SAW chirp filters can also find application as delay-line circuits, where it is necessary to add time delays between diversity antennas to cater to multipath compensation [16].

8.6. REFERENCES

1. C. K. Campbell, Y. Ye and J. J. Sferrazza Papa, "Wide-band linear phase SAW filter design using slanted transducer fingers," *IEEE Trans. on Sonics and Ultrasonics,* vol. SU-29, pp. 224–228, 1982.

2. P. M. Naraine and C. K. Campbell, "Wide band linear phase SAW filters using apodized slanted fingers," *Proc. 1983 IEEE Ultrasonics Symp.*, vol. 1, pp. 113–116, 1983.

3. N. J. Slater and C. K. Campbell, "Improved modelling of wide-band linear phase SAW filters using transducers with curved fingers," *IEEE Trans. on Sonics & Ultrasonics*, vol. SU-31, pp. 46–50, 1984.

4. H. Yatsuda, "Design techniques for SAW filters using slanted finger interdigital transducers," *"Proc. IEEE Trans. Ultrason., Ferroelec., Freq. Contr.*, vol. 44, 1997 (To be published).

5. E. V. Bausk and I. B. Yakovkin, "Withdrawal weighted fan-shaped SAW transducers," *"Proc. IEEE Trans. Ultrason., Ferroelec., Freq. Contr.*, vol. 4, pp. 164–167, March 1995.

6. K. Yamanouchi, T. Matsudo and M. Takeuchi, "Wide bandwidth SAW resonators and VCO using slanted interdigital transducers," *Proc. 1992 IEEE Ultrasonics Symp.*, vol. 1, pp. 57–60, 1992.

7. J. D. Maines, "Surface Wave Devices For Radar Equipment," in H. Matthews (ed.), *Surface Wave Filters*,, Wiley, New York, Chapter 13, 1977.

8. E. Brookner, *Radar Technology*, Artech House, Dedham, pp. 175–180, 1977.

9. R. C. Williamson, "Reflection Grating Filters," in H. Matthews (ed.),*Surface Wave Filters*,, Wiley, New York, Chapter 9, 1977.

10. R. C. Williamson and H. I. Smith, "The use of surface-elastic-wave reflection gratings in large time-bandwidth pulse-compression filters," *IEEE Trans. Microwave Theory Tech.*, vol. MTT-21, pp. 195–205, April 1973.

11. H. M. Gerard, "Surface Wave Interdigital Electrode Chirp Filters," in H. Matthews (ed.), *Surface Wave Filters*,, Wiley, New York, Chapter 8, 1977.

12. P. M. Bakken, A. Rønnekleiv and B. R. Andersen, "Frequency multiplexers and demultiplexers based on the SAW chirp Fourier transformer combined with digital signal processing," *Proc. 1993 IEEE Ultrasonics Symp.*, vol. 1, pp. 137–142, 1993.

13. R. C. Peach, "SAW based systems for communications satellites,", *Proc. 1995 IEEE Ultrasonics Symp.*, vol. 1, pp. 159–166, 1995.

14. H. P. Grassl and H. Engan, "Small-aperture focusing chirp transducers vs. diffraction-compensated beam compressors in elastic SAW convolvers," *IEEE Trans. on Sonics and Ultrasonics*, vol. SU-32, pp. 675–684, September 1985.

15. A. Fauter, L. Reindl, R. Weigel, P. Russer and F. Seifert, "Miniaturized SAW convolver for indoor mobile communication," *Proc.1993 IEEE Ultrasonics Symp.*, vol. 1, pp. 73–77, 1993

16. S. G. Gopani, J. H. Thompson and R. Dean, "GHz SAW delay line for direct sequence spread spectrum, CDMA in-door communication system," *Proc. 1993 IEEE Ultrasonics Symp.*, vol. 1, pp. 89–93, 1993.

17. C. S. Hartmann, D. T. Bell, Jr. and R. C. Rosenfeld, "Impulse model design of acoustic surface-wave filters," *IEEE Trans. Microwave Theory and Techniques*, vol. MTT-21, pp. 162–175, April 1973.

18. D. P. Morgan, *Surface Wave Devices For Signal Processing*, Elsevier, New York, 1985.

19. J. Burnsweig, US Patent No. PD 71403, 1972.

20. V. S. Dolat and R. C. Williamson, "A continuously variable delay-line system," *Proc. 1976 IEEE Ultrasonics Symposium*, pp. 419–423, 1976.

—9—

IDT Finger Reflections
and Radiation Conductance

9.1. Introduction

Surface acoustic wave reflections from the edges of the metallic electrode fingers of an IDT can have a significant influence on its operation, as well as on overall SAW device response. These reflections can either be associated with Rayleigh waves (i.e., "true" SAW) or pseudo-SAW ones. (In the remainder of this chapter the designation "SAW" will be used to apply to either wave type.) These reflections can either be desirable or undesirable, depending on the application. As discussed in Chapter 6, an undesirable feature can be degradation of both the amplitude and phase response in some filter designs, unless minimized, such as through the use of split electrodes. In recent years, however, finger reflections have been used to advantage in the design of RF front-end filters, antenna duplexers, and RF interstage filters employing LSAW substrates.

SAW reflections from the finger edges of an IDT affect its operation in two important ways. First, they are accompanied by a reduction in the average SAW velocity under the IDT, which will also reduce its center frequency. More importantly, however, they give rise to a distortion of its radiation conductance. In a uniform IDT, for example, the distortion will result in an asymmetric conductance with a peak at a lower frequency than for the unperturbed $|(\sin X)/X|^2$ response considered so far. For typical metallization thicknesses, this distortion may often be neglected in the design of filters operating below about 100 MHz. Because the influence of finger reflections increases with frequency, however, they can have a drastic effect on both IDT response and overall filter response. As already mentioned here, such finger reflections are paramount to the operation of leaky-SAW RF resonator-filters and antenna-duplexer ladder filters; they will be examined in Part 2 of this book.

The *film-thickness ratio* h/λ for the metallization of an IDT is an important SAW design parameter, particularly for devices on LSAW substrates.

This chapter examines this parameter and its influence on the radiation conductance of an IDT. In addition, the Mason equivalent circuit model—which was introduced in Chapter 4—will be further examined for incorporation of finger-reflection effects.

9.2. IDT Film-Thickness Ratio and Radiation Conductance

In SAW filter fabrication, thin-metal films are deposited to form the IDTs. While these metal films define IDT geometry and provide electrical contact, they also need to be light enough so that they do not dampen the surface wave excessively. Because of its low density, aluminum (or an aluminum alloy) is the preferred metal for such fabrications. Its thickness h is usually in the range $h = 500$–2000 Å (1 Å $= 10^{-8}$ cm), to provide both a good electrical contact and low resistance.

The film-thickness ratio h/λ, (where $\lambda =$ acoustic wavelength), is an important design parameter when finger reflections are of concern. (This ratio is a small-valued one so it is often expressed as a percentage $h/\lambda \times 100\%$.) The effects of electrode metallization on the IDT frequency response can often be neglected when $h/\lambda \ll 1\%$. While this may be typical when operating frequencies are below about 100 MHz, it can have a significant influence on filter response at gigahertz frequencies, particularly in the current 1.8–1.9 GHz spectral range for cordless telephones and PCS.

A large film-thickness ratio will affect the response characteristics of an IDT in two ways. First, the average SAW velocity under the IDT will be reduced from the free-surface value and will result in a reduction in its center frequency. Moreover, acoustic reflections from—and between IDT—fingers will give rise to a change in its radiation conductance. In a uniform IDT, for example, the radiation conductance will no longer have the idealized $|(\sin X)/X|^2$ frequency response considered in Chapter 4.

Although not strictly feasible, for illustrative purposes it is assumed here that IDT radiation conductance can be described in *three* variant forms, as an aid to modelling an IDT frequency response. These are identified here with the nomenclature $G_a(f)$, $G_{am}(f)$, and $G_{amf}(f)$, where

(a) $G_a(f)$ is the radiation conductance in the absence of any impedance or velocity perturbations and the SAW velocity has the free-surface value v_o throughout;

(b) $G_{am}(f)$ relates to an average SAW velocity v_a under both metallized and unmetallized regions of the IDT, but neglects finger reflections. This response is, therefore, just a replica of $G_a(f)$, which is downward-shifted in frequency.

(c) $G_{amf}(f)$ incorporates the effects of both average velocity shift v_a and finger reflections, and results in a radiation conductance with an asymmetric frequency response.

In reality, of course, $G_{af}(f)$ and $G_{amf}(f)$ cannot be separated as simplistically as considered here.

9.3. Reflections from IDTs with Single-Electrode Geometries

The IDTs examined here are those for a uniform (i.e., nonchirped) SAW filter, with the single-electrode (i.e., solid-electrode) geometry of Fig. 9.1 and metallization ratio $\eta = 0.5$. Let us consider the receiver IDT by itself, and determine what spurious effects it will have on the passage of surface waves emanating from the transmitter IDT.

First, the actual mass of the metal electrodes will dampen the surface wave. This damping can be reduced by using thin-film electrodes made of a light metal, indicating why aluminum is normally preferable to gold for the thin-film metallization of IDTs. Because minimum damping of the surface wave is the desired goal, metal electrode films should be as light as possible; operationally, however, their electrical resistance must not be significant or it will contribute to circuit-factor loading $C(f)$ and insertion loss. In consequence, typical aluminum- film thicknesses used in IDT fabrication are in the range of 500–2000 Å ($1\,\text{Å} = 10^{-8}\,\text{cm.}$)

To first order, the fractional velocity decrease $(dv/v)_m$ due to this mass-loading in the metallized regions can be expressed in terms of the film-thickness ratio h/λ, such that

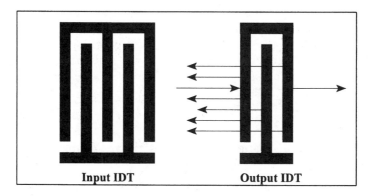

| Input IDT | Output IDT |

FIG. 9.1. Illustrating SAW reflections at output IDT. Interelectrode reflections not shown.

$$\left.\frac{dv}{v}\right|_m \approx 2\pi F\left(\frac{h}{\lambda}\right) + \frac{0.5K^2}{\left(1 + 0.5K^2 + 1/\varepsilon_r\right)}, \qquad (9.1)$$

for thin films with $h/\lambda < 0.01$, where $h =$ metal-film thickness, $\lambda =$ acoustic wavelength, $K^2 =$ electromechanical coupling constant, $\varepsilon_r =$ substrate relative permittivity (dielectric constant), and $F =$ a constant for the metal employed, with values $F = 0.037$ for lithium niobate and $F = 0.01$ for quartz [1], [2]. Unless otherwise stated, the metallization is taken to be aluminum or an aluminum alloy.

Also considered here is an additional perturbation that relates to the fractional velocity decrease $(dv/v)_p$ of the SAW under metallized regions due to the "shorting" of surface piezoelectric fields given by

$$\left.\left|\frac{dv}{v}\right|\right._p = \frac{K^2}{2}. \qquad (9.2)$$

As well as causing mass-loading, however, the deposited metal film will also change the effective "stiffness" of the propagating surface [3]. The fractional velocity change $(dv/v)_s$ accompanying this stiffness perturbation will depend on both the metal and the piezoelectric substrate employed. It can actually cause the SAW velocity to increase, in competition with the change due to mass loading. The total fractional velocity perturbation (dv/v) at and under the metal fingers may therefore be expressed as

$$\frac{dv}{v} = \left.\frac{dv}{v}\right|_p + \left.\frac{dv}{v}\right|_m + \left.\frac{dv}{v}\right|_s. \qquad (9.3)$$

The first term on the right-hand side of Eq. (9.3) will be the dominant one for piezoelectric substrates such as lithium niobate, with relatively large values of K^2. For ST-X quartz, on the other hand, velocity perturbations will be due largely to the other two terms. These perturbations will cause a shift of the center frequency of the IDT. For ST-X quartz it has also been found that recessed aluminum transducers are very effective in reducing acoustic reflections, with much lower acoustic losses than with other configurations [4].

9.4. Impedance Discontinuities in Equivalent SAW Transmission Line

If propagation of the forementioned surface waves is modelled in terms of an equivalent SAW transmission line, it will be appreciated that velocity changes at such IDT surface discontinuities will also change the equivalent electrical impedance parameters. In determining these, caution must be exercised in applying standard RF transmission-line concepts to the modelling of equivalent SAW transmission lines. This can be particularly misleading when determining the signs of the reflection coefficient terms. This results from the complex nature of the E-fields at the piezoelectric surface due to the SAW, in contrast to the relatively simple transverse electromagnetic (TEM) wave propagating in an RF coaxial transmission line. With this in mind, apply the relationship for (voltage) reflection coefficient ρ to a SAW transmission line discontinuity imposed by deposition of metal fingers or strips (with $\eta = 0.5$), as sketched in Fig. 9.2(a). At the boundary between Regions 1 and 2 in Fig. 9.2(b), the (voltage) reflection coefficient ρ_{12} at the leading metal edge, for surface waves incident from the left, is

$$\rho_{12} = \frac{Z_2 - Z_1}{Z_2 + Z_1}, \tag{9.4}$$

where $Z_1 =$ equivalent characteristic impedance of the "unperturbed" SAW transmission line, and Z_2 is the perturbed value under the metallized layer. Likewise, the reflection coefficient ρ_{21} at the back edge is

$$\rho_{21} = \frac{Z_1 - Z_2}{Z_1 + Z_2}, \tag{9.5}$$

so that $\rho_{12} = -\rho_{21}$. Because the metallization ratio $\eta = 0.5$ considered here corresponds to a finger width of $\lambda/4$ at center frequency, the sign of reflection coefficient ρ_{21} will be the same as that for ρ_{12} at this frequency, when

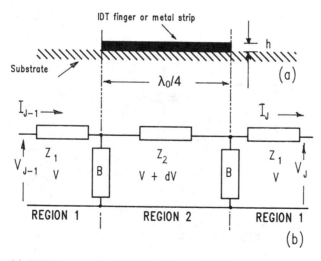

Fig. 9.2. (a) IDT finger, or metal strip, with metallization ratio $\eta = 0.5$; (b) section of equivalent SAW transmission-line model, with characteristic impedances Z_1 and Z_2. B is an equivalent susceptance, relating to energy storage at discontinuities.

both reflection coefficient phasors are referred to the region 1–2 boundary. Similarly, all other metal fingers in the receiving IDT will be in-phase at center frequency, so that the reflected SAW signal components will be additive, as sketched in Fig. 9.3. The reflections will be maximized at center frequency. Away from center frequency the two reflection phasor components are no longer in alignment, so that the resultant phasor magnitude decreases as a function of frequency.

Because these reflected waves can be further reflected from the input IDT, spurious passband ripples of amplitude and phase can ensue in the same manner as for triple-transit interference. Multiple reflections can also occur between fingers of the same IDT, giving rise to spurious resonances.

The impedance perturbation in the metallized region will be small if the metal fingers are thin and have low density (e.g., aluminum or copper-aluminum alloy), or if they are buried in quartz [4]. For this condition set $Z_1 \approx Z_2$ in the denominator of Eqs. (9.4) and (9.5). In the numerators of these equations, however, the sign of the reflection coefficients will be dependent on whether Z_1 or Z_2 is larger for the device under study. In extending this transmission line analogy, recall that the characteristic impedance Z_o of a lossless electrical transmission line is

$$Z_o = \sqrt{\frac{L}{C}} \quad (ohms), \tag{9.6}$$

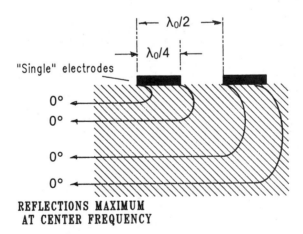

**REFLECTIONS MAXIMUM
AT CENTER FREQUENCY**

FIG. 9.3. Reflections from fingers of a solid-electrode IDT with metallization ratio $\eta = 0.5$ are maximum at centre frequency. Phase references are shown relative to $0°$.

where L = distributed series inductance/unit length (H/m), while C = distributed shunt capacitance/unit length (F/m). Further, the phase velocity v is given by

$$v = \frac{1}{\sqrt{LC}} \quad (m/s).$$ (9.7)

Equations (9.6) and (9.7) may be combined to eliminate L, giving

$$Z_o = \frac{1}{vC}.$$ (9.8)

The product term vC in Eq. (9.8) dictates the value of Z_o.

In translating this result to an equivalent SAW transmission line, it is seen that the characteristic impedance Z_2 in the metallized region of Fig. 9.2(a) will increase or decrease relative to that for the unperturbed regions, depending on how the product vC changes as a result of the metallization. Due to the anisotropic nature of piezoelectric substrates, the interrelationship between the two will, of course, be a complex one. As a result, the impedance discontinuity parameters are often inferred from experiment. For open aluminum strips on YZ-lithium niobate, for example, $Z_2 > Z_1$ in Fig. 9.2, the sign of the reflection coefficient ρ_{12} in Eq. (9.4) is positive at the leading finger edge, yielding

$$\rho_{12} = \frac{\Delta Z}{2Z_1},$$ (9.9)

where $\Delta Z = Z_2 - Z_1$. When the reflection coefficient at the back edge of the electrode is included, the resultant reflection coefficient ρ_s for the single electrode (with $\eta = 0.5$) at center frequency is obtained as

$$\rho_s = 2\rho_{12} = \frac{\Delta Z}{Z_o},$$ (9.10)

if we neglect multiple reflections between IDT fingers.

To a good approximation, the ratio of reflected SAW power P_r to incident power P_i on an array of N single electrodes in Fig. 9.1 at center frequency is

$$\frac{P_r}{P_i} = 10 \log_{10} \tanh^2 |N\rho_s| \quad (dB),$$ (9.11)

for metallization ratio $\eta = 0.5$. Figure 9.4 shows a plot of Eq. (9.11) for representative values of ρ for single-electrode IDTs on YZ-lithium niobate and ST quartz. For high-K^2 piezoelectric substrates, such as lithium niobate,

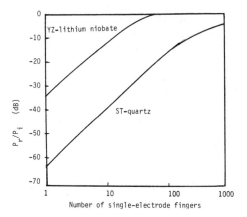

FIG. 9.4. Power reflection from solid-electrode IDTs on Y-Z lithium niobate and ST-quartz, for $\eta = 0.5$.

it can be deduced that significant acoustic power will be reflected for only 10 electrodes, compared with 100 electrodes on ST-X quartz for the same power value.

As considered in Part 2 of this book, the operation of surface wave reflection gratings is based on reflections of surface waves from piezoelectric-surface discontinuities. Such reflections are also *vital* to the desired operation of many leaky-SAW RF resonator-filters and leaky-SAW ladder filters for antenna duplexers, employing IDTs with large values of film-thickness ratio.

9.5. Self-Coupling and Mutual-Coupling Coefficients

9.5.1. SELF-COUPLING COEFFICIENT K_{11}

Equation (9.3), relating the SAW change under the metallized surface of a piezoelectric substrate, is usually reformulated as a second-order function of the film-thickness ratio (h/λ), such that

$$\left[\frac{dv}{v}\right] = \left[\frac{\Delta v}{v}\right]_p + \left[\frac{\Delta v}{v}\right]_m \bullet \left(\frac{h}{\lambda}\right) + \left[\frac{\Delta v}{v}\right]_s \bullet \left(\frac{h}{\lambda}\right)^2. \qquad (9.12)$$

Parameter values in Eq. (9.12) are inferred from velocity measurements on a given piezoelectric substrate, as functions of film-thickness ratio (h/λ). From this, the self-coupling coefficient κ_{11} is defined as

$$k_{11} = k_o \left|\frac{dv}{v}\right| = \left|k_{11p} + k_{11m} + k_{11s}\right|, \qquad (9.13)$$

where k_o = wave vector (i.e., phase constant β_o) at center frequency. Note that the magnitude of (dv/v) is employed in Eq. (9.13), because (dv/v) is negative, and k_{11} is taken as a positive term. This is also called a *velocity-shift* coefficient, as it relates to the decrease in the center frequency of an IDT due to surface loading. Self-coupling coefficient k_{11} is often given in frequency-normal*ized* fashion as k'_{11}, so that

$$k'_{11} = \frac{k_{11}}{k_o} = \left| k'_{11p} + k'_{11m} + k'_{11s} \right|. \tag{9.14}$$

Table 9.1 lists k'_{11} design parameters for illustrative piezoelectrics for Rayleigh-wave and leaky-SAW propagation.

9.5.2. MUTUAL-COUPLING COEFFICIENT K_{12}

Table 9.1 also lists the components of the normalized *mutual-coupling* coefficient k'_{12}. This coefficient relates the coupling between oppositely directed surface waves in a reflection grating or in an IDT. It is employed in coupling-of-modes (COM) analyses of SAW transducers and reflection gratings, as introduced in Eq. (4.53) of Chapter 4. The frequency-

TABLE 9.1

DESIGN DATA FOR RAYLEIGH-WAVE AND LEAKY-SAW PIEZOELECTRIC SUBSTRATES (Aluminum Metallization, with $\eta = 0.5$ and Short-Circuit Gratings)

Wave Type	Rayleigh	Rayleigh	Rayleigh	Leaky-SAW	Leaky-SAW
Substrate	ST-X Quartz [5], [6]	YZ-LiNbO$_3$ [5], [6]	128° YX-LiNbO$_3$ [5], [6]	64° YX-LiNbO$_3$ [7], [8]	36° YX-LiTaO$_3$ [7], [8]
SAW velocity v_o (m/s)	3158	3488	3997	4742	4212
Propagation axis	X	Z	X	X	X
Electromechanical coupling coefficient K^2	0.0016	0.045	0.056	0.113	0.047
c_o Capacitance/ finger pair/unit length (pF/cm)	0.503	4.5	5.0	—	—
k'_{11p}	0.0004	0.018	0.022	0.052	0.0076
k'_{11m}	0.02(h/λ)	0.30(h/λ)	0.091(h/λ)	0.18(h/λ)	−0.0011(h/λ)
k'_{11s}	7.9(h/λ)^2	—	—	1.4(h/λ)^2	3.6(h/λ)^2
TCD (ppm/°C)	≈0	94	75	70	35
k'_{12p}	0.0001	0.0054	0.0064	0.0091	0.0069
k'_{12m}	0.16(h/λ)	0.08(h/λ)	0.14(h/λ)	0.48(h/λ)	0.12(h/λ)
k'_{12s}	—	—	—	—	2.8(h/λ)^2

normalized mutual-coupling coefficient k'_{12} may also be expressed as a function of film-thickness ratio, namely,

$$k'_{12} = \frac{k_{12}}{k_o} = \left| k'_{12p} + k'_{12m} + k'_{12s} \right|.$$ (9.15)

9.6. Effects on IDT Radiation Conductance

The average shifted-velocity v_a under an IDT due to metallization can be expressed in the form

$$v_a = v_o \left(1 - k'_{11} \right),$$ (9.16)

where v_o = free-surface SAW velocity. This gives the average shifted center frequency f_a of a uniform IDT as

$$f_a = \frac{v_a}{\lambda_o},$$ (9.17)

where λ_o = physical wavelength (period) of the mechanical IDT structure.

In the absence of any velocity perturbations, the radiation conductance will be given by Eq. (4.31) of Chapter 4 as

$$G_a(f) = 8N^2 G_o \left| \frac{\sin\left[N\pi \left(f - f_o \right) / f_o \right]}{\left[N\pi \left(f - f_o \right) / f_o \right]} \right|^2$$ (4.31)

where $N = N_p$ = number of finger pairs, f_o = center frequency = v_o/λ_o and v_o = free-surface SAW velocity. The effects of shifted center frequency f_a may be incorporated to give the associated radiation conductance $G_{am}(f)$ as

$$G_{am}(f) = 8N^2 G_o \left| \frac{\sin\left[N\pi \left(f - f_a \right) / f_a \right]}{\left[N\pi \left(f - f_a \right) / f_a \right]} \right|^2.$$ (9.18)

Note carefully that v_a represents an average velocity shift for the metallized and nonmetallized regions of an IDT or reflection grating test structure . This velocity cannot be used in instances where we specifically need to know the actual perturbed velocity v_m under a metallized region or IDT finger. As will be shown, such a situation arises for generalized Mason transmission-line modelling of the interdigital transducer under the influence of acoustic finger reflections.

Example 9.1 Velocity Shift Due to Metallization—IDT on ST-X Quartz Substrate. An IDT is fabricated on an ST-X quartz substrate for use as an IF filter in a mobile communications receiver [9]. If the IDT was originally designed to operate at a center frequency $f_o = 251.9$ MHz with aluminum metallization of negligible thickness, what will be the average shifted velocity v_a and shifted centre frequency f_a if an aluminum film-thickness ratio $h/\lambda = 0.02$ is subsequently employed? ■

Solution. At centre frequency the acoustic wavelength is $\lambda_o = v/f_o = 3158/(251.9 \times 10^6) = 12.5367 \times 10^{-6} = 12.54 \mu m$. From Table 9.1, the normalized self-coupling coefficient $k'_{11} = (0.0004) + (0.02 \times 0.02) + (7.9 \times 0.02^2) = 0.004$. From Eq. (9.16) the average shifted velocity $v_a = v_o \times (1 - k'_{11}) = 3158 \times (1 - 0.004) = 3145$ m/s. The shifted center frequency is $f_a = v_a/\lambda_o = 3145/(12.54 \times 10^{-6}) \approx 250.8$ MHz. ■

Example 9.2 Shifted Radiation Conductance of IDT Due to Velocity Shift. A uniform IDT with $N_p = 200$ finger pairs is fabricated on an ST-X quartz substrate with a finger-pair period of $\lambda_o = 12.54 \mu m$. Aluminum metallization is employed with film-thickness ratio $(h/\lambda_o) = 0.02$. The acoustic aperture is $W = 20 \lambda_o$. Determine: (a) the "unperturbed" IDT center frequency f_o; (b) the average shifted SAW velocity v_a due to the metallization; and (c) the average shifted center frequency f_a. In addition, (d) determine and plot the radiation conductance function in $G_{am}(f)$ Eq. (9.18). ■

Solution. (a) The "unperturbed" center frequency $f_o = v_o/\lambda_o = 3158/(12.54 \times 10^{-6}) = 251.9$ MHz. (b) From Table 9.1 and Eq. (9.15), and using $(h/\lambda_o) = 0.02$, the normalized self-coupling coefficient $k'_{11} = 0.0004 + (0.02 \times 0.02) + 7.9 \times (0.02)^2 = 0.004$. From Eq. (9.16) the averaged shifted SAW velocity due to metallization is $v_a = v_o \times (1 - k'_{11}) = 3158 \times (1 - 0.004) = 3145$ m/s. (c) The average shifted center frequency (excluding the effect of finger reflections is $f_a = v_a/\lambda_o = 3145/(12.54 \times 10^{-6}) = 250.9$ MHz. (d) From Eq. (4.16) the free-surface characteristic conductance is $G_o = 1/Z_o = K^2 C_s f_o$, where $K^2 = $ electromechanical coupling constant and $C_s = C_o W = $ static capacitance of *one* periodic section. From Table 9.1 for ST-X quartz, $K^2 = 0.0016$, $C_s = (0.503 \times 20 \times 12.54 \times 10^{-6}) = 1.261 \times 10^{-14}$ F, so that $G_o = 0.0016 \times (1.261 \times 10^{-14}) \times (251.9 \times 10^6) = 5.083 \times 10^{-9}$ mho. From Eq. (9.18) the shifted radiation conductance is $G_{am}(f)$; it is plotted for $N = N_p = $ number of finger pairs $= 200$ in Example 9.2 that directly follows.

Fundamentals of Surface Acoustic Waves and Devices

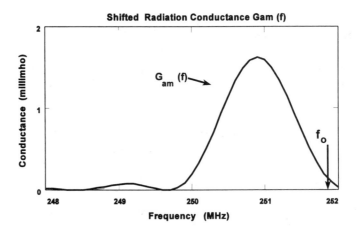

EXAMPLE 9.2. Shifted radiation conductance $G_{am}(f)$, with peak at 250.9 MHz, due to velocity shift alone. This is just a shifted replica of conductance around f_o, with no finger reflections. In reality, $G_{am}(f)$ cannot be truly separated from $G_{amf}(f)$, which also incorporates IDT finger reflections. Vertical scale in millimhos. ■

9.7. The Equivalent Circuit for an IDT Section with Negligible Finger Reflections

Chapter 4 introduced the three-port admittance matrix [Y] and the crossed-field model for an interdigital transducer. In this initial treatment it was tacitly assumed that acoustic finger reflections were negligible, so that the free-surface SAW velocity v_o applied to both metallized and unmetallized regions, with acoustic characteristic impedance Z_o. Figure 9.5 shows the associated equivalent circuit[1] for this simplified situation, applied to a half-wavelength section around a single electrode, with metallization ratio $\eta = 0.5$ [10]. Distance $L = \lambda_o/2$ where $\lambda_o = v_o/f_o =$ wavelength at unperturbed IDT center frequency f_o. The acoustic transmission line segment between ports 1 and 2 is represented by a lumped equivalent-T network, with purely reactive element values $Z_a = Z_b = jZ_o \tan(\theta/2)$ and $Z_c = jZ_o \csc(\theta/2)$. Electrical transit angle $\theta = \pi(f/f_o) = 2\pi(L/\lambda)$, where $f = v_o/\lambda$. The transformer, with turns ratio $\phi:1$ represents the electromechanical transduction from the excited electrode, where $\varphi = W(d/s)$ and $W =$ acoustic aperture, while d and s are appropriate piezoelectric and compliance constants [11]. In what follows, Port 3 is considered to be a completely electrical one with a normalized turns ratio of $1:1$.

---

[1] This equivalent circuit can be designated as either the Mason equivalent circuit or the Smith equivalent circuit.

The individual crossed-field admittance matrix elements for a single solid-electrode section in Fig. 9.5 are given by [10], [11]

$$\begin{pmatrix} i_1' \\ i_2' \\ i_3' \end{pmatrix} = [y] \begin{pmatrix} v_1' \\ v_2' \\ v_3' \end{pmatrix} \tag{9.19}$$

using *small-y* admittance notation for a single section, where

$$[y_a] = \begin{pmatrix} y_{11} & y_{12} & y_{13} \\ y_{21} & y_{22} & y_{23} \\ y_{31} & y_{32} & y_{33} \end{pmatrix} = \frac{j}{Z_o} \begin{pmatrix} -\cot\theta & \operatorname{cosec}\theta & -\tan\dfrac{\theta}{2} \\ \operatorname{cosec}\theta & -\cot\theta & -\tan\dfrac{\theta}{2} \\ -\tan\dfrac{\theta}{2} & -\tan\theta/2 & 2\tan\dfrac{\theta}{2} + \omega Z_o C \end{pmatrix} \tag{9.20}$$

In Eq. (9.20), the 2×2 acoustic sub-matrix $[y_a]$ given by

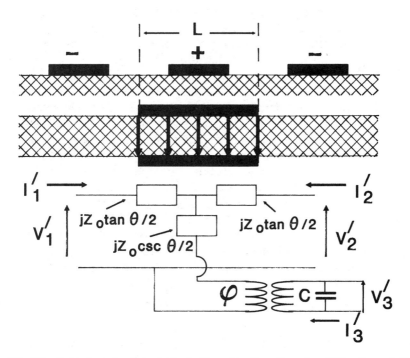

FIG. 9.5.   Equivalent circuit model for half-wavelength section of an IDT, with negligible acoustic reflections from finger discontinuities.

$$[y_a] = \begin{pmatrix} y_{11} & y_{12} \\ y_{21} & y_{22} \end{pmatrix} = \frac{j}{Z_o} \begin{pmatrix} -\cot\theta & \operatorname{cosec}\theta \\ \operatorname{cosec}\theta & -\cot\theta \end{pmatrix} \qquad (9.21)$$

represents an equivalent SAW transmission-line section for an IDT finger.

The preceding admittance relationships were derived for an IDT single finger section where finger reflections were considered to be negligible. In such a situation, the characteristic impedances of both metallized and unmetallized regions of an electrode section have the same value, leading to the simplified Mason equivalent circuit and acoustic transmission line representation in Fig. 9.5. In the following section modifications are considered that apply when finger reflections are nonnegligible, such that metallized and unmetallized regions have differing characteristic impedance. For ease of illustration it is assumed here that the effects of energy storage under discontinuities can be reglected.

9.8. The Equivalent Circuit for an IDT Section with Finger Reflections

Figure 9.6 shows the Mason equivalent circuit for a single-electrode section of an IDT, for the case where finger reflections are nonnegligible [12]–[14]. Again, the metallization ratio is taken as $\eta = 0.5$. From transmission line concepts it is known that wave reflections from an obstacle are caused by impedance discontinuities. The same situation prevails for SAW reflections from a metallization strip. Thus, both the SAW velocity and transmission line characteristic impedance have differing values in the metallized and nonmetallized regions. In modelling terms this means that three transmission line equivalent circuit representations must be considered for each finger section instead of just one as in Fig. 9.5.

The SAW velocity v_m (not v_a here!) under a metallized region of a single electrode will be reduced by the metal overlay. From Eq. (9.8) this implies that the characteristic impedance $Z_{om} = 1/G_{om}$ of the metallized region will be greater than Z_o, the free-surface value. The associated SAW velocity v_m under the metal strip can be derived from a knowledge of the values of both v_a, the *average* shifted velocity in the IDT structure, and f_a the average shifted center frequency. Because f_a will be known, and given by

$$f_a = \frac{1}{2}\left(\frac{\lambda_o}{4v_m} + \frac{\lambda_o}{4v_o} \right)^{-1}, \qquad (9.22)$$

a value for the metallization velocity v_m can be derived. This gives the shifted center frequency f_m for the metallized strip as $f_m = v_m/\lambda_o$.

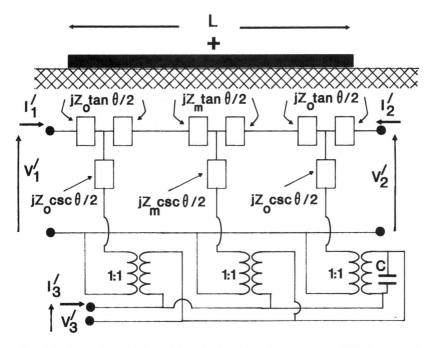

FIG. 9.6. Equivalent circuit model for half-wavelength section of an IDT, incorporating effects of acoustic reflections from finger discontinuities.

Next, and in order to calculate the overall response of the three acoustic transmission submatrices in Fig. 9.6, *ABCD* matrix manipulations must be employed, because Z or Y matrices cannot be directly multiplied. Figure 9.7 recalls the *ABCD* matrix representation of a transmission line section of length d, with characteristic impedance Z_o and complex propagation constant $Y = \alpha + j\beta$. Figure 9.8 illustrates its application to cascaded two-port networks. Returning to Fig. 9.6, the transit angle θ is obtained for each of the two unmetallized regions *of* characteristic impedance Z_o as

$$\theta = \frac{\pi}{4}\frac{f}{f_o}, \tag{9.23}$$

while the transit angle for the metallized strip region of characteristic impedance Z_{om} is

$$\theta_m = \frac{\pi}{2}\frac{f}{f_m}. \tag{9.24}$$

$$\begin{bmatrix} A & B \\ C & D \end{bmatrix} = \begin{bmatrix} \cosh(\gamma d) & Z_o \sinh(\gamma d) \\ \dfrac{1}{Z_o} \sinh(\gamma d) & \cosh(\gamma d) \end{bmatrix}$$

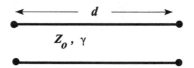

FIG. 9.7. The *ABCD* matrix representation of a transmission-line section of length d, with characteristic impedance Z_o and propagation constant γ.

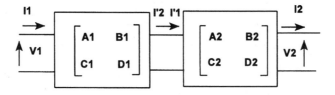

$$\begin{bmatrix} V1 \\ I1 \end{bmatrix} = \begin{bmatrix} A1 & B1 \\ C1 & D1 \end{bmatrix} \begin{bmatrix} A2 & B2 \\ C2 & D2 \end{bmatrix} \begin{bmatrix} V2 \\ I2 \end{bmatrix}$$

FIG. 9.8. The *ABCD* matrix evaluation of cascaded two-port networks.

If an unmetallized region is assumed to be a lossless transmission line segment, it can be represented mathematically by an *ABCD* matrix $[R_u]$, where

$$[R_u] = \begin{bmatrix} A_u & B_u \\ C_u & D_u \end{bmatrix} = \begin{bmatrix} \cosh(j\theta) & Z_o \sinh(j\theta) \\ \dfrac{1}{Z_o} \sinh(j\theta) & \cosh(j\theta) \end{bmatrix}. \tag{9.25}$$

The corresponding matrix $[R_m]$ for a lossless metallized-strip region may be represented by

$$[R_m] = \begin{bmatrix} A_m & B_m \\ C_m & D_m \end{bmatrix} = \begin{bmatrix} \cosh(j\theta_m) & Z_{om}\sinh(j\theta_m) \\ \dfrac{1}{Z_{om}}\sinh(j\theta_m) & \cosh(j\theta_m) \end{bmatrix}. \quad (9.26)$$

From these, the total submatrix $[R_t]$ for the three cascaded sections of transmission line in Fig. 9.6 is obtained as

$$[R_t] = [R_u][R_m][R_u]. \quad (9.27)$$

9.9. Admittance Matrix for Entire IDT with Finger Reflections

Equation (9.27) gives, as $[R_t]$, the effective *ABCD* matrix for the three segments of equivalent transmission line for the single-electrode region of Fig. 9.6. To obtain the total 2×2 *ABCD* matrix [Q] for the equivalent transmission line of the complete IDT with N_p finger pairs, the matrix in Eq. (9.27) is cascaded to obtain

$$[Q] = \begin{bmatrix} Q_{11} & Q_{12} \\ Q_{21} & Q_{22} \end{bmatrix} = [R_u]^{2N_p}. \quad (9.28)$$

Working backwards, and employing *ABCD*-to-*Y* matrix conversions, the acoustic submatrix $[Y_a^f]$ for the entire IDT is

$$[Y_a^f] = \begin{bmatrix} Y_{11}^f & Y_{12}^f \\ Y_{21}^f & Y_{22}^f \end{bmatrix} = \begin{bmatrix} \dfrac{Q_{22}}{Q_{12}} & \dfrac{-1}{Q_{12}} \\ \dfrac{-1}{Q_{12}} & \dfrac{Q_{11}}{Q_{12}} \end{bmatrix}, \quad (9.29)$$

for the general case. Equation (9.29) may now be incorporated into the total 3×3 admittance matrix for the IDT to determine its radiation conductance, which can be in the form $G_a(f)$, $G_{am}(f)$ or $G_{amf}(f)$.

9.10. IDT Radiation Conductance Obtained from 3×3 Admittance Matrix

So far in the considerations of IDT admittance-matrix relationships here, a lossless system has been assumed, as evidenced by the fact that all the matrix elements in Eq. (4.19) involve *trigonometric* (not hyperbolic) functions. Rewriting these lossless relationships here for convenience gives

$$[Y] = \begin{pmatrix} Y_{11} & Y_{12} & Y_{13} \\ Y_{12} & Y_{11} & -Y_{13} \\ Y_{13} & -Y_{13} & Y_{33} \end{pmatrix}$$

$$= \begin{pmatrix} -jG_o \cot(N\theta) & jG_o \operatorname{cosec}(N\theta) & -jG_o \tan\left(\dfrac{\theta}{4}\right) \\[2ex] jG_o \operatorname{cosec}(N\theta) & -jG_o \cot(N\theta) & jG_o \tan\left(\dfrac{\theta}{4}\right) \\[2ex] -jG_o \tan\left(\dfrac{\theta}{4}\right) & jG_o \tan\left(\dfrac{\theta}{4}\right) & j\omega C_T + j4NG_o \tan\left(\dfrac{\theta}{4}\right) \end{pmatrix}$$

$$(9.30)$$

Acoustic reflections from IDT fingers can lead to internal resonances and associated losses. Dealing with such losses requires inclusion of an attenuation coefficient α in a general $[Y^f]$-matrix representation for the IDT with finger reflections. In this book, the author has approximated this as

$$\begin{bmatrix} I_1 \\ I_2 \\ I_3 \end{bmatrix} \approx [Y^f] \begin{bmatrix} V_1 \\ V_2 \\ V_3 \end{bmatrix}, \qquad (9.31)$$

where

$$[Y^f] \approx \begin{pmatrix} Y_{11}^f & Y_{12}^f & -jG_{om}\tanh(\alpha + j\theta_m) \\[1.5ex] Y_{21}^f & Y_{22}^f & jG_{om}\tanh(\alpha + j\theta_m) \\[1.5ex] -jG_{om}\tanh(\alpha + j\theta_m) & jG_{om}\tanh(\alpha + j\theta_m) & j\omega C_T + j4NG_{om}\tanh(\alpha + j\theta_m) \end{pmatrix}$$

$$(9.32)$$

and 2×2 acoustic submatrix terms are given by Eq. (9.29). Solving Eqs. (9.31) and (9.32) for the radiation conductance $G_{amf}(f)$, with the application of matched boundary conditions such that $I_1 = -V_1/Z_o = -V_1 G_o$ and $I_2 = -V_2/Z_o = -V_2 G_o = I_1$ yields

$$G_{amf}(f) = Real\left[\frac{I_3}{V_3}\right] \approx Real\left[\frac{2Y_{31}^2}{\left(G_o + Y_{11}^f - Y_{12}^f\right)}\right]. \qquad (9.33)$$

Example 9.3 illustrates the asymmetric frequency response of $G_{amf}(f)$ that results from finger reflections. In examining this response, the reader may note that the peak of $G_{amf}(f)$ will occur at frequency $f_m = v_m/\lambda_o$. This is as far below the average centre frequency f_a as the unperturbed centre frequency f_o is above it!

Example 9.3 Radiation Conductance of IDT with Finger Reflections. The 200 finger-pair uniform IDT in Example 9.1 has an acoustic aperture $W = 20\lambda_o$. Apply the forementioned Mason equivalent circuit relations to obtain: (a) the shifted radiation conductance $G_{amf}(f)$ with finger reflections, with attenuation coefficient $\alpha = 0.00075$; (b) the shifted radiation conductance $G_{am}(f)$ without finger reflections. ∎

Solution. (a) From Table 9.1 for ST-X quartz, $K^2 = 0.0016$ and $C_o = 0.503 \times 10^{-10}$ F/finger pair/unit length. This gives $C_s = C_o W = 1.261 \times 10^{-14}$ F. For unshifted center frequency $f_o = 251.9$ MHz. Characteristic conductance $G_o = 1/Z_o = K^2 C_s f_o = 5.083 \times 10^{-9}$ mho. For $v_o = 3158$ m/s and $k'_{11} = 0.004$, obtain $v_a = v_o \times (1 - k'_{11}) = 3145$ m/s in Eq. (9.16). From Eq. (9.22), obtain $v_m = 3134$ m/s and $f_m = 250$ MHz. Solve Eq. (9.33) using $\alpha = 0.00075$, and plot $G_{amf}(f)$. Because $G_{amf}(f)$ has a much larger peak value than $G_{am}(f)$ in this example, it is plotted as $G_{amf}(f)/10$ here.

(b) To get $G_{am}(f)$ set $f_o = f_a$ in Eq. (9.23) and Eq. (9.25). Set $f_m = f_a$ in Eqs. (9.24) and (9.26). Solve Eq. (9.33) for the lossless case with $\alpha = 0$. Plot $G_{am}(f)$ and check that this is the same response function as in Example 9.2.

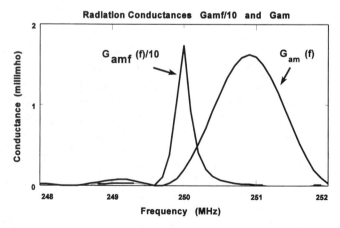

EXAMPLE 9.3. Plot of radiation conductances $G_{am}(f)$ and $G_{amf}(f)$. Vertical scale in millimhos. Vertical scale in millimhos is reduced by a factor of 10 for $G_{amf}(f)$. Compare $G_{am}(f)$ with that given in Example 9.2. ∎

9.11.   Restrictions on Frequency Response Computations

In Chapter 4, a lossless model for a Mason circuit representation of an IDT was used to obtain an admittance-parameter equation for the voltage response of a SAW filter. This was given as

$$H(f) = \frac{V_L}{V_\varepsilon} = \frac{y_{ab}R_L}{\left(1 + y_{aa}R_s\right)\left(1 + y_{bb}R_L\right) - y_{ab}^2 R_s R_L}, \tag{4.29}$$

with the admittance derived from unperturbed $G_a(f)$ radiation conductance functions appropriate to input and output IDTs. On first inspection it might seem that Eq. (4.29) could be converted directly to treat finger reflections by substituting the related $G_{amf}(f)$ functions in Eqs. (4.30), (4.32) and (4.34). While such substitutions would handle finger reflections *within* IDTs, however, this still would not deal with finger reflections *between* IDTs.

Depending on IDT separation, the transmission line section between the IDTs can act as an acoustic resonant cavity, with the IDT electrodes serving as end reflectors for the cavity. This could considerably alter the filter frequency response from that predicted by a modified equation (4.29). Indeed, such a technique has been employed to realize an additional resonance pole in the design of a 250-MHz SAW IF filter, for application to a mobile communications circuit [9].

9.12.   Illustrative Application to Design of 250-MHz IF Filter for Mobile Radio

An example of the *desirable* use of significant IDT finger reflections is illustrated in the frequency response of Fig. 9.9 [15] . This related to the design of a low-loss (~2 dB) 250-MHz IF filter on ST-quartz, as required in a mobile radio application [7]. Input/output IDTs each had 200 electrode finger pairs, with film-thickness ratio $h/\lambda = 2\%$. A distorted radiation conductance $G_{amf}(f)$ was required here to contribute to the passband shaping of the narrowband (BW ~0.18%) resonator-filter. Figure 9.9 illustrates the *desired* effects of the distorted radiation conductance on the frequency response of the input/output IDTs in the absence of end reflection gratings.

9.13.   Summary

This chapter has presented an introductory treatment on the influence of acoustic reflections within IDTs on the radiation conductance of the IDT. The resultant distortion of the radiation conductance was related to impedance discontinuities and acoustic velocity shifts imposed by the IDT. Impor-

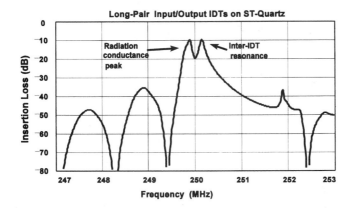

FIG. 9.9. Predicted frequency response of 250-MHz IF filter on ST-quartz, with 200 finger pairs in input/output. Film-thickness ratio h/λ = 2%. Acoustic aperture w = 20 λ. As used in the design of a three-pole IF resonator-filter for mobile phone system. Lower-frequency peak due to $G_{amf}(f)$. Higher-frequency peak due to resonance between input/output IDTs. [After Reference 15].)

tant design parameters examined were the film-thickness ratio h/λ , and frequency-normalized self-coupling and mutual-coupling coefficients k'_{11} and k'_{12}, respectively. The *ABCD* matrices were used to illustrate the modelling of an equivalent acoustic transmission line with impedance discontinuities and losses. A knowledge of such behaviour is of great importance to device design in terms of shaping the device frequency response. This includes several types of SAW/leaky-SAW RF and IF filters with low insertion loss (< ~ 3 dB) for mobile communications circuitry.

9.14. REFERENCES

1. R. H. Tancrell, "Principles of Surface Wave Filter Design", in H. Matthews (ed.), *Surface Wave Filters*, John Wiley and Sons, New York, Chapter 3, 1977.
2. M. B. Schulz and J. H. Matsinger, " Rayleigh wave electromechanical coupling constants," *Applied Physics Letters*, vol. 20, p. 367, 1972.
3. S. Datta, *Surface Acoustic Wave Devices* , Prentice-Hall, Englewood Cliffs, p. 78, 1986.
4. W. J. Tanski, "Developments in resonators on quartz," *Proc. 1977 IEEE Ultrasonics Symp.*, pp. 900–904A, 1977.
5. C. B. Saw, *Single-phase Unidirectional Transducers for Low-loss Surface Acoustic Wave Devices*, Ph.D. Thesis in Electrical Engineering, McMaster University, Hamilton, Ontario L8S 4L7, Canada, 267 pages, July 1988.
6. D-P Chen and H. A. Haus, "Analysis of metal strip SAW gratings and transducers," *IEEE Trans. Sonics and Ultrasonics*, vol. SU-32, pp. 395–408, May 1985.
7. T. Morita, Y. Watanabe, N. Tanaka and Y. Nakazawa, "Wideband low loss double mode SAW filters," *Proc. 1992 IEEE Ultrasonics Symp.*, pp. 95–104, 1992.

8. K. Yamanouchi and M. Takeuchi, "Applications for piezoelectric leaky surface waves," *Proc. 1990 IEEE Ultrasonics Symp.*, pp. 11–18, 1990.

9. Y. Yamamoto and R. Kajihara, "SAW composite longitudinal mode resonator (CMLR) filters and their application to new synthesized resonator filters," *Proc. 1993 IEEE Ultrasonics Symp.*, pp. 47–51, 1993.

10. W. R. Smith, H. M. Gerard, J. H. Collins, T. M. Reeder and H. J. Shaw, "Analysis of interdigital surface wave transducer by use of an equivalent circuit model," *IEEE Trans. Microwave Theory and Techniques*, vol. MTT-17, pp. 856–864, 1969.

11. G. W. Farnell and E. A. Adler, "An overview of acoustic surface wave technology," Final Report to Communications Research Center, Ottawa, Canada, DSS Contract 6001-3-4406, 12 August 1973.

12. W. R. Smith, H. M. Gerard and W. R. Jones, "Analysis and design of dispersive interdigital surface-wave transducers," *IEEE Trans. Microwave Theory and Techniques*, vol. MTT-20, pp. 458–471, July 1972.

13. W. S. Jones, C. S. Hartmann and T. D. Sturdivant, "Second order effects in surface wave devices," *IEEE Trans. Sonics and Ultrasonics*, vol. SU-19, pp. 368–377, July 1972.

14. W. R. Smith, "Experimental distinction between crossed-field and in-line three-port circuit models for interdigital transducers," *IEEE Trans. Microwave Theory and Techniques*, vol. MTT-22, pp. 960–964, 1974.

15. C. K. Campbell, "Scattering and transmission matrix analysis of SAW resonator filters with long-pair IDTs and triple composite longitudinal modes on quartz," *Proc. 1994 IEEE Ultrasonics Symp.*, vol. 1, pp. 309–312, 1994.

PART 2

Techniques, Devices and Mobile/ Wireless Applications

Definition
Device. **A piece of equipment or a mechanism designed to serve a special purpose or perform a specific function**

—10—

Overview of Systems and Devices

10.1. Merits of SAW and Pseudo-SAW Devices

Current mobile-telephone and wireless communications systems around the world operate in frequency bands from about 200 MHz to about 2 GHz. This includes those bands for analog and digital cellular phones and analog and digital cordless phones, as well as those for wide area networks (WAN) and wireless local-area networks (WLAN). The dramatic increase in consumer use of such products and systems has led to a corresponding upsurge in the production of SAW-device products for these systems.

SAW devices are small in size, rugged, and lightweight. Moreover, they can be designed to give exceptional electrical merits of low-loss, high-frequency and high-power capability, coupled with superior passband and stopband characteristics. Leaky-SAW resonator-filters and antenna duplexers are now increasingly dominant in the RF[1] filter stages of mobile phone circuitry, offering strong competition with dielectric filters and resonators in such applications [1]–[3]. Up to six surface-wave filters are now being used in the design of some mobile-phone circuits [4].

10.2. Frequency Bands For Mobile Communications

10.2.1. ANALOG CELLULAR COMMUNICATIONS

For reference, Table 10.1 lists *some* current RF channel allocations for *analog cellular* phone standards [5], [6]. These include the Advanced Mobile Phone System (AMPS), the Total Access Communication System (TACS), the Nippon Total Access Communications System (NTACS), the Nordic Mobile Telephone (NMT), and the Radio Telephone Mobile System (RTMS). As indicated in Table 10.1, analog cellular phones employ narrowband frequency modulation (FM), with frequency-division multiple

---

[1] The generic term "radio frequency (RF)" as used here is intended to include the VHF (30–300 MHz), UHF (300 MHz to 3 GHz), and SHF (3 GHz plus) frequency ranges.

TABLE 10.1

SOME ANALOG CELLULAR PHONE STANDARDS (After Reference [5])

Region	System	Mobile TX (MHz)	Mobile RX (MHz)	Channel Spacing (kHz)	Number of Channels
Americas & Australia	AMPS	824–849	869–894	30	832
Europe	TACS	890–915	935–960	25	1000
Europe	NMT450	453–458	463–468	25	180
Europe	NMT900	890–915	935–960	12.5	1999
France	Radiocom 2000	215.5–233.5	207.5–215.5	12.5	640
		414.8–418	424.8–428	12.5	256
Germany	C-450	450–455.74	460–465.74	10	573
Great Britain	ETACS	871–904	916–949	25	1240
Japan	NTT DoCoMo	925–940	870–885	25/6.25	600/2400
	NTT IDO	915–918.5	860–863.5	6.25	560
	NTT IDO (interleaving)	922–925	867–870	6.25	480
Japan	JTACS/NTACS, DDI	915–925	860–870	25/12.5	400/800
	JTACS/NTACS, DDI, IDO	896–901	843–846	25/12.5	120/240
	JTACS/NTACS, IDO (Interleaving)	918.5–922	863.5–867	12.5	280

access (FDMA), and a small channel separation (e.g., 30 kHz for the 800–900 MHz AMPS cellular band)[2].

10.2.2. ANALOG CORDLESS COMMUNICATIONS

Table 10.2 illustrates some pertinent specifications for the analog European Cordless Telephone, (CT1/CT1+), operating in the 800- to 900-MHz band, and for the analog Japanese Cordless Telephone (JCT) operating in the 200- to 300-MHz band. Frequency modulation is again employed, together with frequency-division duplexing. As is shown, channel spacing is 25 kHz or less.

---

[2] The RF carrier spread dictates the RF filter bandwidth requirement, while the channel spacing dictates the IF filter bandwidth requirement (in a noninterlaced system). Thus, the AMPS system in Table 10.1 has an RF bandwidth of $(849 - 824) = 25$ MHz.

TABLE 10.2

ANALOG CORDLESS PHONE STANDARDS

	Japan Japanese Cordless Phone (JCT)	Europe Cordless Telephone CT1/CT1+
Mobile frequencies	254/380 MHz	CT1: 914/960 CT2: 885/932
Multiple Access Method	FDMA	FDMA
Duplex Method	FDD	FDD
Number of channels	89	CT1: 40 CT1+: 80
Carrier channel spacing	12.5 kHz	25 kHz
Modulation type	FM	FM

10.2.3. DIGITAL CELLULAR COMMUNICATIONS

Table 10.3 gives pertinent specifications for illustrative digital cellular standards. These are: 1) the U.S. Digital Cellular (USDC) system; 2) the Japan Digital Cellular (JDC/PDC) system; and 3) the Global System for Mobile Communications (GSM). Access schemes for digital cellular systems include time-division multiple access with frequency-division multiplexing (TDMA/FDM). This European system employs an access scheme with eight users/channel, with Gaussian minimum-shift-keying (GMSK) modulation employing a 0.3 Gaussian filter.

The North American and Japanese systems shown in Table 10.3 employ a three user/channel TDMA/FDM access scheme, with π/4-shifted differential quadrature phase-shift keying (π/4-DQPSK) digital modulation. This modulation technique is based on standard QPSK modulation techniques, with the modification that each bit is shifted by π/4, to avoid zero-crossing problems. Figure 10.1(a) shows a vector signal analyzer [7] measurement on a North American Digital Cellular (NADC) π/4-DQPSK system, giving the constellation diagram points at symbol times. Figure 10.1(b) shows the measured vector diagram for this TDMA system, with the continuous trajectory of the carrier between symbol times, and resultant amplitude/phase loci. For data-burst operation with a 1/3 duty cycle for the full-burst mode, the maximum and minimum excursion points in Fig. 10.1(b) correspond to 1.6 W (or 32 dBm), and 0.4 W, respectively. The average antenna power within a burst is 0.8 W.[3]

---

[3] In logarithmic terms, 1 mW = 0 dBm, while 32 dBm = 1.6 W, etc.

TABLE 10.3

SOME DIGITAL CELLULAR PHONE STANDARDS

	Japan (JDC/PDC)	North America (USDC) IS-54	North America (USDC) IS-95	Europe (GSM)	Europe (DCS-1800) PCN
RX (MHz)	810–826	869–894	869–894	925–960	1805–1880
TX (MHz)	940–956	824–849	824–849	890–915	1719–1785
RX (MHz)	1429–1453				
TX (MHz) (Mobile range)	1477–1501				
Number of channels	1600	832	20	124	374
Carrier channel spacing (kHz)	25	30	1250	200	25
Users/channel	3	3	798	8	8
Access scheme	TDMA/FDM	TDMA/FDM	CDMA/FDM	TDMA/FDM	TDMA/FDM
Modulation type	π/4-DQPSK	π/4-DQPSK	QPSK/OQPSK	Gaussian MSK (0.3 Gaussian filter)	Gaussian MSK (0.3 Gaussian filter)

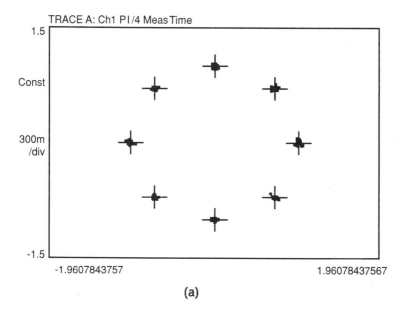

(a)

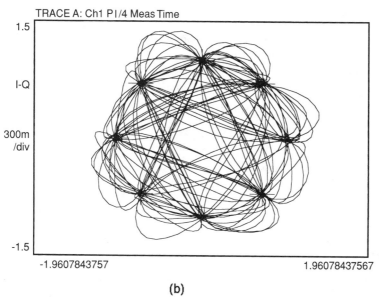

(b)

FIG. 10.1. (a) Vector signal analyzer measurement of constellation diagram at symbol times for NADC $\pi/4$ DQPSK-TDMA system. (b) Measurement of its vector diagram, showing continuous trajectory of carrier between symbol times, and amplitude and phase loci. (Courtesy of Hewlett-Packard Company.)

10.2.4.   DIGITAL CORDLESS COMMUNICATIONS

Some *digital cordless* telephone systems and relevant parameters are listed in Table 10.4. This list includes the Personal Handiphone System (PHS) in Japan, as well as Personal Access Communications Services (PACS) in the USA, and the Digital European Cordless Telecommunications (DECT) system in Europe (all operating in the 1.8- to 1.9-GHz band. Other systems include the Cordless Telephone (CT2/CT2+) system, operating in the 800- to 900-MHz band. Modulation for the PHS, CT2 and DECT systems also uses TDMA/FDM for multiple access with TDD used for the duplex method. The Japanese PHS system uses $\pi/4$-DQPSK modulation, while that for the CT2 and DECT phones uses Gaussian-filtered frequency-shift-keying (GFSK), with a 0.5 Gaussian filter.

10.2.5.   POWER AND RANGE CAPABILITIES FOR CELLULAR AND CORDLESS SYSTEMS

General distinguishing features of—and between—cellular and digital phone architectures are summarized as follows:

Cellular Phones:

- Maximal bandwidth efficiency and frequency reuse in macrocellular high-speed fading environment
- Increased complexity in terminal and base stations
- Large cell size (0.5 to 30 km)
- High-mobility capability (up to 250 km/hr)

TABLE 10.4

SOME DIGITAL CORDLESS PHONE STANDARDS (After Reference [5])

System	PHS/PHP	PACS	DECT
Region	Japan	USA	Europe
Frequency bands (MHz)	1895–1915	1850–1910/1930–1990	1880–1900
Carrier spacing (kHz)	300	300/300	1728
Number of carriers	77	16 pairs/10 MHz	10
Duplexing	TDD	FDD	TDD
Channels per carrier	4	8/pair	12
Modulation	$\pi/4$ DQPSK	$\pi/4$ DQPSK	GFSK
Handset TX power-average (mW)	10	25	10
Handset peak TX power (mW)	80	200	250

- Average handset transmit power of up to ~600 mW for AMPS, with peak powers up to 1.6 W in packet data systems, and up to 2 W for GSM
- FM modulation with frequency-division duplexing

Cordless Phones:

- High-quality speech in a quasi-static environment
- Low-complexity, low-power equipment
- Small cell size (50 to 500 m)
- Average handset transmit power of up to about 50 mW
- Modulation by time-division duplexing

10.3. Spread-Spectrum Code-Division Multiple Access (CDMA)

As consumer-communication demand continue to increase, the use of spread-spectrum techniques for mobile communications is considered to be one of the upcoming key technologies for relieving congestion of signal-traffic channels. Frequency division multiple access (FDMA) is a technique that is implemented by dividing an allocation RF spectrum into different radio channels; TDMA is a multiple access scheme that divides a radio channel into many time slots, where each carries a traffic channel. In contrast, code-division multiple access (CDMA) is a system that uses code sequences as traffic channels in a common radio channel, so that many users may use the same frequency at the same time. The CDMA licensed systems with a large bandwidth per channel (e.g., USA = 1.23 MHz within the cellular 800- to 900-MHz band) are finding increased applications for indoor mobile communications in large buildings. Unlicensed low-power spread-spectrum techniques are being increasingly employed for synchronous or asynchronous voice/data systems employing SAW convolvers [8].

Mobile spread-spectrum radio links operate at low RF power levels, with severe restrictions on such RF power levels, depending on the region involved. In Japan, for example, regulations for spread-spectrum mobile radios operating at frequencies of less than 322 MHz require their field strengths to be less than 500 μV/metre at a distance of 3 metres from the unit.

As an aid to appreciating the differences between the preceding TDD and FDD schemes, Fig. 10.2 gives artistic representations of "frequency-gap" and "time-gap" separations for licensed CDMA access—as well as for TDMA, FDMA, TDD, and TDD/TDMA.

Frequency and Time-Division Duplexing

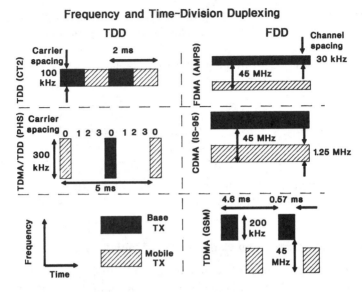

Fig. 10.2.   Illustrative FDD and TDD techniques for some analog and cordless phone systems. (After Reference [5].)

10.4. Wireless Data Systems

Wireless data systems are designed for packet-switched operation. They include mobile wide area networks (WANs) and wireless local area networks (WLANs). Wide area licensed networks include Advanced Radio Data Information Service (ARDIS), MOBITEX, and Cellular Digital Packet Data (CDPD) with mobile units operating with up to 4W of radiated power, and channel transmission rates up to about 20kbit/s.

The WLAN systems include unlicensed low-power spread-spectrum systems[4] for high-data rates (generally greater than 1 Mb/s, with radiated powers less than 1mW) and short-range indoor applications, where penetration through building floors is desirable. Their use has grown dramatically in recent years, as computing resources become more decentralized. In the United States there are a number of products available that operate in unlicensed Industrial, Scientific, and Medical (ISM) frequency bands. Table 10.5 gives some illustrative parameters for two WAN systems, as well as for European and Japanese WLAN networks operating in the 2.4-GHz band. As another example of the miniaturization offered by the use of SAW

---

[4] Chapter 11 examines the use of SAW resonator devices in unlicensed low-power wireless bands that do not employ spread-spectrum techniques.

TABLE 10.5

EXAMPLES OF WAN AND WLAN DATA-TRANSFER SYSTEMS

	RAM — Mobitex (WAN)	Ardis-RD-LAP (WAN)	IEEE 802.11 (WLAN)
Mobile frequencies	(North America) RX: 935-941 MHz TX: 896-902 MHz (Europe/Asia) 403-470 MHz	RX: 851-869 MHz TX: 806-824 MHz	(North America & Europe) 2400-2483 MHz (Japan) 2470-2499 MHz
Number of channels	480	720	FHSS: 79 DSSS: 7
Carrier channel spacing	12.5 kHz	25 kHz	FHSS: 1 MHz DSSS: 11 MHz
Multiple Access Method	TDMA/FDM	TDMA/FDM	CSMA
Duplex Method	FDD	FDD	TDD
Channel bit rate	8 kb/s	19.2 kb/s	1 or 2 MB/s
Modulation	GMSK (0.3 Gaussian filter)	FSK (2 and 4 level)	FHSS: GFSK (0.5 Gaussian filter) DSSS: DBPSK (1 MB/s) DQPSK: (2 MB/s)

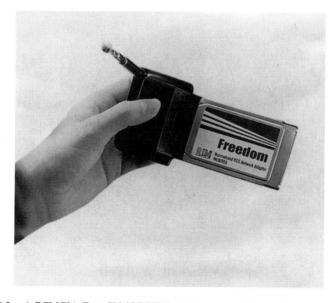

FIG. 10.3. A PCMCIA Type II MOBITEX WAN packet-data 2-W transceiver operating at 900 MHz, with leaky-SAW RF front-end filters in receiver. (Courtesy of Research in Motion, Limited, Waterloo, Ontario, Canada.)

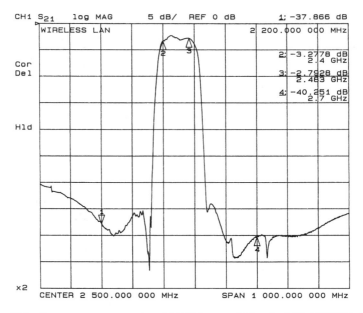

FIG. 10.4. Frequency response of an LSAW front-end filter for 2400–2483 MHz WLAN filter. Horizontal scale: Center at 2.5 GHz, span 1 MHz. Vertical scale: 5 dB/div. Insertion loss: 3.2778 dB at 2.4-GHz Marker 2; 2.7928 dB at 2.483-GHz Marker 3. Out-of-band loss: 40.251 dB at 2.7 GHz Marker 4. (Courtesy of Fujitsu Ltd., Kawasaki, Japan.)

technology, Fig. 10.3 shows a plug-in Personal Computer Memory Card International Association (PCMCIA) Type II card for a MOBITEX WAN packet-data 2-W transceiver operating at 900 MHz, with leaky-SAW RF front-end filters in the receiver. Figure 10.4 shows the frequency response of an LSAW front-end WLAN filter for the 2400- to 2483-MHz band, such as for North American and European wireless systems.

10.5. Architecture for Mobile and Wireless Systems

10.5.1. CELLULAR RECEIVER PERFORMANCE REQUIREMENTS

Here, the important requirements are reviewed for efficient receiver operation in a mobile phone system, such as an analog cellular one with the basic double heterodyne circuitry, as shown in Fig. 10.5 for a narrow-channel AMPS transceiver. The electrical and audio performance of the receiver will be governed both by 1) receiver sensitivity, and 2) receiver selectivity. While the overall receiver sensitivity will depend on the cumulative noise performance of individual components, it will be dominated by the noise

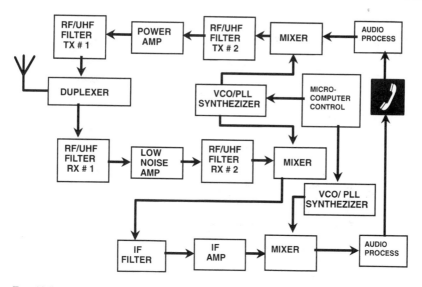

Fɪɢ. 10.5. Basic block diagram of double-heterodyne transceiver, such as for an 800-MHz band AMPS analog cellular system.

performance of the RF front-end stages. Contributions to the overall receiver noise performance will come from thermal noise, amplifier noise, filter noise, image-product noise, mixer noise, detector noise, and local-oscillator noise.[5] These will dictate the minimum required signal-to-noise ratio (SNR) at a specified point in the receiver for an acceptable quality of audio output signal.

Table 10.6 lists typical noise figures (in decibels) to be found in a receiver circuit of the type shown in Fig. 10.5. For illustrative purposes here, the antenna duplexer is considered to incorporate both the front-end transmitter and receiver filters. Note that filter insertion losses translate into equivalent noise figures (in decibels).

The maximum signal power generated at the antenna in an 800-MHz AMPS cellular system is about 600 mW. For losses in the transmit stage, this means that the TX#1 filter in the AMPS system must handle power levels of 1 W (30 dBm). Moreover, and with reference to Fig. 10.7, the front-end receiver filter RX#1 in the AMPS system would have to operate at a power level of about 28 dBm at the crossover point in its lower stopband.

---

[5] Local oscillator noise will be considered in Chapter 18.

TABLE 10.6

TYPICAL NOISE FIGURES IN A DOUBLE-HETERODYNE RECEIVER
(After Reference [9])

Component	Gain or Insertion Loss (dB)	Noise Figure (dB)
RF filter #1	−2	2
Low noise amplifier	12	3.5
RF filter #2	−2.5	2.5
Fist mixer	−8	8.3
First IF filter	−6	6
IF amplifier	20	4.0
Second (active) mixer	12	12
Audio detector		15

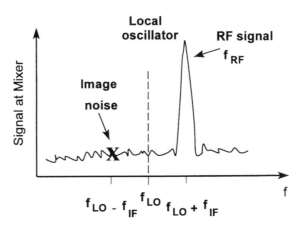

FIG. 10.6.   Image noise with low-side injection. (After Reference [9].)

10.5.2.   FILTER REQUIREMENTS IN THE CELLULAR MOBILE RECEIVER

Let us consider the role that the RF filters and IF filters play in the operation of the receiver portion of Fig. 10.5. In "conventional" antenna duplexers the signal-path separation is provided by microwave ceramic (MWC)
filters, with insertion losses of about 3.0 dB for two-pole MWCs, and losses
of about 5 dB for three-pole MWCs. For illustration here, however, the
antenna duplexer incorporates the transmit and receiver filters TX#1 and
RX#1, respectively, as would be the case for a leaky-SAW ladder-filter
duplexer.

AMPS Analog Cellular Radio

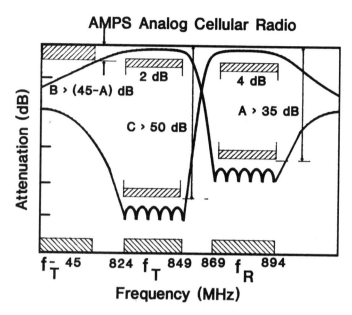

FIG. 10.7. Passband and stop-band specifications for 800-MHz band AMPS cellular radio system. (After Reference [27].)

The receiver preselector RF filter RX#1 must have multifunction capability. It should be both a low-loss and a highly selective filter to limit the bandwidth of spectra entering the follow-up low-noise amplifier (LNA). It must also suppress interference caused by the local oscillator. It is also required to suppress spurious image signals. These requirements must serve to ensure linear operation, with minimum intermodulation distortion, for a dynamic range of about 120 dB. The follow-up RF filter RX#2 is required additionally to suppress second-harmonic signals, and image-frequency noise, which could degrade the performance of the first mixer. Leaky-SAW RF resonator-filters are also becoming dominant in these follow-up stages.

The *shape factor* of the front-end transmit and receive filters becomes increasingly important as frequencies for mobile-communications operation move to ever-higher bands to handle increased subscriber capacity. (Recall that shape factor (SF) is normally defined as $SF = \Delta f_r / \Delta f$, where Δf_r is the maximum filter rejection bandwidth, and Δf is the bandwidth measured at 1-dB points.) For systems operating in the 1.8- to 1.9-GHz band the transition between transmit and receive passband edges is only about 1.1%. This requires a filter shape factor SF < ~1.5 in such systems [10].

10.5.3. IMAGE-FREQUENCY REJECTION

The IF stages are required to be very selective ones, with modest values of insertion loss, for suppressing adjacent user-channel interference, as well as higher-order intermodulation distortion. They determine the modulation bandwidth that can be received, as well as the amount of noise reaching the detector. First-IF filters are now routinely implemented with Rayleigh-wave devices on ST-quartz, with center frequencies ranging from about 45 to over 200 MHz, depending on the concept of the system[6]. Lower-frequency second-IF stages typically employ non-SAW monolithic crystal filters (MCF).

Image-frequency responses entering the IF filter stages result largely from the inherent nonlinear operation of the preceding mixer stage. The desirable IF output from the first mixer in Fig. 10.5 is the difference frequency $f_{IF} = |f_{RF} - f_{LO}|$, between the incoming RF message frequency and that of the first local oscillator signals As sketched in Fig. 10.6, however, the mixer output can also contain image noise within the signal image frequency if the latter is present at the RF input port at the mixer. Additional spurious IF responses may be introduced through the RF or local oscillator signals to a degree depending on the mixer nonlinearity and its *intercept points*[7]. These will be at frequencies f_m given by

$$f_m = n \cdot f_{LO} \pm m \cdot f_{RF}, \qquad (10.1)$$

where m and n are integers from zero upwards [9], [11]. The problem that arises from such mixing occurs when the mixer output frequency corresponds to the desired first-IF frequency f_{IF}. This will occur for spurious signals at incoming RF frequencies f_{RF1} and f_{RF2}, and for integer values of m and n in Eq.(10.1), such that [9]

$$f_{RF1} = \frac{n \cdot f_{LO} - f_{IF}}{m} \qquad (10.2)$$

$$f_{RF2} = \frac{n \cdot f_{LO} + f_{IF}}{m}. \qquad (10.3)$$

Depending on the system concept, the local oscillator signal applied to the mixer can be injected above or below the RF signal frequency. For high-side injection, for example, the two most common spurious responses entering the first-IF stage in Fig. 10.5 could be due to the image signal at $f_{LO} + f_{IF}$

---

[6] Some combined GSM/DCS1800 cellular phones employ an IF of 459 MHz.

[7] The intercept point is related to system nonlinearity and represents a fictitious level at which desired and undesired signal levels have the same amplitude.

(for $m = n = 1$ in Eq. (10.3)), as well as the " half-IF" signal at $f_{LO} - f_{IF}/2$ (for $m = n = 2$). Low-frequency first-IF stages can also be particularly susceptible to higher-order spurious RF responses where integers m and n differ only by 1 [9]. From the preceding considerations, it can be deduced that the first-IF center frequency should be such as to locate the half-IF value well outside the RF band allocated for the system design. Thus, for an 800-MHz AMPS receiver with a bandwidth allocation of 25 MHz, the first-IF frequency would normally be above 50 MHz.

The second-mixer IF output can also be degraded by spurious inputs due to *second-image* signals around the first-IF frequency. As a result, the performance of a double-heterodyne receiver may be dictated by the second-image rejection capability rather than by the adjacent-channel selectivity.

10.5.4. FILTER REQUIREMENTS IN THE CELLULAR MOBILE TRANSMITTER

The RF transmission power levels for cellular mobile phones are typically in the range of from 500 mW to 2 W, while those for cordless phones are in the 10-mW range. In order to save battery power, the RF power amplifier in the transmit stage of Fig. 10.5 may be designed to operate in Class-C mode. Thus the front-end filter TX#1 will be required to suppress both higher-harmonics and noise generated by Class-C amplifier operation that could otherwise appear in the receiver band. Figure 10.7 illustrates passband and stopband specifications for the TX#1 and RX#1 filter stages for the AMPS cellular radio system, operating in the 800-MHz band. The purpose of the interstage RF filter TX#2 is to suppress close-in noise [6].

10.5.5. SINAD SPECIFICATIONS FOR A UHF MOBILE RADIO RECEIVER

The mobile radio receiver of Fig. 10.5 must satisfy specifications for in-channel and out-of-channel performance, in terms of SINAD parameters [12]. A SINAD determination, which can be measured using a SINAD audio analyzer, is given as the ratio of {Signal(S) + Noise(N) + Distortion(N)} at the receiver output to {Noise(N) + Distortion(D)} at the same output level, so that

$$SINAD = 20 \log_{10} \left(\frac{(S + N + D)}{(N + D)} \right) \quad (in\ decibels). \quad (10.4)$$

The North American Electronics Industry Association (EIA) FM standard defines usable sensitivity as the input RF level which produces 12-dB SINAD at greater than 50% of rated audio output power. A SINAD audio analyzer acts first as a broadband voltmeter and measures the total output of the receiver. A filter in the analyzer then notches out the audio modulation—and the resultant noise plus distortion is measured.

For adequate out-of-channel interference suppression in such systems, the selectivity should be at least 70 to 90 dB for channel separations of 12.5 or 25 kHz. Specifications are typically met with the RF signal reduced to -109 dBm, corresponding to $\approx 0.8\,\mu$V across 50 Ω [13].

10.5.6. ARCHITECTURES FOR DIGITAL CELLULAR RADIOS

In conventional analog cellular radios, the radio transceiver simultaneously transmits and receives RF signal information. In digital cellular radios employing TDMA, however, the transmit bursts do not overlap the receive bursts. This leads to less demanding requirements on the rejection levels of the TX#1 and RX#1 filters, at their mutual frequency bands. Additionally, some receivers for narrowband digital receivers may only require a single IF down-conversion stage [14].

Separation between transmit and receive signal paths can be obtained with PIN diode switching [9], as simplistically sketched in Fig. 10.8(a). One alternative is to use leaky-SAW duplexer techniques, as sketched in Fig. 10.8(b) [15]. For such a SAW-based duplexer, the main objectives in designing the associated front-end filters relate to attainment of: 1) a low insertion loss; 2) large off-band rejection at image and harmonic frequencies; and 3) high impedance presented at the mutual frequency bands, as depicted in Fig. 10.8(c).

The trade-off in digital communications is between the bit-error rate (BER) and the transmission bandwidth. Figure 10.9 gives the basic functional blocks for a GSM data transceiver, which can represent a good compromise in the selection of BER and transmission bandwidth.

10.5.7. TIME-DIVERSITY RECEIVERS FOR WIRELESS COMMUNICATIONS

The "ASH" receiver sketched in Fig. 10.10 represents a radically new wireless receiver design. In contrast to the frequency-diversity receivers outlined in the preceding, the ASH receiver operates in a time-diversity mode. As a result, it dispenses with local-oscillator circuitry [16].

In the circuit of Fig. 10.10, the incoming RF signal passes through a bandpass filter, for application to the first RF amplifier. This first RF amplifier is switched on and off by a pulse generator to sample the incoming signal. The signal samples are stored in a follow-up SAW delay line, which is chosen to accept hundreds of samples per incoming data bit in a delay period of typically 0.5 μs. As this sampled signal emerges from the delay

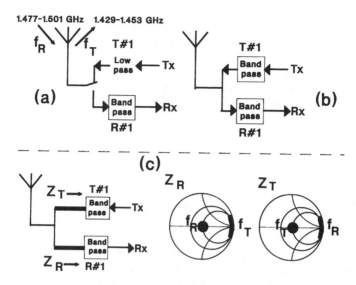

FIG. 10.8. (a) Basics of antenna duplexer stage using diode-switching of TDMA sequence. (b) Duplexer using LSAW bandpass filters in transmit and receive stages. (c) Smith chart Impedance specifications for (b)at mutual frequency bands. (After Reference [15].)

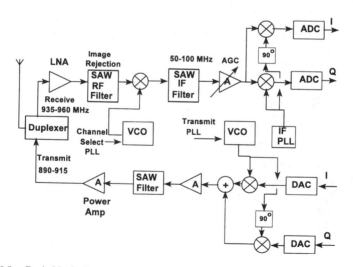

FIG. 10.9. Basic block diagram of a GSM transceiver, with I-Q modulation/demodulation.

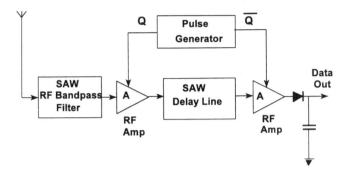

Fig. 10.10.   Block diagram of an ASH time-diversity wireless receiver, employing a low-loss SAW delay line for data-storage and transfer. (After Reference [16].)

line, the second RF amplifier is switched on while the first is switched off. In that way, there is no feedback to the first amplifier, which could otherwise cause the circuit to become unstable. Thereafter, the gating-signal is removed from the data bit using a lowpass filter after the detector. In this way, this time-diversity receiver has a sensitivity comparable to that of a single-conversion receiver, with a capability for about 100-dB rejection of undesired out-of-band signals.

10.6.   Applications of Surface Wave Devices

10.6.1.   SAW RF Resonator-Filters and Duplexers in Mobile-Phone Circuitry

To date, competing technologies for implementation of RF front-end filters are those for dielectric types involving microwave ceramic (MWC) filters and those based on SAW technology. While MWC filters exhibit very low losses, they have the disadvantage of relatively large size. In contrast, leaky-SAW resonator-filters can also enjoy low loss, coupled with better shape factors (SF) and higher close-in stopband rejection. A variety of resonator filters currently exist for application to both RF front-end and interstage filtering in transceivers. Typical fractional-bandwidth requirements range from 2 to 5%.

One notable RF resonator-filter type is that for the longitudinal-coupled dual-mode resonator-filter [17]. This filter type is fabricated on leaky-SAW substrates with large values of K^2 in order to achieve strong coupling between adjacent resonant modes. Because of limited out-of-band rejection this filter type is normally series-connected for low-loss four-pole resonator-filter operation. With their uniquely flat passband response, these filters find ready application in RX#1 and TX#1 front-end RF filtering stages for AMPS, GSM, PCN, and DECT.

Nonresonator-type RF filters include the low-loss interdigitated interdigital (IIDT) structures, which consist of a series of IDTs alternately connected to the input and output IDTs [18]. Because almost all of the acoustic power generated by the input IDT collection can be delivered to the load, if properly matched, insertion losses can be as low as 1 dB. Moreover, as there is no directly reflected signal as with simple bidirectional IDTs, this filter type will have a small passband ripple value. Because it can be difficult to obtain out-of-band rejections greater than 35 dB, however, this type is used primarily as noise-reducing filters in antenna duplexers as well as in interstage RF filters TX#2 and RX#2 in Fig. 10.5.

In conventional (non-SAW) antenna duplexers in cellular systems, both the front-end transmit and receive filters in the three-port structure are fabricated using sections of semirigid coaxial resonators employing high-permittivity ceramic dielectrics for miniaturization. In general, $4 \sim 7$ resonant sections are employed in each filter. The relatively large size of these structures can become significant, however, in the design of transceiver circuitry. To combat this, much smaller leaky-SAW-based antenna duplexers for 300- to 400-MHz transceivers have been developed with minimum insertion losses of 2 dB, with package sizes as small as $7.5 \times 8.5 \times 2.5 \, \text{mm}^3$ [19]. In addition, package sizes of $14 \times 6 \times 2 \, \text{mm}^3$ have been produced for 1.5-GHz leaky-SAW duplexers, with minimum loss of 0.8 dB for the TX#1 front-end stage [15].

10.6.2. SAW IF Filters for Mobile Phones

Low-loss ($< 5 \, \text{dB}$) and midloss ($< 10 \, \text{dB}$) IF filters are now standard components in mobile phones, with tolerance constraints favoring Rayleigh wave designs on ST-quartz. They may be employed in single- or double-conversion heterodyne receivers. Current IF centre frequencies range from 45 to over 250 MHz. The centre-frequency selection may be dictated in part by the preceding RF filter response [20]. With QPSK digital modulation, upper IF frequencies are limited by the speed of the follow-up A/D converters. With MSK or GMSK modulation, however, gigahertz-range center frequencies can be employed, using SAW injection-locked oscillators for carrier and NRZ data recovery [21], [22].

Current principal SAW IF filter designs include low-loss single-phase unidirectional transducers (SPUDTs), in-line longitudinally coupled resonator filters, waveguide-coupled (transverse mode) resonator-filters, and midloss Z-path filters. The SPUDT design aim is to minimize the passband distortion, rather than achieve very-low insertion loss. The SPUDT filters can employ open or shorted reflection grating elements, floating electrodes of open and shorted strips, or with interlaced reflectors; these filters are also the key elements in time-diversity (ASH) wireless receivers [16]. However, the SPUDT structure has a length drawback at lower IF center frequencies.

To combat size restrictions, Z-path SAW IF filters employ weakly inclined width-weighted reflectors (4°), such that the Z-path transit reduces the chip length for low frequencies. At 45 MHz, for example, a 16-DIP package can be employed with 3-dB bandwidth of 180 kHz and insertion loss of 10 dB.

In-line two-pole or cascaded four-pole SAW IF resonator filters can be employed above ~150 MHz, with a bandwidth capability of ~0.13% and insertion loss of 2.5 dB. Because of their weak mode-coupling, waveguide-coupled SAW resonator filters have small fractional bandwidths (BW < ~0.1%), coupled with steep shape factors, low loss (~1 to 4 dB) and ultimate rejection of about 70 dB, which makes them especially suitable for pager applications.

10.6.3. CODING AND CONVOLVERS IN MOBILE AND
WIRELESS COMMUNICATIONS

Pulse compression techniques with PN fixed-code SAW tapped delay lines, or arbitrarily coded SAW convolvers are suitable for combined CDMA-TDMA (TCDMA) systems, such as the 1.9-GHz PCS system in the United States. For digital voice this uses a bandwidth of ~30 MHz and a time-slot of ~5 μs for low BER and high cordless-telephone (CT) capacity. Such coding is especially suitable in buildings with small cell size (30 to 50 metres maximum), and high CT-subscriber capacity. An advantage of CDMA is in its ability to distinguish between multipaths with time delays greater than the reciprocal of the CDMA bandwidth. To combat multipath interference, broadband SAW delay lines (up to 2 μs) can be added, in conjunction with distributed antennas. As will be seen in Chapter 16, a variety of coding schemes can be used with the fixed-code PN structures, including Barker codes, quadraphase codes, and minimum-shift keying (MSK). Moreover, miniature 350-MHz SAW convolvers with an integration time $T = 3 \mu$s, bandwidth $B = 50$ MHz ($TB = 150$), and convolution efficiency $\eta_c = -70$ dBm have been developed for indoor mobile CT communications in cell sizes of up to 50 metres [23].

As described in Chapter 17, asynchronous spread-spectrum systems have also been developed that employ efficient layered-type SAW convolvers. These can be applied to: 1) operation in the United States licence-free 900-MHz spread-spectrum band (902 to 928 MHz, 1 W), using a direct-sequence code-shift-keying (DS/CSK); 2) duplex operation up to 100 metres in Japan in the very low-power licence-free spread-spectrum band, where the allowed transmitter field-strength is less than $500\,\mu V/m$ at a distance of 3 metres, in the frequency range below $322\,MHz$—corresponding to a transmitted power of less than $0.05\,\mu W/120\,kHz$; and 3) full-duplex operation at $2.45\,GHz$, in the 2-GHz United States spread-spectrum band (2.40000 to $2.4835\,GHz$, $1\,W$) [24].

10.6.4. LOCAL OSCILLATOR REQUIREMENTS

The SAW/SSBW/STW oscillators have the benefits of small size and weight, low power drain, low vibration sensitivity and aging, and frequency capabilities at selected frequencies into the gigahertz range. Depending on the application, they can operate in free-running, VCO, hybrid VCO, comb-frequency hopping, or injection-locked modes, as sketched in Fig. 10.11. Feedback elements include high-Q structures such as resonators, resonator-filters, or SPUDTs. While the phase noise of the injection-locked oscillator is primarily dependent on the phase noise of the injection signal, the phase noise of the other oscillators is dictated primarily by oscillator-loop parameters. In Rayleigh-wave oscillators on ST-quartz (limited to $< \sim 15\,dBm$), phase noise below the half-loop bandwidth is due primarily to flicker frequency, with $30\,dB/decade$ roll-off. In single-pole resonators, phase noise at 10-kHz offset can be low enough (-120 to $-150\,dBc/Hz$) to avoid adjacent-channel interference. Typical noise floors are -160 to $-170\,dBc/Hz$, with ultraprecision floors of $-184\,dBc/Hz$. In commercial

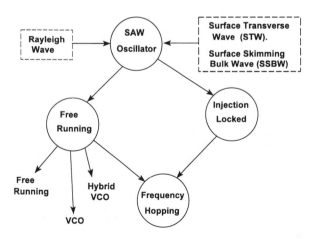

FIG. 10.11. Operational modes for free-running and controlled surface wave oscillators.

devices, the set-frequency tolerance ranges from ±30 to ±200 ppm, depending on cost. Vibration sensitivities are given as $\Delta f/f = 1 \times 10^{-9}/g$. Two- and four-pole oscillators can be designed for a 1-dB bandwidth group delay, which is flat over 400 ppm in order to handle 5-year aging [25].

The STW/SSBW oscillators on 90°-rotated ST-quartz can operate at a factor of 1.6 higher frequency, with the SAW-quartz IDT, at up to about 25 dBm. Triple-mode STW/SSBW oscillators with two STW modes and one SSBW mode have been designed with a stability of ±1.4 ppm over –45°C to +70°C [25].

Injection-locked oscillators have also been employed for carrier-recovery in digital systems operating at 1 GHz, with 80-dB dynamic range for a 2-Mb/s BPSK signal [21], [22].

10.7. Coverage of Part 2

To follow up on the basic coverage of and information on surface acoustic wave material and devices given in Part 1, Part 2 of this book examines the principles of operation of major SAW devices either currently in use in mobile and wireless communications circuitry or under intensive development for such applications.

The coverage begins in Chapter 11 with an examination of the principles of operation of surface-wave reflection gratings that operate in the fundamental mode, and their incorporation into the resonator structures that are employed in RF and IF filtering stages, as well as in local oscillator circuitry. Chapter 11 also deals with Bleustein-Gulyaev-Shimizu (BGS) resonators that have been employed as delay elements in DECT demodulator circuitry.

Chapter 12 proceeds to examine the principles of operation of unidirectional transducers, as applied here to SPUDT designs, which are used widely in mobile circuitry, as well as in time-diversity wireless receivers.

Many different types of RF filter structures are currently in use in front-end, interstage, and local-oscillator circuitry of mobile and wireless circuitry, or are under development for such applications. They are increasingly being incorporated in ASIC packages. Figure 10.12 illustrates a base-station transmitter multichip module for the North American IS-54 dual-mode cellular system. This incorporates an I-Q modulator, followed by a 50-dB output-power amplifier on a silicon bipolar ASIC, together with an LSAW four-pole longitudinally coupled resonator-filter on 64° YX-LiNbO$_3$. [26]. Figure 10.13 illustrates the corresponding RF base-station receiver multichip module, again incorporating a four-pole LSAW longitudinally coupled low-loss (3 dB) resonator-filter [26].

Chapters 13 and 14 deal with representative SAW filter designs for such

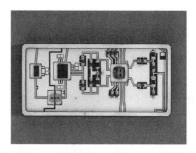

FIG. 10.12. The RF transmitter multichip module for base station in dual-mode IS-54 digital cellular system. Includes I-Q modulator, 50-dB power amplifier on silicon bipolar ASIC, and LSAW four-pole longitudinally coupled resonator-filter. (Courtesy of NORTEL, Northern Telecom, Canada.)

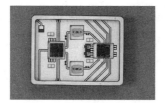

FIG. 10.13. Corresponding RF receiver multichip module for base station in dual-mode IS-54 digital cellular system. Again incorporates a low-loss (3 dB) LSAW four-pole resonator-filter. (Courtesy of NORTEL, Northern Telecom, Canada.)

applications. A possibly greater number of IF filter designs, and their variants, are also in current use in mobile and wireless circuitry.

Chapter 15 deals with a representative selection of such IF-filter designs.

Chapter 16 examines SAW-based techniques applied to fixed-code matching filters for signal extraction in spread-spectrum communications.

Chapter 17 extends this spread-spectrum study to the use of SAW-based convolvers for indoor mobile communications with cell sizes of up to about 50 metres, as well as outdoor systems employing asynchronous correlation with high-efficiency layered SAW convolvers.

Chapter 18 considers a variety of Rayleigh-wave and pseudo-SAW oscillator designs for use in fixed-frequency or VCO applications, as well as for data extraction.

Chapter 19 examines SAW filter technology as applied to representative designs for digital microwave radio, for clock recovery in fiber-optic SONET networks, and for satellite communications systems.

Chapter 20 concludes with a postscript on new techniques and developments that could be significant to the use of surface wave devices in the continuing expansion of global communications.

10.8. REFERENCES

1. K. Wakino, T. Nishikawa, H. Matsumoto and Y. Ishikawa, "Quarter wave dielectric transmission diplexer for land mobile communications," *Proc. 1979 IEEE MTT Symp.*, pp. 278–280, 1979.

2. T. Nishikawa, "RF front end circuit components miniaturized using dielectric resonators for cellular portable telephones, " *IEICE Trans*, vol. E74, pp. 1556–1562, June 1991.

3. T. Ishizaki, M. Fujita, H. Kagata, T. Uwano and H. Miyake, "A very small dielectric planar filter for portable telephones," *IEEE Trans. Microwave Theory Tech.*, vol. 42, pp. 2017–2022, November 1994.

4. H. Yatsuda, T. Horishima, T. Eimura and T. Ooiwa, "Miniaturized SAW filters using a flip-chip technique," *IEEE Trans. on Ultrasonics, Ferroelectrics, and Frequency Control*, vol. 43, pp. 125–130, January 1996.

5. J. E. Padgett, C. G. Günther and T. Hattori, "Overview of wireless personal communications," *IEEE Communications Magazine*, pp. 28–41, January 1995.

6. J. Machui, J. Bauregger, G. Riha and I. Schropp, "SAW devices in cellular and cordless phones," *Proc. 1995 IEEE Ultrasonics Symp.*, vol. 1, pp. 121–130, 1995.

7. Hewlett-Packard Company, "Using Error Vector Magnitude Measurements to Analyze and Troubleshoot Vector-Modulated Signals," Product Note 89400-14, 1996.

8. A. Fauter, L. Reindl, R. Weigel, P. Russer and F. Seifert, "Miniaturized SAW convolver for indoor mobile communication," *Proc. 1993 IEEE Ultrasonics Symp.*, vol. 1, pp. 73–77, 1993.

9. P. Vizmuller, *RF Design Guide*, Artech House, Boston, 1995.

10. K. Hashimoto, M. Ueda, O. Kawachi, H. Ohmori, O. Ikata, H. Uchishiba, T. Nishihara and Y. Satoh, "Development of ladder type SAW RF filter with high shape factor," *Proc. 1995 IEEE Ultrasonics Symp.*, vol. 1, pp. 113–116, 1995.

11. W. H. Hayward, *Introduction To Radio Frequency Design*, Prentice-Hall, New Jersey, 1982.

12. Hewlett-Packard Note, "Application and measurements of low phase noise signals using the 8662A synthesized signal generator," *Hewlett-Packard Application Note #283-1*, November 1981.

13. C. K. Campbell, J. J. Sferrazza Papa and P. J. Edmonson, "Study of a UHF mobile radio receiver using a voltage-controlled SAW local oscillator," *IEEE Trans. Sonics and Ultrasonics*, vol. SU-31, pp. 40–46, Jan. 1984.

14. K. Anemogiannis, J. Bauregger and G. Riha, "Cordless phone system architectures based on SAW filters," *Proc. 1993 IEEE Ultrasonics Symp.*, vol. 1, pp. 85–88, 1993.

15. M. Hikita, N. Shibagaki, K. Asai, K. Sakiyama and A. Sumioka, "New miniature SAW antenna duplexer used in GHz-band digital mobile cellular radios," *Proc. 1995 IEEE Ultrasonics Symp.*, vol. 1, pp. 33–38, 1995.

16. D. L. Ash, "New UHF receiver architecture achieves high sensitivity and very low power consumption," *RF Design*, pp. 32–44, December 1994.

17. T. Morita, Y. Watanabe, M. Tanaka and Y. Nakazawa, "Wideband low loss double mode SAW filters," *Proc. 1992 IEEE Ultrasonics Symp.*, vol. 1, pp. 95–104, 1992.

18. H. Yatsuda, Y. Takeuchi and T. Horishima, "Surface acoustic wave filter composed on interdigitated interdigital transducers and one port resonators," *Japan J. Appl. Phys.*, vol. 33, pp. 2979–2983, May 1984.

19. O. Ikata, Y. Satoh, H. Uchishiba, H. Taniguchi, N. Hirasawa, K. Hashimoto and H. Ohmori, "Development of small antenna duplexer using SAW filters for handheld phones," *Proc. 1993 IEEE Ultrasonics Symp.*, vol. 1, pp. 111–114, 1993.

20. C. C. W. Ruppel *et al.*, "SAW devices for consumer communications applications," *IEEE Trans. Ultrason., Ferroelec., Freq. Control*, vol. 40, pp. 438–452, Sept. 1993.

21. P. J. Edmonson, P. M. Smith and C. K. Campbell, "Injection-locking techniques for a 1-GHz digital receiver using acoustic-wave devices," *IEEE Trans. Ultrason., Ferroelec., Freq. Control*, vol. 39, pp. 631–637, Sept. 1992.

22. P. J. Edmonson, P. M. Smith, C. K. Campbell, "A SAW based carrier recovery scheme for a 915 MHz BPSK wireless system," *Proc. 16th Biennial Conference on Communications*, Queen's University, Kingston, Canada, 1992.

23. A. Fauter, L. Reindl, R. Weigel, P. Russer and F. Seifert, "Miniaturized SAW convolver for indoor mobile communication," *Proc. 1993 IEEE Ultrasonics Symp.*, vol. 1, pp. 73–77, 1993.

24. K. Tsubouchi, "An asynchronous spread-spectrum wireless modem using SAW convolver," *Proc. International Symp. on Surface Acoustic Wave Devices for Mobile Communications*, Sendai, Japan, pp, 215–222, 1992.

25. C. K. Campbell, "SAW oscillators and resonators," *Proc. International Symp. on Surface Acoustic Wave Devices for Mobile Communication*, Sendai, Japan, 1992.

26. J. Saw, M. Suthers, J. Dai, Y. Xu, R. Leroux, J. Nisbet, G. Rabjohn and Z. Chen, "SAW technology in RF multichip modules for cellular systems, " *Proc. 1995 IEEE Ultrasonics Symp.*, vol. 1, pp. 171–175, 1995.

27. M. Hikita, N. Shibagaki, T. Akagi and K. Sakiyama, "Design methodology and synthesis techniques for ladder-type SAW resonator coupled filters," *Proc. 1993 IEEE Ultrasonics Symp.*, vol. 1, pp. 15–24, 1993.

—11—

SAW Reflection Gratings and Resonators

11.1. Introduction

11.1.1. STANDING SURFACE WAVES AND RESONATOR STRUCTURES

The previous chapters have examined applications of *travelling* surface acoustic waves to SAW filter design. This chapter considers resonator applications involving *standing* surface acoustic waves in Rayleigh-wave and BGS structures.

Surface-wave reflection concepts are significantly different from those for microwave resonant cavities. In microwave resonators, total electromagnetic wave reflection at a boundary is realized by use of a metal plate. In contrast, a single reflector element on the surface of a piezoelectric substrate will not lead to total SAW reflection. Because an elliptically polarized Rayleigh wave has a 90° phase shift between horizonal and vertical surface deformations, a single metal strip on the piezoelectric surface can not simultaneously satisfy total reflection conditions for both displacement components. In practice, a SAW reflection grating may employ up to several hundred appropriately placed reflector elements to achieve near-total SAW reflection. Surface acoustic wave (SAW) resonator structures are realized by employing either one or two IDTs bounded by such reflection gratings [1], [2]. One-port SAW resonators employ a single IDT for input and output, in conjunction with two SAW reflection gratings, while two-port SAW resonators are formed using separate IDTs for input and output signals, which are also contained between SAW reflection gratings.

11.1.2. SAW RESONATORS IN WIRELESS COMMUNICATIONS CIRCUITS

One-port single-pole SAW resonators can serve as fixed-frequency replacements for bulk-wave crystal oscillators operating in high-overtone modes [3]. (The term *"pole"* used here relates to a mechanical-resonance condition, analogous to that with a single inductance-capacitance-resistance (LCR) lumped-element tuned circuit of series or parallel construction.) One-pole SAW resonators are also used as constituents of waveguide-coupled multiple-pole resonator filters, for applications in narrowband intermediate-frequency (IF) stages in analog cellular phone systems. Two-port one-pole SAW resonators, with somewhat more flexible design constraints than their one-port counterparts, are also employed in oscillator applications for both fixed-frequency and tuning applications. To date, the majority of surface wave resonator designs have been based on Rayleigh-wave types, fabricated on cuts of quartz for temperature stability.

11.1.3. RESONATORS IN LOW-POWER WIRELESS APPLICATIONS

A current major market for one-port and two-port SAW resonators has been created by their use in oscillator circuitry, for low-power unlicensed wireless applications (e.g., remote control security, automobile keyless entry, etc.) Such unlicensed systems typically transmit less than 1 mW of power, and operate over distances of from 5 to 100 metres. Table 11.1 lists some frequency-band allocations for unlicensed wireless systems in different countries. Power emission specifications for both the fundamental emission and spurious harmonics are governed by the regulatory authority of

TABLE 11.1

SOME FREQUENCIES FOR UNLICENSED LOW-POWER WIRELESS
COMMUNICATIONS (After Reference [4])

Country	Frequency (MHz)
United States	260 to 470 902 to 928
Canada	260 to 470 902 to 928
Japan	Below 322
United Kingdom	418 Automotive only: 433.92
Germany	433.92
France	223.5 to 225 Automotive only: 433.92
South Africa	403.916 and 411
Australia	303.825 and 318
Hong Kong	314

the country in question. In Japan, for example, regulations for spread-spectrum mobile radios operating at frequencies less than 322 MHz require their field strengths to be less than 500μV/metre at a distance of 3 metres from the unit.

Applications of narrowband low-power wireless systems include: a) a keyless entry systems for automobiles; b) remote controls for garage doors, keyless locks, remote lighting, etc.; and c) handheld bar-code readers for wireless. These low-power wireless systems all use digital modulation techniques. Transmission rates for these systems, with long bit sequences, are normally in the range from 150–1200 bits/s for control signals, while data rates range from 1200–9600 bits/s [4]. At the other data-rate extreme, low-power spread-spectrum data transmission at 1 Mb/s has been reported, using with coded matched filters [5].

One-port SAW resonators are normally employed in Colpitts oscillators in these low-power systems, as sketched in the elementary circuit of Fig. 11.1(a), while two-port SAW resonators (or resonator-filters) are used in Pierce oscillator circuits as sketched in Fig. 11.1(b). These SAW-based superheterodyne low-power wireless receivers can also enjoy significant improvements in sensitivities compared with conventional LC-receiver

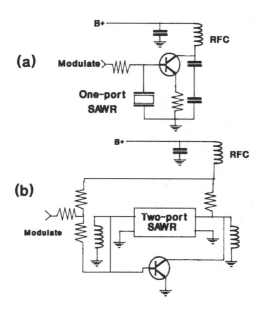

FIG. 11.1. (a) Elementary Colpitts oscillators employing a one-port SAW resonator. (b) Elementary Pierce oscillator with two-port SAW resonator (or resonator-filter in feedback loop).

types. Typical sensitivities for SAW-based wireless receivers are or the order of $-105\,\text{dBm}$ $(1.26\,\mu\text{V})$ compared to only $-80\,\text{dBm}$ $(22\,\mu\text{V})$ for LC receivers [4].

11.1.4. SCOPE OF THIS CHAPTER

Reflection gratings are essential to the operation of one-pole surface wave resonators as well as multiple-pole resonator-filters for RF and IF filter stages of mobile communications transceivers. In consequence, this chapter first reviews the basics of SAW reflection grating theory, as applied to the design of one-port single-pole SAW resonators, as well as two-port single- and multiple-pole designs. The SAW resonators considered will be those involving Rayleigh-wave propagation. Because surface transverse wave (STW) resonators are employed mainly in oscillator circuitry, consideration of them will be deferred until Chapter 18.

In this chapter two simplifications are applied in illustrating the principles of operation of one-pole Rayleigh-wave resonators. The first is that distortion of the radiation conductance by finger reflections is reglected. This is because the number of IDT fingers is normally small, so that the resonance width will be unaffected by perturbations in the broad radiation conductance. In the second simplification, it is assumed that the design frequencies have already been compensated for velocity shift, as associated with self-coupling coefficient κ'_{11} (introduced in Chapter 9).

Because of the vital importance of surface-wave resonator structures in mobile/wireless circuitry, analytic transmission-matrix and coupling-of-modes techniques are presented in some detail in this chapter, as these can be applied to a range of Rayleigh-wave and LSAW devices in RF filter, antenna duplexer and IF stages.

Somewhat idealized SAW standing-wave patterns are presented as an aid to understanding requirements for optimum placement of reflection gratings. Transmission-matrix theory is employed in deriving the frequency response of these resonators [6]. Spurious multimode resonator response is also examined, together with degradation imposed by higher-order transverse SAW modes.

11.2. SAW Reflections and Reflection Gratings

11.2.1. BASIC CONCEPTS

As depicted in Fig. 11.2, the elements of SAW/pseudo-SAW reflection gratings are composed of periodically spaced discontinuities. These can consist simply of open- or short-circuited thin metal strips (or a combina-

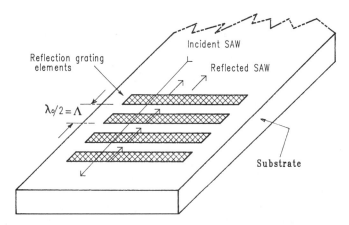

F<small>IG.</small> 11.2. Elements of a surface-wave reflection grating can consist of open- or shorted-metal strips, grooves, or pedestals.

tion of the two), deposited onto the surface of the piezoelectric substrate. Alternatively, they can be formed by etching shallow grooves (or pedestals) into the substrate surface. In each instance, the periodicity of the grating at center frequency yields a cumulative resultant reflection from the *front* edges of the discontinuities at center frequency f_o. In addition, the metallization ratio[1] η is chosen so that SAW reflections from the *back* edges of these discontinuities also add constructively at f_o. Although a metallization ratio $\eta = 0.5$ is normally used, some anomalous behaviour has been reported for open metal strips on lithium niobate [7].

The number of strips (or grooves) required for near-total reflectivity will depend on the reflection mechanisms involved. Where piezoelectric shorting is dominant, as on lithium niobate substrates, near-total SAW reflection can be attained with only a few hundred metal strips. Where mass loading dominates, as in ST-quartz substrates, up to a few thousand grating elements may be required to yield a reflection-coefficient magnitude $|\rho|$ close to unity. To keep packaging sizes realistic, SAW resonator designs on quartz substrates are normally used for frequencies above about 50 MHz.

While the reflection of SAW waves from a reflection grating will be maximal at the (centre) frequency for which all of the individual reflections are additive, there will be a narrow frequency range over which the grating will reflect SAW waves with reduced efficiency. The width of this grating

---

[1] The notations *metallization ratio* and *film-thickness ratio* are also loosely applied to depths or heights of etched grooves or pedestals on a piezoelectric substrate.

"stopband" can be specified as the frequency range over which the grating reflection coefficient magnitude $|\rho|$ exceeds some minimum value. This specification will depend on a number of design parameters including: a) the type of substrate used; b) the type of reflection grating; c) the number of reflector elements; and d) diffraction and other losses in the grating structure.

11.2.2. REFLECTION MECHANISMS IN DIFFERENT SAW REFLECTION GRATINGS

Figure 11.3 depicts three types of SAW reflection gratings employing elements of etched grooves, open-circuited or short-circuited thin-film metal strips on the substrate surface. In some designs the grooves are filled with metal to increase the overall grating reflectivity [8]. Typically, the depth of etched grooves may range from 500 to 2000 Å, with similar values for the thickness of deposited metal films (usually of aluminum or a copper-aluminum alloy).

For correct placement of a SAW reflection grating in a resonator structure, it is essential that the *phase* of the reflected wave at the grating edge be

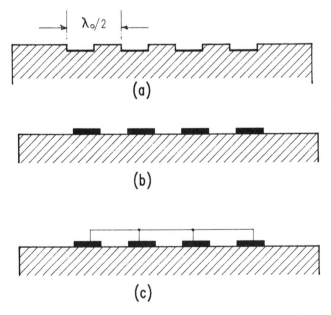

FIG. 11.3. Elements of SAW reflection grating structures, incorporating: (a) shallow grooves etched into the surface of the piezoelectric crystal substrate; (b) open (unconnected) thin metal strips; and (c) shorted (connected) thin metal strips.

correctly identified. Because of the different reflection mechanisms that can compete in a SAW reflection grating, however, its determination may not be straightforward.

Four distinct types of reflection mechanisms have been attributed as contributing to the reflection process in a SAW grating. These are associated with: a) piezoelectric shorting; b) geometric discontinuities; c) electrical regeneration; and d) mass loading [8]. Significant features of these are listed here.

1. *Piezoelectric Shorting.* The reflection effects due to piezoelectric shorting are normally significant only for materials with large values of electromechanical coupling coefficient K^2, such as lithium niobate (LiNbO$_3$) or lithium tantalate (LiTaO$_3$).

2. *Geometric Discontinuity.* A geometric discontinuity occurs on the substrate surface when grooves are etched into it, or when metal strips are deposited. Aluminum strips on quartz or lithium niobate have very similar acoustic propagation properties. As a result, the geometric reflection from an open-circuited aluminum strip of thickness h is similar to that for a groove of the same depth. For the case of very thin metal strips on lithium niobate, this geometric effect is usually smaller than for piezoelectric shorting or regeneration. (This may not be the case, however, if thick aluminum strips are involved, with film-thickness ratio $h/\lambda > \sim 1\%$.) For open metal strips on lithium niobate, the geometric effect subtracts from the net piezoelectric and regenerative effects. For shorted aluminum strips on lithium niobate, the geometric effect adds to the piezoelectric one, so that the latter construction is a more efficient SAW reflector than the open-strip one. With aluminum metal strips on ST-quartz, the geometric effect is the dominant one and is thus the only one of the four mechanisms that need to be considered.

3. *Electrical Regeneration.* Surface waves incident on a grating of open aluminum strips on lithium niobate establish a time-varying electric potential between adjacent strips. In turn, this potential causes regeneration of a surface acoustic wave in addition to the reflected component due to piezoelectric shorting. The regenerated component is larger than, but has opposite phase to, that for piezoelectric shorting. Piezoelectric regeneration and shorting are negligible in ST-quartz.

4. *Mass Loading.* The differences in both density and elastic properties between metal strips and piezoelectric substrate give a mass-loading discontinuity. Loading due to grooves on ST-quartz only induces reflections, comparable to piezoelectric shorting on lithium niobate, if very deep

TABLE 11.2

MEASURED CHARACTERISTICS OF SOME RAYLEIGH-WAVE REFLECTION GRATING TYPES

	Sign of Reflection Coefficient ρ	Velocity Perturbation in Strip ($\Delta v/v$)
YZ-lithium niobate (shorted aluminum strips)	−	−0.018
YZ-lithium niobate (open aluminum strips)	+	−0.006 to −0.010
ST-quartz (shorted aluminum strips)	−	−0.0005
ST-quartz (open aluminum strips)	−	−0.0002
Shallow grooves on ST-quartz and lithium niobate	+	—

grooves are employed ($h/\lambda > 3\%$). To circumvent problems associated with the use of excessively deep grooves, reflection gratings of shallow grooves filled with gold have been used to increase substantially the reflection coefficient magnitude [8].

As more than one mechanism may be contributing at about the same level in a given grating, it is essential to determine whether or not their reflection coefficient phases have the same or opposite sign. This determination can be a difficult one when two reflection mechanisms tend to cancel one another. In some cases, contradictions exist between theory and experiment as to the phase of the grating reflection coefficient.

Table 11.2 lists some parameters for various Rayleigh-wave reflection grating types on YZ-lithium niobate and on ST-quartz. The first column gives the sign of the grating reflection coefficient, the second gives the SAW velocity perturbation in each grating element, for a metallization ratio (or groove equivalent) $\eta = 0.5$ at center frequency. Reflection grating strips on weak piezoelectrics, such as quartz, have the same reflection coefficient *phase* (but different magnitude) for both open- and short-circuited metal strips with metallization ratio $\eta = 0.5$ [9].

11.3. One-Port SAW Resonators

11.3.1. GENERAL CONSIDERATIONS

SAW resonators may be configured electrically as one-port or two-port networks. Their operation is based around the judicious use of SAW reflection gratings to form resonant structures. In the one-port SAW resonator, surface waves emitted from both sides of the excited IDT are constructively reflected at the center frequency by SAW reflection gratings, which give rise to SAW standing waves within the IDT. These can be characterized in terms of amplitude or of piezoelectric surface potential.

The elements of the reflection gratings are both periodically spaced and normal to the SAW propagation direction, to cause narrowband performance. As the SAW gratings reflect surface waves emanating from both sides of an excited bidirectional IDT, the device insertion loss can be less than 6 dB. As will be seen, the operation of SAW resonators is crucially dependent on the separation between IDTs and adjacent reflection gratings, as this controls the standing wave pattern and optimum performance of the resonant structure.

11.3.2. RESONATOR PARAMETERS AND SENSITIVITIES

Those SAW resonators of the one-port or two-port type, with synchronous or optimum grating placement, are normally fabricated on selected cuts of ST-quartz, with differing free-surface velocities v_o, coupling coefficients K^2, and turn-over temperatures T_t. Langasite ($La_3Ga_5SiO_{14}$) with $K^2 = 0.3\%$ and $v_o = 2400$ m/s is currently receiving considerable attention as a possible replacement over quartz due to its wider-bandwidh capability.

The equivalent circuit of the one-port SAW resonator (identical to the BAW one) has a large static-to-motional capacitance ratio C_T/C_r, with small frequency separation between series- and parallel-resonance modes. In oscillator applications this usually limits the one-port SAW resonator to fixed-frequency designs. Two-port designs normally have short cavities for single-mode operation. For SAW operation, quartz has a parabolic temperature-frequency dependence around the inversion temperature T_o, corresponding to a frequency shift of 100 ppm in the temperature range from −30 to +75°C. This turnover temperature (normally $T_o = 25°C$) can be adjusted by the choice of crystal cut and metallization film-thickness ratio. As sketched in Fig. 11.4, the significance of this parabolic characteristic is that the IDT center frequency will shift **downwards** for temperature deviations on either side of T_o.

Resonator aging (dependent on packaging, preparatory "burn-in" temperature, and oscillator power level) is typically specified at ± 10 ppm/year, although values as low as ± 1 ppm/year are indicated for larger oscillator powers. Vibrational sensitivities for ST-cut SAW devices are given as $\Delta f/f \approx 1 \times 10^{-9}$/g [10]–[12]. In commercial resonators, the set-frequency tolerance at 25°C ranges from about ± 30 to ± 200 ppm depending on cost. Using laser-trimming of all-quartz package (AQP) sealed resonators, however, the set tolerance of custom 300-MHz to 1-GHz resonators can be within ± 1 ppm, allowing for dual resonator operation for higher oscillator powers. In the latter type of device, attainable $1/f$ flicker phase noise levels for

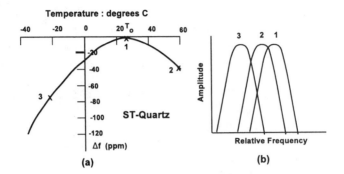

(a)   (b)

FIG. 11.4. (a) Parabolic temperature-frequency response of ST-quartz around inversion
temperature T_o~25°C. (b) Positive or negative temperature changes around T_o will result in
downward shift of IDT center frequency.

500-MHz SAW resonators with $Q_L = 4500$ are given as ~140 dBc/Hz at a
Fourier frequency offset $f_F = 1$ Hz.

A problem that arises with SAW reflection gratings of large aperture is in
the onset of spurious transverse modes that degrade the device Q-factor.
This can be particularly troublesome with anisotropic piezoelectric substrates such as ST-quartz.

11.3.3. Lumped Equivalent Circuit Parameters for One-Port Resonator

Figure 11.5(a) illustrates the geometric structure of a one-port SAW resonator. Here, the reflection gratings and IDT are assumed to have the same
centre frequency, while the IDT metallization ratio is $\eta = 0.5$ in this illustration. Figure 11.5(b) shows its idealized lumped-element equivalent
circuit in the vicinity of resonance. This consists of a series inductance-
capacitance- resistance (LCR) branch shunted by a capacitor. Shunt capacitor C_T represents the IDT capacitance, while elements L_r, C_r, and R_r relate
to equivalent motional parameters for the series-resonance condition. The
capacitance ratio C_T/C_r is an important one in one-port SAW resonator
design, as it is often required that the parallel and series resonance frequencies be offset by a specific amount for optimum performance. Figure 11.6
gives a sample frequency response computation for a 250-MHz resonator
application, using the appropriate parameters from Fig.11.5(b). Figure 11.7
shows the distance-parameter notation for later application to the transmission-matrix analysis of the one-port SAW resonator. Distance d_c represents
a penetration depth into a reflection grating, at which point the wave has

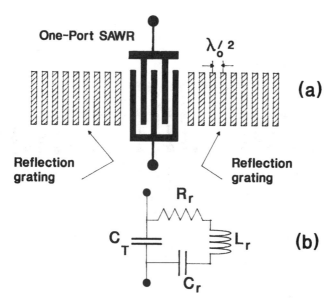

FIG. 11.5. (a) One-port SAW resonator, with open-circuit reflection-grating electrode strips. (b) Idealized LCR lumped-equivalent circuit in vicinity of resonance.

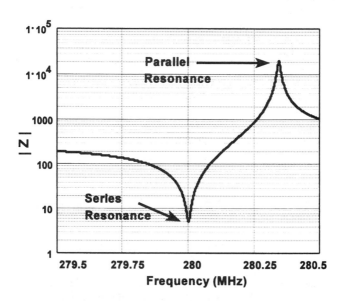

FIG. 11.6. Frequency response computation for LCR circuit of Fig. 11.5(b). Parameters $L_r = 72.64\,\mu\text{H}$, $C_r = 44.8 \times 10^{-16}\,F$, $R_r = 5\,\Omega$, $C_T = 1.813 \times 10^{-12}\,F$. Horizontal scale: 279.5 to 280.5 MHz. Logarithmic vertical scale: 1 to $10^5\,\Omega$.

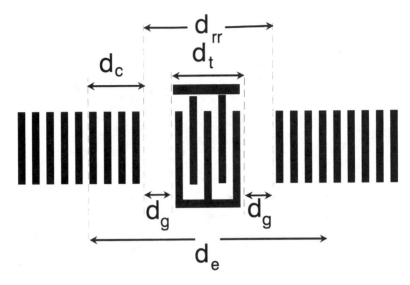

<small>Fɪɢ. 11.7. Important distance parameters in design of a one-port SAW resonator.</small>

decayed to $1/e$ of its value at entry. This penetration distance d_c can be taken to correspond to the location of a fictitious "mirror," from which the SAW wave is completely reflected.

In considering the individual parameters in the lumped LCR equivalent circuit of Fig. 11.5(b), recall from Chapter 4 that the IDT static capacitance C_T is

$$C_T = N_p C_s = N_p C_o W, \tag{11.1}$$

where N_p = number of IDT finger-pairs, C_s = capacitance/finger-pair, C_o = capacitance/finger pair/unit length, and W = acoustic aperture. Values of C_o for various piezoelectric substrates are given in Table 2.1 of Chapter 2.

It may be shown [13] that, if IDT finger reflections are neglected, the equivalent series resistance R_r in Fig. 11.5(b) may be approximated by

$$R_r \approx \frac{1}{G_a(f_o)} \frac{\left(1-|\rho|\right)}{\left(1+|\rho|\right)}, : \quad (ohms) \tag{11.2}$$

where $G_a(f_o) = 8K^2 f_o C_s N_p^2$ = unperturbed radiation conductance at IDT center frequency f_o, as given in Chapter 4. Parameter ρ (with $|\rho| < 1$) in Eq. (11.2) is a dimensionless reflection coefficient, relating the ratio of reflected-to-incident surface waves entering the reflection grating. This parameter ρ is analogous to the (voltage) reflection coefficient of a load termination in a transmission line, given as the ratio of reflected-to-incident voltage waves. For the time being it will be assumed that the reflection grating geometries are such that ρ has a maximum value at IDT center frequency f_o.

As noted, d_c can be defined as the distance into the reflector at which the SAW has decayed to $1/e$ of its value in the cavity. The grating penetration distance d_c into a reflection grating will be dependent on a number of factors, including: a) the piezoelectric substrate employed; b) the impedance discontinuity imposed by grating metallization material and thickness; and c) the operating frequency. Because of the gradual decay of the SAW in the grating, there is no unique way of specifying the effective reflection point [14]. Typically, d_c has been shown to have values on the order of $20\lambda_o$ at resonance for thin aluminum grating strips on Y-Z lithium niobate [15].

The equivalent series inductance L_r in Fig. 11.5(b) may be approximated by [13],[14]

$$L_r \approx \frac{d_e}{\lambda_o} \frac{1}{\left(4f_o G_a(f_o)\right)},: \quad \left(henries\right) \tag{11.3}$$

where d_e = effective cavity length in Fig. 11.7, and $\lambda_o = v/f_o$, in terms of SAW velocity v. From Eq. (11.3) the equivalent series capacitance C_r is obtained as

$$C_r = \frac{1}{4\pi^2 f_o^2 L_r},: \quad \left(farads\right) \tag{11.4}$$

For high-Q resonators, the frequency separation between parallel and series resonant frequencies is

$$f_p - f_s = \frac{f_o}{2} \frac{C_r}{C_T}. \tag{11.5}$$

For grating reflection coefficient magnitudes $|\rho|$ close to unity, the resonant cavity Q is approximated by [13]

$$Q \approx \frac{d_e}{\lambda_o} \frac{2\pi}{\left(1-|\rho|^2\right)}. \tag{11.6}$$

11.3.4. FURTHER DESIGN CONSIDERATIONS

By itself, the design of an efficient SAW reflection grating is a necessary but not sufficient criterion for ensuring that the two-port resonator will operate properly. A second criterion of *paramount importance* relates to the placement of each reflection grating in Fig. 11.7 with respect to its adjacent IDT. An incorrect positioning of either IDT or grating can degrade the electromechanical transduction between input and/or output voltages and acoustic resonance, to the point where SAW resonance action can be lost altogether. SAW reflections from the IDTs can also corrupt the desired response. This latter effect can be reduced, however, by embedding the IDT fingers in grooves on quartz substrates [16].

An understanding as to how the SAW reflection gratings should be positioned can be gained from drawing on analogies with transmission-line representations of electromagnetic resonators. In a conventional microwave resonant cavity, resonance occurs when the total phase shift of E- and H-field components is $2n\pi$ radians (n = integer) after two complete transits of the travelling wave within the cavity bounded by metal reflectors. Interference between oppositely travelling waves results in standing waves of E and H. For a given excitation, their intensities within the resonator will be determined by input/output coupling losses as well as those due to the finite reflectivity ($|\rho| < 1$) of the interior metal boundaries. The degree of coupling loss will depend on the placement of input/output coupling holes with respect to the E- or H-field spatial patterns. In similar fashion, acoustic resonance occurs within the SAW resonator when the total phase shift ϕ of the surface wave is $\phi = 2n\pi$ within the cavity bounded by the two reflection gratings. Whereas energy oscillation in the electromagnetic resonator was between E and H fields, here it is between mechanical stress and strain fields. For a given input excitation, the amplitudes of the SAW standing waves and piezoelectric surface potential will be dependent on electromechanical transduction losses and by those due to the finite reflectivity $|\rho|$ of each reflection grating. The transduction efficiency depends on the positioning of the IDTs with respect to the SAW standing wave. It will be optimum when the IDT metal fingers are positioned exactly above the SAW standing-wave maxima.

11.3.5. STANDING-WAVE PATTERNS IN THE ONE-PORT SAW RESONATOR

The SAW standing-wave patterns at centre frequency in the one-port SAW resonator can be depicted in much the same way as for conventional transmission lines, using amplitude or piezoelectric potential as the variable instead of voltage. In this manner, SAW amplitude distributions within the one-port resonator may be derived from a knowledge of the complex reflection coefficient ρ of the SAW reflection grating. For simplicity, consider that the two SAW gratings have identical acoustic reflectivities. Also assume that there is no mode conversion and or accompanying decrease in reflection efficiency by mode conversion such as due to bulk waves.

The all-important parameter here is the phase of the grating reflection coefficient. Recall how the voltage standing-wave distributions in an electromagnetic transmission line are affected by the voltage reflection coefficient ρ_L of the load termination [17]. For an open-circuit load $\rho_L = +1$, so that the resultant voltage magnitude is a maximum at the load. Conversely, $\rho_L = -1$ for a short-circuit load, so that the resultant voltage is zero at this point. Figure 11.8 depicts the idealized standing-wave patterns derived in this way for a one-port resonator on ST-X quartz using gratings of either open- or short-circuited metal strips. The gratings positions shown are for the optimum placement condition and are the same for both shorted and unshorted grating strips. The idealized standing wave patterns sketched here assume grating reflectivities $|\rho| = 1$.

As the standing-wave patterns are periodic, the positions for both gratings can be "moved out" by an integer number of acoustic half wavelengths from the optimal minimum ones, while preserving the resonance. Such displacements are illustrated in both configurations of Fig. 11.8. While the reflection grating positioning and geometry are the same for shorted and unshorted gratings in the one-port SAW resonator of Fig. 11.7, the actual performance of two such devices will not be the same. This is due to the difference in grating strip reflectivities. The structure of Fig. 11.8(a) is often the preferred one for substrates employing ST-quartz, because of the larger reflection coefficient for shorted grating strips.

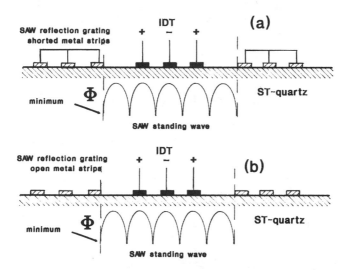

FIG. 11.8.	Standing-wave patterns of Rayleigh-wave amplitude in one-port resonator on ST-quartz. Reflection gratings are placed for optimum response with: (a) shorted-metal grating strips; and (b) open-metal strips. In both examples, gratings are displaced $\lambda/2$ outwards from minimum close-in positions.

11.4. Two-Port SAW Resonators

11.4.1. Transfer Function Requirements

Figure 11.9(a) shows the input IDT, output IDT, and reflection grating placement for a two-port SAW resonator, where the input/output IDTs have equal finger numbers. Figure 11.9(b) is the LCR lumped-equivalent circuit for 0° input/output phase shift at resonance. Figure 11.9(c) is the corresponding equivalent circuit incorporating a 1:−1 ideal transformer, applicable to designs with 180° input/output phase shift at resonance. For circuit implementation of the 0 or 180° phase shift designs, however, it is important to note that these phase shifts at the resonator peak response will only be met if the input/output capacitors are "tuned out" by shunt inductors. Figure 11.10 indicates important distance parameters in the design of a two-port SAW resonator.

The voltage transfer-function response between input and output IDTs may be considered to be a composite of two contributions. In the absence

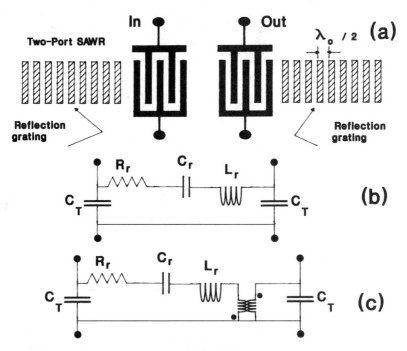

Fig. 11.9. (a) A two-port SAW resonator. (b) Lumped-equivalent circuit for 0° phase shift between input/output IDTs at resonance. (c) Ideal 1:−1 transformer included for 180° phase shift at resonance. In each case the input/output capacitors C_T have to be "tuned out," to meet the 0° or 180° specifications.

of reflection gratings, it would just be that for a SAW filter with uniform, and equal, input and output IDTs, giving a $|(\sin X)/X|^2$ amplitude distribution in the absence of significant IDT finger reflections. With the gratings included, the resonator response around center frequency will be superimposed on this, as illustrated in Fig. 11.11 for a two-pole resonator filter. This

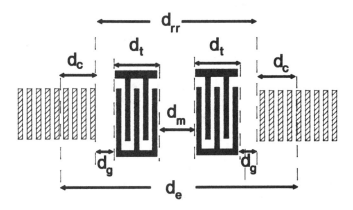

FIG. 11.10. Important distance parameters in design of a two-port SAW resonator.

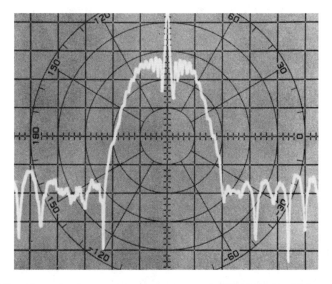

FIG. 11.11. Frequency response of a two-port two-pole 537.5-MHz Rayleigh-wave resonator, with insertion loss of 4.2 dB. (Courtesy of Sawtek Inc., Orlando, Florida.)

underlying IDT response will limit the dynamic range of the resonance component. If, for example, the SAW resonator is assigned an insertion loss of 3 dB and dynamic range of not less than 25 dB, the minimum insertion loss of the uniform filter structure must be *at least* 28 dB. This is achieved by mismatching input and output IDT radiation conductances. Although the IDTs by themselves are then mismatched, the additional resonance action can bring the overall response to near-match at center frequency.

The resonator transfer response will depend on a number of parameters. As in the case of the one-port resonator, one of the most critical parameters relates to the spacing between each reflection grating and its adjacent IDT. Other influential parameters are: a) grating reflectivity and loss; b) separation between reflection gratings; c) number of IDT fingers; d) IDT film thickness and resistivity; e) K^2 and C_o values for the piezoelectric substrate; and f) the operating frequency.

Example 11.1 Dynamic Range of a Two-Port SAW Resonator on Lithium Niobate. A two-port SAW resonator is to be fabricated on YZ-lithium niobate for operation at $f_o = 100$ MHz with 50-Ω source and load impedances. Uniform IDTs with acoustic aperture $W = 80\lambda$ are to be employed in input and output, with $N = 4$ finger pairs in each. If the device is allowed an insertion loss of 3 dB, what will be the dynamic range of the resonator response? (Neglect IDT finger reflections.) ∎

Solution. From Table 2.1, $K^2 = 0.045$, $C_o = 4.6$ pF/cm and $v = 3488$ m/s for YZ-lithium niobate. The acoustic wavelength at 100 MHz is $\lambda_o = 3488/f_o = 3488/(100 \times 10^6) = 34.8 \times 10^{-6}$ m $= 34.8\,\mu$m. This gives $W = 80\lambda_o = 80 \times 34.8 \times 10^{-6} = 2.8 \times 10^{-3}$ m, so that $C_s = WC_o = (2.8 \times 10^{-3}) \times (4.6 \times 10^{-10}) = 1.3 \times 10^{-12}$ F/finger pair. The total IDT capacitance is $C_T = C_s N = 1.3 \times 10^{-12} \times 4 = 5.2$ pF. The unperturbed radiation conductance $G_a(f_o) = 8K^2 f_o C_T N = 8 \times 0.045 \times (100 \times 10^6) \times (5.2 \times 10^{-12}) \times 4 = 7.5 \times 10^{-4}$ mho. The input mismatch value $q = G_a R_s = 7.5 \times 10^{-4} \times 50 = 3.7 \times 10^{-2}$. From Eq. (5.19) for $p = 0$, the input power transmission loss is $-10 \cdot \log_{10} |T_{31}|^2 = 10 \cdot \log_{10}[2q/(1+q)^2] \approx 11$ dB. This gives the total IDT insertion loss for equal IDTs as ≈ 22 dB. A device insertion loss of 3 dB thus allows for a resonator dynamic range of $(22 - 3) = 19$ dB. ∎

Example 11.2 Comparison with a Two-Port SAW Resonator on ST-Quartz. If the two-port resonator were to be fabricated on ST-quartz instead of lithium niobate as in Example 11.1, how many finger pairs would be required in the input and output IDTs for the same level of IDT insertion loss? Assume that the acoustic aperture is to remain at $W = 80\lambda$. (Again neglect IDT finger reflections.) ∎

Solution. From Table 2.1, $K^2 = 0.0016$, $C_o = 0.55\,\text{pF/cm}$ and $v = 3158\,\text{m/s}$ for ST-quartz. Proceeding as in Example 11.1, the acoustic wavelength is $\lambda_o = 3488/f_o = 3158/(100 \times 10^6) = 31.58 \times 10^{-6}\,\text{m} = 31.58\,\mu\text{m}$. The acoustic aperture is $W = 80\ \lambda_o = 80 \times (31.58 \times 10^{-6}) = 2.52 \times 10^{-3}\,\text{m}$. The two radiation conductances must have the same value for the same level of IDT insertion loss. On quartz, therefore, $G_a(f_o) = 8K^2 f_o C_T N = 8K^2 f_o W C_o N^2 = 7.5 \times 10^{-4}\,\text{mho}$. This gives $8 \times 0.0016 \times (100 \times 10^6) \times 2.52 \times 0^{-3} \times (0.55 \times 10^{-10}) \times N^2 = 7.5 \times 10^{-4}$, which reduces to $(1.77 \times 10^{-7}) \times N^2 = 7.5 \times 10^{-4}$, so that $N = 65$ finger pairs for the resonator implementation on quartz. ∎

11.4.2. STANDING-WAVE PATTERNS IN THE TWO-PORT SAW RESONATOR

The optimal placement for reflection gratings in two-port SAW resonators follows that for the one-port resonator considered here. Figure 11.12 depicts the SAW standing-wave patterns derived in this way for a two-port resonator using gratings of either open- or short-circuited metal strips on lithium niobate, using the appropriate reflection coefficient signs in Table 11.2. Reflection grating and IDTs are assumed to have the same centre

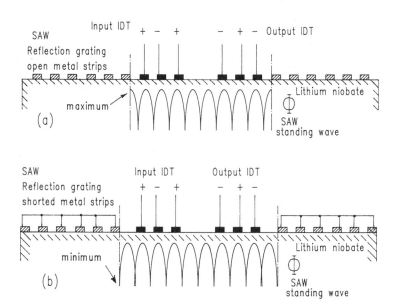

FIG. 11.12. SAW standing-wave patterns in two-port resonator on lithium niobate, with 180° phase shift between input/output at resonance. Optimally placed reflection gratings are positioned $\lambda/2$ out from minimum position. (a) With gratings with open metal strips. (b) With gratings of shorted metal strips.

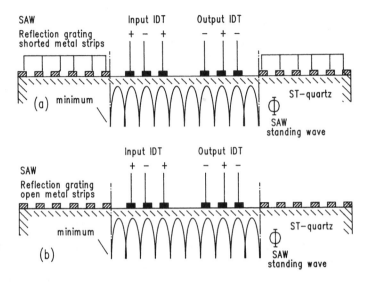

FIG. 11.13.   (a) Positioning gratings with short-circuit electrodes on ST-quartz, for resonator action. Reflection coefficient per aluminum strip is $\rho = -0.0034$ for $h/\lambda = 0.6\%$ and $\eta = 0.5$.
(Reference [9].) (b) Same positioning with open strips, but $\rho = -0.0024$ per strip. (Reference
[9].)

frequency, while the IDT metallization ratio is $\eta = 0.5$ in this illustration.
The idealized standing-wave patterns sketched here assume grating
reflectivities $|\rho| = 1$. Figure 11.13 shows the SAW standing-wave patterns
for open and shorted gratings on ST-quartz. The IDT metallization ratio is
again $\eta = 0.5$.

Depending on the sign of the reflection coefficient for the substrate and
grating connection, the optimal spacings between IDTs and adjacent reflection gratings are either increased or decreased by $\lambda_o/8$ from that for "synchronous" spacing between the two structures. This can be troublesome in
high-frequency lithographic fabrications where spacings between reflection
gratings and adjacent IDTs are to be decreased by $\lambda_o/8$. To counterbalance
this reduction, and because the standing-wave patterns are periodic, the
positions for both gratings can be "moved out" by an integer number of
acoustic half-wavelengths, as illustrated in Figs. 11.12(b) and 11.13.

11.4.3.   INPUT/OUTPUT VOLTAGE POLARITY OF TWO-PORT RESONATOR

Standing-wave representations for the two-port SAW resonator are also
useful in establishing the phase difference between input and output voltages in single-pole devices. This will depend on both the relative polarities
of the input/output leads as well as on the number of half-cycles of standing

waves between phase centres of input and output IDTs. For the same electrical lead polarities, for example, an odd number of half-cycles between IDT phase centers will result in an overall 180° phase shift between input and output at centre frequency, under matched conditions, as illustrated in Figs. 11.12 and 11.13.

11.4.4. A SAW Notch Filter Using a Two-Port Resonator

The fore-mentioned 180° input/output phase shift capability can be useful in single-pole oscillator and other feedback applications. It can also be used to configure a SAW narrowband notch filter. The simplest configuration merely requires the connection of a small resistor between input and output ports, which value is equal to that of a series resistor in an equivalent lumped LCR resonant circuit. Such a notch filter response is illustrated in Fig. 11.14, incorporating a 674-MHz two-port resonator on ST-quartz.

11.4.5. Potential Phase-Shift Problems with the Two-Port Resonator

As already noted, attainment of 0 or 180° phase shifts at resonance in the designs of Fig. 11.9 would require the shunt capacitive reactance of the input/output IDTs to be "tuned out" by appropriate shunt inductors. This becomes a critical matter when these two-port resonators are employed as

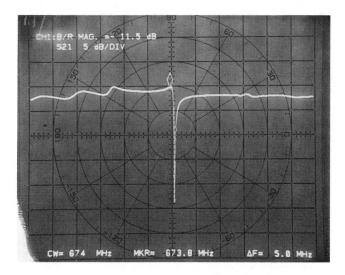

FIG. 11.14. A highly-selective 674-MHz notch filter using a bypass resistor across a 180° two-port SAW resonator. Horizontal scale: 673.8 MHz ± 5 MHz. Vertical scale: 5 dB/div.

oscillator feedback elements. In the absence of such inductors it would be necessary to include an addition phase shift in the loop, by the series addition of a lowpass filter or length of coaxial line.

An alternative method that can be applied to the 0° phase design of Fig. 11.15(a) is to essentially convert it from a two-port network to a one-port network by cross connection of input/output leads as shown in Fig. 11.15(b). As shown in Fig. 11.15(c), the two-port condition is restored by addition of ground-return leads. With this technique the equivalent motional parameters L_r, C_r, and R_r remain unchanged, but two static capacitances C_T reduce to a single effective one with a value of $C_T/2$. In this way one matching inductor is required to maintain the desired 0° phase shift at resonance [4].

In the design of transistor oscillators for low-power transmitters, problems may also arise with excessive phase shift in the transistor stage. This may be as much as about 240° in a UHF transistor. In this instance, the feedback element would have to provide +120 or −140° to meet loop phase-shift requirements. As the phase shift across the 3-dB bandwidth of a one-pole resonator is only ±45°, however, the preceding two-port one-pole SAW resonators would not meet this criterion. In this case—and

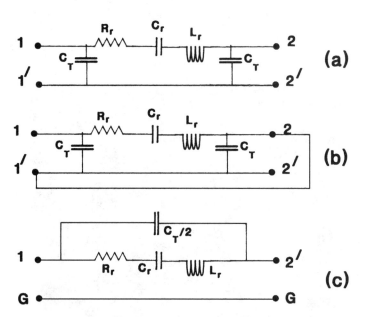

FIG. 11.15.   Reconfiguring a two-port 0°-phase SAW resonator. (a) 0°-phase SAW resonator. (b) Cross connection. (c) Cross connection plus addition of ground return. Reduces effective static capacitance to $C_T/2$, with only one inductor required for matching.

as described in a later chapter—one alternative would be to use a two-pole SAW resonator-filter, with ±90° phase shift across the 3-dB bandwidth.

11.5. Avoiding Multimode Effects in Single-Pole Resonators

The amplitude of the frequency response of the two-port single-pole SAW resonator will be symmetric about center frequency when optimal reflection-grating spacing is employed, and IDT finger reflections are negligible Moreover, this configuration will have minimum insertion loss for a given design. This is illustrated in Fig. 11.16 for the measured amplitude and phase response of a 90-MHz two-port SAW resonator fabricated on lithium niobate. Depending on the overall cavity length, however, spurious longitudinal (i.e., multimode) modes can arise if the cavity lengths d_{rr} or d_e in Fig. 11.10 are large enough to support them. The frequency separation Δf between such multimode responses may be approximated by $\Delta f = v/(2d_e)$, where $v = $ SAW velocity, and $d_e \approx$ effective cavity length [15], [18]. These modes do not appear in the experimental SAW resonator response of Fig. 11.16, because the cavity length was small in this instance.

The multimode situation can often occur in two-port SAW resonator designs on substrates such as ST-X quartz with low K^2 coupling factors. This is because a large number of fingers is normally required to increase the

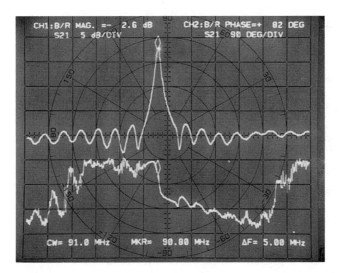

Fɪɢ. 11.16. Measure amplitude and phase responses of 90-MHz two-port one-pole Rayleigh-wave resonator on YZ-LiNbO₃. Horizontal span: ± 5 MHz about center frequency. Vertical scales: 5 dB/div and 90°/div. Compare with theoretical plots in Figs. 11.26 and 11.27.

radiation conductance $G_a(f_o)$ to sufficient levels for impedance matching at input and output. Recall from Chapter 4 that the unperturbed radiation conductance for IDTs with negligible finger reflections is given by

$$G_a(f_o) = 8K^2 f_o C_s N^2, : \quad (mho) \qquad (4.22)$$

where $N =$ number of IDT finger pairs, and $C_s =$ capacitance per finger pair. The radiation conductance can be increased by increasing C_s through the use of wider acoustic apertures W. As will be discussed later in the chapter, large values of acoustic aperture can give rise to response degradation by spurious transverse modes in the reflection gratings, so we can be forced to use larger numbers of IDT finger pairs. Figure 11.17 illustrates a computation for such a multimode response in a long resonator cavity for a 673.9-MHz SAW resonator on ST-quartz, showing spurious longitudinal modes on either side of the main response. (The computation employs the transmission-matrix method considered later in this chapter.)

It has been shown [19] that these spurious multimode responses can be removed if the gratings are "synchronously" spaced with respect to the

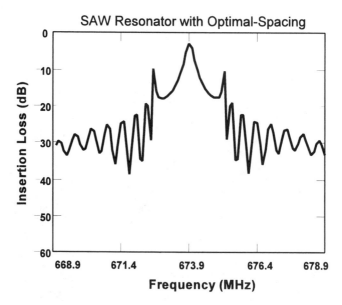

F‌IG. 11.17. Computed frequency response of 673.9-MHz two-port one-pole Rayleigh-wave resonator, with large cavity length, on ST-quartz. Shown for optimal placement of end reflection gratings, giving modes on either side of main peak. Horizontal scale: 668.9 to 678.9 MHz. Vertical scale: 0 to −60 dB.

IDTs. This will result in an asymmetric response about center frequency, with some increase in insertion loss as well as a slight displacement of the response peak. Despite this, resonance Q-values of over 10,000 can be obtained with the synchronous configuration, with excellent out-of-band rejection of spurious multimodes [19]. The computation of Fig. 11.18 shows the modified response resulting from such synchronous spacing, giving a lower resonance frequency as well as a deep notch in the upper stopband. This result may be compared with the experimental amplitude and phase response of Fig. 11.19 for a 674-MHz two-port SAW resonator on ST-quartz with such synchronous spacing.

The preceding technique was developed to circumvent degradation in the sidebands of a two-port SAW resonator by longitudinal multimode action. It should be noted, however, that multimode operation is not always harmful. Indeed, it can be usefully employed in certain other SAW designs, such as for multimode harmonic resonators [20], comb filters [21], and multiple-pole SAW resonators [6].

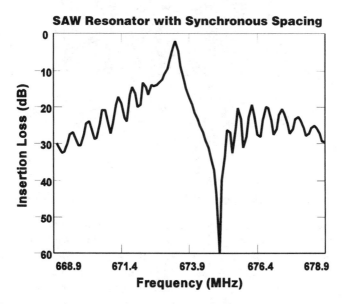

FIG. 11.18. Computed frequency response of same 673.9-MHz Rayleigh-wave resonator as for Fig. 11.17, but using synchronous grating displacement. Note that this causes a downward shift of the resonance response peak.

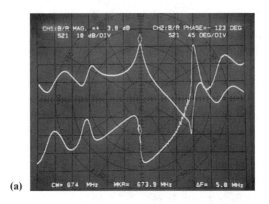

(a)

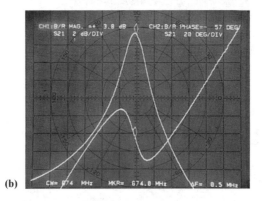

(b)

Fɪɢ. 11.19. Measured amplitude and phase response of a two-port one-pole Rayleigh wave resonator on ST quartz. (a) Horizontal scale: 673.9 ± 5 MHz. Vertical scales: 10 dB/div and 45°/div. (b) Expanded vertical scales 2 dB/div and 20°/div. Compare with computation of Fig. 11.18.

11.6. Quality-Factor Q of a Rayleigh-Wave Resonator

SAW resonator performance is normally expressed in terms of its insertion loss (IL) and quality-factor Q. In the one-port and two-port configurations the insertion loss would ideally be zero if all of the SAW energy were contained within the resonant cavity. In practice, various loss contributions typically constrain the insertion loss to a few decibels. These include acoustic attenuation, diffraction, losses in the reflection gratings, mode conversion to bulk waves, external circuit loading, and IDT electrode resistance.

For SAW resonator structures, the loaded Q, Q_L can be given in terms of contributing loss parameters as

$$\frac{1}{Q_L} = \frac{1}{Q_m} + \frac{1}{Q_d} + \frac{1}{Q_r} + \frac{1}{Q_b} + \frac{1}{Q_c} + \frac{1}{Q_e} \tag{11.7}$$

and

$$Q_L = \frac{f_o}{\Delta f}, \tag{11.8}$$

where f_o = resonator center frequency and Δf = 3-dB bandwidth; such that Q_L is the inverse of the resonator fractional bandwidth $BW_3 = \Delta f / f_o$. Spurious transverse modes can also degrade the response peak and the Q.

In Eq. (11.7) Q_m is referred to as the *material Q*, associated with viscous damping of the SAW on the substrate. This is what limits the ultimate attainable loaded Q. (For ST-quartz, $Q_m \approx 10{,}500$ at 1 GHz.) Above about 200 MHz this acoustic attenuation increases to about f^2, as given in Table 6.4 of Chapter 6. Material Q_m also incorporates frequency-independent loss terms due to air (or inert gas) loading if the device is not encapsulated in vacuum.

The term Q_d in Eq. (11.7) represents losses due to diffraction in the resonator and in the reflection grating itself. It has been shown that Q_d is proportional to $(W/\lambda)^2$, where W = resonator aperture in acoustic wavelengths. Radiation Q_r incorporates losses both in and from the finite SAW reflection gratings, while Q_b relates to energy loss due to mode conversion to acoustic bulk waves. The last two terms in Eq. (11.7) concern losses in the IDT and external circuitry, such that Q_c incorporates external circuit coupling while Q_e relates IDT losses due to the finite resistance of their metal film electrodes; Q_L values as high as $Q_L = 80{,}000$ are reported for SAW resonators operating at 100 MHz; they decrease to about $1/f$ to $3000 \leq Q_L \leq 10{,}000$ at 1 GHz. These Q values may be compared with those for bulk acoustic wave (BAW) resonators with loaded Q values of up to 10,000 at 100 MHz and up to 30,000 at 1 GHz [22].

For off-the-shelf one-port SAW resonators in the range 100 to 1200 MHz, matched insertion losses are in the range 0.5 to 2.5 dB, with $Q_L \sim 1600$ to 7000. Over the same frequency range two-port resonator losses are 1 to 4 dB, and $Q_L \sim 3000$ to 13,000. Two-port SAW resonators have been reported at 2.6 and 3.3 GHz, with IL < 11 dB, $Q_o = 2000$, and IL = 17 dB, $Q_o = 1600$, respectively.

11.7. Matrix Building Blocks for the SAW Resonator

11.7.1. PREAMBLE

Rather than using lumped electrical equivalent circuits to relate the frequency response of a one-port or two-port SAW resonator, a more versatile approach employs the analytical use of transmission matrices. One such technique employs reflection-grating matrix representations derived by the coupling-of-modes (COM) method, originally derived for analysis of thick optical holograms [23] and distributed feedback lasers [24], [25] before being applied to SAW devices [26], [27]. In the transmission-matrix approach employed here, the two-port SAW resonator is represented electrically and acoustically as a composite of three different types of "building blocks" using *complex* transmission matrices [6], [21]. Building blocks for IDTs are 3×3 transmission matrices relating both electrical and acoustic parameters. Those for SAW reflection gratings are 2×2 matrices relating acoustic transmission, reflection and loss. The third building block is a 2×2 transmission line matrix applied to acoustic spaces between IDTs and adjacent gratings. The elements of this last matrix will depend on which segment of the structure it is applied. While at first glance, the matrix relations appear complicated, they can be easily programmed on a personal computer. As will be shown, both the magnitude and phase response of the two-port SAW resonator can be predicted accurately to first-order in this way.

11.7.2. THE THREE TRANSMISSION MATRIX BUILDING BLOCKS

The frequency response of a two-port SAW resonator may be computed by multiplying transmission matrices appropriate to each section of the SAW resonator structure and then applying boundary conditions. The cascaded matrix relationships are derived from three basic "building blocks." The first of these is a 2×2 matrix [G] for the SAW reflection gratings, as derived from COM theory introduced in Chapter 4 [6], that relate their acoustic transmission, reflection and loss performance. The second building block is a 3×3 transmission matrix [T] for the IDTs, involving both acoustic and electric parameters. It may be derived from the 3×3 admittance matrix [Y] in Chapter 4, as applied to the crossed-field model of the IDT [28]–[30]. The third building block is a 2×2 acoustic transmission line matrix, applicable to the various SAW transmission line segments between IDTs and gratings.

11.7.3. THE 2×2 GRATING MATRIX [G]

Matrix [G] is a 2×2 transmission matrix applied to the SAW reflection gratings as derived from COM theory [6]. In view of the complexity of this subject [26], [27], only its highlights are considered here, as they apply to the two-port resonator design.

It is assumed first that SAW propagation in a periodic and uniform reflection grating with element spacing period Λ can be represented by a scalar-wave equation

$$\frac{d^2\Phi}{dx^2} + \left[\frac{\omega^2}{v^2(x)}\right]\Phi = 0, \qquad (11.9)$$

in terms of quasi-static surface electric potential Φ, angular frequency ω, and SAW velocity $v(x)$. Here, velocity $v(x)$ is not constant, but is perturbed sinusoidally about value v_o as the surface wave passes through the grating. The corresponding Bragg frequency f_o for reflection is given by $f_o = v_o/2\Lambda$, in terms of the grating period Λ. In solving Eq. (11.9) to first-order for this velocity perturbation, a pair of coupled-wave equations may be obtained that relate the propagation of "forward" and "backward" SAW waves. In derivative form these are

$$-R' - j\delta R = j\kappa S \qquad (11.10a)$$

$$S' - j\delta S = j\kappa R, \qquad (11.10b)$$

where R and S (and their derivatives R' and S') relate the amplitudes of the forward- and backward-waves, respectively, $\delta = 2\pi(f - f_o)/v_o + \kappa_{11}$ is the frequency deviation from the Bragg frequency f_o and $\kappa = \kappa_{12}$ is the mutual-coupling coefficient in Chapter 9. Using this approach, the grating matrix $[G]$ for a reflection grating of shallow grooves is [6]

$$[G] = \begin{pmatrix} G_{11} & G_{12} \\ G_{21} & G_{22} \end{pmatrix}, \qquad (11.11)$$

with elements

$$[G] = C\begin{pmatrix} \left[\frac{\sigma}{\kappa} + j\left(\frac{\delta - j\alpha}{\kappa}\right)\tanh(\sigma L)\right]e^{j\beta_o L} & je^{-j\theta}\tanh(\sigma L)e^{j\beta_o L} \\ -je^{j\theta}\tanh(\sigma L)e^{-j\beta_o L} & \left[\frac{\sigma}{\kappa} - j\left(\frac{\delta - j\alpha}{\kappa}\right)\tanh(\sigma L)\right]e^{-j\beta_o L} \end{pmatrix}, \qquad (11.12)$$

where

$$\sigma = \left[\kappa^2 - (\delta - j\alpha)^2\right]^{1/2}, \qquad (11.13)$$

$$C = \left(\frac{\kappa}{\sigma}\right)\cosh(\sigma L), \qquad (11.14)$$

and κ = grating mutual-coupling coefficient (m^{-1}), α = grating attenuation constant (m^{-1}), L = grating length (m), $\beta = \omega_o/v_o$ is the "unperturbed" phase constant (rad/m) and θ = reference phase. For a narrowband approximation, the frequency-deviation (detuning) parameter δ is

$$\delta = \frac{2\pi\left(f - f_o\right)}{v} + \kappa_{11}, \tag{11.15}$$

in terms of unperturbed SAW velocity v and self-coupling (velocity-shift) coefficient κ_{11} in Chapter 9. Frequency deviation from resonance is given in terms of a normalized parameter Δ, where

$$\Delta = \frac{\delta}{\kappa} \tag{11.16}$$

so that

$$\sigma = \sqrt{\left(\kappa^2 - \delta^2\right)} = \lambda\sqrt{\left(1 - \Delta^2\right)} : \; \left(m^{-1}\right) \tag{11.17}$$

when the grating attenuation constant $\alpha = 0$.

11.7.4. CHOICE OF REFERENCE AXES

In computing the frequency response of the two-port SAW resonator, it is essential that the reference axes for matrix relationship be identified correctly. In a manner analogous to notation used for incident wave and reflected wave propagation in a transmission line, a "+" sign may be associated with an incident surface acoustic wave and a "–" sign with a component reflected from a discontinuity. Note that while incident or "forward" waves are usually sketched from left to right in transmission line diagrams, they can be in either direction here, when considering reflections from each of the SAW reflection gratings. In matrix form, amplitudes of forward and reverse surface waves at an ith reference axis are

$$\left[W_i\right] = \begin{pmatrix} w_i^+ \\ w_i^- \end{pmatrix}. \tag{11.18}$$

Consider, for example, the choice of reference axes in Fig. 11.20, for manipulations involving grating matrix [G] of Eq. (11.12). Here, the ith reference axis for the forward and backward SAW waves w_i^+ and w_i^- is associated with the ith element of the reflection grating. This is given by

$$\left[W_{i-1}\right] = \left[G_i\right]\left[W_i\right], \tag{11.19}$$

where $[G_i]$ is the 2×2 transmission sub-matrix of the ith element of a reflection grating. From this, a one-dimensional transmission line type of

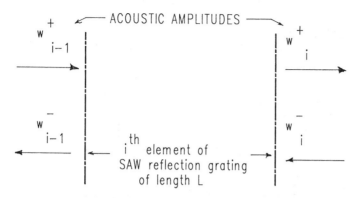

Fig. 11.20. The SAW reference planes at the *i*th element of a SAW reflection grating of length *L*.

equation for the electric potential Φ at the surface of the piezoelectric substrate can be solved as

$$\Phi(x) = w^+(x)e^{-j\beta_o x} + w^-(x)e^{+j\beta_o x} \qquad (11.20)$$

along some SAW propagation axis x, where $\beta_o = \pi/\Lambda$ is the propagation constant (rad/m) of the surface acoustic wave at the Bragg frequency $f_o = v_o/2\Lambda$, when the grating has maximum reflection. This acoustic wave equation is solved and used to relate SAW amplitudes at ends $x = -L$ and $x = 0$ of the reflection grating of length L in Fig. 11.21, *for a wave incident from the left.* An approximate solution of this acoustic wave equation for a narrow frequency range about center frequency f_o yields

$$[W(-L)] = [G][W(0)], \qquad (11.21)$$

for the entire reflection grating. Because this is an integer number of periods in length at centre frequency, the number of grating elements N_g is

$$N_g = \frac{L}{\Lambda}. \qquad (11.22)$$

If there is no incident SAW wave at the plane $x = 0$, then in Fig. 11.21 at reference plane $x = -L$ the reflection coefficient at centre frequency is

$$\rho(\Delta = 0) = \frac{w^-(-L)}{w^+(-L)}. \qquad (11.23)$$

For a reflection grating of shallow grooves, Eqs. (11.12) and (11.23) yield

$$\rho(\Delta = 0) = \frac{G_{21}}{G_{22}}\bigg|_{\Delta=0} = j\tanh(|\sigma L|) \qquad (11.24)$$

for the grating reflection coefficient at centre frequency.

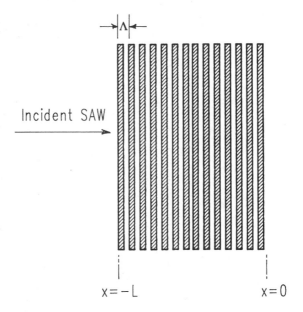

FIG. 11.21. Boundary notation for SAW reflection grating, with surface wave arriving from the left.

The j term in Eq. (11.24) results form the choice of reference axis. Here, it is $\lambda_o/8$ away from the *physical edge* of the first grating element. For the reader familiar with electrical transmission-line theory, a reference shift of $\lambda/8$ corresponds to a rotation of $90°$ on the Smith Chart. If the reference plane is moved to the leading edge of the first element of the reflection grating, the reflection coefficient magnitude at center frequency is approximated as a pure real quantity [31]

$$\rho(\Delta = 0) = (-1)^n \tanh(|\sigma L|). \tag{11.25}$$

In Eq. (11.25) index n is chosen to fit experimental reflection data in Table 11.2. With lithium niobate substrates, $n = 0$ for gratings composed of open metal strips or shallow grooves, while $n = 1$ for shorted metal strips. With ST-quartz substrates, on the other hand, $n = 1$ for gratings of open- and short-circuited metal strips [9].

Figure 11.22(a)(b) illustrates the predicted magnitude and phase response of a lossless SAW reflection grating about a centre frequency $f_o = 76\,\text{MHz}$ on *YZ*-lithium niobate. This was computed from Eqs. (11.12), (11.13) and (11.14) for a grating of length $= 0.96\,\text{cm}$, $\eta = 0.5$, $\kappa = \kappa_{12} =$

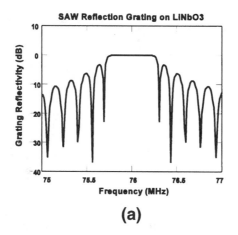

(a)

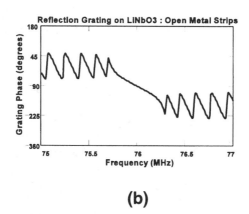

(b)

FIG. 11.22. (a) Reflectivity magnitude for 76-MHz grating on LiNbO₃, with 420 open-metal grating strips, and $\eta = 0.5$, $\kappa = \kappa_{12} = 4.5\,\mathrm{cm}^{-1}$, $\alpha = 0.016\kappa$. Horizontal scale: 75 to 77 MHz. Vertical scale: −40 to +10 dB. (b) Reflection phase −90° at f_o for open strips and $\lambda/8$ reference axis. Vertical scale: −360 to +180 degrees.

$4.5\,\mathrm{cm}^{-1}$, and attenuation constant $\alpha = 0.016\,\mathrm{cm}^{-1}$. Observe the narrow stopband range in Fig. 11.22(a) over which the reflectivity $|\rho|$ is close to unity. If the separation between gratings is too large, the stopband will support multimode resonance rather than the single one desired here. Note also that the stopband width will decrease with increasing grating attenuation, as shown in Fig. 11.23 for the same grating with an exaggerated loss of $\alpha = 1.0\,\mathrm{cm}^{-1}$ to highlight the effect.

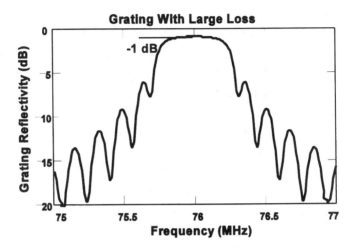

FIG. 11.23. Grating reflectivity with attenuation coefficient increased to $\alpha = 1\,\text{cm}^{-1}$, to emphasize effect. Horizontal scale: 75 to 77 MHz. Vertical scale: −20 to 0 dB.

Example 11.3 SAW Reflection Grating on Lithium Niobate. A SAW reflection grating of length $L = 0.96\,\text{cm}$ is fabricated on YZ-lithium niobate for operation at 100 MHz. The grating is of open metal strips, with grating coefficient $\kappa = 4.5\,\text{cm}^{-1}$. Calculate (a) the grating reflection coefficient ρ at center frequency and (b) the number of elements in the grating. Assume that the grating is a lossless one. ∎

Solution. (a) For a lossless grating at center frequency, Eq. (11.13) reduces to $\sigma = \kappa$. In Eq. (11.25)

$$\rho(\Delta = 0) = \tanh(\sigma L) = \tanh(\kappa L) = \tanh(4.5 \times 0.96) = +0.999.$$

(b) The acoustic wavelength at center frequency is

$$\lambda_o = v/f_o = 3488 \big/ \left(100 \times 10^6\right) \approx 34 \times 10^{-4}\,\text{cm}.$$

The grating period is $\Lambda = \lambda_o/2 = 17 \times 10^{-4}\,\text{cm}$. From Eq. (11.22), the number of elements in the grating is $N_g = L/\Lambda = 0.96/(17 \times 10^{-4}) = 564$ open metal strips. ∎

11.7.5. The 3×3 IDT Transmission Matrix [T]

A 3×3 transmission matrix is required to relate electrical and acoustic parameters for each IDT. The transmission matrix [T] can be equated to these by

$$\begin{pmatrix} w_{i-1}^+ \\ w_{i-1}^- \\ b_i \end{pmatrix} = [T] \begin{pmatrix} w_i^+ \\ w_i^- \\ a_i \end{pmatrix}, \tag{11.26}$$

where b_i and a_i, respectively, denote complex electrical output and input strengths at the i th port. Reference planes for the IDT are as shown in Fig. 11.24.

The IDT matrix elements in Eq. (11.26) are given as [6]

$$[T] = \begin{pmatrix} t_{11} & t_{12} & t_{13} \\ -t_{12} & t_{22} & t_{23} \\ st_{13} & -st_{23} & t_{33} \end{pmatrix}, \tag{11.27}$$

where s is a symmetry parameter and $s = 1$ or $s = -1$ for an IDT with an even or odd number of electrodes N_t.

11.7.6. ACOUSTIC AND ELECTRICAL TERMS IN [T]

In the notation of Reference [6], subscripts 1 and 2 relate to SAW "acoustic ports" and matrix elements within the 2×2 acoustic submatrix in Eq. (11.27). Terms in Eq. (11.27) involving subscript 3 relate to *electrical*

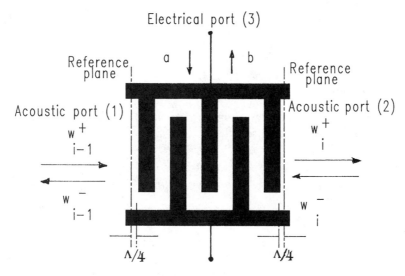

FIG. 11.24. Reference planes at offsets $\lambda/8 = \Lambda/4$, as used here for an IDT, together with designations for acoustic and electrical ports.

transfer parameters. In this way Eq. (11.26) may be split into two equations. One of these relates SAW amplitudes at and within the IDT. The other concerns the electrical output from the IDT. The equation for acoustic waves at the $(i+1)$th and ith reference planes in Fig. 11.24 is

$$[W_{i-1}] = [t_i][W_i] + a_i \cdot [\tau_i], \tag{11.28}$$

where a_i is the input electrical signal at the ith plane, and $[t_i]$ is the acoustic submatrix in Eq. (11.28) given by

$$[t_i] = \begin{pmatrix} t_{11} & t_{12} \\ -t_{12} & t_{22} \end{pmatrix}_i, \tag{11.29}$$

while $[\tau_i]$ is the column matrix relating to input coupling so that

$$[\tau_i] = \begin{pmatrix} t_{13} \\ t_{23} \end{pmatrix}_i. \tag{11.30}$$

The electrical signal b leaving the IDT in Fig. 11.24 is given by

$$[b_i] = [\tau_i'][W_i] + a_i \bullet [t_{33}]_i, \tag{11.31}$$

where the symbol "$\bullet$" in Eqs. (11.28) and (11.31) represents the scalar (dot) product, while $[\tau']$ is an output coupling column matrix

$$[\tau_j'] = s \bullet \begin{pmatrix} t_{13} \\ -t_{23} \end{pmatrix}_i \tag{11.32}$$

in terms of symmetry parameter s.

11.7.7. INDIVIDUAL TERMS IN IDT MATRIX [T]

From a knowledge of the scattering matrix of an IDT [6], [30], the elements of Eq. (11.27) are obtained as

$$[T] = \begin{pmatrix} s(1+t_o)e^{j\theta_t} & -st_o & t_{13} \\ st_o & s(1-t_o)e^{-j\theta_t} & t_{13}e^{-j\theta_t} \\ st_{13} & -st_{13}e^{-j\theta_t} & t_{33} \end{pmatrix}, \tag{11.33}$$

where

$$t_o = \frac{G_a(R_s + Z_e)}{1 + j\theta_e} \tag{11.34}$$

$$t_{13} = \frac{\sqrt{2G_a Z_e}}{1 + j\theta_e} e^{j\theta_t/2} \qquad (11.35)$$

$$t_{33} = 1 - \frac{2j\theta_c}{1 + j\theta_e} \qquad (11.36)$$

$$s = (-1)^{N_t} \qquad (11.37)$$

is a symmetry parameter for the IDT with N_t electrodes, such that transit angle

$$\theta_t = N_t \Lambda \delta. \qquad (11.38)$$

In addition,

$$\theta_c = \omega C_T (R_s + Z_e) \qquad (11.39)$$

and

$$\theta_e = (\omega C_T + B_a)(R_s + Z_e). \qquad (11.40)$$

The total IDT capacitance C_T is

$$C_T = \frac{(N_t - 1)C_s}{2} \qquad (11.41)$$

for C_s = static-capacitance/electrode pair, Z_e = load resistance or source resistance, N_t = number of IDT electrodes (not pairs), R_s = combined IDT metal and lead resistance.

Assuming that finger reflection effects can be neglected within the narrowband resonator response, radiation conductance G_a can be represented by the unperturbed sinc function expression,

$$G_a = G_o (N_t - 1)^2 \left[\sin\left(\frac{\theta_t}{2}\right) / \left(\frac{\theta_t}{2}\right) \right]^2, \qquad (11.42)$$

where

$$G_o = 8K^2 C_s f_o \qquad (11.43)$$

and K^2 = electromechanical coupling constant. The corresponding radiation susceptance B_a is

$$B_a \approx 2G_o (N_t - 1)^2 \left[\frac{(\sin(\theta_t) - \theta_t)}{\theta_t^2} \right]. \qquad (11.44)$$

11.7.8. The Acoustic Transmission Line Matrix [D]

The remaining matrix for the two-port SAW resonator design is a 2×2 one corresponding to each acoustic transmission line separating IDTs and SAW reflection gratings. This is

$$[W(d)] = [D][W(0)], \tag{11.45}$$

where the elements of complex matrix [D] are

$$[D] = \begin{pmatrix} e^{j\beta d} & 0 \\ 0 & e^{-j\beta d} \end{pmatrix} \tag{11.46}$$

in terms of phase constant $\beta = 2\pi/\lambda$ and acoustic line length d between appropriate reference planes.

11.7.9. Overall Acoustic Matrix [M] for the Two-Port SAW Resonator

As shown in Fig. 11.25, the overall acoustic matrix [M] of the constituents of the two-port SAW resonator is now obtained as the product of the composite building blocks, such that [M] is a 2×2 complex acoustic matrix

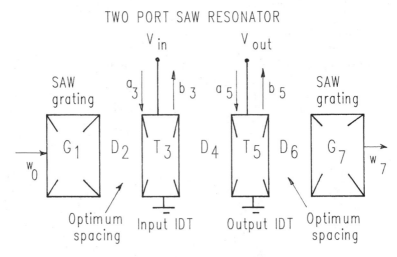

Fig. 11.25.   Representation of a two-port, one-pole SAW resonator in terms of its constituent matrix building blocks. The notation use here follows Reference [6].

$$[M] = [G_1][D_2][t_3][D_4][t_5][D_6][G_7].$$ (11.47)

The subscript notation follows Cross and Schmidt [6]. Here, $[G_1]$ and $[G_7]$ relate to the two SAW reflection gratings, $[D_2]$ and $[D_6]$ are the spacings between the gratings and adjacent IDTs, while $[D_4]$ gives the separation between IDTs. The remaining 2×2 matrices $[t_3]$ and $[t_5]$ are the acoustic submatrices within IDT transmission matrix [T]. In Figure 11.25 the terms a and b alongside IDTs T_3 and T_5 relate the incident and reflected electrical signal amplitudes, respectively, at these terminals.

11.7.10. APPLICATION OF BOUNDARY CONDITIONS

The frequency response of the two-port SAW resonator cannot be computed until boundary conditions are applied and source and load impedances are included. From Eq. (11.28), however, the SAW amplitudes associated with transducer T_3 in Fig. 11.25 are given by

$$[W_2] = t_3[W_3] + a_3\tau_3,$$ (11.48)

which represents two equations with four unknowns $W_2^{\pm}$ and $W_3^{\pm}$. Two more equations can be obtained by applying boundary equations. As there are no surface waves externally incident on the reflection gratings in the two-port resonator, set

$$w_0^+ = w_7^- = 0.$$ (11.49)

Applying boundary conditions to the reference axes of transducer T_3 gives

$$[W_0] = ([G_1][D_2])[W_2]$$ (11.50)

and

$$[W_3] = ([t_5][D_6][G_7])[W_8].$$ (11.51)

On combining Eqs. (11.48), (11.50) and (11.51), the outward propagating SAW waves w_7^+ and w_0^- are related to the input voltage a_3 by

$$\begin{pmatrix} 0 \\ W_0^- \end{pmatrix} = [M] \begin{pmatrix} W_7^- \\ 0 \end{pmatrix} + a_3[G_1][D_2][\tau_3],$$ (11.52)

where the overall 2×2 acoustic matrix [M] is given in Eq. (11.47). In addition, a choice of matched conditions at input and output yields voltage values

$$b_3 = a_7 = 0.$$ (11.53)

At the output transducer T_5,

$$[W_5] = [D_6][G_7][W_7].$$ (11.54)

Finally, the electrical output voltage V_{out} is derived from the scalar product

$$V_{out} = b_5 = [\tau'] \bullet [W_5],$$ (11.55)

while the overall phase response $\phi(f)$ is

$$\phi(f) = \tan^{-1}\left(\frac{Imaginary(V_{out})}{Real(V_{out})}\right).$$ (11.56)

11.7.11.   ILLUSTRATIVE FREQUENCY RESPONSE COMPUTATION

Figure 11.26 shows an illustrative FORTRAN computation of the magnitude response of a two-port SAW resonator on YZ-lithium niobate, using the forementioned relationships. Figure 11.27 shows the associated phase response. Parameters used here were: source and load resistances $= 50\,\Omega$; reflection grating length $L = 0.96\,\text{cm}$; total fingers (not pairs) in each IDT $N_t = 5$; self-coupling coefficient $\kappa_{11} = 0$; grating mutual-coupling coefficient $\kappa = \kappa_{12} = 4.5\,\text{cm}^{-1}$; grating loss coefficient $\alpha = 0.045\,\text{cm}^{-1}$; separation between

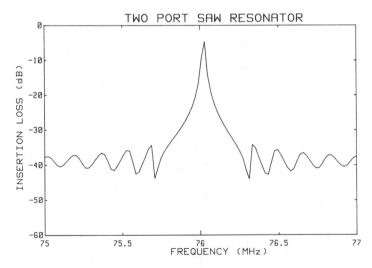

FIG. 11.26.   FORTRAN computation for the amplitude response of a 76-MHz, two-port one-pole SAW resonator. Grating lengths $L = 0.96\,\text{cm}$, $N_t = 5 =$ number of fingers in each IDT, $\kappa = 4.5\,\text{cm}^{-1}$, $\alpha = 0.045\,\text{cm}^{-1}$. IDT separation $= 10\,\lambda_o$. Contact resistance $= 7\,\Omega$. Compare with the experimental response of Fig. 11.16.

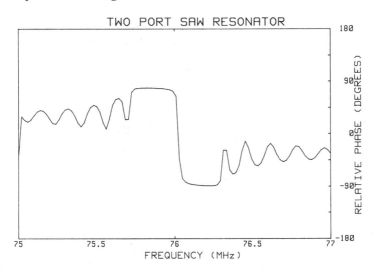

Fig. 11.27. FORTRAN computation of the phase response for the 76-MHz, two-port SAW resonator. Compare with the experimental response of Fig. 11.26.

IDTs = 10 wavelengths at center frequency $f_o = 76$ MHz; and IDT film and lead resistance = 7 Ω. These predicted responses may be compared with the experimental ones of Fig. 11.16. The excellent agreement between experimental and theoretical responses supports the use of an unperturbed radiation conductance here.

11.8. Transverse Modes in SAW Resonator Gratings

In Section 11.5, it was shown that longitudinal multimodes can degrade the desirable frequency response of a two-port SAW resonator. Another source of degradation (both in one-port and two-port SAW resonators) is due to spurious *transverse modes* that occur *inside* the SAW reflection gratings.

Spurious longitudinal multimode responses that occur between gratings contribute to out-of-band amplitude and phase ripple on both sides of the resonator response peak [15], [20]. These can be minimized in a number of ways, using design trade-offs of varying severity. One technique already considered in this chapter involves the use of synchronous placement of IDTs and adjacent reflection gratings, rather than the optimal one for minimum insertion loss. Such multimode responses can readily be computed using the transmission-matrix techniques already detailed herein.

Diffraction and beam-steering of a SAW wave in a uniform reflection grating of finite width can give rise to spurious higher-order *transverse* modes of grating vibration. These spurious modes appear on the high-frequency side of the grating stopband, and can degrade the Q of single-pole, one-port and two-port SAW resonators [16], [18], [32], [33] as well as multiple-pole, resonator-filters such as waveguide-coupled ones [34]. In addition, transverse modes can corrupt the voltage-frequency response of a voltage-controlled SAW oscillator employing a SAW resonator in the feed-back loop.

Transverse modes [35], [36] can be reduced or suppressed by several techniques, including: 1) reducing the acoustic aperture of the IDTs and reflection gratings in the resonator, to shift the transverse modes to higher frequencies, 2) using tapered reflection gratings [37]; or 3) using reflection gratings comprising of both electrical-shorted and open-reflection grating strips, with positive and negative reflectivities [38], [39]. It should be noted, however, that transverse modes are less likely to occur in SAW resonators on strongly autocollimating substrates such as YZ-lithium niobate, due to a resultant decrease in diffraction or beam steering in these substrates.

11.9. Bleustein-Gulyaev-Shimizu (BGS) PZT Ceramic Resonators

As discussed in Chapter 2, BGS waves on PZT piezoelectric ceramic sub-strates can experience high reflectivity at the propagation edges of the substrate [40]–[42]. A polarization parallel to the plane of the PZT sub-strate is essential for BGS propagation.

Using BGS wave propagation, one-port resonators can be fabricated without the use of end reflection gratings, thereby achieving a considerable savings in device size. Because conventional PZT piezoelectric ceramics can have large variations in wave propagation velocity, not all such ceramics are suitable for this application. The PZT-1 and PZT-2 ceramics have been developed for resonator action: PZT-1 ceramics have a high K^2 value ($K^2 = 22\%$) and small TCD (9 ppm/°C), and are suitable for low-Q resonator applications and PZT-2 ceramics have small K^2 ($K^2 = 4\%$) and small TCD (7 ppm/°C), and are suitable for high-Q resonator applications [40]–[42].

As illustrated in Fig. 11.28 for a device with N finger pairs, inner IDT electrodes have finger widths of $\lambda/4$, while the end electrodes at substrate edges have widths of $\lambda/8$. As for a waveguide resonant cavity, this will give a resonance condition when the distance between the two edges is set at integer multiples of $\lambda/2$. As shown in Fig. 11.28, and provided edge defini-tion is precise, all of the possible even-mode spurious responses at ($2N \pm 2$, $4, \ldots$) will be eliminated as they occur at nulls of the IDT frequency response [40]–[42]. Only the $2N$ mode at center frequency remains for

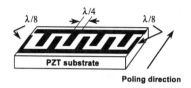

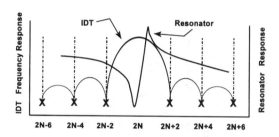

FIG. 11.28. (Upper) BGS one-port resonator showing poling direction. (Lower) shows cancellation of all modes except the central one, when edge-to-edge dimensions are exactly $n\lambda/2$ for integer n. (After Reference [42].)

excitation, provided that the edge-to-edge distance deviation is less than about $\lambda/20$. Figure 11.29 is a photograph of a section of a PZT wafer and BGS resonator fabrications, showing elements after dicing and packaged components [42]. Figure 11.30(a) shows the impedance magnitude and phase response of a 109-MHz BGS resonator designed for the DECT demodulator in Fig. 11.30(b). As this resonator application required a wideband frequency response as well as reduced capacitance, it was designed with a divided IDT. The resonance frequency was 106.1 MHz, with a frequency difference of 4.44 MHz between resonance and anti-resonance frequencies ($\Delta f/f_r = 4.2\%$). The sensitivity of this demodulator was -88 dBm, with a peak-to-peak voltage output of 800 mV, and a temperature coefficient of 18 ppm/°C. This performance compared most favorably with an equivalent LC resonator-based demodulator with a sensitivity of -83 dBm, peak-to-peak voltage output of 400 mV, and a temperature coefficient of 220 ppm/°C [42].

11.10. Summary

This chapter has reviewed the theory and operational characteristics of one-port and two-port Rayleigh-wave resonators and reflection gratings, as well as one-port BGS resonators. By themselves, single-pole, Rayleigh-wave

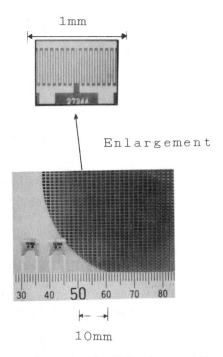

FIG. 11.29. Photograph of a section of a PZT wafer and BGS resonator fabrications, showing elements after dicing and packaged components. (Reprinted with permission from Morozumi, Kadota and Hayashi, [42], © IEEE, 1996.)

resonators find a major market in circuitry for low-power wireless communications. The BGS resonators have found application in the feedback-delay stages of DECT demodulators. Additionally, and as will be examined in Chapter 13, one-port LSAW resonators are building blocks for ladder filters in front-end RF filters and antenna duplexers, while two-port LSAW resonator-filters are also employed for RF front-end and interstage filtering. As will be demonstrated in Chapter 15, two-port Rayleigh-wave resonators are constituents for multiple-pole IF filters for mobile phones.

Because of the vital importance of surface-wave resonator structures in mobile/wireless circuitry, analytic transmission-matrix and coupling-of-modes techniques were presented in some detail, because these can be applied to a range of Rayleigh-wave and LSAW devices. The all-important locations of the SAW reflection gratings for optimum performance were examined in relation to SAW standing wave distributions in the resonant cavity. Resonator Q-parameters and device stabilities were considered for various oscillator applications.

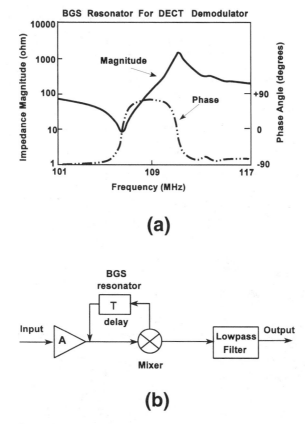

(a)

(b)

Fig. 11.30. (a) Impedance magnitude and phase response of 109-MHz BGS resonator for DECT demodulator. Horizontal scale: 101 to 117 MHz. Vertical scales: 1 to 10 kΩ and −90° to +90°. (b) BGS resonator in delay circuit of DECT demodulator. (After Reference [42].)

11.11. REFERENCES

1. E. A. Ash, "Surface wave grating reflectors and resonators," *Proc. IEEE G-MTT International Microwave Symp.*, pp. 385–386, 1970.
2. C. S. Hartmann and R. C. Rosenfeld, U.S. Patent No. 3,886,504, May 1985.
3. M. E. Mierzwinski and M. E. Terrien, "280-MHz production SAWR," *Hewlett-Packard Journal*, vol. 32, pp. 15–16, Dec. 1981.
4. RF Monolithics, *1995 Product Data Book*, RF Monolithics Inc., Dallas, Texas, 1995.
5. T. Shiba, A. Yuhara, M. Moteki, Y. Ota, K. Oda, K. Tsubouchi, "Low loss SAW matched filters with low sidelobe sequences for spread spectrum communications," *Proc. 1995 IEEE Ultrasonics Symp.*, vol. 1, pp. 107–112, 1995.
6. P. S. Cross and R. V. Schmidt, "Coupled surface-acoustic-wave resonators," *Bell System Technical Journal*, vol. 56, pp. 1447–1481, 1977.

7. E. G. S. Paige, A. G. Stove and R. C. Woods, "SAW reflection from aluminium strips on LiNbO3," *Proc. 1981 IEEE Ultrasonics Symposium*, vol. 1, pp. 144–147, 1981.
8. C. Dunnrowicz, F. Sandy and T. Parker, "Reflection of surface waves from periodic discontinuities," *Proc. 1977 IEEE Ultrasonics Symp.*, pp. 386–390, 1976.
9. D-P Chen and H. A. Haus, "Analysis of metal-strip SAW gratings and transducers," *IEEE Trans. Sonics and Ultrasonics,* vol. SU-23, pp. 395–408, May 1985.
10. T. E. Parker and G. K. Montress, "Precision surface-acoustic-wave (SAM) oscillators," *IEEE Trans. on Ultrasonics, Ferroelectrics, and Frequency Control*, vol. 35, pp. 342–364; May 1988.
11. J. A. Greer, G. K. Montress, and T. E. Parker and, "Applications of laser-trimming for all quartz package, surface acoustic wave devices," *Proc. 1989 IEEE Ultrasonics Symp.*, pp. 179–184, 1989.
12. T. E. Parker and G. K. Montress, "Frequency stability of high performance SAW oscillators," *Proc. 1989 IEEE Ultrasonics Symp.*, pp. 37–45, 1989.
13. E. J. Staples, J. S. Schoenwald, R. C. Rosenfeld and C. S. Hartmann, "UHF surface acoustic wave resonators," *Proc. 1974 IEEE Ultrasonics Symp.*, pp. 245–252, 1974.
14. E. A. Ash, "Fundamentals of signal processing devices," in A. A. Oliner (ed.), *Acoustic Surface Waves*, Topics in Applied Physics, vol. 24, Springer-Verlag, New York, p. 113, 1978.
15. W. H. Haydl, B. Dischler and P. Hiesenger, "Multimode SAW resonators—a method to study optimum performance," *Proc. 1976 IEEE Ultrasonics Symp.*, pp. 287–296, 1976.
16. W. J. Tanski, "Developments in resonators on quartz," *Proc. 1977 IEEE Ultrasonics Symposium*, pp. 900–904A, 1977.
17. R. A. Chipman, *Transmission Lines,* Schaum's Outline, McGraw-Hill, New York, 1968.
18. W. J. Tanski and H. van der Vaart, "The design of SAW resonators on quartz with emphasis on two ports," *Proc. 1976 IEEE Ultrasonics Symp.*, pp. 260–265, 1976.
19. P. S. Cross and W. A. Shreve, "Synchronous IDT SAW resonators with Q above 10,000," *Proc. 1979 IEEE Ultrasonics Symp.*, pp. 824–829, 1979.
20. C. K. Campbell, "Narrow-band filter design using a staggered multimode SAW resonator," *IEEE Trans. Sonics Ultrasonics,* vol. SU-32, pp. 65–70, January 1985.
21. C. K. Campbell and C. B. Saw, "Analysis and design of low-loss SAW filters using single-phase unidirectional transducers," *IEEE Trans. Sonics and Ultrasonics, Ferroelectrics and Freq. Control*, pp. 357–367, May 1987.
22. J. J. Gagnepain, "Rayleigh Wave Resonators and Oscillators," in E. A. Ash and E. G. S. Paige (eds), *Rayleigh Wave Theory and Application*, Springer-Verlag, Berlin, pp. 151–172, 1985.
23. H. Kogelnik, "Coupled wave theory for thick hologram gratings," *Bell System Technical Journal*, vol. 48, pp. 2909–2947, November 1969.
24. H. Kogelnik and C. V. Shank, "Coupled-wave theory of distributed feedback lasers," *Journal Applied Physics*, vol. 43, pp. 2327–2335, May 1972.
25. H. Haus and C. V. Shank, "Antisymmetric taper of distributed feedback lasers," *IEEE J. Quantum Elec*, vol. QE-12, pp. 532–538, Sept. 1976.
26. H. A. Haus and P. V. Wright, "The analyses of grating structures by coupling-of-modes theory," *Proc. 1980 IEEE Ultrasonics Symp.*, vol. 1, pp. 277–281, 1980.
27. P. V. Wright, "A coupling-of-modes analysis of SAW grating structures", Ph.D. Thesis in Electrical Engineering, Massachusetts Institute of Technology, MA, April 1981.
28. W. R. Smith, Jr., "Studies of microwave acoustic transducers and dispersive delay lines," Ph.D. Thesis in Applied Physics, Stanford University, CA, December 1969.
29. W. R. Smith, H. M. Gerard, J. H. Collins, T. M. Reeder and H. J. Shaw, "Analysis of interdigital surface wave transducers by use of an equivalent circuit model," *IEEE Trans. Microwave Theory and Techniques*, vol. MTT-17, pp. 856–864, November 1969.

30. S. G. Joshi and P. Sudhakar, "Scattering parameters of interdigital surface acoustic wave transducers," *IEEE Trans. Sonics and Ultrasonics*, vol. SU-24, pp. 201–206, May 1977.
31. L. A. Coldren and R. L. Rosenberg, "Scattering matrix approach to SAW resonators," Proc. 1976 IEEE Ultrasonics Symp., pp. 266–271, 1976.
32. C. K. Campbell, "Transverse modes in one-port SAW resonators," *IEEE Trans. on Ultrasonics, Ferroelectrics and Frequency Control*, vol. 39, pp. 785–787, Nov. 1992.
33. E. J. Staples and R. C. Smythe, "Surface acoustic wave resonators on ST-quartz," *Proc. 1975 IEEE Ultrasonics Symp.*, pp. 307–310, 1975.
34. C. K. Campbell, P. M. Smith and P. J. Edmonson, "Aspects of modelling the frequency response of a two-port waveguide-coupled SAW resonator-filter," *IEEE Trans. on Ultrasonics, Ferroelectrics and Freq. Control*, vol. 39, pp. 768–773, Nov. 1992.
35. H. A. Haus, "Modes in SAW grating resonators," *J. Applied Physics,* vol. 48, pp. 4955–4961, December 1977.
36. H. A. Haus and K. L. Wang, "Modes of grating waveguide," *J. Applied Physics*, vol. 49, pp. 1061–1069, May 1978.
37. P. S. Cross, "Surface acoustic wave resonator- filters using tapered gratings," *IEEE Trans. on Sonics and Ultrasonics*, vol. SU-25, pp. 313–319, Sept. 1978.
38. M. Takeuchi and K. Yamanouchi, "Self-suppression effects of spurious transverse modes in SAW reflectors and resonators with a positive and negative reflectivity," *Proc. 1985 IEEE Ultrasonics Symp.*, pp. 266–269, 1985.
39. M. Takeuchi and K. Yamanouchi, "New types of SAW reflectors and resonators consisting of reflecting elements with positive and negative reflection coefficients," *IEEE Trans. on Ultrasonics, Ferroelectrics, and Freq. Control*, vol. 33, pp. 369–375, July 1986.
40. M. Kadota, K. Morozumi, T. Ikeda, and T. Kasanami, "Ceramic resonators using BGS waves," *Japanese Journal of Applied Physics*, vol. 31, Supplement 31-1, pp. 219–221, 1992
41. K. Morozumi, M. Kadota, and S. Hayashi, "Characteristics of Bleustein-Gulyaev-Shimizu wave resonators on ceramics at high frequencies," *Japanese Journal of Applied Physics*, vol. 35, Part 1, No. 5B, pp. 2991–2993, May 1996.
42. K. Morozumi, M. Kadota, and S. Hayashi, "Characteristics of BGS wave resonators using ceramic substrates and their applications," *Proc. 1996 IEEE Ultrasonics Symp.*, vol. 1, pp. 81–86, 1996.

—12—

Single-Phase Unidirectional Transducers For Low-Loss Filters

12.1. Introduction

Single-phase unidirectional transducers (SPUDTs) are employed for low-loss IF and RF filtering in mobile and wireless communications. SPUDT development for IF filtering applications received added impetus from the use of low-loss IF filter stages in cordless phone systems to reduce both power and noise. These SPUDT filters are capable of operating with insertion losses of less than 3 dB. Filter fractional bandwidths can range from about 0.1 to 5%, although SPUDT filters have been designed for 10% bandwidth operation with some trade-off in insertion loss [1]. The IF frequencies for mobile phones depend on system design, and currently vary from about 45 MHz to over 450 MHz, with 459 MHz being applied to some combined GSM/DCS1800 phones.

There are a variety of applications for SPUDT-based filters. As will be examined in more detail in Chapter 15, they are employed in IF circuitry stages of GSM transceivers with 200 kHz channel spacing, as well as in CDMA mobile systems with a large bandwidth per channel (e.g., 1.23 MHz for the IS-95 North American Digital Cellular architecture). IF filter stages in mobile-phone receivers must provide adequate out-of-band suppression of spurious interference. In European GSM phone systems, for example, while the main power is concentrated within about 160 kHz in a 200-kHz channel, the signal spectrum can extend into adjacent channels [2] as illustrated in Fig. 12.1. In this GSM system, the IF bandwidth at 20 to 25 dB insertion-loss levels usually has to be less than 200 kHz, with an ultimate rejection of 50 to 80 dB. These isolation levels can be readily attained with single or cascaded SPUDT-filter designs [3]. Where temperature stability is critical, SPUDT IF filters are normally fabricated on ST-quartz. Also, and to be examined further in Chapter 15, composite SPUDT/reflection grating IF filters are employed in European DECT cordless phones with channel spacings of 1.728 MHz.

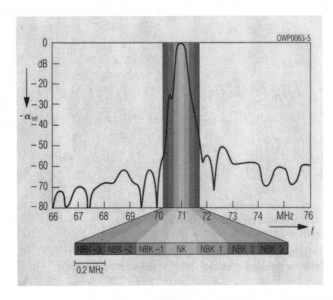

FIG. 12.1.   Typical frequency response of IF filter for GSM phone with 200 kHz channels, illustrating permissible spread into adjacent channels. (Courtesy of Siemens-Matsushita Components GmbH & Co. KG, Munich, Germany.)

The use of SPUDT-based filters in mobile/wireless communications is, however, not solely restricted to IF filter functions. As one radically different example, low-loss SPUDT delay lines are employed in RF stages of time-diversity ASH wireless receivers (Fig. 10.10), with center frequencies ranging from 180 to 450 MHz. The target specifications of these time-diversity receivers include a sensitivity of −100 dBm at 1.0 kb/s data rate, and a sensitivity of −85 dBm for data rates of 56 kb/s. Ultimate out-of-band rejections are prescribed as 80 dB [4]. Low-loss SPUDT-based delay lines with 1.5-MHz bandwidths are employed [5] for incoming data storage and transfer. Figure 12.2 illustrates the response of one of these 0.5-μs SAW delay lines to a 0.5-μs RF pulse input.

As still another example of their circuit-application diversity, SPUDT structures can also be designed to form low-loss comb filters for use in frequency-hopping (FH) oscillators [6], [7]. Hopping oscillators can find application for WLAN spread-spectrum operation in ISM bands, employing direct-sequence (DS) code expansion in conjunction with frequency hopping [8].

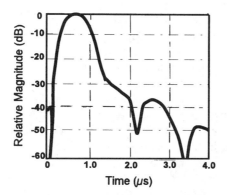

Fig. 12.2.   Typical delay response of a $0.5\,\mu s$ SAW delay line, due to RF impulse stimulus at delay-line frequency. (After Reference [4].)

12.1.1. Scope of This Chapter

Various SPUDT configurations can be employed to implement low-loss SAW filters. The original SPUDT structure employed split-electrode IDTs fabricated in a two-stage metallization process [9], [10]. A second SPUDT filter design introduced in 1983 employed a single-metallization comb-type structure [11]. Other SPUDT designs include those of the floating-electrode unidirectional transducer (FEUDT)[1] [12], as well as those using natural orientations of SAW substrates [13]. Due to its design versatility, the comb-type SPUDT design is the focus of study in this chapter. Chapter 15 will examine further aspects of SPUDT-based designs for low-loss IF filter stages in mobile-phone receivers.

This chapter first reviews the principles of operation of low-loss SAW filters employing SPUDTs. In presenting these, it is convenient to consider first comb filter structures before progressing to single passband responses. This is because both "standard" SAW comb filters and the comb-type SPUDT design employ ladder-type interdigital transducer constructions, where the IDT fingers are clustered in periodically spaced groups or "rungs." The SPUDT construction additionally employs sets of SAW reflection gratings, judiciously interspersed within these finger groupings. By selection of the relative periodicities of the rungs in the input and output IDTs, the SPUDT device can be designed for operation as either (not both)

---

[1] The FEUDT structure on high-coupling leaky-SAW substrates is normally employed for RF filtering. This structure will be reviewed in Chapter 14.

a low-loss *narrowband* filter, or as a low-loss *wideband* comb filter [14]. Both the SPUDT-based narrowband filter and the comb filter can be analyzed using the transmission-matrix techniques of Chapter 11 [15]–[17], or from a simple understanding of SAW standing-wave patterns in these structures. Illustrative examples are given to demonstrate these principles. This is followed by sample SPUDT-based low-loss comb-filter designs. As will be demonstrated in Chapter 18, SPUDT-based comb filters can find application in multimode oscillator circuits for frequency-agile spread-spectrum communications systems [6], [7]. The chapter concludes with highlights of the double-metallization SPUDT structure, and the electrode-width-controlled SPUDT (EWC-SPUDT) commonly used in IF filtering for mobile phones, as well as on the natural SPUDT (NSPUDT). The NSPUDT structure uses conventional split-electrode IDTs, but takes advantage of unconventional piezoelectric crystal-substrate orientations. In this way the effective locations of the transduction centres can be varied with respect to the centres of reflection, in order to attain unidirectionality [13].

12.2.   Basic SAW Comb Filters Using a Tapped IDT Delay Line

Basic SAW comb filters are readily implemented with the tapped delay line IDT structure of Fig. 12.3(a). In this elementary configuration, the output transducer is split into two equal groups of unapodized IDT fingers (or "rungs" as in a ladder) with N_p finger pairs. Each group has the same finger period. Moreover, their phase centers are separated by an integer number W of acoustic wavelengths at the synchronous center frequency f_o set by the finger period. The input IDT is a wideband one with constant acoustic aperture and few finger pairs and with the same finger period as the output IDT.

In Fig. 12.3(a), the ladder grouping is arbitrarily shown as employed in the output IDT. The device is reciprocal, however, so input and output IDTs can be interchanged without harm to the operational principles.

To gain an understanding of the operation of the structure of Fig. 12.3(a), consider that the input IDT has but a few finger pairs. As a result, its transfer function amplitude response $|(\sin X)/X|_i$ is assumed to be wide enough, so that the overall comb filter response is dominated by the output IDT. The transfer function magnitude $|H(f)|$ of this comb filter can be deduced in a straightforward manner, using voltage phasor relationships. The output voltage phasor is the vector sum of component voltages induced in the two rungs by the incident surface wave. Because the two rungs are separated by an integer number W of acoustic wavelengths at center fre-

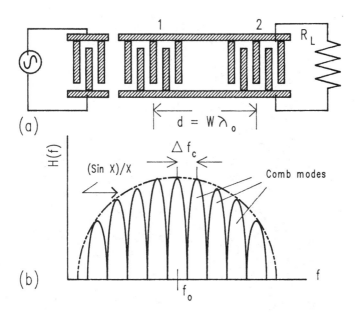

Fɪɢ. 12.3.   (a) SAW comb filter employing an output tapped delay line with two IDT "rungs." (b) Idealized frequency response of (a). Broadband response of input IDT is neglected here, for illustrative simplicity.

quency f_o, the two voltage phasors will add constructively at f_o to yield a maximum resultant voltage phasor. As the excitation frequency is varied above or below centre frequency, however, the differential phase shift $\Delta\phi$ between the rungs will deviate from the constructive amount $\Delta\phi = 2W\pi$ at f_o. The output voltage magnitude will be reduced to zero when the two phasors are 180° out of phase. These voltage magnitude variations are periodic with frequency. When superimposed on the transfer function response $|(\sin X)/X|_o$ of the unapodized comb output IDT, the result is the amplitude response depicted in Fig. 12.3(b). (The broader $|(\sin X)/X|_i$ amplitude response of the input IDT is neglected here for pictorial simplification.) If each rung in the output IDT has few finger pairs N_p, then many comb modes will be contained within its overall $|(\sin X)/X|_o$ amplitude envelope.

   The transfer function of this basic SAW comb filter can be readily derived. Consider that rungs 1 and 2 have an equal number of finger pairs N_p and that the separation d between their phase centers corresponds to SAW delay time τ. In addition, designate the individual voltages induced into IDT rungs 1 and 2 as V_1 and V_2, respectively. As functions of time and frequency these voltage components are

$$V_1 = \frac{(\sin X)}{X}\bigg|_o e^{-j2\pi ft} \tag{12.1a}$$

and

$$V_2 = \frac{(\sin X)}{X}\bigg|_o e^{-j2\pi f(t-\tau)}, \tag{12.1b}$$

where $X = N_p\pi(f-f_o)/f_o$ and N_p = the (equal) number of finger pairs in each output rung. When these individual phasors are added at convenient reference time $t = 0$, the magnitude of the resultant output voltage $|V_{out}|$ is obtained using simple trigonometric relations as,

$$|V_{out}| = |V_1 + V_2| = 2\left|\frac{(\sin X)}{X}\right|_o |\cos(\pi f\tau)|. \tag{12.2}$$

Note that the second term in Eq. (12.2) varies more rapidly with frequency than the first, thereby yielding the comb mode response within the $|(\sin X)/X|_o$ envelope of the output IDT.

The frequency separation Δf_c between adjacent comb mode peaks may be deduced from Eq. (12.2). This will depend on the excursions of the cosine term. In evaluating this, it is convenient to replace the time parameter τ in Eq. (12.2) by the equivalent expression $\tau = d/v$ where $d = W\lambda_o$ is the distance between phase centers of adjacent rungs and $v = f_o\lambda_o$. The term $\cos(\pi Wf/f_o)$ has unity magnitude when the argument is 0 or π. This gives the separation between adjacent comb mode peaks as

$$\Delta f_c = \frac{f_o}{W}. \tag{12.3}$$

Moreover, the transfer function magnitude $|H(f)|$ is

$$|H(f)| \propto |\cos(\pi f\tau)|\left|\frac{(\sin X)}{X}\right|_o \quad where \quad X = \frac{N_p\pi(f-f_o)}{f_o}, \tag{12.4}$$

as sketched in Fig. 12.3(b), which neglects the response fall-off due to the wideband input IDT.

The number of comb modes (i.e., comb "teeth") within the $|(\sin X)/X|_o$ response of Eq. (12.4) is estimated by proceeding as follows. First, from Chapter 3, the 4-dB bandwidth BW_4 of $(\sin X)/X|_o$ is given by $BW_4 \approx f_o/N_p$ where N_p = number of finger pairs in each output IDT rung. From Eq. (12.3) the separation between comb peaks is $\Delta f_c = f_o/W$, where W = number of acoustic wavelengths (at center frequency) between phase centers of adjacent rungs. As a result, the number of comb modes M within the $|(\sin X)/X|_o$ main lobe is approximately

$$M \approx \frac{BW_4}{\Delta f_c} + 1 \approx \frac{W}{N_p} + 1, \tag{12.5}$$

where N_p = number of finger pairs in each output IDT rung and W = number of acoustic wavelengths (at center frequency) between rungs.

From Eq. (12.5) it is seen that a large number of comb modes M requires a large separation $W\lambda_o$ between the two rungs, as well as a small number of finger pairs N_p in each. As the number of finger pairs N_p is reduced, however, the insertion loss of the device will increase. Typical filters of this type have insertion losses in the 20–40 dB range. A consequence is that this type of SAW comb filter can only be used in circuit stages where the signal level is high (i.e., millivolts or higher). This precludes its use in receiver "front end" applications. This limitation can be overcome, however, by the use of a low-loss SAW comb filter incorporating SAW reflection gratings.

Example 12.1 SAW Comb Filter Design Calculations. A UHF SAW comb filter with the geometry of Fig. 12.3(a) is required to operate at a center frequency $f_o = 400$ MHz. Each of the two rungs in the output IDT is composed of $N_p = 2$ finger pairs. The phase centers of the two rungs are spaced $40\lambda_o$ apart. Determine: (a) the 4-dB bandwidth of the overall response; (b) the frequency separation Δf_c between comb mode peaks; (c) the approximate number of comb modes M (i.e., comb "teeth") within the overall 4-dB bandwidth of the $(\sin X)/X|_o$ amplitude response envelope; (d) the 3-dB bandwidth of each comb mode; and (e) the percentage bandwidth of the comb mode at center frequency f_o. ■

Solution.   (a) Neglecting the assumed wideband response of the input IDT, the overall 4-dB bandwidth is $BW_4 = f_o/N_p = (400 \times 10^6)/2 = 200$ MHz. (b) The frequency separation Δf_c between comb modes is $\Delta f_c = f_o/W = (400 \times 10^6)/40 = 10$ MHz. (c) From Eq. (12.5), the number of comb modes within the overall 4-dB response is $M \approx (W/N_p) + 1 \approx (40/2) + 1 \approx 21$. (d) For this simple comb filter with only two rungs in the output IDT, the 3-dB bandwidth BW_3 of each comb mode will be one-half of the null-to-null bandwidth BW_{nn} set by the cosine roll-off term in Eq. (12.2). For this two-rung IDT obtain $BW_{nn} = \Delta f_c$, where Δf_c is the frequency separation between comb mode peaks. The 3-dB bandwidth of each comb mode is $BW_3 = 0.5 \times 10^7 = 5$ MHz. (e) The percentage bandwidth of the comb mode at f_o is $(BW_3/f_o) \times 100\% = 100 \times (5 \times 10^6)/(400 \times 10^6) = 1.25\%$. ■

12.3. SAW Comb Filters with More Complex IDT Structures

12.3.1. SAW Comb Filter with a Three-Rung IDT

Consider that the output tapped delay line has three IDT rungs, as sketched in Fig. 12.4(a). Here, a broadband input IDT illuminates the output IDT containing three rungs with N_p finger pairs in each and separations $d = W\lambda_o$ (W = integer). The transfer function of this structure will differ significantly from that of the two-rung device of Fig. 12.3. In the two-rung device the amplitudes of adjacent comb modes are essentially equal (neglecting the broadband roll-off). This is not the case with the SAW comb filter employing a three-rung IDT.

While the transfer function for the three-rung device can be derived by extending Eq. (12.1) to include three voltage terms instead of two, it is again instructive to relate this by the voltage phasor approach sketched in Fig. 12.4(b), for a physical separation between rungs of $d = W\lambda_o$ at filter center frequency f_o. Under SAW excitation, the *magnitudes* of the component voltages induced in each rung are equal, so that $|V_1| = |V_2| = |V_3| = |V|$. The magnitude of the resultant output voltage phasor $|V_{out}|$ has a maximum value $|V_{out}| = |V_1| + |V_2| + |V_3| = 3|V|$ at

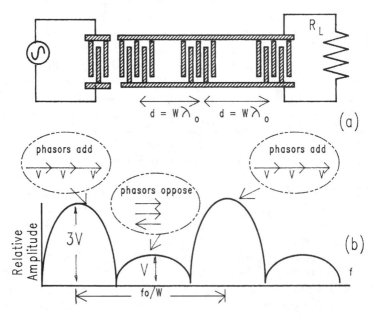

Fig. 12.4. (a) SAW comb filter using a three-rung tapped delay line. (b) Use of voltage phasors to illustrate the comb pattern. (Sin X)/(X) amplitude response excluded for simplicity.

center frequency. As before, the value of $|V_{out}|$ will vary as the excitation frequency is varied. This time, however, it will reduce to zero when the differential acoustic phase shift $\Delta\phi$ between adjacent rungs is only $\pm120°$. This means that the bandwidth of comb modes in this device is reduced from that of the two-rung filter. Moreover, $|V_{out}|$ will have the reduced value $|V_{out}|=|V|$ when the phase shift $\Delta\phi=180°$. (This reduced response may be considered to be a sidelobe.) As the frequency is varied further, the phasors will start to align so that $|V_{out}|$ again takes the value $|V_{out}|=3|V|$ when $\Delta\phi=\pm360°$.

Two points of significance emerge from an examination of these phasor relationships. The first is that the frequency separation between *principal* comb mode peaks is still $\Delta f_c = f_o/W$. The second is that the 3-dB bandwidth of each comb mode will be less than that in the two-rung structure.

12.3.2. SAW Comb Filter with Four (Or More) Rungs in One IDT

Figure 12.5 shows the comb frequency response of a four-rung output IDT, as may be derived from the forementioned voltage-phasor approach. Here, the only comb responses of maximum magnitude $4|V|$ are those that occur

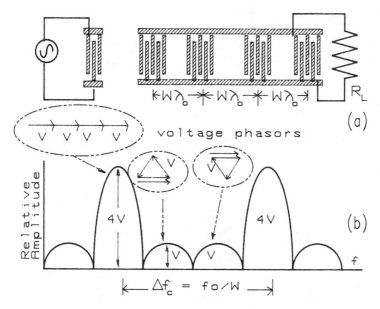

Fig. 12.5. (a) SAW comb filter using a four-rung tapped delay line. (b) Use of voltage phasors to illustrate the comb pattern. $(\sin X)/(X)$ amplitude response again excluded for simplicity.

at frequency separations $\Delta f_c = f_o/\text{W}$. Sidelobe peak magnitudes are again limited to $|V|$.

When these concepts are extended to the general case of an output IDT comprising R rungs, each with equal finger pairs N_p and separation $d = W\lambda_o$, the null-to-null bandwidth BW_{nn} of each principal peak can be deduced as

$$BW_{nn} = \frac{2f_o}{RW} \quad \left(only\ one\ IDT\ with\ rungs\right) \qquad (12.6)$$

when the broadband response of the input IDT is neglected. Here, f_o = center frequency of the overall response, R = number of rungs in the output IDT and W = integer number of wavelengths λ_o between phase centers of adjacent rungs. The 3-dB bandwidth of a single comb mode is given by

$$BW_3 \approx \frac{f_o}{RW} \quad \left(only\ one\ IDT\ with\ rungs\right). \qquad (12.7)$$

12.3.3. SAW Comb Filter with Rungs In Input and Output IDTs

Increased comb mode selectivity and bandwidth reduction is obtained if both input and output IDTs have identical rung structures, as sketched in Fig. 12.6. For this construction, the overall transfer function magnitude is $|H(\omega)| = |H_1(\omega).H_2(\omega)| = |H_1(\omega)|^2$, because the individual transfer function responses of the two transducers overlap. If the magnitude response of each

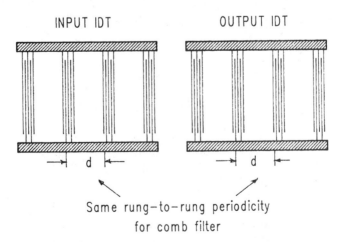

INPUT IDT   OUTPUT IDT

Same rung–to–rung periodicity for comb filter

FIG. 12.6.   The SAW comb filter formed by identical rung structures in input/output IDT.

comb mode in *one* IDT is approximated as having a cosine roll-off about center frequency, as in Eq. (12.4), the SAW comb filter will have an overall *cosine-squared* amplitude roll-off for each comb mode.

Recall that the 3-dB bandwidth level corresponds to the half-power points (or the 0.707 amplitude points), relative to the maximum response of the filter. For a cosine roll-off in amplitude, these points correspond to parameter values $\cos(\pm45°) = 0.707$. For a cosine-squared roll-off, however, the 3-dB points correspond to $\cos^2(\pm32.75°) = 0.707$. From the ratio of these arguments, the 3-dB bandwidth relation in Eq. (12.7) is replaced by

$$BW_3 \approx \frac{f_o}{1.37RW} \approx \frac{0.72f_o}{RW} \quad (equal\ rungs\ in\ input\ output). \quad (12.8)$$

for the structure of Fig. 12.6.

Example 12.2 400-MHz SAW Comb Filter Parameters.   A SAW comb filter is to be designed for operation around a center frequency $f_o = 400\,\text{MHz}$ on an ST-X quartz substrate [6]. Equal input and output IDTs are to be employed, with $R = 50$ rungs in each. The spacing d between phase centers of rungs at center frequency $f_o = v/\lambda_o$ corresponds to $d = W\lambda_o = 40\lambda_o$. In addition, each rung contains $N_p = 2$ finger pairs. Determine: (a) the 3-dB bandwidth of each comb mode; (b) the physical spacing d between phase centers of adjacent rungs; (c) the time delay τ between these phase centers; and (d) the frequency separation Δf_c between principal peaks of the response. ∎

Solution.   (a) From Eq. (12.8) $BW_3 \approx 0.72f_o/(RW) = (0.72 \times 400 \times 10^6)/(50 \times 40) = 144\,\text{kHz}$. (b) The SAW velocity on ST-X quartz is $v = 3158\,\text{m/s}$. This gives the acoustic wavelength $\lambda_o = v/f_o = 3158/(400 \times 10^6) = 7.89 \times 10^{-6}\,\text{m} = 7.89\,\mu\text{m}$, so that the spacing between phase centers of rungs is $d = 40\lambda_o = 40 \times 7.89 = 315.6\,\mu\text{m}$. (c) The time delay τ between adjacent rungs is $\tau = d/v = (315.6 \times 10^{-6})/3158 = 0.1\,\mu\text{s} = 100\,\text{ns}$. (d) The frequency spacing Δf_c between principal peaks is $\Delta f_c = 1/(0.1 \times 10^{-6}) = 10\,\text{MHz}$. ∎

12.4.   SAW Filters with Single-Phase Unidirectional Transducers (SPUDTs)

12.4.1.   THE SINGLE-METALLIZATION SPUDT

Figure 12.7 sketches four different types of single-phase unidirectional transducers (SPUDTs). Figure 12.7(a) shows the single-metallization comb-type SPUDT. Figure 12.7(b) shows the double-metallization SPUDT [9].

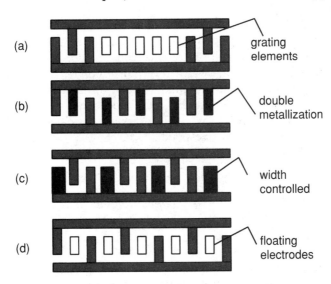

Fɪɢ. 12.7. Four illustrative SPUDT designs. (a) Comb-type SPUDT analyzed in this chapter. (b) Double-metallization SPUDT. (c) EWC-SPUDT. (d) Floating-electrode SPUDT.

Figure 12.7(c) shows the EWC-SPUDT [18], while Fig. 12.7(d) depicts the structure of the floating-electrode SPUDT [12].

The single-metallization SPUDT of Fig. 12.7(a) now examined in more detail consists of groups of SAW reflection gratings judiciously interspersed between the "rungs" of a SAW comb filter. Figure 12.8 gives a sketch of an elemental SPUDT-based SAW filter, using two IDT rungs and two SAW reflection gratings in input and output IDTs. In practice, these SPUDTs may incorporate many rungs with interspersed short-length reflection gratings of metal strips or grooves. In designs on lithium niobate, for example, each IDT rung may contain around five single-electrode or split-electrode fingers, while each constituent reflection grating may contain 10 or more metal strips. It has been found that individual metal strips on lithium niobate have SAW reflectivities on the order of 1% [19], [20]. A high degree of SAW directivity can be attained on lithium niobate with a hundred or more metal strips distributed between the reflection gratings in each SPUDT.

12.4.2. Selecting a Single or a Comb Filter Response

Each IDT rung in Fig. 12.7(a) can be configured with an even or odd number of fingers, of single- or split-electrode geometry. For constructive interference, reference planes for individual IDT rungs must be separated

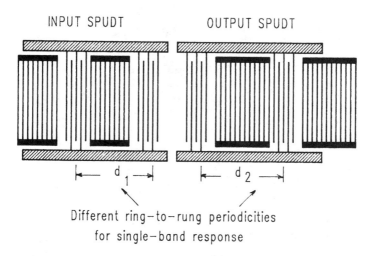

Fɪɢ. 12.8. Elemental single-passband SPUDT filter in Fig. 12.7(a). Input/output SPUDTs each employ two IDT rungs and two shorted-electrode reflection gratings. Input/output IDTs have differing rung periodicities for single passband response. Shorted-grating displacement shown here for use on LiNbO₃ substrate.

by an integer number of acoustic wavelengths at center frequency. As there is a multiplicity of rungs in both input and output ports, however, this periodicity gives rise to a second data sampling; this is in addition to that imposed by the IDT fingers themselves. Unless compensated for, this additional sampling produces a multiplicity of comb transfer function responses. A low-loss SAW comb-filter transfer function is realized if identical input and output SPUDT configurations are employed. On the other hand, a single low-loss narrowband transfer function response can be obtained by the use of differing rung periodicities in input and output SPUDTs, so that only one of the multiple comb responses of the input SPUDT will overlap with one in the output SPUDT.

12.4.3. Positioning the Reflection Gratings within the SPUDT

The single-metallization SPUDT design exploits the fact that the centre of reflection of a SAW wave is at the edge of a metallic reflector strip, while the centre of transduction in an excited IDT finger is at its centre. In a single-electrode IDT with a metallization ratio of $\eta = 0.5$, this manifests as a $\lambda/8$ acoustic wavelength difference between the two locations [10].

The transmission-matrix analytical techniques applied to SAW resonator design in Chapter 11 may be readily applied to the SPUDT structure of Fig. 12.7(a), once the correct positions of the constituent SAW reflection

INPUT SPUDT

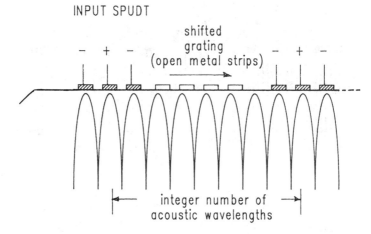

FIG. 12.9. The SAW standing-wave pattern in elemental section of excited comb-type SPUDT. Displacement of open-strip reflection grating shown here is appropriate for an input comb-type SPUDT on LiNbO₃.

TABLE 12.1

DIRECTIONS OF λ/8 GRATING DISPLACEMENT IN INPUT AND OUTPUT COMB-TYPE SPUDTs

	Open Metal Strips		Shorted Metal Strips	
	Input	Output	Input	Output
Y-X and *Y-Z* lithium niobate	⟶	⟵	⟵	⟶
ST-quartz	⟵	⟶	⟵	⟶

gratings are determined with respect to their adjacent IDT rungs. This positioning of the SAW reflection gratings should be approached with caution, as the direction of the required λ/8 displacement from the grating midposition depends on the substrate used, as well as on whether open- or shorted-metal strip reflection gratings are employed. Table 12.1 lists the relative directions of this λ/8 shift in input and output SPUDTs on lithium niobate and ST-quartz. The grating shifts in each instance relate to the appropriate reflection coefficient sign in Table 11.2 of Chapter 11.

Figure 12.8 thus shows the required λ/8 shift for shorted metal gratings in input and output SPUDTs on lithium niobate. Figure 12.9, on the other hand, illustrates the required λ/8 shift of a reflection grating composed of open metal grating strips in an input SPUDT on lithium niobate, such that the "front edges" of grating elements are located at SAW standing wave maxima, for maximum directivity.

12.5. Illustrative Design of SPUDT-Based SAW Filter

12.5.1. IDENTIFYING THE REFERENCE PLANES

For simplicity, the coupling-of-modes algorithmic design technique is demonstrated here for the elemental SPUDT-based SAW filter of Fig. 12.8, employing two IDT "rungs" and two SAW reflection gratings in both input and output SPUDTs. In practice, a typical design would have considerably more IDT rungs and SAW reflection gratings in input and output transducers.

By multiplying together an appropriate sequence of transmission matrices, closed-form expressions can be obtained for the passband response of the comb-type SPUDT of Fig. 12.7(a). The notation and parameters used in this chapter for complex transmission matrices $[G]$, $[T]$ and $[D]$ applied to SPUDT design are exactly the same as in the derivations for the SAW resonator in Chapter 11.

Consider the SPUDT filter of Fig. 12.8 to be represented by the matrix block diagram representation of Fig. 12.10, where the individual building blocks are numbered consecutively from left to right (as in the notation of Reference [16]). In Fig. 12.10 the input signal is applied to transducers T_3 and T_7 connected in parallel, while the output voltage is obtained from transducers T_9 and T_{13} connected in parallel.

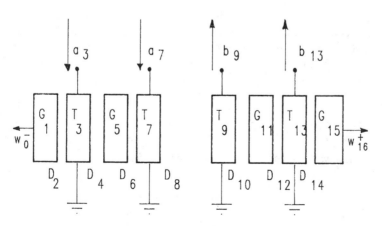

FIG. 12.10. Notation used here for transmission matrices appropriate to comb-type SPUDT-based filter geometry of Fig. 12.8.

The reference planes used here for an IDT are given in Fig. 12.11, while from Chapter 11 the 3×3 transmission matrix for each constituent IDT within a SPUDT is

$$[T] = \begin{bmatrix} t_{11} & t_{12} & t_{13} \\ -t_{12} & t_{22} & t_{23} \\ st_{13} & -st_{23} & t_{33} \end{bmatrix}, \qquad (12.9)$$

where s is a symmetry parameter given as $s = 1$ or $s = -1$ for an IDT with an even or odd number of electrodes N_t, respectively. Following the previous notation, subscripts 1 and 2 for the transmission matrix elements in Eq. (12.9) apply to "acoustic ports" and to a 2×2 acoustic submatrix, while terms with subscript 3 involve electrical ports. Acoustic and electrical parameters at the two reference planes of the IDT are related to the transmission matrix $[T]$ by

$$\begin{bmatrix} w_{i-1}^+ \\ w_{i-1}^- \\ b_i \end{bmatrix} = [T] \begin{bmatrix} w_i^+ \\ w_i^- \\ a_i \end{bmatrix}, \qquad (12.10)$$

where w^+ and w^- are forward and reverse surface wave amplitudes, respectively. Moreover, a_i and b_i are complex electrical input and output voltages, respectively, at the ith port; the reference planes are in Fig. 12.11.

12.5.2. Simplified Vector Notation

A simplifying matrix notation may be used to group forward and reverse surface acoustic waves components of complex amplitude w^+ and w^- at a given reference plane. The two components at the ith plane, for example, may be grouped as

$$[W_i] = \begin{bmatrix} w_i^+ \\ w_i^- \end{bmatrix}. \qquad (12.11)$$

In this way, forward and reverse complex amplitudes of surface acoustic wave components at input and output planes of a SAW reflection grating of length L are linked by

$$[W(-L)] = [G][W(0)]. \qquad (12.12)$$

In these formulations it is important to note that acoustic responses are computed at "input ports"; given specifications are at "output ports." Thus,

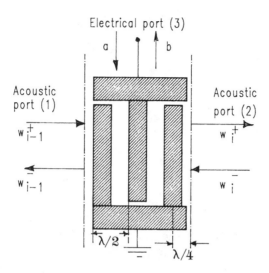

FIG. 12.11.   Reference axes used here for an IDT rung in comb-type SPUDT, with $\lambda/8$ shift from outer-electrode edges.

terms w_{i-1}^+ and w_{i-1}^- relate complex amplitudes of forward and reverse acoustic waves, respectively, at the $(i-1)$th plane due to an acoustic stimulus at the ith plane.

12.5.3.   APPLICATION TO A SAMPLE SPUDT DESIGN

Using the preceding notation, the surface acoustic wave amplitudes at the $(i-1)$th and ith reference planes for the IDT in Fig. 12.11 are related by

$$[W_{i-1}] = [t_i][W_i] + a_i[\tau_i], \qquad (12.13)$$

where a_i is the input electrical signal at the ith plane and $[t_i]$ is an acoustic submatrix within Eq. (12.9) given by

$$[t_i] = \begin{pmatrix} t_{11} & t_{12} \\ -t_{12} & t_{22} \end{pmatrix}_i. \qquad (12.14)$$

In addition, $[\tau_i]$ is the column matrix (vector) that relates to input coupling

$$[\tau_i] = \begin{bmatrix} t_{13} \\ t_{23} \end{bmatrix}_i. \qquad (12.15)$$

Likewise, the electrical signal b leaving the IDT in Fig. 12.11 is given by

$$b_i = [\tau_i'] \bullet [W_i] + a_i(t_{33})_i, \tag{12.16}$$

where the symbol $\bullet$ in Eq. (12.16) gives the scalar (dot) product and $[\tau_i']$ is an output coupling vector

$$[\tau_i'] = s \begin{bmatrix} t_{13} \\ -t_{23} \end{bmatrix}_i, \tag{12.17}$$

in terms of symmetry parameter s.

Using Eqs. (12.13) and (12.14), the SAW amplitude associated with transducer T_3 in Fig. 12.10 is

$$[W_2] = [t_3][W_3] + a_3[\tau_3]. \tag{12.18}$$

If there are no external surface waves incident on either end of the structure, the SAW boundary parameters are

$$w_o^+ = w_{16}^- = 0. \tag{12.19}$$

In addition, a choice of matched conditions in input and output SPUDTs gives

$$b_3 = b_7 = a_9 = a_{13} = 0, \tag{12.20}$$

for incident and reflected complex voltage components a and b, respectively. In terms of the three "building blocks" for the SPUDT, the two matrix equations that relate acoustic amplitudes at the reference planes for transducer T_3 are

$$[W_0] = [G_1][D_2][W_2], \tag{12.21}$$

$$[W_3] = [D_4][G_5][D_6][t_7][D_8][t_9][D_{10}][G_{11}][D_{12}][t_{13}][D_{14}][G_{15}][W_{16}]$$
$$+ a_7[D_4][G_5][D_6][\tau_7]. \tag{12.22}$$

By combining Eqs. (12.18), (12.21) and (12.22), the "outward" acoustic waves w_{16}^+ and w_0^- are obtained in terms of electrical input stimuli a_3 and a_7, as

$$\begin{bmatrix} 0 \\ w_o^- \end{bmatrix} = [M] \begin{bmatrix} w_{16}^+ \\ 0 \end{bmatrix} + a_3[G_1][D_2][\tau_3] + a_7[G_1][D_2][t_3][D_4][G_5][D_6][\tau_7], \tag{12.23}$$

where $[M]$ is the overall 2×2 acoustic transmission matrix, given by

$$[M] = [G_1][D_2][t_3][D_4][G_5][D_6][t_7][D_8][t_9][D_{10}][G_{11}][D_{12}][t_{13}][D_{14}][G_{15}].$$
(12.24)

The components of acoustic matrix $[W_{13}]$ are found from $[W_{16}]$ as

$$[W_{13}] = [D_{14}][G_{15}][W_{16}].$$
(12.25)

In this example the complex component b_{13} of the electrical output signal is

$$b_{13} = [\tau_{13}^l] \bullet [W_{13}],$$
(12.26)

while the second component b_9 of electrical output signal is obtained from

$$[W_9] = [D_{10}][G_{11}][D_{12}][t_{13}][D_{14}][G_{15}][W_{16}]$$
(12.27)

as

$$b_9 = [\tau_9^l] \cdot [W_9].$$
(12.28)

The total electrical output signal V_{out} in this example is obtained as the complex sum,

$$V_{out} = b_9 + b_{13}$$
(12.29)

for electrical input $V_{in} = a_3 = a_7 = 1$, while the overall phase response ϕ_f of the filter is obtained from Eq. (12.29) as

$$\phi_f = \tan^{-1}\left(\frac{Imaginary(V_{out})}{Real(V_{out})}\right).$$
(12.30)

The recursive algorithm applied to the foregoing illustrative example can be readily extended to SPUDT configurations of Fig. 12.7(a) with considerably more IDT "rungs" in both input and output sections.

12.6. Experimental Performance of a SPUDT-Based SAW Filter

12.6.1. ILLUSTRATIVE EXAMPLE

Figure 12.12(a) shows the measured amplitude response of a 79-MHz SPUDT-based design of Fig. 12.7(a), as fabricated on 128° *Y-X* lithium niobate. Single-electrode IDT geometry was employed here. Figure 12.12(b) shows the predicted transfer function response using the preceding analytic method. The slight dip in the passband in Fig. 12.12(b) around center frequency is attributed to regenerative triple-transit-interference. The more pronounced one in the experimental response is attributed to a

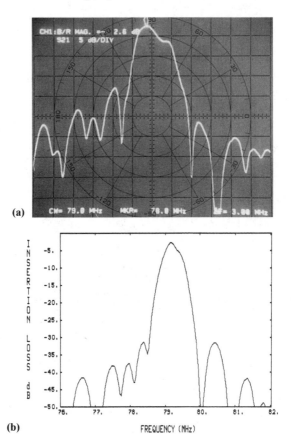

FIG. 12.12.   (a) Measured frequency response of 79-MHz comb-type SPUDT-based SAW filter on 128° YX-LiNbO₃, using ten IDT rungs and gratings in each SPUDT. Minimum matched insertion loss 2.6 dB at 78.9-MHz marker. Horizontal scale: 76–82 MHz. Vertical scale: 5 dB/div. (b) Predicted response. (Reprinted with permission from Campbell and Saw [17], © IEEE, 1987.)

combination of regenerative TTI and finger reflections from the single-electrode IDTs. Finger reflection effects were not incorporated in the version of the model described here.

In this example, both input and output transducers had 10 IDT rungs, with five single-electrode fingers in each, as well as acoustic apertures of $70\lambda_o$. The input and output SPUDTs had different periodicities, as already discussed in Section 12.4.2. To achieve this, the input SPUDT used 10 open aluminum strips in each reflection grating, while the output SPUDT used 14 open aluminum ones. In each instance the metallization ratio was $\eta = 0.5$.

With the use of open metal strips, the reflection gratings in each SPUDT were shifted $\lambda_o/8$ towards one another as indicated in Table 12.1. Separations between edges of adjacent IDT and reflection gratings were $0.375\lambda_o$ and $0.625\lambda_o$. With this design, the matched forward insertion loss was 2.6 dB at 79 MHz with a 3-dB fractional bandwidth of 1%, together with lower and upper sideband rejection of 16 and 25 dB, respectively.

The directivity of the SPUDTs was obtained by measuring the amplitude response in a second design where the same input and output SPUDTs were turned "back-to-back." With this reverse configuration the insertion loss at centre frequency was typically ~ 35 dB; giving a forward-to-reverse directivity ratio of ~ 32 dB, or 16 dB per SPUDT.

12.6.2. EXAMPLE OF DESIGN VERSATILITY

This type of SPUDT has design versatility in that trade-offs can readily be applied between bandwidth, insertion loss and close-in sidelobe suppression. As an example of this, Fig. 12.13(a) shows a network analyzer measurement of the amplitude response of a modified SPUDT structure. In this example, the two separations between adjacent IDT and reflection grating edges in input and output SPUDTs were reduced from $0.375\ \lambda_o$ to $0.25\ \lambda_o$ and from $0.625\ \lambda_o$ to $0.5\ \lambda_o$, compared with the values used for Fig. 12.12. In this second example a synchronous center frequency $f_o = 82$ MHz was employed. The forward matched insertion loss for this second device was 3.4 dB at f_o, with 3-dB fractional bandwidth of $\sim 1.5\%$. Lower and upper sidelobe suppression is now 22 and 25 dB, respectively.

Figure 12.13(b) shows the predicted response of the filter as derived from the foregoing analytic technique, while Fig. 12.13(c) is the "corrected" experimental response after applying time-gating to the impulse response. Here, both electromagnetic feedthrough and triple-transit-interference were "removed" from the experimental response, for comparison with theory.

12.6.3. INCREASING THE CLOSE-IN SIDELOBE SUPPRESSION

The SPUDT-based SAW filter designs examined in the preceding have employed input and output transducers with constant acoustic aperture in each rung of the ladder structure. By applying a suitable window function to cause aperture weighting of these SPUDTs, however, it is possible to increase the close-in sidelobe suppression at the expense of some trade-off in insertion loss.

The window function aperture-weighting technique applied to the SPUDT of Fig. 12.7(a) differs somewhat from that considered in Chapter 3.

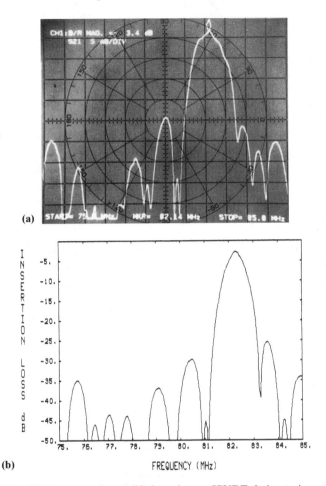

(a)

(b)

FREQUENCY (MHz)

Fɪɢ. 12.13. (a) Response of a modified comb-type SPUDT design to increase sidelobe suppression. Minimum insertion loss 3.4 dB at 82.14-MHz marker. Close-in sidelobe suppression is now greater than 20 dB. Horizontal scale: 75 to 85 MHz. Vertical scale: 5 dB/div. (b) Predicted response on same scales. (Reprinted with permission from Campbell and Saw [17], © IEEE, 1987.)

In the bidirectional transducers considered in Chapter 3, the aperture-weighting window function was applied to *all* of the electrodes in such IDTs. In the method applied to this SPUDT, however, the acoustic apertures of the IDTs in each rung are kept constant. The aperture weighting results in IDT *rungs* with differing acoustic apertures within the SPUDT. One window function that can be applied to the SPUDT is the Dolph-Chebyshev one used in antenna design [1], [21].

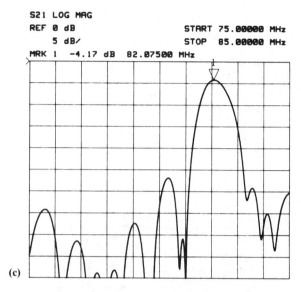

S21 LOG MAG
REF 0 dB
5 dB/
MRK 1 -4.17 dB 82.07500 MHz

START 75.00000 MHz
STOP 85.00000 MHz

(c)

FIG. 12.13, Continued. (c) "Corrected" experimental response of Fig. 12.13(a) after time-gating, to "remove" effects of TTI and electromagnetic feedthrough. Same scales as (a). (Reprinted with permission from Campbell and Saw [17], © IEEE, 1987.)

An experimental example of the use of Dolph-Chebyshev aperture-weighting in SPUDT design is shown in the network analyzer measurement of Fig. 12.14. Here, the input SPUDT incorporated 10 aperture-weighted IDT rungs, using a 35-dB Dolph-Chebyshev weighting. Each rung had five fingers with single-electrode geometry. The rung-to-rung period was $W = 8\lambda_o$. Ten unweighted SAW reflection gratings were used with 10 metal strips in each. The output SPUDT had 6 unweighted IDT rungs, with five single electrodes per rung. The rung period in the output SPUDT was $W = 10\lambda_o$, for selection of only a single narrowband filter response. Moreover, the six reflection gratings in the output SPUDT incorporated 14 metal strips per grating bank. For this 82-MHz design example on lithium niobate, the close-in sidelobe suppression is about 40 dB. This was at the expense of a slight increase in insertion loss to 4.3 dB at $f_o = 82$ MHz [1].

12.6.4. USE OF CASCADED SPUDT-BASED SAW FILTERS

A considerable increase in sidelobe suppression can be attained by the used of cascaded SPUDT SAW filters. It may be anticipated that both the overall insertion loss and close-in sidelobe suppression should double if a good

INPUT SPUDT

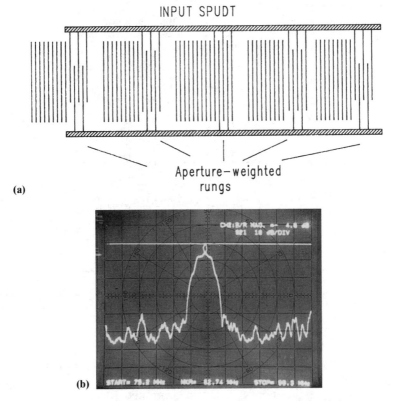

Aperture—weighted
rungs

(a)

(b)

Fɪɢ. 12.14.   (a) Aperture-weighting of input comb-type SPUDT. Open grating strips shown for input SPUDT on LiNbO₃. (b) Measured response of aperture-weighted design, with 40-dB sidelobe suppression. Insertion loss is 4.3 dB at 82.74-MHz marker. Horizontal scale: 75.2 to 90.3 MHz. Vertical scale: 10 dB/div. (Reprinted with permission from Saw and Campbell [1], © IEEE, 1987.)

match is achieved between the two filters. Using SPUDT-based filters, however, the overall increased insertion loss may still be at an acceptable level. Figure 12.15 shows the result of cascading two aperture weighted SAW filters with the individual responses as shown in Fig. 12.14. The overall insertion loss in Fig. 12.15 is ≈ 8.3 dB at center frequency with matching between the two filters; the close-in sidelobe suppression is about 70 dB. These cascaded designs on ST-quartz can be employed to meet the template specifications for the 200-kHz GSM mobile phone specification, as shown in Fig. 12.1.

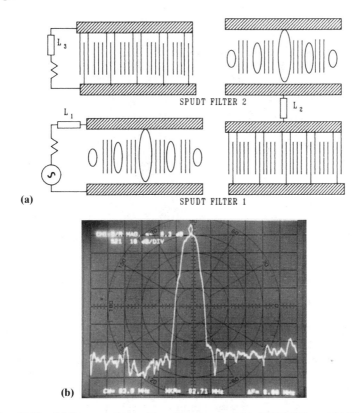

(a)

SPUDT FILTER 2

SPUDT FILTER 1

(b)

FIG. 12.15. (a) Two cascaded and tuned aperture-weighted SPUDT-based SAW filters. (b) Measured response of cascaded filters, with close-in sidelobe suppression of about 70 dB. Insertion loss is 8.3 dB at 82.71-MHz marker. Horizontal scale: 83 MHz ± 8 MHz. Vertical scale: 10 dB/div.

12.7. Low-Loss SAW Comb Filters Using Unidirectional Transducers

As considered in section 12.4.2, low-loss wideband SAW comb filters can be implemented using identical SPUDTs in input and output transducers. Although it might seem puzzling that wideband action can be realized with transducers employing narrowband SAW reflection gratings, it results from signal sampling by both the periodically spaced SAW reflections gratings and the IDT rungs [22].

Figure 12.16 illustrates the configuration of a low-loss SAW comb filter, employing identical SPUDTs in input and output stages. As may be determined from Table 12.1, this particular design is appropriate for use with a

INPUT SPUDT　　　　　　　OUTPUT SPUDT

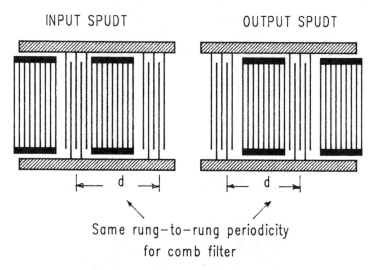

Same rung–to–rung periodicity
for comb filter

FIG. 12.16.　Low-loss comb filter uses identical SPUDTs in input and output transducers.

lithium niobate substrate. Figure 12.17(a) shows a network analyzer measurement of the amplitude response of an 80-MHz low-loss comb filter, with minimum insertion loss ≈ 3.7 dB. Five comb modes are spaced 10 MHz apart from 60 to 100 MHz. This filter employed $R = 10$ rungs in each SPUDT, with spacing $d = W\lambda_o = 8\lambda_o$ between phase centers of adjacent rungs at center frequency $f_o = 80$ MHz. Each rung contained $N = 5$ electrodes. In addition, the 10 SAW reflection gratings in each SPUDT consisted of 10 unshorted thin-film aluminum strips. Figure 12.17(b) shows the predicted response using the transmission-matrix design technique [14]. Related parameters for the design are derived in Example 12.3.

Example 12.3　Design of SPUDT-Based Low-Loss SAW Comb Filter.　A low-loss SAW tapped delay line comb filter employs identical comb-type SPUDTs in input and output transducers. It is designed for operation around a center frequency $f_o = 80$ MHz and fabricated on lithium niobate. Each SPUDT has $R = 10$ IDT rungs spaced $d = W\lambda_o = 8\lambda_o$ apart at center frequency, while the electrodes in each rung contain $N_p =$ two finger pairs. Experimentally, the overall response is found to contain five principal comb modes spaced 10 MHz apart. In addition, the insertion loss and 3-dB bandwidth of a single comb mode, as measured on a network analyzer, are 3.7 dB and 660 kHz, respectively. Calculate (a) the frequency separation between adjacent comb mode peaks and (b) the number of comb modes within the 4-dB level of the overall response. In addition, (c) compare the measured

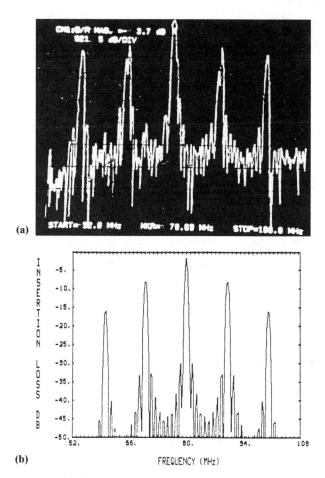

(a)

(b)

FREQUENCY (MHz)

Fɪɢ. 12.17. (a) Experimental response of low-loss comb filter using identical SPUDTs in input and output. Comb spacing is 10 MHz. Insertion loss is 3.7 dB at 80-MHz center frequency. Horizontal scale: 50 to 100 MHz. Vertical scale: 5 dB/div. (b) Predicted response. Vertical scale: 10 dB/div. (Reprinted with permission from Saw, Campbell, Edmonson and Smith [14], © IEEE, 1986.)

bandwidth result with that calculated for a "conventional" tapped delay line comb filter without grating reflectors, but with the same values of rungs and rung separations. ■

Solution. (a) From Eq. (12.3) the separation between adjacent comb peaks is $\Delta f_c = f_o/W$, where W = integer number of wavelengths between adjacent rungs. This gives $\Delta f_c = (80 \times 10^6)/8 = 10$ MHz, in agreement with experiment. (b) From Eq. (12.5) the number of modes M within the $(\sin X)/X$

overall response of each SPUDT is $M \approx (W/N_p) + 1 = (8/2) + 1 = 5$. (c) From Eq. (12.8) the 3-dB bandwidth of each comb mode peak for a "conventional" (i.e., non-SPUDT) comb filter with rungs in input and output IDTs would be $BW_3 \approx 0.72.f_o/(RW) = (0.72 \times 80 \times 10^6)/(10 \times 8) = 720 \, \text{kHz}$. This last result may be compared with the measured value of 660 kHz. ∎

12.8. Highlights of Other SPUDT Structures

12.8.1. DOUBLE METALLIZATION SPUDT

Figure 12.7(b) outlines the basic double-metallization SPUDT. Here, the transducer employs a split-electrode ($\lambda/8$) structure, with an additional metallization layer applied to selected electrodes [9], [13]. Normally, gold or chromium is used as the second overlay over the initial aluminum layer. It is noted, however, that SPUDT action should be realizable by using two thicknesses of the same metal film [23]. The ensuing unidirectionality finds its basis in the fact that while the single-level split-electrodes have no localized reflections at centre frequency, the effect of the additional metal layer on selected electrodes is to locate the centers of reflection at the centres of these thicker electrodes. Because the centres of transduction remain at the centres of the gaps between split electrodes, this establishes the required $\lambda/8$ shift between centres of transduction and reflection.

12.8.2. ELECTRODE WIDTH-CONTROLLED SPUDT (EWC-SPUDT)

One commonly used noncomb SPUDT geometry for IF filter designs in GSM phones is that shown in Fig. 12.7(c), which can be considered to be modified split-finger IDTs. This structure is known as the electrode width-controlled SPUDT (EWC-SPUDT) [18]. Its interlaced reflection strips are $\lambda/4$ wide, and withdrawal-weighting of reflectivity is achieved by replacing the $\lambda/4$ electrodes by two grounded $\lambda/8$ electrodes with $\lambda/8$ spacing. Relative positions between split-fingers and reflector strips determine the shift between centres of transduction and centres of reflection. The reflector elements can be chosen so that reflection and regeneration effects cancel over a broad frequency range when the IDTs are properly matched [24]. This structure is normally fabricated on ST-X quartz, and passband response shaping is achieved by selective withdrawal-weighting of electrodes. In designs for GSM/PCN IF filters, the general aim is to minimize the passband distortions due to spurious signals, rather than to achieve very-low insertion loss. Complex matching networks may be required for enhanced operation [18].

12.8.3. Natural SPUDT (NSPUDT) Structure

The NSPUDT structure is just a conventional split-electrode single-metallization structure. By exploiting the anisotropy of the underlying piezoelectric crystal substrate, however, orientations can be obtained whereby the centers of reflection can be displaced spatially from the centres of the electrodes to obtain the required unidirectionality. To date, useful NSPUDT orientations have been obtained for quartz, lithium niobate, lithium tantalate, and gallium arsenide [13]. Two basic quartz crystal cuts support NSPUDT action. The first is a singly rotated *y*-cut, with wave propagation $\pm 25°$ off the *X*-axis. The second quartz cut is the doubly rotated 111 cut, with higher electromechanical coupling coefficient [25]. NSPUDT operation has also been reported for lithium borate ($Li_2B_4O_7$) with Euler angles (0°, 18°, 90°) and (0°, 51°, 90°), as well as the (0°, 78°, 90°) orientation with zero temperature coefficient of delay [26].

A design complication that arises with NSPUDT structures is that the unidirectionality arises from the asymmetry of the piezoelectric crystal, rather than from the symmetric IDT configuration. As a result, all devices with the same metallization will have the same directivity direction. Because of this, a transducer with reverse directionality can only be obtained by changing the metallization [13].

12.9. A Reminder On TTI and Insertion Loss in SPUDT-Based Filters

As illustrated in Fig. 5.13 of Chapter 5, SPUDT-based filters do not obey the same rules for TTI and insertion loss relationships as do filters with conventional bidirectional IDTs. As shown in the SPUDT example of Fig. 5.13, both the passband TTI ripple and insertion loss generally decrease as matching at the electrical ports is increased. And TTI ripple decreases as the effects of acoustic reflection and piezoelectric regeneration begin to cancel out [27].

12.10. Summary

Low-loss SPUDT filters find many applications in communications circuitry, as highlighted by three diverse examples relating to: a) IF channel selection in GSM mobile phones; b) low-loss delay lines in time-diversity wireless receivers; and c) frequency-hopping SPUDT comb filters for spread-spectrum communications. As illustrated, a variety of different fabrication techniques can be used to implement a particular SPUDT design. The most versatile of these is the comb-structure design, which can be applied to realize either a comb-filter response for frequency hopping, or a single passband one for channel selection.

The transmission-matrix analytic techniques applied to SAW resonators in Chapter 11 were again applied here to illustrate both the comb or single passband capability of this particular SPUDT design.

12.11. REFERENCES

1. C. B. Saw and C. K. Campbell, "Improved design of single-phase unidirectional transducers for SAW filters," *Proc. 1987 IEEE Ultrasonics Symp.*, vol. 1, pp. 169–172, 1987.
2. J. Machui, J. Bauregger, G. Riha and I. Schropp, "SAW devices in cellular and cordless phones," *Proc. 1995 IEEE Ultrasonics Symp.*, pp. 121–130, 1995.
3. M. Solal, P. Dufilie and P. Ventura, "Innovative SPUDT based structures for mobile radio applications," *Proc. 1994 IEEE Ultrasonics Symp.*, pp. 17–22, 1994.
4. D. L. Ash, "New UHF receiver architecture achieves high sensitivity and very low power consumption," RF Design, pp. 32–44, December 1994.
5. Peter Wright, "Group single-phase unidirectional transducers with 3/8 and 5/8 sampling," Patent No. 5,073,763, December 1991.
6. M. F. Lewis, "Practical frequency source for use in agile radar," *Electronics Letters*, vol. 21, pp. 1017–1018, October 1986.
7. C. B. Saw, P. M. Smith, P. J. Edmonson and C. K. Campbell, "Mode selection in a multimode SAW oscillator using FM chirp mixing injection," *IEEE Trans. Ultrasonics, Ferroelectrics and Frequency. Control*, vol. UFFC-35, May 1988.
8. K. Pahlavan, "Wireless data communications," *Proc. IEEE*, vol. 82, pp. 1398–1430, Sept. 1994.
9. C. S. Hartmann, P. V. Wright, R. J. Kansy and E. M. Garber, "An analysis of SAW interdigital transducers with internal reflections and the application to the design of single-phase unidirectional transducers," *Proc. 1982 IEEE Ultrasonics Symp.*, vol. 1, pp. 40–45. 1982.
10. P. V. Wright, "A coupling-of-modes analysis of SAW grating structures," *Ph.D. Thesis in Electrical Engineering*, Massachusetts Institute of Technology, Massachusetts, April 1981.
11. M. F. Lewis, "Low loss SAW devices employing single stage fabrication," *Proc. 1983 IEEE Ultrasonics Symp.*, vol. 1, pp. 104–108, 1983.
12. K. Yamanouchi and H. Furuyashiki, "Low-loss SAW filter using internal reflection types of new single-phase unidirectional transducers," *Proc. 1984 IEEE Ultrasonics Symp.*, vol. 1, pp. 68–71, 1984.
13. P. V. Wright, "The natural single-phase unidirectional transducer: a new low-loss SAW transducer," *Proc. 1985 IEEE Ultrasonics Symp.*, vol. 1, pp. 58–63, 1985.
14. C. B. Saw, C. K. Campbell, P. J. Edmonson and P. M. Smith, "Multi-frequency pulsed injection-locked oscillator using a novel low-loss SAW comb filter," *Proc. 1986 IEEE Ultrasonics Symp.*, vol. 1, pp. 273–276, 1986.
15. H. A. Haus and P. V. Wright, "The analyses of grating structures by coupling-of-modes theory," *Proc. 1980 IEEE Ultrasonics Symp.*, vol. 1, pp. 277–281, 1980.
16. P. S. Cross and R. V. Schmidt, "Coupled surface-acoustic-wave resonators," *Bell System Technical Journal*, vol. 56, pp. 1447–1481, 1977.
17. C. K. Campbell and C. B. Saw, "Analysis and design of low-loss SAW filters using single-phase unidirectional transducers," *IEEE Trans. Ultrasonics, Ferroelectrics and Frequency Control*, vol. UFFC-34, pp. 357–367, May 1987.

18. C. S. Hartmann and B. P. Abbott, "Overview of design challenges for single phase unidirectional SAW filters," *Proc. 1989 IEEE Ultrason. Symp.*, vol. 1, pp. 79–89, 1989.

19. E. G. S. Paige, A. G. Stove and R. C. Woods, "SAW reflections from aluminium strips on LiNbO3," *Proc. 1981 IEEE Ultrasonics Symp.*, vol. 1, pp. 144–147, 1981.

20. C. Dunnrowicz, F. Sandy and T. Parker, "Reflections of surface waves from periodic discontinuities," *Proc. 1976 IEEE Ultrasonics Symp.*, pp. 386–390, 1976.

21. C. B. Saw, "Single-phase unidirectional transducers for low-loss surface acoustic wave devices," *Ph.D. Thesis in Electrical Engineering*, McMaster University, Hamilton, Ontario, L8S 4L7, Canada June 1988.

22. C. B. Saw and C. K. Campbell, "Sampling effect of distributed reflector arrays with a single-phase unidirectional SAW transducer," *IEEE Trans. Ultrason. Ferroelec. and Freq. Control*, vol. 37, pp. 115–117, March 1990.

23. P. V. Wright, private communication.

24. C. C. W. Ruppel *et al.*, "SAW devices for consumer communication applications," *IEEE Trans. Ultrason. Ferroelec. and Freq. Control*, vol. 40, pp. 438–452, Sept. 1993.

25. D. F. Thompson and P. V. Wright, "Wide bandwidth NSPUDT coupled resonator filter," *Proc. 1991 IEEE Ultrasonics Symp.*, vol. 1, pp. 181–188, 1981.

26. M. Takeuchi, H. Odagawa and K. Yamanouchi, "Crystal orientations for natural single phase unidirectional transducer (NSPUDT) on $Li_2B_4O_7$," *Electronics Letters*, vol. 30, pp. 2081–2082, November 1994.

27. C. S. Lam, D-P Chen, B. Potter, V. Narayanan and A. Vishwanathan, "A review of the applications of SAW filters in wireless communications," *International Workshop on Ultrasonics Application*, Nanjing, China, Sept. 1996.

—13—

RF and Antenna-Duplexer Filters for Mobile/Wireless Transceivers

13.1. Introduction

Surface wave RF front-end filters have insertion losses comparable with dielectric filters. Their additional electrical merits relate to their "rectangular" passband capability, and higher levels of close-in stopband suppression. They also have physical attributes of small size and ruggedness. To date, most surface wave filter designs for RF front-end and interstage applications in mobile communications are fabricated on leaky-SAW substrates. As given in Table 2.2 of Chapter 2, these include 64° Y-X LiNbO$_3$, 41° Y-X LiNbO$_3$, and 36-42° Y-X LiTaO$_3$. Acoustic propagation velocities on LSAW piezoelectric-substrate cuts are higher than for Rayleigh-wave counterparts. Indeed, using the same Rayleigh-wave transducer lithography, they can be designed for an increase of up to 60% in fundamental-frequency operation. Because of the increased wave penetration of the LSAW within the piezoelectric substrate, filters based on such wave mechanisms can handle higher powers before the onset of nonlinearities or destructive breakdown. As LSAW propagation is subsurface, device operation can also be less sensitive to surface contamination. Some LSAW substrate cuts, such as the LST-cut on quartz, have excellent temperature stability [1].

The characteristics of LSAW RF resonator-filters are highly dependent on IDT metallization thickness (normally quoted in terms of film-thickness ratio h/λ). There are also differences between the behaviour of LSAW and Rayleigh-wave propagation in reflection gratings, as will be examined in the material that follows.

The RF filter structures examined in this chapter are those employing reflection gratings and/or IDTs with significant acoustic reflections from electrodes. The structures examined here relate principally to those employing LSAW substrates, and include:

- Leaky-SAW longitudinal-mode two-pole and four-pole RF interstage resonator-filters
- RF front-end ladder filters and antenna duplexers employing cascaded one-port LSAW resonators
- RF balanced-bridge filters using one-port resonators fabricated on either Rayleigh-wave, LSAW, or STW substrates

Following this coverage, Chapter 14 will deal with some aspects of RF filters employing nonresonator or mixed structures.

As a necessary prelude to examining the operation of these filter types, differences that can arise between the operation of Rayleigh-wave and LSAW reflection gratings, as well as in interdigital transducers must be considered.

13.2. Leaky-SAW Propagation under Reflection Gratings

13.2.1. BEHAVIOR OF REFLECTION GRATINGS WITH
LARGE FILM-THICKNESS RATIO

Before considering leaky-SAW reflection gratings with large values of met-allization film-thickness ratio (h/λ), some aspects of the propagation and reflection characteristics of Rayleigh waves in reflection gratings will be reviewed. In this connection, it has been demonstrated that the frequency response of thick-film reflection gratings with shorted grating elements will depend on the relative contributions of the piezoelectric and mechanical components κ_{12p} and κ_{12m}, respectively, of the mutual-coupling coefficient κ_{12} given in Eq. (9.15) of Chapter 9. For example, for shorted reflection gratings on 128° Y-X LiNbO$_3$ $|\kappa_{12}| = |\kappa_{12p}| \sim |\kappa_{12m}|$, while for open-strips $|\kappa_{12}| = |\kappa_{12p}| + |\kappa_{12m}|$. This situation can, however, reverse itself depending on the piezoelectric employed, as well as on the electrode metal and its thickness h [2]. For Rayleigh-wave propagation in shorted reflection gratings on such substrates, it has been found that the magnitude of the grating reflectivity decreases with increasing values of h/λ, while the grating sidelobes become asymmetric about center frequency [2].

13.2.2. LSAW PROPAGATION AND ATTENUATION IN REFLECTION GRATINGS

With LSAW propagation in reflection gratings, the forementioned anomalous responses obtained with Rayleigh waves may be additionally compounded by attenuation, velocity dispersion, and scattering [3]–[5]. For example, LSAW propagation on 64° Y-X LiNbO$_3$ has a stronger coupling to electric fields, with $|\kappa_{12p}| \approx |\kappa_{12m}|$ for aluminum (Al) electrodes with $h/\lambda \sim 5\%$.

While the resultant large value of LSAW grating reflectivity will determine the width of the grating stopband, the reflectivity itself can be corrupted by bulk-wave scattering. Fast-shear bulk-waves that skim the substrate surface will have a velocity close to that of LSAW on a free surface. This will cause a significant attenuation of the grating reflectivity above the center of the stopband (i.e., above the Bragg frequency) when the metallization film-thickness ratio h/λ is large (e.g., $h/\lambda \sim 1$ to 5%). With the minimum metallization thickness h required for good electrical conductivity on the order of 500–2000 Å, it will be appreciated that h/λ values can be large at the frequencies of concern for RF filters in mobile-communications circuitry.

13.2.3. PHASE RESPONSE OF LSAW REFLECTION GRATINGS

The phase response of an LSAW reflection grating around the Bragg frequency can differ from that for Rayleigh-wave reflection. In relating this, recall from Chapter 11 that the reference-axis notation used here for specifying the grating phase response is $\lambda_o/8$ away from the first grating element. (This corresponds to a 90° rotation on the Smith Chart.) From closed-form dispersion relations for shear surface waves, including LSAW, in periodic gratings, it has been shown [5] that the phase of the midband response can differ significantly from this 90° value. From experimental observations on reflection gratings with large h/λ on 64° *Y-X* LiNbO$_3$, the author has incorporated an empirical phase-shift correction of 45° in the analytic design of dual-mode LSAW RF resonator-filter designs [6], [7].

For a Rayleigh-wave reflection grating, the frequency-dependence of its reflection coefficient $\rho(f)$ may be derived from the grating matrix $[G]$ of Eq. (11.12) in Chapter 11, as $\rho(f) = G_{21}/G_{11}$, so that

$$\rho(f) = \rho_S(f) = \frac{-je^{j\theta}\tanh(\sigma L_G)}{\left[\dfrac{\sigma}{k_{12}} - \left(\dfrac{\delta - j\alpha}{k_{12}}\right)\tanh(\sigma L_G)\right]}, \tag{13.1}$$

where $\sigma = [\kappa_{12}^2 - (\delta - j\alpha)^2]^{1/2}$ (m^{-1}), κ_{12} = mutual-coupling coefficient (m^{-1}), α = grating attenuation coefficient (m^{-1}), $\delta = 2\pi(f - f_{go})/v + \kappa_{11}$ = detuning parameter, where f_{go} = grating Bragg frequency, L_G = grating length (m), v = velocity, and κ_{11} = self-coupling (i.e., velocity-shift) coefficient (m^{-1}). As given in Table 11.2 of Chapter 11, phase reference angles $\theta = 0$ and $\theta = \pi$ for open- and shorted-metal grating strips, respectively, on Rayleigh-wave substrates with strong piezoelectric coupling, while $\theta = \pi$ for both open- and shorted-metal strips on quartz.

In deriving an expression for the reflection coefficient $\rho_{LG}(f)$ of an LSAW reflection grating with open-metal strips on 64° Y-X LiNbO$_3$, and based on experiments, the author has empirically modified Eq. (13.1) to obtain

$$\rho_{LG} \approx \rho_S A_L e^{j(\pi/4)}$$

(13.2)

for large values of film-thickness ratio h/λ, where A_L = an attenuation factor relating to the increase in attenuation α above the Bragg frequency, due to bulk-wave scattering of the LSAW, while the exponential term is the difference between SAW and LSAW reflection phase shifts at the Bragg frequency, as inferred from experiment [6], [7]. Figure 13.1 gives a computation example of Eq. (13.2) applied to a LSAW reflection grating with open-metal strips on 64° Y-X LiNbO$_3$. Note the absence of ripple in the reflectivity response above the stopband. This observation can be useful, experimentally, in determining whether or not an unknown filter with reflection gratings is operating as a SAW or a leaky-SAW one.

This discussion has been applied to reflection gratings with *open*-metal strips. The analysis may be much more complex, however, when *shorted*-metal strips are employed. This is because anomalous dispersion (i.e., velocity-frequency) characteristics have been reported for LSAW reflection gratings with shorted-metal strips on 36° Y-X LiTaO$_3$, with considerable discrepancy between theory and experiment; as well as a stopband width only about one-half of that predicted [8]. It may be postulated that, for shorted LSAW reflection gratings, the anomalous responses result from either an additional phase-shift term in the shorted elements or a change in the effective grating penetration depth d_e given in Fig. 11.7 of Chapter 11.

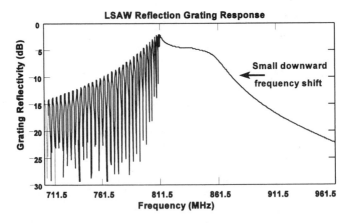

FIG. 13.1.   Illustrative predicted frequency response of LSAW reflection grating, with attenuation above grating stopband. Horizontal scale: 711.5 to 961.5 MHz. Vertical scale: −30 to 0 dB.

Significant levels of acoustic reflections from IDT fingers will also cause these to act, in part, as LSAW reflection gratings. From the author's experience of such LSAW resonator-filter fabrications, the reader should note that it may be necessary to fabricate a trial number of such structures in order to obtain an empirical determination of the desired optimum-spacing between LSAW reflection gratings and adjacent IDTs, for a given film-thickness ratio h/λ on the selected substrate.

13.3. Approximating the Radiation Conductance in Thick-Electrode IDTs

13.3.1. RADIATION CONDUCTANCE OF RAYLEIGH-WAVE IDTS WITH FINGER REFLECTIONS

Chapter 9 dealt with the effects of velocity-shift and finger reflections on the radiation conductance of a uniform Rayleigh-wave IDT. Depending on the perturbation mechanisms considered, this led to three useful formulations for its radiation conductance: 1) $G_a(f)$, the sinc-function type conductance at centre frequency f_o when both velocity shift and finger reflections were neglected; 2) $G_{am}(f)$, the sinc-function radiation conductance shifted downwards to average center frequency f_a by SAW velocity perturbations; and 3) $G_{amf}(f)$, the distorted and shifted radiation conductance peaking at frequency f_m, due to both SAW velocity shift and finger reflections. This distorted conductance could be derived from the general form of the Mason equivalent circuit as

$$G_{amf}\left(f\right) = Real\left[\frac{I_3}{V_3}\right] \approx Real\left[\frac{2Y_{31}^2}{\left(G_o + Y_{11}^f - Y_{12}^f\right)}\right], \qquad (9.33)$$

where the peak of $G_{amf}(f)$ occurs at frequency $f_m = v_m/\lambda_o$, which is as far below the average centre frequency f_a as the unperturbed centre frequency f_o is above it. In another approach to formulating the radiation conductance of an IDT with finger reflections, $G_{amf}(f)$ is obtained from a precise analysis [9] as the product of the velocity-shifted sinc-function radiation conductance $G_{am}(f)$ and a distortion parameter $U(f)$ such that, for metallization ratio $\eta = 0.5$,

$$G_{amf} \approx G_{am}\left(f\right)\left|U\left(f\right)\right|^2, \qquad (13.3)$$

where

$$U\left(f\right) = \frac{1+P}{e^{j\phi} + Pe^{-j\phi}}, \qquad (13.4)$$

where $P = [\beta - (\delta + \kappa_{11})]/\kappa_{12}$, $\beta = [(\delta + \kappa_{11})^2 - \kappa_{12}^2]^{1/2}$, $\delta = 2\pi(f - f_o)/v = $ detuning parameter, $\kappa_{11} = $ self-coupling (velocity-shift) coefficient (m^{-1}), $\kappa_{12} = $ mutual-coupling coefficient (m^{-1}). Phase term $\phi = N_p \lambda_o/2$, where $N_p = $ number of IDT electrode pairs.

13.3.2. RADIATION CONDUCTANCE FOR LEAKY-SAW IDT WITH FINGER REFLECTIONS

An expression employed by the author to approximate the radiation conductance of LSAW IDT structures with finger reflections on 64° Y-X LiNbO$_3$, involves the velocity-shifted radiation conductance and the reflection coefficient ρ_{LT} of the LSAW IDT [6]. Parameter ρ_{LT} is calculated in the same manner as for Eq. (13.1) and Eq. (13.2), but with IDT length L_T replacing grating length L_G. The IDT leaky-SAW radiation admittance $Y_L(f)$ is expressed as [6]

$$Y_{LT}(f) = G_{LT}(f) + jB_{LT}(f)$$

$$\approx G_{am}(f)\left[1 - \rho_{LT}\exp\left(j\frac{\pi}{2}\right)\right]^2 \qquad (13.5)$$

Equation (13.5) applies to an electrode metallization ratio $\eta = 0.5$. The additional phase-shift term $\pi/2$ accounts for LSAW reflection from finger

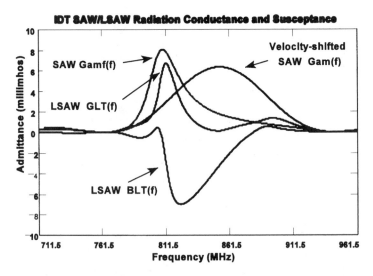

FIG. 13.2. Example comparing SAW/LSAW IDT distorted radiation conductances due to finger reflections. Horizontal scale: 711.5 to 961.5 MHz. Vertical scale: −10 to +10 mmhos.

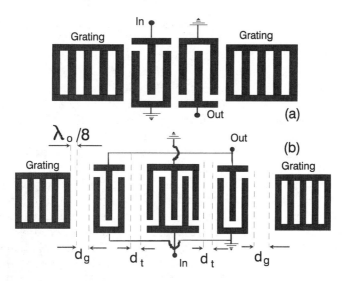

FIG. 13.3.   (a) IDT structure of longitudinal-mode LSAW resonator-filter, for operation in first and second modes. (b) Symmetric IDT structure for operation in first and third longitudinal modes.

edges, whereas IDT transduction is from the centre of an electrode finger[1].

Figure 13.2 is an illustrative computation for the radiation conductances of a sample LSAW IDT, showing the "equivalent" Rayleigh-wave radiation conductance $G_{amf}(f)$ of Eq. (13.3), as well as the LSAW radiation conductance $G_{LT}(f)$ and susceptance $B_{LT}(f)$ computed using Eq. (13.5).

13.4.   Longitudinally Coupled LSAW Resonator-Filters

13.4.1.   BASIC CONSIDERATIONS

Figure 13.3(a)(b) shows the IDT patterns employed for two types of RF low-loss leaky-SAW longitudinally coupled resonator-filters on 64° Y-X LiNbO₃ [10]. The two-port structure of Fig. 13.3(a) looks deceptively like the two-port Rayleigh-wave resonators considered in Chapter 11, but Its operation is significantly different. This is because these LSAW resonator-filters exploit the use of IDTs with significant acoustic reflections from fingers, in conjunction with piezoelectric substrates with large values of electromechanical coupling coefficients K^2. These cause the coupling of

---

[1] As will be examined in Chapter 14, the transduction center shifts from the electrode center in asymmetric IDTs.

adjacent longitudinal modes resonances in the cavity formed by the end reflection gratings.

Both filter designs of Fig. 13.3 operate as multimode structures with standing waves (see Chapter 11). The design of Fig. 13.3(a) has one input and one output IDT, and employs coupling of adjacent longitudinal modes of opposite symmetry (i.e., opposite input/output polarity). Its two coupled modes are designated as the first and second modes. The structure of Fig. 13.3(b) has one input IDT and two parallel-connected output IDTs. The symmetric placement of output IDTs in this latter design ensures coupling of two longitudinal modes of the same symmetry, designated as first and third modes. To ensure adequate out-of-band rejection, these filters can be cascaded to form four-pole structures. Because a four-pole design based on Figure 13.3(a) normally requires an intermediate shunt filter to maintain a flat passband response [10], the more versatile filter type of Fig. 13.3(b) will be considered. This can realize fractional bandwidths from 2.7% to about 3.7%, which are needed to deal with temperature-range specifications in systems such as the TACS 900-MHz cellular band.

Figure 13.4(a) is a simplistic inductance-capacitance-resistance (*LCR*) lumped-element equivalent circuit for approximating the passband response of the circuit in Fig. 13.3(a), in which the two longitudinal modes (the first and second modes) are coupled through an ideal transformer with $1:-1$ transformer ratio. In Fig. 13.4(a), C_{T1} and C_{T2} are the input and output IDT capacitances, respectively, while R_{g1} and R_{g2} are the corresponding radiation resistances. The series resonant elements $L_{r1,2}$, $C_{r1,2}$, and $R_{r1,2}$ are the equivalent "motional" parameters for the acoustic resonances of the two modes. These motional parameters can be approximated by adapting the technique for obtaining Rayleigh-wave resonator equivalent parameters, as given in Chapter 11. Figure 13.4(b) shows a corresponding lumped-equivalent circuit with motional parameters $L_{r1,3}$ $C_{r1,3}$, and $R_{r1,3}$ approximating the passband response of the circuit of Fig. 13.3(b) with adjacent symmetric modes. Other resonator branches can be added in each case, to include out-of-band responses due to "quasi-modes" [11].

Figure 13.5 illustrates an experimental frequency response of a single two-pole LSAW resonator-filter of this type, on 64° *Y-X* LiNbO$_3$ [10]. The 1-dB bandwidth of this 836.5-MHz resonator-filter is about 30 MHz, with a passband insertion loss ~ 1 dB. The relatively poor cut-off response in the upper stopband is characteristic of this LSAW filter type, and may be attributed to high-frequency remnants of the distorted LSAW radiation conductance in Fig. 13.2.

To compensate for this response degradation in the upper stopband, the filters of Fig. 13.3(b) are usually cascaded to give four-pole operation [10]. Figure 13.6 gives the measured response of an 836-MHz device. In Fig. 13.6,

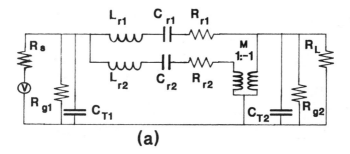

(a)

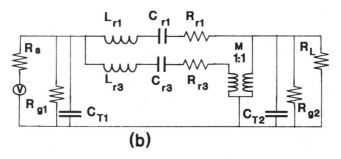

(b)

FIG. 13.4. (a) One possible LCR lumped-equivalent circuit representation for first/second mode RF LSAW resonator-filter with structure of Fig. 13.3(a). (b) Possible LCR lumped-equivalent network for first/third mode filter with structure of Fig. 13.3(b). Note polarities of coupling transformers in both instances.

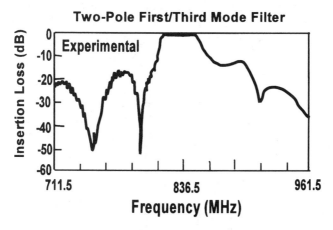

FIG. 13.5. Experimental frequency response of 836.5-MHz two-pole RF LSAW resonator-filter on 64° Y-X LiNbO$_3$, with longitudinally coupled first/third modes. Horizontal scale: 711.5 to 961.5 MHz. Vertical scale: 0 to −60 dB. (After Reference [10].)

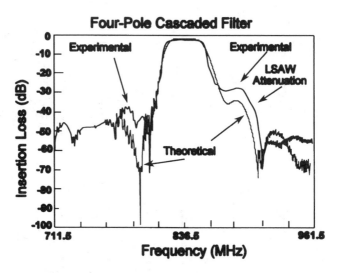

Four-Pole Cascaded Filter

Fig. 13.6. Experimental response of 836.5-MHz four-pole LSAW resonator-filter frequency response, using two cascaded two-pole filters of Fig. 13.5. For AMPS Tx interstage
filtering. Compare with theoretical response from Eq. (13.11a,b). Horizontal scale: 711.5 to
961.5 MHz. Vertical scale: 0 to −100 dB.

the 1-dB bandwidth is about 25 MHz, with a midband insertion loss of about
2.2 dB. The close-in, out-of-band rejection is ~ 28 dB. The far-out rejection,
however, is close to 60 dB, making this design type suitable for interstage
RF filtering and image rejection in the TX#2 and RX#2 RF stages in the
analog cellular circuit of Fig. 10.5. Figure 13.7 shows the response of a
similar design in a DCC6 package $(3.8 \times 3.8 \text{mm}^2)$, employed as a GSM
Rx-interstage filter for image, LO and Tx suppression. In Fig. 13.7 the
insertion loss is 3 dB, with a 1.5-dB bandwidth of 38.5 MHz.

In Chapter 11 it was mentioned that the frequency separation between
longitudinal modes in a two-port SAW resonator was $\Delta f = v/(2d_e)$, where d_e
is the effective cavity length in Fig. 11.10 [12]. This implies that the mode
separation will become larger as the cavity length is reduced. The same
situation holds for an LSAW resonator-filter. For maximum bandwidth in
the structure of Fig. 13.3(b), the input and output IDTs should be moved
as close together as possible. In practice they are moved towards each
other until they touch. For this reason, and as sketched in Fig. 13.3(b), care
must be taken to ensure that the end-electrodes of adjacent IDTs have the
same polarity. Also observe that the output IDTs in Fig. 13.3(b) are connected in parallel. This means that for input/output matching to 50-Ω source
and load impedances, the two output IDTs will be required to have fewer

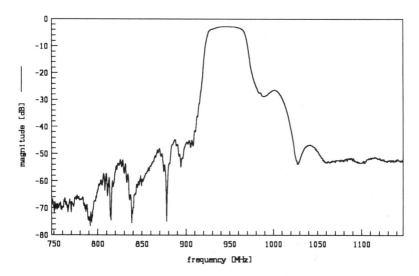

FIG. 13.7. Response of four-pole longitudinal-mode LSAW resonator-filter. For GSM Rx-interstage and suppression of image, LO and Tx. Insertion loss: 3 dB. 1.5 dB bandwidth: 38.5 MHz. In DCC6 package (3.8×3.8 mm^2). Horizontal scale: 750 to 1150 MHz. Vertical scale: −80 to 0 dB. (Courtesy of Siemens-Matsushita Components GmbH & Co. KG, Munich, Germany.)

fingers than for the input IDT, as their radiation conductances will be in parallel.

This resonator-filter type normally employs LSAW reflection gratings with shorted grating elements, which are synchronously-placed with respect to adjacent IDTs. As discussed in Section 13.2.3, shorted gratings can give rise to anomalous dispersion responses. To treat this, an empirical correction factor $C = 0.375\lambda_o$ is included by the author in the admittance relationships for LSAW gratings in these particular designs on 64° Y-X LiNbO$_3$, as derived from experimental observations [6].

13.4.2. ADMITTANCE RELATIONSHIPS

The analytic technique employed by the author for obtaining the response of this two-pole LSAW resonator-filter incorporates the two-port admittance parameters given in Chapter 4. Susceptance parameters are ignored in the driving-point admittance terms. With the output IDTs connected in paralle, individual two-port networks formed by the input IDT and one output IDT can be considered. For each, in isolation, the output voltage response V_L is

$$V_L = \frac{y_{ab}R_L V_{in}}{\left(1 + y_{aa}R_s\right)\left(1 + y_{bb}R_L\right) - y_{ab}^2 R_s R_L}, \tag{13.6}$$

where y_{aa} and y_{bb} are short-circuit input/output driving-point radiation admittances, respectively, and $y_{ab} = y_{ba}$ are short-circuit transfer admittances.

Recall that reference axes for IDTs and gratings are taken here as being $\lambda_o/8$ in front of the electrodes, as shown in Fig. 13.3(b). When input/output IDTs just touch, distance parameter d_t in Fig. 13.3(b) becomes $d_t = -\lambda_o/4$, so that distance D between IDT geometric midpoints is

$$D = \left[\frac{\left(N_{in} + N_{out}\right)}{2}\lambda_o - \frac{\lambda_o}{4}\right], \tag{13.7}$$

where N_{in} and N_{out} are the number of finger-pairs in the input and output IDTs, respectively. Next, the short-circuit transfer admittances $y1_{ab}$ and $y3_{ab}$ for the first and third modes are defined here as

$$y1_{ab} = y3_{ab} \approx \sqrt{Gin_{amf}(f)}\sqrt{Gout_{amf}(f)}\, e^{j\beta\frac{D}{2}}, \tag{13.8}$$

where $Gin_{amf}(f)$ and $Gout_{amf}(f)$ are the radiation conductances of input and output IDTs, respectively, with velocity-shift and finger reflections incorporated. Susceptance terms due to IDT capacitances are ignored in Eq. (13.8), as well as Hilbert transforms of radiation conductances. With the same simplifications, the driving-point admittances y_{aa} and y_{bb} are

$$y_{aa} \approx Gin_{amf}(f) \tag{13.9a}$$

$$y_{bb} \approx Gout_{amf}(f) \tag{13.9b}$$

13.4.3. GRATING REFLECTION COEFFICIENTS FOR FIRST AND THIRD LONGITUDINAL MODES

In obtaining the overall frequency response of this resonator-filter type by the preceding method, it is necessary to compute the frequency responses of the two longitudinal modes separately, and then combine them. In the design illustrated here, the LSAW reflection gratings with shorted metal strips are synchronously placed with respect to adjacent IDTs. The grating distance parameter d_g in Fig. 13.3(b) is set to $d_g = d1_g = 0$, for the calculation of the lower-order longitudinal mode. To get phase conditions between the gratings to satisfy the higher-frequency third-mode excitation, we set $d_g = d3_g = \lambda_o/4$ for both end gratings for this case. (Note that the grating position is mechanically unchanged. The $d3_g$ parameter just creates an electrical phase change to satisfy phase shift conditions for the third mode.)

For application to the two modal conditions, the reflection coefficient ρ_{LG} in Eq. (13.2) for an LSAW reflection grating is replaced by coefficients $\rho 1_{LG}$ and $\rho 3_{LG}$ incorporating the distance parameters $d1_g$ and $d3_g$, respectively. In addition, an empirical correction factor $C = 0.375\lambda_o$ is included in the reflection coefficient parameters for the shorted-element LSAW reflection gratings, to account for anomalous dispersion. This was derived from experimental observations on such gratings on $64°$ Y-X LiNbO$_3$ [6], [7], [13]. The two grating reflection coefficients become

$$\rho 1_{LG} \approx \rho_S A_L e^{j\pi\frac{1}{4}} e^{j\beta\frac{3}{8}\lambda_o} : \quad \left(for \quad d1_g = 0\right) \tag{13.10a}$$

$$\rho 3_{LG} \approx \rho_S A_L e^{j\pi\frac{1}{4}} e^{j\beta\frac{3}{8}\lambda_o} e^{j\beta\frac{1}{4}\lambda_o} : \quad \left(for \quad d3_g = \lambda_o/4\right) \tag{13.10b}$$

where ρ_S is the SAW grating reflection coefficient in Eq. (13.1) and $\beta = 2\pi/\lambda = $ phase constant. For the shorted reflection gratings used here, reference phase $\theta = \pi$ is set in Eq. (13.1), with $\theta = 0$ for the reflection coefficients of the IDTs. These reflection coefficients are at the reference axes for the gratings. To obtain the voltage relationships with this admittance parameter method, the coefficients in Eq. (13.10) have to be phase-shifted by an amount $[\exp(-j\beta N_{out}/2)]$ to the centre of the output IDTs. From Eqs. (13.8), (13.9), and (13.10), and for unit input voltage, the voltages responses $V1_L$ and $V3_L$ for the first and third modes are obtained as

$$V1_L \approx \frac{y1_{ab} R'_L \left[1 \pm \left(\rho 1_{LG} e^{-j\beta\frac{N_{out}}{2}}\right)\right]}{\left(1 + y_{aa}\left[R_s + R_t\right]\right)\left(1 + y_{bb}\left[R'_L + R_t\right]\right) - y1_{ab}^2 R_s R'_L} \tag{13.11a}$$

$$V3_L \approx \frac{y3_{ab} R'_L \left[1 \pm \left(\rho 3_{LG} e^{-j\beta\frac{N_{out}}{2}}\right)\right]}{\left(1 + y_{aa}\left[R_s + R_t\right]\right)\left(1 + y_{bb}\left[R'_L + R_t\right]\right) - y3_{ab}^2 R_s R'_L} \tag{13.11b}$$

where resistance R_t is included for IDT and lead resistance. In these designs, source and load resistances are typically $R_s = R_L = 50\,\Omega$, while $R'_L = 2R_L$ in Eq. (13.11) — as the output IDTs are connected in parallel. The resultant output voltage is $V_{out} = (V1_L + V3_L)/\sqrt{2}$.

The number of fingers in input/output IDTs can be critical to the design. Designs considered here apply to input/output IDTs with an *odd* number of fingers in each. To further preserve phase relationships, the numerator *plus* sign in Eq. (13.11) is used when $(N_{out} - 0.5)$ is an odd integer, with the *minus* sign employed when ($N_{out} - 0.5$) is an even integer, where N_{out} is the number of finger *pairs* in the output IDT.

13.4.4. PREDICTED FREQUENCY RESPONSES FOR THE FOUR-POLE STRUCTURES

Figure 13.6 shows the predicted frequency responses of a sample cascaded four-pole 836.5-MHZ LSAW resonator-filter on 64° *Y-X* LiNbO₃, as applied to input and output IDTs with $N_{in} = 15.5$ and $N_{out} = 9.5$ finger pairs, respectively. The IDT metallization thickness was set as $h/\lambda_o = 5\%$ for aluminum, the acoustic aperture was $W = 40\lambda_o$. End reflection gratings each contained 250 shorted-metal strips, and were placed synchronously (i.e., $d_g = 0$) with adjacent output IDTs. To deal with the distortion of the IDT radiation conductances due to finger reflections, the Bragg-frequency of the end gratings was set 19 MHz below the IDT midband frequency. Input/output termination resistances were $R_s = R_L = 50\,\Omega$, with $R_t = 6\,\Omega$ for IDT and wiring losses. The 1-dB bandwidth is $BW_1 \sim 25$ MHz, with passband insertion loss IL ~ 2.5 dB. The midband group delay of this four-pole structure was computed as ~ 22 ns.

Example 13.1 Filter Bandwidth Allowing for Operational Temperature Range. An RF filter is to be fabricated on a 64° *Y-X* LiNbO₃ LSAW substrate, for application in an AMPS cellular-phone circuit. It is required to operate over a 25 MHz bandwidth from 869 to 894 MHz, in an operating environment with a temperature variation from −20°C to +50°C. Determine the additional bandwidth to be included in the filter design, to deal with this temperature variation. ■

Solution. From Table 9.1, the temperature coefficient of delay (TCD) for LSAW 64° *Y-X* LiNbO₃ substrates is TCD = −70 ppm/°C. This gives a delay variational range of $70 \times 70 = 4900$ ppm. A variation in delay corresponds to a variation in LSAW velocity, and a shift in device centre frequency. Taking the midband filter design frequency as $869 + 12.5 = 881.5$ MHz, this gives the temperature-dependent frequency shift as $4900 \times 881.5 = 4.32$ MHz. The total bandwidth required is therefore $25 + 4.32 = 29.32$ MHz, allowing for $(4.32)/2 = 2.16$ MHz to be added at each band edge. ■

Example 13.2 Filter Bandwidth That Includes Fabrication Tolerances. How much additional bandwidth should be included in the RF filter design of Example 13.1 to allow for lithographic fabrication tolerances? ■

Solution. A "rule-of-thumb" estimate often used to allow for fabrication tolerances on substrates such as 64° *Y-X* LiNbO₃ is to increase further the bandwidth requirement by 50% of the estimated TCD value. Using this estimate, the additional bandwidth required would be $0.5 \times 4.32 = 2.16$ MHz, giving a total required bandwidth of $29.32 + 2.16 = 31.48$ MHz. ∎

Example 13.3 RF Filter Fractional Bandwidth. Determine the numerical value of the filter fractional bandwidth for the 881.5-MHz RF filter of Example 13.2. ∎

Solution. The filter fractional bandwidth will be $(31.48 \times 100)/881.5\%$, or $\sim 3.6\%$. ∎

13.5. Leaky-SAW Ladder Filters for Antenna Duplexers in Mobile Radios

13.5.1. LOSS AND PERFORMANCE REQUIREMENTS FOR DUPLEXERS

Because of their power-handling and low-loss capabilities, leaky-SAW RF filters are finding increased application in front-end antenna-duplexer stages of analog and digital mobile radio systems. An added bonus is that these antenna-duplexer modules can be significantly smaller than conventional dielectric-resonator counterparts [14]. Their required filtering characteristics are dependent, however, on whether their applications is intended for analog or digital cellular systems.

In analog cellular radios employing frequency-division multiple access (FDMA), such as the AMPS phone circuitry sketched in Fig. 10.5, the radio transceiver must simultaneously transmit and receive RF signals. This requires high rejection-level capability for the TX#1 and RX#1 stages in Fig. 10.5, with very high impedance values at mutual frequency bands. That is to say, the impedance of the transmitter filter must be highly mismatched at the *receive* frequency, while the impedance of the receiver filter must be highly mismatched at the *transmit* frequency. Figure 10.7 shows the passband and stopband specifications for the 800-MHz band AMPS analog cellular radio system. In this instance, the transmit filter must have a power handling capability of up to 1 W, with low insertion loss (less than 2 dB); the receive filter must also have low insertion loss (less than 4 dB) with high sidelobe suppression capability. They must also have small values of second- and third-order nonlinearities, which can otherwise degrade the sensitivity of the transceivers when operating under such high transmit power values.

Such stringent stopband specifications do not apply to the digital cellular transceivers. As they operate under time division multiple access (TDMA), the transmit sequences do not overlap with receive sequences. In the 1.5-GHz digital cellular band, however, insertion loss and impedance specifications for LSAW antenna duplexers are critical factors. Moreover, temperature shifts in LSAW filters at 1.5 GHz can be about twice as large as in those that have been fabricated for 800-MHz operation.

Consider, for example, antenna-duplexer operation at 1.5 GHz in the Japanese digital cellular (JDC) band. Figure 10.8(a) of Chapter 10 shows a basic sketch of an antenna duplexer stage using diode switching of the TDMA sequence burst between transmit and receive stages. In such antenna-switching circuitry the diode switch and the lowpass filter in the transmitter stage can each have an insertion loss of 0.5 to 0.7 dB, for a total insertion loss of up to 1.4 dB [14]. In contrast, the LSAW antenna duplexer basic representation of Fig. 10.8(b) employs bandpass filters in both transmit and receive front-end stages. Apart from low insertion loss values that compete with those of Fig. 10.7(a), the LSAW front-end filters require high rejection at the image frequency and at second- and third-harmonic frequencies. As depicted in Fig. 10.8(c), they also require high impedances at mutual frequency bands, although these are not as high as those for the analog FDMA systems. For digital cellular 1.5-GHz transceivers, losses as

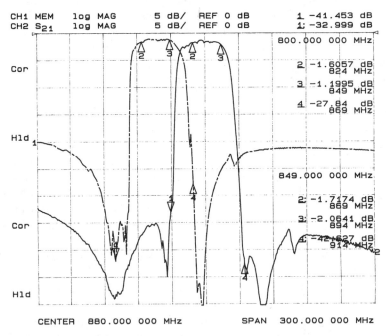

FIG. 13.8.   Low-loss LSAW Tx and Rx filters on LiTaO₃, for AMPS standards. Tx-band loss: 1.60 to 1.19 dB, with 27.84 dB out-of band attenuation at 869 MHz. Rx-band loss: 1.71 to 2.06 dB. Horizontal scale: Center 880 MHz with span 300 MHz. Vertical scale: 5 dB/div. (Courtesy of Fujitsu Ltd., Kawasaki, Japan.)

low as 0.8 dB and 1.6 dB for the transmitter and receiver front-end filters, respectively, have been reported for devices fabricated on Y-X LaTiO$_3$ substrates. While 36° Y-X LaTiO$_3$ substrates have mostly been used to date, recent fabrications have employed 42° Y-X LaTiO$_3$ substrates, as shown in Fig. 2.17.

Figure 13.8 shows an example of the frequency response of low-loss Tx and Rx front-end filters for cellular AMPS specifications (Tx: 1.99 dB at 849 MHz; Rx: 2.06 dB at 894 MHz), with 27.84 dB out-of-band attenuation at 869 MHz. Figure 13.9 shows an alternative design (Tx: 2.24 dB at 849 MHz; Rx: 2.26 dB at 894 MHz) with enhanced out-of-band attenuation of 48.9 dB at 849 MHz.

13.5.2. RESONATOR ELEMENTS FOR LEAKY-SAW LADDER FILTERS

One-port resonator sections are used as "building blocks" in the construction of LSAW ladder filters for front-end transmit and receive filters and duplexers in mobile/wireless transceivers. Depending on the filter

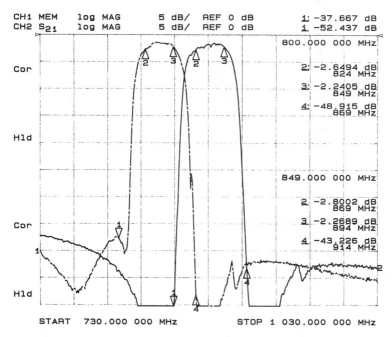

FIG. 13.9. Alternative Tx/Rx design with enhanced out-of-band attenuation. Tx-band loss: 2.64 to 2.24 dB, with 48.9 dB out-of-band attenuation at 869 MHz. Rx-band loss: 2.80 to 2.26 dB. Horizontal scale: 730 to 1030 MHz. Vertical scale: 5 dB/div. (Courtesy of Fujitsu Ltd., Kawasaki, Japan.)

application, the number of ladder elements in a design may range from two to sixteen or more, in various cascaded configurations. For 800-MHz applications the ladder filters can be constructed on 64° Y-X LiNbO$_3$ or 36° Y-X LiTaO$_3$, or on a combination of both substrates [15]. For ladder-filter applications in the 1.5-GHz band, however, single substrates of Y-X LiTaO$_3$ are employed due to their better temperature stability (see Table 2.2 in Chapter 2) because the relative frequency shifts in this band are about twice those for the 800-MHz ladder filters [14]. Figure 13.10 illustrates a commercial LSAW-based ladder-type GSM antenna-duplexer module, employing LSAW one-port resonator elements. Figure 13.11 shows the very small packages attainable for antenna-duplexer modules employing these LSAW ladder-filter designs. Figure 13.12 gives the frequency responses of the transmit/receive filters in such a GSM antenna duplexer, while Fig. 13.13

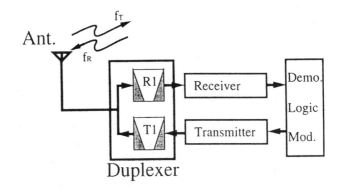

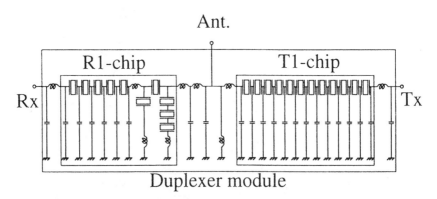

Fig. 13.10. Layout of a commercial LSAW-based GSM antenna-duplexer module employing LSAW one-port resonator elements in ladder-filter design. (Courtesy of Hitachi Ltd., Central Research Laboratory, Tokyo, Japan.)

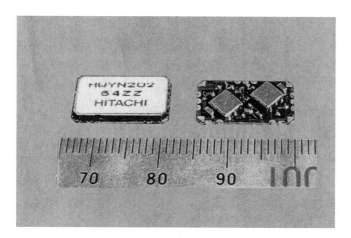

FIG. 13.11. Millimeter scale emphasizes the very small package sizes attainable with LSAW-based antenna duplexers. Antenna-duplexer here is for GSM digital cellular system. Tx and Rx front-end filters are fixed on glass epoxy substrate. (Courtesy of Hitachi Ltd., Central Research Laboratory, Tokyo, Japan.)

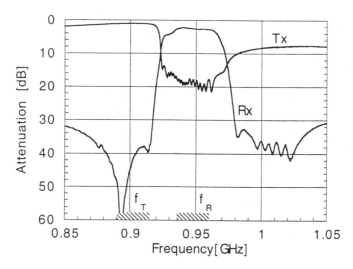

FIG. 13.12. Frequency responses of LSAW-based antenna duplexer for GSM application. Horizontal scale: 850 to 1050 MHz. Vertical scale: 0 to 60 dB. (Courtesy of Hitachi Ltd., Central Research Laboratory, Tokyo, Japan.)

shows the wideband response with large out-of-band attenuation at second and third harmonics.

Recall that one-port Rayleigh-wave resonators were considered in Chapter 11. As shown in Fig. 11.5(b), the lumped-element equivalent circuit

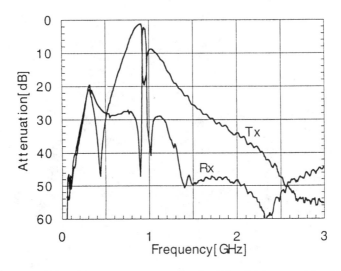

FIG. 13.13.   Wideband frequency responses for the GSM duplexer module of Fig. 13.12, showing out-of-band attenuation performance. Horizontal scale: 0 to 3 GHz. Vertical scale: 0 to 60 dB. (Courtesy of Hitachi Ltd., Central Research Laboratory, Tokyo, Japan.)

of this one-port resonator comprises a series resonant path with impedance minimum $|Z_s|$ for motional parameters R_r, L_r and C_r, ascribing SAW vibrational modes and losses. This gave the series impedance minimum shown in Fig. 11.6. The maximum impedance $|Z_p|$ at a higher frequency in Fig. 11.6 related the parallel-resonance condition established by IDT quasi-static capacitance C_T.

There are two significant fabrication differences between the Rayleigh-wave one-port resonators of Chapter 11, and the leaky-SAW RF ones considered here. First, the LSAW designs can call for a significantly distorted radiation conductance $G_{amf}(f)$ to enhance the passband response. Additionally, LSAW substrates, with their greater depths of subsurface acoustic wave penetration, are required to handle RF power levels of up to about 2 W for packet-data networks. Both conditions can be satisfied by the use of LSAW substrates with large electromechanical coupling coefficients K^2 such as 64° Y-X LiNbO$_3$ and 36–42° LiTaO$_3$ (Table 2.2), together with metallization film-thickness ratios $h/\lambda \sim 3$ to 5%.

Depending on the design application, LSAW one-port resonators for 800 MHz and above may be constructed by using either a large number of IDT fingers with strong reflections (as shown in Fig. 13.14(b)), or additional LSAW end reflection gratings as shown in Fig. 13.14(a). Figure 13.14(c) gives the idealized lumped LCR equivalent circuit for both. This is of the

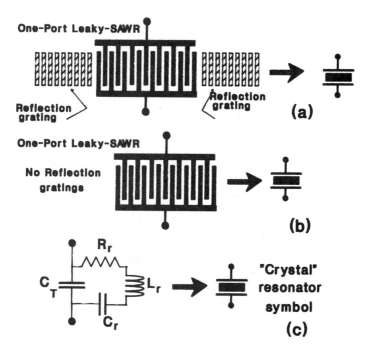

FIG. 13.14. (a) One-port SAW or LSAW resonator with reflection gratings. (b) Resonator action depending only on significant IDT finger reflections. (c) LCR lumped-equivalent circuit and symbol.

same network form as for the Rayleigh-wave one-port resonator. Because it is also the same as for a conventional bulk-wave quartz-crystal resonator [16] (where IDT capacitance C_T takes the place of the crystal holder and shunt capacitance of electrodes in parallel), the "crystal" symbol may be employed as shown.

Figure 13.15(a)(b) gives the computed impedance responses for two sample 950-MHz LSAW designs on 64° Y-X LiNbO$_3$, using an aluminum film thickness ratio $h/\lambda = 3.7\%$, and acoustic aperture $W = 60\,\mu$m. These were derived using transmission-matrix computations and LSAW reflection-grating relationships. The plot in Fig. 13.15(a) gives the impedance magnitude response for an IDT with $N_p = 70$ finger pairs, together with LSAW reflection gratings composed of 15 shorted-metal strips. The effect of these end gratings is to give a deeper notch for $|Z_s|$, the series impedance resonance [17]. As demonstrated in Fig. 13.15(b), this notch depth could be maintained in this instance by increasing the number of IDT finger pairs to $N_p = 100$, and omitting the end reflection gratings.

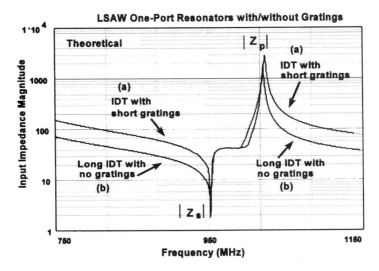

FIG. 13.15. Impedance magnitude response computations for 950-MHz, one-port resonators of Fig. 13.13. (a) 70 IDT finger pairs and reflection gratings with 15 shorted strips. (b) 100
IDT finger pairs only. Horizontal scale: 750 to 1100 MHz. Vertical scale: 1 to 10,000 Ω.

13.5.3. THE ELEMENTAL LADDER FILTER

Now consider the elemental ladder-filter stage of Fig. 13.16, made up of two
of the foregoing lumped-equivalent *LCR* resonator circuits, with their
"crystal" equivalents LE1 and LE2. Motional parameters in element LE1
are designated $L1_r$, $C1_r$, $R1_r$, with IDT capacitance $C1_T$. Likewise, motional
parameters in element LE2 are designated $L2_r$, $C2_r$, $R2_r$, with IDT capacitance $C2_T$.

What are the conditions to be imposed on the parameters of these series-
and shunt-connected elements that will yield a desired bandpass response?
First, the series-resonance frequencies $f1_s$, $f2_s$ of the motional parameters in
elements LE1 and LE2, respectively, are

$$f1_s = \frac{1}{2\pi\sqrt{L1_r C1_r}},$$

$$f2_s = \frac{1}{2\pi\sqrt{L2_r C2_r}} \tag{13.12}$$

The corresponding parallel-resonance frequencies $f1_p$ and $f2_p$, (also known
as *antiresonance frequencies*) are [16]

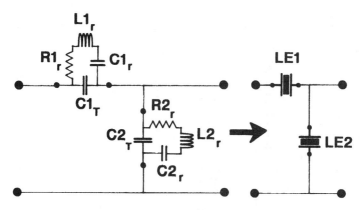

FIG. 13.16. Elemental ladder filter stage employing two LCR lumped-equivalent resonators, together with "crystal-type" symbolic representation.

$$f1_p = f1_s \left(1 + \frac{C1_r}{2C1_T}\right)$$
$$f2_p = f2_s \left(1 + \frac{C2_r}{2C2_T}\right)$$

(13.13)

From Eq. (13.13) it is seen that the ratio between motional and quasi-static capacitances determines the separation between series-resonance and parallel-resonance frequencies in each ladder element. The ratio between the IDT quasi-static impedances $C1_T$ and $C2_T$ will determine both the insertion loss of the filter and its stopband rejection characteristics, necessitating some design trade-offs [18].

In obtaining an LSAW bandpass response using the preceding ladder elements LE1 and LE2, the aim is to adjust parameters so that at midband frequency f_o the impedance of LE1 is minimized and the impedance of LE2 is maximized. This gives $f1_s \approx f2_p$, as sketched in Fig. 13.17. The sources of the "notches" that characterize a ladder filter now become apparent. The notch at the lower cutoff is due to the shunt effect of $Z2_s$, while that at the higher cutoff is due to $Z1_p$. Figure 10.4 in Chapter 10 is a typical example of this ladder-filter response characteristic.

There will be a trade-off between insertion loss and stopband rejection. This will be determined, in part, by the ratio of the IDT quasi-static capacitances $C2_T/C1_T$ in Fig. 13.16. Even if this ratio is kept constant, however, the filter input/output impedances can be varied by changing the *absolute* values of $C2_T$ and $C1_T$ [18]. One technique for determining input/output impedance matching conditions employs classic *constant-k* filter synthesis

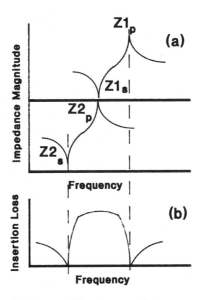

Fɪɢ. 13.17. Adjusting the "one-port" parameters of the resonators in Fig. 13.16 to give desired passband response.

techniques [19]. For ease of illustration, consider the basic half-section lowpass filter of Fig. 13.18, with input/output image impedances Z_T and Z_Π; Z_T is the impedance looking into terminals A-A, with Z_Π connected across terminals B-B. Conversely, Z_Π is the impedance looking into terminal B-B, with Z_T connected across A-A. This yields the impedance relationships [19]

$$Z_T = \left(Z_1 Z_2\right)^{1/2}\left(1 + Z_1/4Z_2\right)^{1/2} \qquad (13.14a)$$

$$Z_\Pi = \left(Z_1 Z_2\right)^{1/2}\frac{1}{\left(1 + Z_1/4Z_2\right)^{1/2}}, \qquad (13.14b)$$

giving

$$Z_T Z_\Pi = Z_1 Z_2 \qquad (13.15a)$$

$$Z_T Z_\Pi = R^2 = k^2, \qquad (13.15b)$$

for the constant-k type filter with equal source and load resistances $R_g = R_L = R$.

Now apply these image concepts to the bandpass circuit of Fig. 13.16. If motional resistances $R1_r$ and $R2_r$ are neglected in Fig. 13.16, the series impedance $Z1(\omega)$ of ladder element LE1 and the shunt impedance $Z2(\omega)$ of

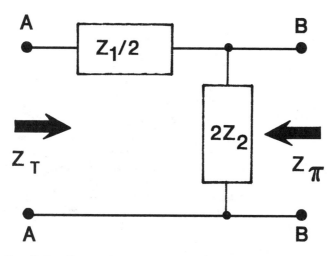

FIG. 13.18.   Constant-k filter section with image impedances Z_T and Z_Π.

ladder element LE2 can be approximated using pole-zero frequencies. In this way

$$Z1(\omega) \approx \frac{2}{C1_T} \frac{(\omega - \omega1_s)}{(\omega - \omega1_p)} \qquad (13.16)$$

$$Z2(\omega) \approx \frac{0.5}{C2_T} \frac{(\omega - \omega2_s)}{(\omega - \omega2_p)}, \qquad (13.17)$$

is obtained so that

$$Z1(\omega)Z2(\omega) \approx R^2 \qquad (13.18)$$

over the frequency range of interest. At the midband frequency $f_o = \omega_o/2\pi$, Eq. (13.18) may be approximated by

$$R^2 \approx \frac{1}{\omega_o^2 C1_T C2_T}, \qquad (13.19)$$

in terms of the IDT quasi-static capacitances $C1_T$ and $C2_T$ in the series and shunt arms, respectively. Figure 13.19 illustrates a typical plot of $C1_T$ versus $C2_T$, with the bounded region for 50-Ω input/output impedance matching with voltage standing-wave ratios (VSWRs) typically VSWR <2 in the passband. Here, $C1_T = L.C2_T$ where L is a proportionality constant. Typically, $L = 11.7$ for ladder filter designs for 900-MHz ladder filters, and $L = 2.8$ for 1.9-GHz designs [18].

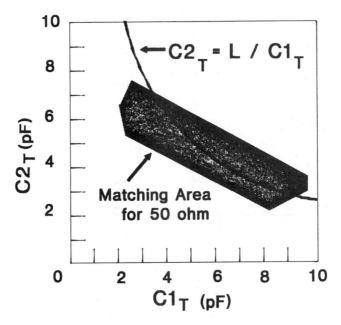

Fɪɢ. 13.19.   Adjusting the series and shunt capacitances in ladder filter section for 50-Ω input/output impedance matching. (After Reference [18].)

13.5.4.   Pʀᴀᴄᴛɪᴄᴀʟ LSAW Lᴀᴅᴅᴇʀ Fɪʟᴛᴇʀꜱ

Depending on the application, LSAW ladder filters may contain from 2–16 or more cascaded LSAW one-port resonator elements. For wideband application it is desirable to maximize the frequency separation $(\omega_p - \omega_s)$ between parallel- and series-resonance impedance points $|Z_p|$ and $|Z_s|$, respectively, in a ladder element in Fig. 13.16. It has been shown [20] that this frequency separation $(\omega_p - \omega_s)$ is approximately proportional to the electromechanical coupling coefficient K^2 of the piezoelectric substrate. Current LSAW piezoelectrics with large K^2 values include 41° Y-X LiNbO$_3$ ($K^2 = 17.2\%$), 64° Y-X LiNbO$_3$ ($K^2 = 11.3\%$) or 36–42° Y-X LiTaO$_3$ ($K^2 = 4.7\%$), as given in Table 2.2.

Antenna duplexers for some mobile phone systems must satisfy particularly exacting response requirements. As indicated in Table 10.1, the British ETACS cellular phone standard has a 33-MHz bandwidth assigned to transmit and receive frequencies, with only 12-MHz spacing between these bands. For these specifications, some commercial LSAW-based antenna-duplexer modules employ ladder-filter constructions on both 64° Y-X LiNbO$_3$ and 36° Y-X LiTaO$_3$ substrates, with differing ladder arrays for transmit and receive filters [20]. These are connected in cascade in one package [21]. In this way the 64° Y-X LiNbO$_3$ section caters for wideband and low insertion-loss aspects, while the 36° Y-X LiTaO$_3$ section provides good temperature stability and sharp cut-off characteristics[2]. For 1.5-GHz band applications, frequency stability and drift requirements normally restrict fabrications to designs on 36° Y-X LiTaO$_3$, with better temperature stability [14]. Figure 13.20 shows an LSAW-based antenna-duplexer module for an analog cellular system (ETACS, EAMPS, etc.), using a two-chip configuration on 36° Y-X LiTaO$_3$ and 64° Y-X LiNbO$_3$ for desired frequency-response characteristics. Figure 13.21 shows the frequency response of this module in an ETACS cellular system. Figure 13.22 shows the wideband frequency response and large out-of-band attenuation at second and third harmonics.

Depending on the frequency and application, LSAW ladder filters may employ one-port resonator elements in both series and shunt paths, as in Fig. 13.23(a). Alternatively, the shunt elements may be only capacitance-to-ground of resonator-resonator coupling strips, as indicated in Fig. 13.23(b),

FIG. 13.20. The LSAW-based antenna-duplexer module for analog cellular system (ETACS, EAMPS, etc.). Employs two-chip configuration on 36° Y-X LiTaO$_3$ and 64° Y-X LiNbO$_3$ for desired frequency response. (Courtesy of Hitachi Ltd., Central Research Laboratory, Tokyo, Japan.)

---

[2] 41° Y-X LiNbO$_3$ is not used here, as its propagation loss and bulk-wave radiation is large in these structures [20].

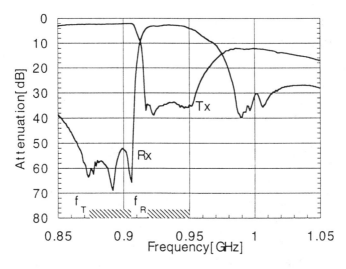

Fɪɢ. 13.21.   Frequency responses of LSAW-based antenna-duplexer module in Fig. 13.20, in ETACS cellular system. Horizontal scale: 850 to 1050 MHz. Vertical scale: 0 to 80 dB. (Courtesy of Hitachi Ltd., Central Research Laboratory, Tokyo, Japan.)

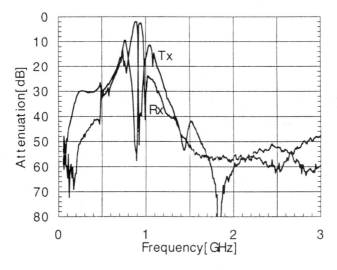

Fɪɢ. 13.22.   Wideband frequency responses of antenna duplexer of Fig. 13.21, showing out-of-band attenuation. Horizontal scale: 0 to 3 GHz. Vertical scale: 0 to 80 dB. (Courtesy of Hitachi Ltd., Central Research Laboratory, Tokyo, Japan.)

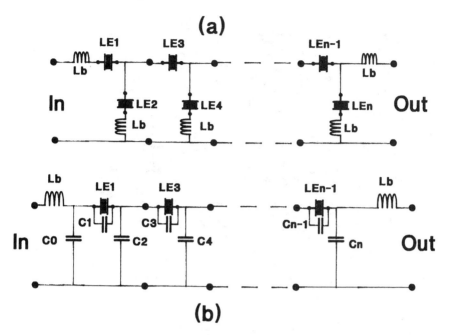

FIG. 13.23. (a) The LSAW ladder filter design on 36° *Y-X* LiTaO$_3$ with one-port resonators in series and shunt paths, in conjunction with lead inductances. (b) LSAW ladder filter with series-connected one-port resonators, and shunt "floating" capacitances between electrode finger edges.

in conjunction with "floating" capacitances between electrode finger edges (see Fig. 13.10). This gives the simplified ladder construction on 36° *Y-X* LiTaO$_3$, as depicted in Fig. 13.24, using 300 ~ 400 finger pairs in each of the series-connected one-port resonator elements[3] [22], [23].

13.5.5. POWER-HANDLING CAPABILITY OF LSAW LADDER FILTERS

The LSAW ladder filters for transmitter front-end filters T#1 in 1.5-GHz digital cellular radios have been reported with insertion loss levels as low as 0.8 dB and with insertion losses of 1.6 dB for corresponding receiver front-end filters R#1 [14]. In the π/4-DQPSK modulation schemes for such systems, each bit is shifted by π/4 to avoid zero-crossing problems. In digital cellular systems in the United States and Japan, signal bursts have a dura-

---

[3] The inductance L_b of bonding wires can affect significantly the frequency response characteristics of LSAW ladder filters, and should be included in design computations.

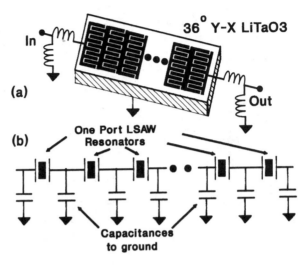

FIG. 13.24.   Layout of IDTs in ladder filter of Fig. 13.23(b) on 36° *Y-X* LiTaO₃.

tion of 6.7 ms, with a duty cycle of 1/3 for a full-rate mode, as well as a duration of 3.3 ms for the half-rate mode. Maximum and minimum powers are 1.6 W and 0.4 W, respectively, with an average power in the antenna of 0.8 W during a signal burst. (See the TDMA constellation display in Fig. 10.1.) Thus, the transmit filter T#1 must be able to handle 1.6 W at the filter output, while simultaneously presenting a high impedance at the receiver frequency, to protect the receiver stages against destruction and/or saturation. Bulk-wave radiation can have a detrimental effect on the filter impedance at receiver-band frequencies. This is found to be strongly dependent on electrode metallization thickness in substrates such as 36° *Y-X* LiTaO₃, but can be kept to minimal values with electrode thicknesses on the order of 1400 Å in 1.5-GHz ladder filters [14].

The temperature stability of the LSAW piezoelectric also becomes of critical importance for designs above 1 GHz. At this rotational angle of 36°, the LSAW wave reduces to a shear-horizontal (SH) type surface wave with propagation characteristics that are greatly affected by the existence of a ferroelectric inversion layer below the substrate surface. A method that has been reported for reducing the TCD of 36° *Y-X* LiTaO₃ involves creating such an inversion layer below the substrate surface by proton-exchange and heat treatment. In this way temperature coefficients of delay TCD = 12.6 ppm/°C have been obtained for the metallized surface case, as compared with TCD ∼ 35 ppm/°C without such treatment [24].

Because the relative transitional frequency between transmit and receive passband edges is only about 1.1% in 1.5-GHz transceivers, it is extremely important that the filter shape factor (SF) be less than 1.5 in the front-end filters. In this chapter the shape factor is defined as

$$SF = \frac{Bandwidth\ for\ minimum\ loss\ of\ 20\,dB}{Bandwidth\ for\ minimum\ loss\ of\ 3\,dB}. \tag{13.20}$$

A reduction in the shape factor requires a reduction in the separation between the series-resonance frequency f_s and the parallel (antiresonance) frequency f_p in the LSAW one-port resonator. With reference to Eq. (13.13) this may be expressed in the form

$$\frac{f_p}{f_s} = \sqrt{1 + \frac{1}{2\gamma}}, \tag{13.21}$$

where $\gamma = C_T/C_r =$ capacitance ratio is inversely proportional to electrome-chanical coupling coefficient K^2. Another fabrication method for controlling K^2 and γ in $36°$ Y-X LiTaO$_3$ involves argon-ion implantation with a dosage of 5×10^{13}/cm^2, in an accelerating voltage of 180 keV. In this way shape factors $SF = 1.4$ were obtained for LSAW ladder filters [25], [26].

Example 13.4 Capacitance Ratio γ in Resonator-Filter Design. (a) Is it desirable to have a large or a small capacitance ratio γ for maximum bandwidth in a SAW/LSAW resonator-filter design? (b) How does capacitance ratio γ relate to the electromechanical coupling coefficient K^2 of the piezoelectric substrate to be employed? (c) Other factors being equal, and given the choice between two piezoelectrics, would $36°$ Y-X LiTaO$_3$ or $64°$ Y-X LiNbO$_3$ be used for maximum bandwidth? ∎

Solution. (a) The capacitance ratio is $\gamma = C_T/C_r$, where $C_T =$ IDT static capacitance and $C_r =$ equivalent capacitance in the resonance circuit. For such an application maximum frequency separation is required between the series-resonance frequency f_s and the parallel (antiresonance) frequency f_p in the lumped LCR equivalent circuit of Fig. 13.14(c). From Eq. (13.21), a large frequency separation corresponds to a small value of capacitance ratio γ. (b) Capacitance ratio is inversely proportional to electromechanical coupling coefficient K^2. (c) $64°$ Y-X LiNbO$_3$ has the higher value of K^2. ∎

13.6. One-Port LSAW Resonators in Balanced-Bridge Filters

13.6.1. DESIGN CONSIDERATIONS

One *potential* drawback to the foregoing use of RF ladder filters employing one-port LSAW resonators relates to their stopband suppression capability. Outside their passband, the IDTs of the one-port resonators do not excite surface waves. Instead, they serve merely as simple coupling-capacitors [17], [27]. Because of this, their out-of-band frequency response is largely dependent on the capacitance ratios in the series and parallel arms.

The balanced-bridge circuit of Fig. 13.25 is designed to deal with capacitance feedthrough between input and output terminals [17]. It consists of two pairs of LSAW one-port resonators, also known as "impedance elements." In Fig. 13.25 the series-resonance frequency of one pair (IE1) of impedance elements is chosen to coincide with the parallel antiresonance frequency of the second pair (IE2). In the passband frequency range the low impedance of elements IE1 connect the input to the output, with insertion loss dictated by the resistive loss in the series-resonance paths. At the same time this coupling is supported by the large antiparallel impedance of elements in IE2. By using the appropriate design, the static capacitances of all four impedance elements can be made essentially equal, so that the feedthrough voltages across them cancel out, thereby increasing stopband suppression. The frequency shift between elements in IE1 and IE2 will determine the passband width of the filter. The permissible amount of this frequency shift will also be dependent on the electromechanical coupling coefficient K^2 of the substrate employed.

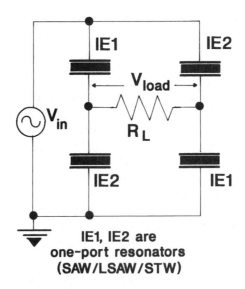

FIG. 13.25. A single-section balanced-bridge RF bandpass filter circuit employing pairs of one-port resonators of SAW, LSAW, or STW type.

13.6.2. PROPOSED MERITS OF BALANCED BRIDGE FILTERS

Those merits reported to date [17] of the type of balanced bridge filters employed as one-port resonators include:

- The four one-port resonators do not interact acoustically, so they can be placed on the same piezoelectric substrate.
- The resonators normally employ IDTs with a large number of finger pairs. The correspondingly large number of current paths thus tends to reduce the degradation caused by electrode resistivity. Moreover, the IDT parasitic capacitances can be minimal in comparison with the IDT static capacitance.
- The frequency response of the constituent one-port resonators (with or without end reflection gratings) can be accurately modelled.
- Operation of the one-port resonators can be based on Rayleigh wave, leaky-SAW (LSAW) or surface transverse wave (STW) propagation, as well as with a variety of piezoelectric substrates.
- Because the acoustic waves are *not* being transmitted between IDTs, the size of the balanced bridge filter can be made to be much smaller than for a coupled-resonator filter operating at the same frequency.
- Balanced-bridge filters offer the potential for wider achievable passbands than for SAW or LSAW resonator-filter structures.
- Bulk-wave radiation interference for balance-bridge filters employing LSAW impedance elements is not as critical as for coupled-resonator filters.
- The frequency response characteristics of the one-port resonators can be predicted accurately using computer-aided design (CAD) and modelling techniques [27]–[33].

One potential fabrication drawback to the use of the balanced-bridge circuit of Fig. 13.25 relates to the ground connections. As noted, if the input source is grounded, the output must be floating (and vice-versa).

13.7. Summary

This chapter first reviewed the characteristic features of LSAW propagation in reflection gratings compared to those for Rayleigh waves, including the effects, and significance, of significant finger reflections on the distortion of the radiation conductance of an IDT. This was followed by an examination of the design and applications of LSAW two-pole and four-pole longitudinally coupled resonator-filters for use in RF interstage circuitry in mobile-phone systems, Next, the design of some LSAW ladder filter types was highlighted, as applied to RF front-end and antenna-duplexer applications, employing LSAW one-port resonators as building blocks. The chapter concluded with a brief review of the operation and merits of surface-wave balanced bridge filters employing one-port resonators.

13.8. REFERENCES

1. Y. Shimizu and M. Tanaka, "A new cut of quartz for SAW devices with extremely small temperature by leaky surface wave," *Electronics and Communications in Japan*, Part 2, vol. 69, pp. 48–56, 1986.
2. Y. Suzuki, H. Shimizu, M. Takeuchi, K. Nakamura and A. Yamada, "Some studies of SAW resonators and multiple-mode filters," *Proc. 1976 IEEE Ultrasonics Symp.*, pp. 297–302.
3. C. S. Hartmann and V. Plessky, "Propagation, attenuation, reflection and scattering of leaky waves in Al gratings on 41-, 52-, and 64-LiNbO$_3$," *Proc. 1993 IEEE Ultrasonics Symp.*, vol. 2, pp. 1247–1250, 1993.
4. V. P. Plessky and C. S. Hartmann, "Characteristics of leaky SAWs on 36-LiTaO$_3$ in periodic structures of heavy electrodes," *Proc. 1993 IEEE Ultrasonics Symp.*, vol. 2, pp. 1239–1242, 1993.
5. V. P. Plessky, "A two-parameter coupling-of-modes model for shear horizontal type SAW propagation in periodic gratings," *Proc. 1993 IEEE Ultrasonics Symp.*, vol. 1, pp. 195–200, 1993.
6. C. K. Campbell, "Longitudinal-mode Leaky SAW resonator filters on 64° Y-X lithium niobate," *IEEE Trans. Ultrasonics, Ferroelectrics, Frequency Control*, vol. 42, pp. 883–888, September 1995.
7. P. J. Edmonson and C. K. Campbell, "Radiation conductance and grating reflectivity weighting parameters for dual mode leaky-SAW resonator filter design," *Proc. 1994 IEEE Ultrasonics Symp.*, vol. 1, pp. 75–79, 1995.
8. K. Y. Hashimoto, M. Yamaguchi and H. Kogo, "Interaction of high-coupling leaky SAW with bulk waves under metallic-grating structure on 36° YX-LiTaO$_3$," *Proc. 1985 IEEE Ultrasonics Symp.*, vol. 1, pp. 16–21, 1985.
9. T. Uno and H. Jumonji, "Optimization of quartz SAW resonator structure with groove gratings," *IEEE Trans. Sonics & Ultrasonics*, vol. SU-30, pp. 299–310, Nov. 1982.
10. T. Morita, Y. Watanabe, M. Tanaka and Y. Nakazawa, "Wideband low loss double mode SAW filters," *Proc. 1992 IEEE Ultrasonics Symp.*, vol. 1, pp. 95–104, 1992.
11. D. P. Chen, S. Jen and C. S. Hartmann, "Resonant modes in coupled resonator filters and the unique equivalent circuit representation," *Proc. 1995 IEEE Ultrasonics Symp.*, vol. 1, pp. 59–62, 1995.
12. W. J. Tanski and H. van der Vaart, "The design of SAW resonators on quartz with emphasis on two ports, " *Proc. 1976 IEEE Ultrasonics Symp.*, pp. 260–265, 1976.
13. P. J. Edmonson, "Coupling-of-modes studies of surface acoustic wave oscillators and devices," Ph.D. Thesis in Electrical Engineering, McMaster University, Hamilton, Ontario, Canada, pp. 152, February 1995.
14. M. Hikita, N. Shibagaki, K. Asai, K. Sakiyama and A. Sumioka, "New miniature SAW antenna duplexer used in GHz-band digital mobile cellular radios," *Proc. 1995 IEEE Ultrasonics Symp.*, vol. 1, pp. 33–38, 1995.
15. N. Shibagaki, T. Akagi, K. Hasegawa, K. Sakiyama, M. Hikita, "New design procedures and experimental results of SAW filters for duplexers considering wide temperature range," *Proc. 1994 IEEE Ultrasonics Symp.*, vol. 1, pp. 129–134, 1994.

16. M. E. Frerking, *Crystal Oscillator Design and Temperature Compensation*, Van Nostrand Reinhold Co., New York, 1978.
17. J. Heighway, S. N. Kondratiev and V. P. Plessky, "Balanced bridge SAW impedance element filters," *Proc. 1994 IEEE Ultrasonics Symp.*, vol. 1, pp. 27–30, 1994.
18. O. Ikata, T. Miyashita, T. Matsuda, T. Nishihara and Y. Satoh, "Development of low-loss band-pass filters using SAW resonators for portable telephones," *Proc. 1992 IEEE Ultrasonics Symp.*, vol. 1, pp. 111–115, 1992.
19. *Reference Data for Radio Engineers, Fifth Edition*, Howard W. Sams & Co., New York, chapter 7, 1972.
20. M. Hikita, N. Shibagaki, T. Akagi and K. Sakiyama, "Design methodology and synthesis techniques for ladder-type SAW resonator coupled filters," *Proc. 1993 IEEE Ultrasonics Symp.*, vol. 1, pp. 15–24, 1993.
21. M. Hikita, T. Tabuchi, N. Shibagaki, T. Akagi and Y. Ishida, "New high-performance and low-loss SAW filters used in ultra-wideband cellular radio systems," *Proc. 1991 IEEE Ultrasonics Symp.*, vol. 1, pp. 225–230, 1991.
22. M. Hikita, Y. Ishida, T. Tabuchi and K. Kurosawa, "Miniature SAW antenna duplexer for 800-MHz portable telephone used in cellular radio systems," *IEEE Trans. Microwave Theory Tech.*, vol. 36, pp. 1047–1056, June 1988.
23. M. Hikita, Y. Ishida, T. Tabuchi and K. Kurosawa, "Miniature SAW antenna duplexer for 800-MHz portable telephone used in cellular radio systems," *IEEE Trans. Microwave Theory Tech.*, vol. 36, pp. 1047–1056, June 1988.
24. K. Nakamura and A. Tourlog, "Effect of a ferroelectric inversion layer on the temperature characteristics of SH-type surface acoustic waves on $36°$ YX-LiTaO$_3$ substrates," *IEEE Trans. Ultrasonics, Ferroelectrics, and Frequency Control*, vol. 41, pp. 872–875, Nov. 1994.
25. K. Hashimoto, M. Ueda, O. Kawachi, H. Ohmori, O. Ikata, H. Uchishiba, T. Nishihara and Y. Satoh, "Development of ladder type SAW RF filter with high shape factor," *Proc. 1995 IEEE Ultrasonics Symp.*, vol. 1, pp. 113–116, 1995.
26. O. Ikata, Y. Satoh, H. Uchishiba, H. Taniguchi, N. Hirasawa, K. Hashimoto and H. Ohmori, "Development of small antenna duplexer using SAW filters for handheld phones," *Proc. 1993 IEEE Ultrasonics Symp.*, vol. 1, pp. 111–114, 1993.
27. C. K. Campbell, "Applications of surface acoustic and shallow bulk acoustic wave devices," *Proceedings of IEEE,* vol. 77, pp. 1453–1484, October 1989. (Invited paper with 322 references).
28. S. M. Ritchie, B. P. Abbott and D. C. Malocha, "Description and development of a SAW filter CAD system," *IEEE Trans. Microwave Theory Tech.*, vol. 36, pp. 456–466, 1988.
29. A. Reddy, "Design of SAW bandpass filters using new window functions," *IEEE Trans. Ultrason, Ferroelec, Freq. Cont.*, vol. 35, pp. 50–56, Jan 1988.
30. D. C. Malocha and C. D. Bishop, "The classical truncated cosine series functions with applications to SAW filters," *IEEE Trans. Ultrason, Ferroelec, Freq. Cont.*, vol. UFFC-34, pp. 75–85, Jan 1987.
31. C. S. Hartmann, "Weighting interdigital surface wave transducers by selective withdrawal of electrodes," *Proc. 1973 IEEE Ultrasonics Symp.*, pp. 423–426, 1973.
32. M. Yamaguchi, K. Y. Hashimoto and H. Kogo, "A simple method of reducing sidelobes for electrode-withdrawal SAW filters," *IEEE Trans. Sonics Ultrasonics*, vol. SU-26, pp. 334–339, Sept. 1979.
33. K. R. Laker *et al.*, "Computer-aided design of withdrawal-weighted SAW bandpass filters," *IEEE Trans. Circuits Syst.*, vol. CAS-25, pp. 241–251, 1978.

—14—

Other RF Front-End and Interstage Filters for Mobile/Wireless Transceivers

14.1. Scope of This Chapter

Chapter 13 dealt with illustrative LSAW-based RF interstage resonator filters with longitudinal-mode coupling, as well as front-end ladder filters and antenna-duplexer structures employing one-port LSAW resonators as impedance-element building blocks. This chapter illustrates the operation of two geometrically different leaky-SAW RF filter types for application in front-end and/or interstage sections—such as in RX2 and TX2 locations in the analog cellular circuit of Fig. 10.5.

The first of these low-loss designs relates to electrode-structures employing interdigitated interdigital transducers (IIDTs). While they do not use "lumped" reflection gratings as such, they can be interpreted as resonant structures due to reflections between their interlaced transducers. Also, the high-frequency stopband suppression of them can be enhanced in some designs by combining the IIDTs with LSAW one-port resonators at both the input and output terminals. Additional shaping of the sideband levels of an IIDT RF filter for mobile phone circuitry may be required, however, to attain a prescribed level of sidelobe attenuation. One shaping method involves the use of an IDT electrode-weighting technique known as *withdrawal-weighting*. While a simple approximation of the frequency response of an IDT employing withdrawal-weighting may be obtained using delta-function modelling as described in Chapter 4, a detailed computation requires a modelling method employing mixed-unit scattering matrices. This P-matrix method will be outlined in this chapter.

The second low-loss RF filter design technique incorporates variants of the single-phase unidirectional transducer (SPUDT) geometry already presented in Chapter 12. The variant examined here belongs to the class of unidirectional transducers known as floating-electrode unidirectional transducers (FEUDTs). Each cell in the FEUDT structure (between oppositely polarized active electrodes) contains "floating" electrodes in both open and

395

internally shorted configurations. Depending on the particular geometry used, the FEUDT filters can be designed for both fundamental and second-harmonic operation.

14.2. RF Filters Employing Interdigitated Interdigital Transducers (IIDTs)

14.2.1. GENERAL CHARACTERISTICS AND LIMITATIONS

IIDTs are parallel-connected and interleaved IDTs used to design SAW filters with insertion losses less than the 6-dB limit imposed by use of single input/output IDTs [1]. As illustrated in Fig. 14.1, each input IDT (except for the end ones) is surrounded by an output IDT to maximize bidirectional SAW "capture." The inherent insertion loss of the structure is governed by the level of SAW energy "escaping" from the end IDTs. This total SAW energy loss should therefore decrease as the number of interleaved IDTs increases. For an IIDT filter composed of S_{in} input and S_{out} output IDT sections, the ideal insertion loss (IL) is given by [1]

$$IL = 10 \cdot \log\left(\frac{S_{in}}{S_{out}}\right) : (dB).$$
(14.1)

Attainable fractional bandwidths with these structures are in the order of 2%, with bandwidths of up to 10% reported with the use of slanted-finger IDTs [2].

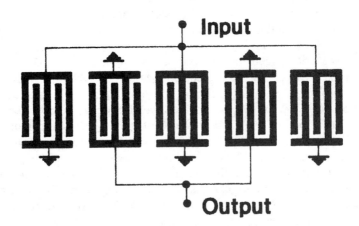

FIG. 14.1. Basic construction of an interdigitated interdigital transducer (IIDT).

These structures are much more complicated to design than would appear to be the case at first glance. Factors that can contribute to the degradation of IIDT filter response are as follows:

- The ideal insertion loss given in Eq. (14.1) applies to a perfectly matched IIDT filter. Because this matching can only be achieved at one frequency, the passband amplitude response will not be flat. Passband amplitude ripple can be severe (e.g., 1 dB), unless there is compensation.

- The IIDT structure corresponds to that of a sampled-data system as considered in Chapter 7. It can exhibit the inherent characteristic of a comb filter, with a multiplicity of comb-frequency responses.

- As the excited IIDT structure will support standing acoustic waves between adjacent IDTs, it can have some resonator-filter characteristics. One result of this is that the placement of adjacent IDTs must be precisely controlled if low insertion loss is to be achieved [2].

- Unless compensated for by methods such as withdrawal-weighting of IDT fingers, the stopband suppression can be relatively poor (e.g., 20 dB).

Despite these design difficulties and inherent limitations, some remarkably good IIDT filters have been produced commercially for RF filter applications in mobile-phone systems. Although these IIDT-based filters were originally applied to Rayleigh-wave substrates, recent commercial filters for use in 800-MHz to 2-GHz bands have exploited the capabilities of leaky-SAW substrates such as 36° Y-X LiTaO$_3$, to obtain filter insertion losses in the 1.6- to 3-dB range, with stopband attenuations of 45 dB [3]–[4].

14.2.2. RATIONALE FOR USING IIDTs IN MOBILE PHONE CIRCUITRY

Performance requirements for surface-wave RF front-end filters in duplexer stages of current mobile phone systems operating into the 2-GHz band include: a) low insertion loss (e.g., less than ~3 dB); b) fractional bandwidths in the range from 1 to 4%; and c) power-handling capabilities up to about 2 W. The power capability will depend on the surface area of the transducer structure, as well as acoustic power density under the electrodes. Because they have a relatively large electrode-metallization area, the use of IIDTs fabricated on LSAW/SAW substrates makes them most suitable for attaining required power-handling specifications. As an LSAW has a greater penetration depth into the piezoelectric substrate than a Rayleigh wave, the commensurate reduction in acoustic power density over

Rayleigh-wave devices—for the same structure, frequency, and operating conditions—makes them increasingly favoured (or necessary) in front-end applications. Even with the use of an LSAW substrate, however, the RF filter or duplexer structure must still have sufficient electrode metallization area to handle the total power requirements. While some designs of LSAW ladder filters satisfy this criterion, this is not necessarily the case for the longitudinally coupled LSAW resonator-filters considered in Chapter 13.

As an added advantage, the operation of the IDTs in the composite IIDT filter is *not* based on the use of a large film-thickness ratio (h/λ). Because of this, the IIDT center-frequency shift due to processing variations in electrode line widths is less serious than for structures with large (h/λ) values. A second feature is that even with thin-film metallization and a corresponding increase in electrode resistivity, the insertion loss of the IIDT is not significantly degraded by such resistivity, as this is counter-balanced by the large number of IDTs connected in parallel [3], [4].

14.2.3. WITHDRAWAL-WEIGHTED INTERDIGITAL TRANSDUCERS

As mentioned in Chapter 3, the first amplitude sidelobes of an unapodized IDT are only about 13 dB below the central response peak. This relatively poor sidelobe response may not be satisfactory for many filter applications in mobile communications circuitry. While improvements to the sidelobe response of two-port SAW filters considered in previous chapters concerned the *apodization* of the electrode fingers in an IDT, this finger-weighting technique is unsuitable for use with the IIDTs of Fig. 14.1. This is because, with apodization, the desired frequency response would be corrupted by transducer interactions with acoustic wave components of differing beamwidth. An alternative electrode-weighting technique that maintains a constant acoustic beamwidth is in the use of the *withdrawal-weighting* of electrodes [5]–[8]. In this technique, introduced in early days of Rayleigh-wave filter design [5], electrode fingers are selectively "withdrawn" from a uniform IDT with constant finger overlap, as depicted in Fig. 14.2, to implement the prescribed transducer response. From previous considerations of linear-phase response of IDTs with uniformly spaced electrodes in Chapter 3, electrodes must be withdrawn symmetrically about the centre axis of the IDT if linear-phase response is to be preserved in such designs.

The electrodes selected for weighting need not actually be "withdrawn" from the IDT structure . Instead, they may merely be "grounded" in order to preserve the acoustic aperture of the propagating wave against degradation by beam steering and other effects. For "standard" two-port filters with

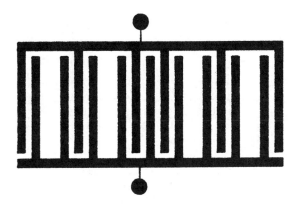

Fɪɢ. 14.2. Withdrawal weighting must be applied symmetrically to the IDT, if linear phase is to be preserved.

single IDTs at input and output ports, this allows one IDT to be apodized and the other to be withdrawal-weighted without resorting to the use of a multistrip coupler (MSC). As an example of this, a 335-MHz filter on ST-quartz, with 0.34% bandwidth and 15-dB insertion loss, yielded a sidelobe suppression of 63 dB using a combination of overlap- and withdrawal-weighting [8]. This is an attractive route to follow for IDT designs on low-coupling piezoelectric substrates (e.g., quartz) where the use of an MSC is normally impractical. Because the filter approximation will deteriorate if too many electrodes are "withdrawn," however, this technique is usually restricted to narrowband filter applications where the number of IDT electrodes is large. It can readily be implemented for IDTs using either solid- or split-electrode geometries. These electrodes can be removed individually or in finger pairs.

To outline the procedure for designing a withdrawal-weighted transducer, consider that the desired array-factor frequency response of an IDT is as sketched in Fig. 14.3(a). The Fourier transform of this array factor can be computed to obtain an approximate sampled impulse response as sketched in Fig. 14.3(b). The aim of the withdrawal-weighted technique is then to devise a suitable constant-amplitude impulse response, as shown in Fig. 14.3(c), whose Fourier transform approximates the desired array-factor response. This technique is not as simple as it might seem, however, as there is no unique withdrawal-weighting pattern commensurate with the desired design. If the IDT has N electrodes, the number of possible withdrawal-weighting combinations is 2^N, as each electrode can be present or absent. Because the number of permutations for IDTs with large N can be prohibitively large, a practical design technique is to consider electrode-groups of

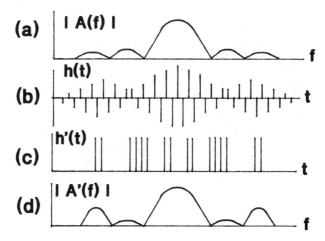

FIG. 14.3.   Illustrating the application of withdrawal-weighting. (a) Desired frequency response. (b) Its Fourier transform. (c) One example of constant amplitude impulse response derived from (b). (d) Frequency response for impulse response in (c).

eight or more, and withdraw each group separately in deriving the best-fitting IDT pattern [9].

14.2.4. SCATTERING PARAMETERS FOR COMPONENT IDTs

Now evaluation of the frequency response of an IDT with withdrawal-weighting or aperture-weighting of the constituent electrodes is considered. The approach here involves obtaining the overall acoustic and acoustoelectric contributions in transmission-matrix form. Unfortunately, the closed-form IDT transmission-matrix relations of Chapter 11 cannot be used as these relate to uniform unweighted IDTs. An alternative approach, as will now be pointed out, is to employ the mixed-unit 3×3 P-matrix formula introduced in Chapter 4 that allows for the cascading of individual IDT finger contributions, including those governed by withdrawal-weighting and/or aperture-weighting [10]. Recall that the equivalent circuit of an IDT is represented as a three-port network with two acoustic ports and one electrical port. Following the notation of Chapter 11 and Fig. 14.4, the propagation of acoustic waves along positive- and negative-reference directions between acoustic ports is designated in terms of W^+ and W^- components, respectively, while the electrical-port parameters relate to current and voltage. This gives the mixed-unit scattering-matrix relationships as

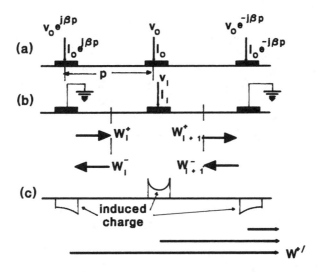

Fig. 14.4. (a) Current and voltage excitation of IDT array with single electrodes. (b) Excitation of single electrode in ith cell. (c) Charge distribution for (b). (After Reference [10].)

$$
\begin{bmatrix} W_{i-1}^- \\ W_i^+ \\ I_{i-1} \end{bmatrix} = \begin{bmatrix} P_{11} & P_{12} & P_{13} \\ P_{21} & P_{22} & P_{23} \\ P_{31} & P_{32} & P_{33} \end{bmatrix} \begin{bmatrix} W_{i-1}^+ \\ W_i^- \\ V_i \end{bmatrix}, \tag{14.2}
$$

where suffixes 1 and 2 refer to acoustic ports while suffix 3 applies to the electrical port. As in Chapter 11, wave amplitude terms W^+ and W^- are evaluated at references axes $\lambda/8$ away from electrode-finger edges. Thus, W_i^+ is the acoustic-wave propagation in the positive-reference direction (with open-surface potential) as it enters the ith electrode. Similarly, W_{i+1}^- is the acoustic-wave propagation in the negative-reference direction, while I_i is the induced current. In Eq. (14.2), parameters $P_{11} = P_{22}$, $P_{12} = P_{21}$, and $P_{13} = P_{23}$ are dimensionless, while $P_{31} = P_{32}$ and P_{33} have the units of mhos. Acoustic-aperture (tap) weighting is obtained by scaling the P_{3i} and P_{i3} terms, as required.

In the computational method [10], and with reference to Fig. 14.4, a voltage is applied to one electrode, while the rest are considered to be grounded. This gives the charge distribution shown in Fig. 14.4(c). The electroacoustic transfer function ($P_{13} = P_{23}$) of the ith electrode in a grounded array assumes that the total acoustic wave is generated in that segment alone. This is repeated for all electrodes of differing polarity as shown in Fig. 14.4(a), giving the total wave at summing point $W^{+/}$. Each contribution can be phase-shifted to refer to the centre of the ith electrode, and summed to give the total electroacoustic transfer function.

In this analysis it is assumed that weak-coupling approximations hold. This implies that the acoustic wave amplitude is constant while traversing the electrode. Under these conditions, the foremential mixed-unit matrix parameters in $[P]$ (corrected for energy conservation and reciprocity) may be derived [2], [10]. For the acoustic-dependent terms

$$P_{11} = P_{22} = m_o \cdot e^{j(\theta_{12} \pm \pi/2)} e^{-j\pi f/f_o} \tag{14.3}$$

where

$$m_o = j \left| \frac{\Delta v}{v} \right| \cdot 2\pi(s + N) \cdot \sum_{m=0}^{N} \alpha_m \left[L_{2N+2s-m}(\cos \Delta) \right.$$
$$\left. + \frac{(-1)^{N-m} L_{N-m+s}(-\cos \Delta) L_{N+2s-1}(\cos \Delta)}{L_{s-1}(-\cos \Delta)} \right] \tag{14.4}$$

$$P_{12} = P_{21} = \left(1 - |m_o|^2 \right)^{1/2} e^{j\theta_{12}} e^{-j\pi f/f_o} \tag{14.5}$$

and

$$P_{31} = P_{32} = j\omega(\varepsilon_p + \varepsilon_o) W \cdot 2\pi(s + N) e^{-j\pi f/2 f_o}$$
$$\bullet \sum_{m=0}^{N} \alpha_m \left[L_{N-m+s}(\cos \Delta) \right.$$
$$\left. + \frac{(-1)^{N-m} L_{s-1}(\cos \Delta) L_{N-m+s}(-\cos \Delta)}{L_{s-1}(-\cos \Delta)} \right] . \tag{14.6}$$

The voltage-dependent matrix parameters are

$$P_{13} = P_{23} = j \left| \frac{\Delta v}{v} \right| \cdot \frac{2(\sin \pi s) L_N(\cos \Delta)}{L_{s-1}(-\cos \Delta)} \cdot e^{-j\pi f/2 f_o} \tag{14.7}$$

and

$$P_{33} = \frac{2 P_{13} P_{31}^*}{Z_o} + j\omega C_p , \tag{14.8}$$

where $P_{31}^\star$ is the complex conjugate of P_{13}. Parameters in the foregoing equations are:

L_n = nth Legendre polynomial

$\Delta v/v$ = half of the piezoelectric coupling coefficient

N = integer value of $(2f/f_o)$

f_o = open surface velocity/electrode spacing, $= v/2p$

p = distance between centers of (uniformly spaced) adjacent electrodes

Δ = $\pi\,\eta$, where η = electrode metallization ratio

s = $2f/f_o - N$

ω = normalized radian frequency

ε_o = permittivity of free space (8.85×10^{-12} F/m)

ε_p = $\varepsilon_r \varepsilon_o$ = piezoelectric substrate (ε_r = relative permittivity)

W = acoustic aperture, which may be set to $W=0$ for a "withdrawn" electrode

θ_{12} = phase angle correction for transmitted wave vector

C_p = capacitance of an electrode pair (equal to the capacitance of one electrode in an alternatively connected IDT array)

Z_o = characteristic electroacoustic impedance (given by $|A|^2/2P_a$, where A represents the open surface electric potential, and P_a is total acoustic power flow)

Equations (14.3)–(14.8) do not include the circuit-loading effects of source impedance R_s and load impedance R_L. For the case where $R_s = R_L = 1/Y_L$, superposition methods may be applied to obtain a modified mixed-unit scattering matrix $[P']$, with elements given by [2], [10]

$$P'_{11} = P_{11} - P_{13}P_{31}\Big/\big(Y_L + P_{33}\big) \tag{14.9}$$

$$P'_{12} = P_{12} - P_{31}P_{32}\Big/\big(Y_L + P_{33}\big) \tag{14.10}$$

$$P'_{13} = P_{13} \cdot Y_L\Big/\big(Y_L + P_{33}\big) \tag{14.11}$$

$$P'_{21} = P_{21} - P_{23}P_{31}\Big/\big(Y_L + P_{33}\big) \tag{14.12}$$

$$P'_{22} = P_{22} - P_{23}P_{32}\Big/\big(Y_L + P_{33}\big) \tag{14.13}$$

$$P'_{23} = P_{23} \cdot Y_L\Big/\big(Y_L + P_{33}\big) \tag{14.14}$$

$$P'_{31} = P_{31} \tag{14.15}$$

$$P'_{32} = P_{32} \tag{14.16}$$

$$P'_{33} = P_{33} \cdot Y_L\Big/\big(Y_L + P_{33}\big). \tag{14.17}$$

In order to simplify the computation of cascaded transmission-matrices, the
P-matrix of Eq. (14.2) can be transformed into a transmission matrix for
one electrode finger to yield

$$
\begin{bmatrix} W_{i-1}^{+} \\ W_{i-1}^{-} \\ I_{i-1} \\ V_{i-1} \end{bmatrix} = \begin{bmatrix} t_{11} & t_{12} & 0 & t_{14} \\ t_{21} & t_{22} & 0 & t_{24} \\ t_{31} & t_{32} & 1 & t_{34} \\ 0 & 0 & 0 & 1 \end{bmatrix} \begin{bmatrix} W_{i}^{+} \\ W_{i}^{-} \\ I_{i} \\ V_{i} \end{bmatrix},
\tag{14.18}
$$

where parameters of the transmission-matrix terms are

$$
\begin{array}{ll}
t_{11} = 1/P_{21}' & t_{12} = -P_{22}'/P_{21}' \\
t_{14} = -P_{23}'/P_{21}' & t_{21} = P_{11}'/P_{21}' \\
t_{22} = P_{12}' - P_{11}'P_{22}'/P_{21}' & t_{24} = P_{13}' - P_{11}'P_{23}'/P_{21}' \\
t_{31} = P_{31}'/P_{21}' & t_{32} = P_{32}' - P_{22}'P_{31}'/P_{21}' \\
t_{34} = P_{33}' - P_{23}'P_{31}'/P_{21}' &
\end{array}
\tag{14.19}
$$

From these, the total transmission matrix $[T]$ for a single IDT within the
IIDT structure is finally obtained as

$$
\begin{bmatrix} W_{L}^{+} \\ W_{L}^{-} \\ I_{in} \\ V_{in} \end{bmatrix} = \begin{bmatrix} T_{11} & T_{12} & 0 & T_{14} \\ T_{21} & T_{22} & 0 & T_{24} \\ T_{31} & T_{32} & 1 & T_{34} \\ 0 & 0 & 0 & 1 \end{bmatrix} \begin{bmatrix} W_{o}^{+} \\ W_{o}^{-} \\ 0 \\ V_{in} \end{bmatrix},
\tag{14.20}
$$

where T_{ij} are the individual terms, W_{o}^{+} and W_{o}^{+} are the incoming and outgoing acoustic waves at one end of the IDT, and W_{L}^{+} and W_{L}^{-} are the incoming
and outgoing acoustic wave components at the other end; V_{in} and I_{in} correspond to the applied (or induced) voltages and currents at the individual
transmitting and receiving IDTs.

Delay line sections between IDTs are the same as given in Chapter 11,
namely

$$
[W(d)] = [D][W(0)],
\tag{11.45}
$$

where the elements of complex matrix $[D]$ are

$$
[D] = \begin{pmatrix} e^{j\beta d} & 0 \\ 0 & e^{-j\beta d} \end{pmatrix}
\tag{11.46}
$$

in terms of phase constant $\beta = 2\pi/\lambda$ and acoustic line length d between appropriate reference planes. The overall frequency response of the IIDT may then be computed using these matrix building blocks [2], [10].

14.2.5. EXAMPLE OF A HIGH-PERFORMANCE RF IIDT FILTER

One significant advantage of withdrawal-weighting of IDT electrodes over apodization is that the withdrawal-weighting can be applied to more than one IDT, thereby overcoming the problems with input/output IDT apodization discussed in Chapter 4. To this end, Fig. 14.5 shows the filter configuration of a high-performance commercial 878-MHz IIDT filter fabricated on 36° Y-X LiTaO$_3$ substrate, employing withdrawal-weighted IDTs as well as input/output one-port LSAW resonators. Shunt inductances for matching were employed in both input and output stages. This filter had an insertion loss of 2.5 dB in the passband from 870 to 887 MHz, and a sidelobe attenuation > 45 dB in the upper stopband region from 925 to 942 MHz. The IIDT incorporated 13 interlaced IDTs with an acoustic aperture of 0.05 mm

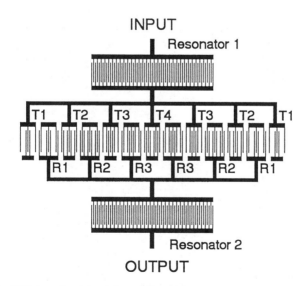

FIG. 14.5. Withdrawal-weighted Interdigitated interdigital transducer (IIDT) for 878-MHz RF front-end filter on 36° Y-X LiTaO$_3$. Incorporates one-port LSAW resonators in input/output, to improve matching and enhance out-of-band sidelobe suppression. Reprinted from Yatsuda, Takeuchi and Horishima, Reference [4]. (Courtesy of *Japanese Journal of Applied Physics*.)

TABLE 14.1

ELECTRODE POLARITIES FOR 878-MHz IIDT IN FIG. 14.5, FOR ONE-HALF OF EACH SYMMETRIC IDT (After Reference [4].)

Total number of interlaced IDTs: 13

Number of finger pairs in each transmitting (T) IDT: 24

Number of finger pairs in each receiving (R) IDT: 34

Acoustic aperture: 0.05 mm

Operating wavelength: 4.72 μm

T1:

T2:

T3:

T4:

R1:

R2:

R3:

and IDT operational wavelength of 4.72 µm. These consisted of seven input IDTs with 24 initial electrode pairs and four differing withdrawal-weighting geometries (with finger positions changed and/or removed). The six inter-laced output IDTs were designed with 34 initial electrode pairs, with three types of withdrawal-weighting. In order to maintain a linear-phase response the withdrawal-weighting of individual IDTs was applied symmetrically about the IDT midpoint. Table 14.1 lists the relative polarities of the weighted IDTs over one-half of each symmetric structure.

The application of this particular filter required an attenuation of 45 dB in a specified upper stopband region from 925 to 942 MHz. For this reason, the one-port resonators at input and output were designed to have differing parallel (antiresonance) frequencies in this band. Both one-port resonators contained 200 electrode finger pairs with 0.05 mm aperture. As shown in Fig. 14.6, in order to spread the antiresonance frequencies within the 925- to 942-MHz stopband region, the electrode-electrode (half-wavelength) pitch for one resonator IDT was set at $\Lambda 1 = 2.28\,\mu$m, while that for the second was set at $\Lambda 2 = 2.25\,\mu$m.

Figure 14.7 illustrates the measured frequency responses for the composite filter. For the computations, and as derived from experiment, a propaga-

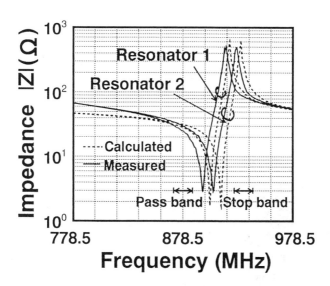

FIG. 14.6. Staggered impedance responses of input/output one-port LSAW resonators in 878-MHz IIDT filter of Fig. 14.5 to spread the antiresonance frequencies within the 925- to 942-MHz stopband region. Reprinted from Yatsuda, Takeuchi and Horishima, Reference [4]. (Courtesy of *Japanese Journal of Applied Physics*.)

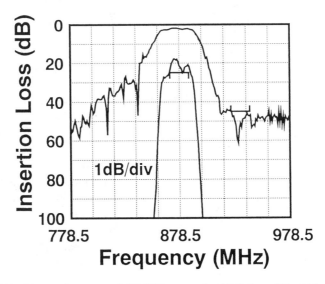

FIG. 14.7.   Measured response of 878-MHz composite IIDT filter of Fig. 14.5, with 2.5-dB insertion loss and specified 45-dB stopband attenuation. Horizontal scale: 778.5 to 978.5 MHz. Vertical scales 10 dB/div and 1 dB/div. Reprinted from Yatsuda, Takeuchi and Horishima, Reference [4]. (Courtesy of *Japanese Journal of Applied Physics*.)

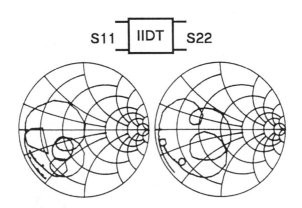

FIG. 14.8.   50-Ω Smith Chart input/output impedances, S11 and S22, of IIDT of Fig. 14.5, without matching or one-port resonators. Frequency range 678.5 to 1078.5 MHz. Reprinted from Yatsuda, Takeuchi and Horishima, Reference [4]. (Courtesy of *Japanese Journal of Applied Physics*.)

tion loss of 0.003 neper/wavelength was employed, in conjunction with an acoustic velocity of 4120 m/s for the IIDT filter by itself, and a velocity of 4140 m/s for the one-port resonators. Figure 14.8 shows the input and output impedance responses S_{11} and S_{22}, respectively, for the 50-Ω input/output

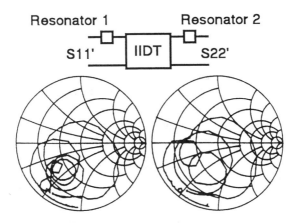

FIG. 14.9.   50-Ω Smith Chart input/output impedances, S11 and S22, of IIDT of Fig. 14.5, with one-port resonators included. Frequency range 678.5 to 1078.5 MHz. Reprinted from Yatsuda, Takeuchi and Horishima, Reference [4]. (Courtesy of *Japanese Journal of Applied Physics.*)

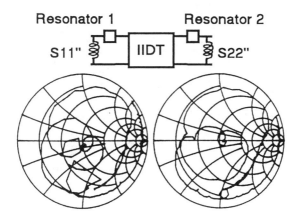

FIG. 14.10.   50-Ω Smith Chart input/output impedances, S11 and S22, of IIDT of Fig. 14.5, with one-port resonators and tuning inductors included. Frequency range 678.5 to 1078.5 MHz. Reprinted from Yatsuda, Takeuchi and Horishima, Reference [4]. (Courtesy of *Japanese Journal of Applied Physics.*)

design, as measured over the frequency range from 678.5 to 1078.5 MHz. Those shown in Fig. 14.8 apply to the IIDT by itself. Figure 14.9 shows the matching obtained with the IIDT in conjunction with the two one-port resonators, while Fig. 14.10 gives the results for the IIDT in conjunction with resonators and matching shunt inductances.

14.3. The Floating-Electrode Unidirectional Transducer (FEUDT)

14.3.1. GENERAL CONSIDERATIONS

The floating-electrode unidirectional transducer (FEUDT) represents a variant of the single-phase unidirectional transducer (SPUDT) introduced in Chapter 12. However, unlike the general SPUDT of Chapter 12, which employs banks of reflection gratings interspaced *between* IDT "rungs," the FEUDT employs single distributed reflectors of both the open- and short-circuit type *within* each IDT. The FEUDT structures reported to date can only be configured for a single-mode response. Originally developed for operation on Rayleigh-wave substrates (128° Y-X LiNbO$_3$) [11], these FEUDT structures have since been applied to LSAW-substrates and shear-horizontal (SH) wave propagation [12]. The operation of RF filters employ-

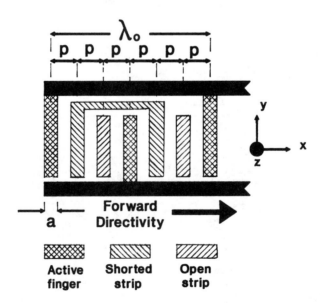

FIG. 14.11.   Example of a floating electrode unidirectional transducer (FEUDT) with minimum electrode width of $\lambda/12$. (After Reference [12].)

ing such FEUDT structures has been demonstrated for operation in the 4-GHz band, with minimum insertion loss of 6.2 dB [13].

While a number of FEUDT structures of differing geometries has been demonstrated [12], their basis of operation is illustrated here with reference to the example shown in Fig. 14.11, which depicts one unit "cell" of the overall structure, containing six electrodes per wavelength. Because a single floating electrode will only reflect a small portion of the incident acoustic energy (i.e., ~ 1%), 20 or more connected cells may be required to provide the optimum overall FEUDT structure for a given application.

The unit FEUDT-cell shown in Fig. 14.11 contains two IDT electrodes—two connected (shorted) floating electrodes and two unconnected (open) floating electrodes. These are shown here with equal widths a, pitch p, metallization ratio $\eta = a/p = 0.5$, over an acoustic wavelength $\lambda_o = 6p$. The floating electrodes will dictate the directivity of the FEUDT in two ways. First, they will govern the amplitude and phase of acoustic reflections within the cell, such that contributions from IDT fingers, shorted floating electrodes, and open floating electrodes add constructively to give an enhanced forward directivity. Second, their asymmetric placement will result in a shift of the centre of transduction in an IDT finger.

With proper design, and attainment of a high forward-to-reverse directivity ratio (e.g., 15 dB) two-port low-loss filter operation is implemented using FEUDTs in input/output stages (and configured so that their forward-directivities are towards each other).

14.3.2. Highlights of Analytical Results

As a detailed analysis of FEUDT structures can be quite formidable, only the general procedures and highlights will be given here as an aid to promoting an understanding of their operation. Such an analysis has been given [11], employing coupling-of-modes (COM) theory to describe the distributed coupling between forward- and reverse-coupling of interacting SAW. With respect to the directional notation in Fig. 14.11, SAW mode amplitudes are $A^+(x)$ and $A^-(x)$, respectively, where $|A^\pm(x)|^2$ represents the power per unit area along acoustic-aperture axis y. Voltage V is applied to the active electrodes, with current flow $I(x)$ flowing from these electrodes; V and $I(x)$ represent peak values of voltage and current, respectively. Distributed coupling may be expressed by the coupled-mode equations [11], [14]

$$\frac{dA^+(x)}{dx} = -j\kappa_{11}A^+(x) - j\kappa_{12}e^{j2\delta x}A^-(x) + j\zeta e^{j\delta x}V \qquad (14.21a)$$

$$\frac{dA^-(x)}{dx} = -j\kappa_{12}^* e^{-j2\delta x} A^+(x) + j\kappa_{11} A^-(x) - j\zeta^* e^{-j\delta x} V \qquad (14.21b)$$

$$\frac{dI(x)}{dx} = -4j\zeta^* e^{-j\delta x} A^+(x) - 4j\zeta e^{j\delta x} A^-(x) + j\omega C V, \qquad (14.21c)$$

where κ_{11} = self-coupling coefficient (m^{-1}) in Chapter 9, relating to velocity shift, while κ_{12} = mutual-coupling coefficient (m^{-1}), ζ = transduction coefficient (m^{-1}), C = electrode capacitance/unit length/unit width. Asterisks (*) in Eq. (14.21b) indicate complex-conjugate quantities. Also included in Eq. (14.21) is a phase-mismatch parameter δ (m^{-1}), where

$$\delta = sk_o - \frac{2\pi}{\lambda_o}, \qquad (14.22)$$

in which sk_o = an uncoupled SAW propagation constant, $k_o = 2\pi/p$ = wavenumber (i.e., phase constant) of the equivalent grating comprising the fixed and floating electrodes, p = electrode-electrode pitch, $\lambda_o = 6p$ = acoustic wavelength at midband, s = dimensionless parameter, $f_o = v_o/\lambda_o$ = Bragg frequency of equivalent grating, v_o = free-surface wave velocity, and $s \approx 1/6$ near the Bragg frequency.

A most important outcome to the solutions of Eq. (14.21) relates to both the mutual-coupling coefficient k_{12} and the transduction coefficient ζ. Thus, while k_{12} and ζ are *real* quantities in a "conventional" IDT, they are both *complex* quantities for asymmetric IDTs such as the FEUDT. They may be defined as

$$\zeta = |\zeta| e^{j\psi}, \quad \kappa_{12} = |\kappa_{12}| e^{2j\phi}. \qquad (14.23)$$

The phase angle term ϕ in Eq. (14.23) relates to the x-direction displacement x_R of the centre of reflection of active electrodes, while phase angle term ψ gives the displacement x_T of the centre of transduction,

$$x_R = \frac{\lambda_o}{2\pi} \phi, \quad x_T = \frac{\lambda_o}{2\pi} \psi. \qquad (14.24)$$

For "conventional" symmetric IDT structures such as SAW resonators and SPUDTs the optimum condition for acoustic unidirectionality occurs when the centre of transduction x_T in an active IDT electrode is at its midpoint, while the centre of reflection x_R is at an electrode edge. As a result, for an electrode in a symmetric IDT with metallization ratio $\eta = 0.5$ and centre-frequency width of $\lambda/4$, the shift between the two centres is $\lambda_o/8$, [corresponding to $x_R - x_T = \lambda_o/8$ and $\phi - \psi = \pi/4$ in Eq. (14.24)]. By contrast, for the asymmetric FEUDT of Fig. 14.11 with metallization ratio $\eta = a/p = 0.5$, it is shown [11] that the shift of the transduction center is $-11°$. Since the electrode width is $\lambda_o/12 = 30°$, this shift moves the centre of transduction almost to the edge of the IDT.

Two other significant features of the excited FEUDT structure of Fig. 14.11 concern: a) its standing-wave patterns; and b) its radiation conductance [11]. With respect to these, an analysis of standing-wave patterns—corresponding to the surface potential—reveals that those at the upper stopband edge frequency f_U differ in phase by 90° from those at the lower stopband edge f_L. Since *each* standing wave is composed of two

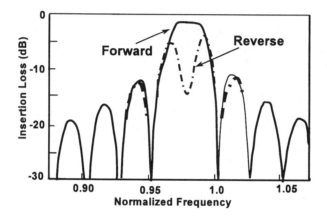

Fig. 14.12. Typical forward- and reverse-directivity of an FEUDT.

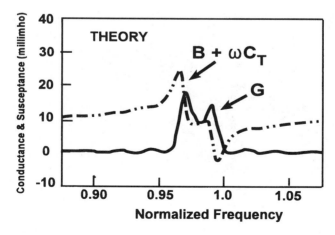

Fig. 14.13. Radiation admittance of FEUDT. Note that radiation conductance has a double peak. (After Reference [12].)

oppositely directed travelling waves, the four resultant travelling waves with different phases combine to give one forward-propagating wave, which provides the FEUDT unidirectionality as illustrated in Fig. 14.12 for this structure. Observe the characteristic null in the back-to-back directivity, which is characteristic of many unidirectional transducer types. Furthermore, and as depicted in Fig. 14.13 the radiation conductance has a dual-peak structure whose peaks are close to f_U and f_L.

14.3.3. OTHER FEUDT TRANSDUCER CONFIGURATIONS

While the FEUDT of Fig. 14.11 was initially developed for SAW applications, it can also be employed with LSAW substrates. As noted, one drawback with this structure is a lithographic one, relating to its minimum electrode or strip width of $\lambda_o/12$. As an improved version of this FEUDT, both in lithography and in performance, Fig. 14.14 shows a unit cell of a narrow-gap FEUDT (NGFEUDT), with minimum electrode width of $\lambda/5$ [12] . in this design, the extremely narrow gap between electrodes can be readily etched using anodic oxidation techniques.

A significant feature of the NGFEUDT is that since its alternating excitation electrodes have differing widths, its harmonic response is no longer constrained to odd-harmonic ones. Even harmonic responses are accentu-

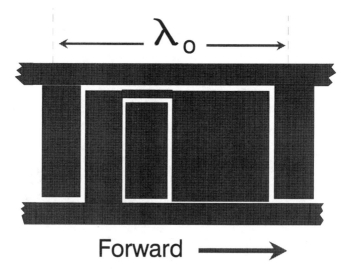

FIG. 14.14.   Narrow-gap FEUDT (NGFEUDT) with minimum electrode width of $\lambda/5$. This structure can have even and odd harmonic responses, since the alternating electrodes have unequal widths. (After Reference [12].)

ated by the use of the narrow electrode-gap. Indeed, an RF filter employing this structure on a 41° Y-X LiNbO$_3$ LSAW substrate has been reported with an insertion loss of 2.6 dB at the fundamental frequency of 360 MHz, together with an insertion loss of 2.8 dB at the 700-MHz second harmonic [12].

14.4. Summary

This chapter has highlighted some types of low-loss RF filters for wireless/mobile applications that do not employ "lumped" reflection gratings to yield their low insertion loss. Those based on the IIDT structure are available commercially, while FEUDT structures are under active development.

14.5. REFERENCES

1. M. F. Lewis, "SAW filters employing interdigitated interdigital transducers, IIDT," *Proc. 1982 IEEE Ultrasonics Symp.*, vol. 1, pp. 12–17, 1982.
2. P. M. Smith, "Studies of surface acoustic wave interdigitated transducers," Ph.D. thesis in electrical engineering, McMaster University, Hamilton, Ontario, Canada, October 1987.
3. H. Yatsuda, T. Inaoka, Y. Takeuchi, and T. Horishima, "IIDT type low-loss SAW filters with improved stopband rejection in the range of 1 to 2 GHz," *Proc. 1992 IEEE Ultrasonics Symp.*, vol. 1, pp. 67–70, 1992.
4. H. Yatsuda, Y. Takeuchi, and T. Horishima, "Surface acoustic wave filter composed of interdigitated interdigital transducers and one-port resonators," *Japanese J. Applied Physics*, vol. 33, Part 1, No. 5B, pp. 2979–2983, May 1994.
5. C. S. Hartmann, "Weighting interdigital surface wave transducers by selective withdrawal of electrodes," *Proc. 1973 IEEE Ultrasonics Symp.*, pp. 423–426, 1973.
6. M. Yamaguchi, K. Y. Hashimoto, and H. Kogo, "A simple method of reducing sidelobes for electrode-withdrawal SAW filters," *IEEE Trans. Sonics Ultrasonics*, vol. SU-26, pp. 334–339, Sept. 1979.
7. K. R. Laker *et al.*, "Computer-aided design of withdrawal-weighted SAW bandpass filters," *IEEE Trans. Circuits Syst.*, vol. CAS-25, pp. 241–251, 1978.
8. A. J. Slobodnik, JR., K. R. Laker, T. L. Szabo, W. J. Kearns, and G. A. Roberts, "Low sidelobe SAW filters using overlap and withdrawal-weighted transducers," *Proc. 1977 IEEE Ultrasonics Symp.*, pp. 757–762, 1977.
9. S. Datta, *Surface Acoustic Wave Devices*, Prentice-Hall, New Jersey, pp. 113–115, 1986.
10. C. M. Panasik and B. J. Hunsinger, "Scattering matrix analysis of surface acoustic wave reflectors and transducers," *IEEE Trans. Sonics Ultrasonics*, vol. SU-28, pp. 79–91, March 1981.
11. M. Takeuchi and K. Yamanouchi, "Coupled-mode analysis of SAW floating electrode type unidirectional transducer," *IEEE Trans. Ultrasonics, Ferroelectrics, Frequency Control*, vol. 40, pp. 648–658, November 1993.
12. M. Takeuchi, K. Yamanouchi, K. Murata, and K. Doi, "Floating-electrode-type SAW unidirectional transducers using leaky surface waves and their applications to low-loss filters," *Electronics and Communications in Japan*, Part 3, vol. 76, pp. 99–110, 1993.

13. K. Yamanouchi, C. H. S. Lee, K. Yamamoto, T. Meguro, and H. Odagawa, "GHz-range low-loss wide band filters using new floating electrode type unidirectional transducers," *Proc. 1992 IEEE Ultrasonics Symp.*, vol. 1, pp. 138–142, 1992.
14. C. S. Hartmann, P. V. Wright, R. J. Kansy, and E. M. Garber, "An analysis of SAW interdigital transducer with internal reflections and the application to the design of single-phase interdigital transducers," *Proc. 1982 IEEE Ultrasonics Symp.*, pp. 40–45, 1982.

—15—

SAW IF Filters for Mobile Phones and Pagers

15.1. Introduction

15.1.1. GENERAL REQUIREMENTS ON IF FILTERS FOR MOBILE CIRCUITRY

Intermediate-frequency (IF) filters play an essential and multipurpose role in mobile receivers. Some of these are highlighted as follows:

- In conjunction with the local oscillator[1], the first IF filter is required to set the level of adjacent-channel selectivity—which can be stringent in narrowband receivers,

- It should suppress image signals, as well as any strong interference that could otherwise saturate the follow-up IF amplifier, which is usually a high-gain one.

- In dual-conversion receivers, as sketched in the basic outline of Fig. 15.1, the first IF filter should attenuate the second image, and suppress close-in intermodulation (IM) distortion signals. This latter requirement may also require impedance isolation from the first mixer [1]. In digital cellular/cordless mobile systems, as well as wireless ones, the IF filters also are required to have a high degree of linearity, coupled with a low value of both group-delay ripple and passband amplitude ripple, in order to minimize signal corruption.

- The shape factor required of the first IF filter will depend on the modulation system employed. In narrowband analog cellular receivers employing FM modulation, the desirable shape factor will be that for a highly rectangular bandpass response. The same situation will hold for wideband CDMA receivers. A different situation exists, however, for

---

[1] The influence of local oscillator phase noise on adjacent-channel selectivity is examined in Chapter 18.

417

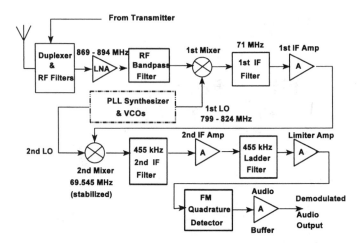

FIG. 15.1. Basic 800-MHz receiver for dual-heterodyne analog cellular mobile telephone, with sample IF frequencies shown for illustrative purposes. The first IF filter can be a waveguide-coupled Rayleigh-wave resonator-filter on ST-quartz.

the first IF filter response in GSM or DECT receivers employing Gaussian Minimum Shift Keying (GMSK) modulation[2]. Because of the power spectral distribution of MSK signals [2], the main power density within a channel will be within a frequency band of about 160 kHz, and its spectrum will extend into adjacent channels. To minimize subscriber interference, the permissible levels of the three adjacent channels (as well as the ultimate rejection), are essential design factors in GSM and DECT systems. As an example of this power containment, Fig. 15.2 shows the frequency response of an illustrative 71-MHz commercial SAW IF filter for a GSM phone in a DCC-14 package. This particular example has an insertion loss of 11.5 dB, with a group delay ripple of 0.7 μs. Its typical adjacent-channel selectivity for first, second and third adjacent channels are 5, 23 and 36 dB respectively, with a blocking level of 50 dB.

- Although the overall noise performance of a receiver will be governed largely by the noise contributions of the front-end high-gain amplifier stage, the equivalent noise bandwidth of the IF circuitry will determine the amount of noise appearing at the detector. It will also affect the attainable modulation bandwidth. For these reasons it is desirable to

---

[2] Gaussian filtering is applied to pre-smooth the binary message bits so that there are no abrupt "1," "0" transitions. The result of this will be to narrow the spectral bandwidth of the signal for increased containment within the prescribed channel.

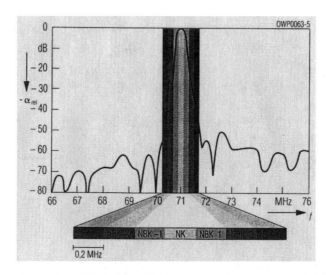

Fig. 15.2.   Illustrative 71-MHz SAW IF channel filter for European GSM phone, with insertion loss 11.5 dB and group delay 0.7 μs. Adjacent channel selectivity: 1st = 5 dB; 2nd = 23 dB; 3rd = 36 dB, with blocking = 50 dB. Horizontal scale : 66–76 MHz. Vertical scale: 0 to −80 dB. NK is desired channel, NBK is adjacent channel. (Courtesy of Siemens Matsushita Components GmbH and Co. KG, Munich, Germany.)

employ low-loss or mid-loss IF filter components. This can serve to reduce the power consumption in handheld phones.

- IF filters for mobile phones and pagers must also be compact, low-cost devices with temperature-stable operation over a given design range. In some systems the temperature-range specification is from − 20 to + 75°C. More severe specifications can, however, cover the range from − 30 to + 85°C.

At this time, center frequencies for mobile IF frequencies do not fall into any specific category. For the most part, the center-frequency allocations depend on the system concept. To minimize image-frequency problems, the minimum frequency employed is usually not less than about 45 MHz. The upper- frequency limit can be 400 MHz or more in some designs, with this limit constrained by the speed of the follow-up analog/digital (A/D) converters in digital architectures. For example, European cellular phones are being developed to support both GSM and DCS1800/PCN standards. The chip set in these phones would receive 900- and 1800-MHz RF inputs, and down convert these to a high IF frequency. Figure 15.3 shows an example of packaged 459-MHz IF filter for such a combined system, as mounted in a leadless 7 × 5 mm chip carrier.

FIG. 15.3. Packaged 459-MHz IF filters for combined European GSM/DCS1800 cellular phone. Contained in a leadless 7 × 5 mm chip carrier. Shown against a German 1 pfennig coin. (Courtesy of VI TELEFILTER, Teltow, Germany.)

As noted in the foregoing, the IF-filter bandwidth (and shape) in a mobile phone or pager will be dictated by the channel width and by the modulation scheme. The actual bandwidth required in an installed IF filter will tend to be greater than the design specification, as the actual IF component will have to take into consideration fabrication tolerances in the manufacturing process, as well as for shifts of the center frequency due to environmental temperature changes.

Table 15.1 lists representative channel widths, channel numbers and modulation schemes for a selection of analog cellular/cordless mobile phone systems. Table 15.2 lists these parameters for a selection of digital cellular/cordless systems. Advantages of time-domain multiple access (TDMA) or spread-spectrum code-division multiple access (CDMA) in digital cellular systems include flexibility for mixed voice/data communications[3]. In CDMA-based systems such as North American IS-95 in Table 15.2, the basic user-channel data rate is 9.6 kb/s. Its spectrum is spread by a chip rate of 1.2288 Mchip/s, with a different spreading process used for forward and return paths. In order to reduce interference between

---

[3] Spread-spectrum concepts and techniques are examined in Chapters 16 and 17.

TABLE 15.1

ANALOG MOBILE CELLULAR/CORDLESS TELEPHONE CHANNEL WIDTHS

Concept	System	Channel Width (kHz)	Number of Channels	Modulation
ANALOG CELLULAR	AMPS Advanced Mobile Phone Service	30	832	FM
TELEPHONES	NMT Nordic Mobile Telephone	NMT-900: 12.5 NMT-450: 25	NMT-900: 1999 NMT-450: 200	FM
	TACS Total Access Communication System	ETACS: 25 NTACS: 12.5	ETACS: 1240 NTACS: 400	FM
	JTACS/NTACS (Japan) (interleaved)	25/12.5 25/12.5 12.5	400/800 120/240 280	FM
	NTT (Japan) (interleaved)	25/6.25 6.25 6.25	600/2400 560 480	FM
ANALOG CORDLESS	CT0 Cordless Telephone 0	1.7, 20, 25, 40	10, 12, 15, 20, 25	FM
TELEPHONES	JCT Japanese Cordless Telephone	12.5	89	FM
	CT1/CT1+ Cordless Telephone 1	25	CT1: 40 CT1+: 80	FM

subscribers, all signals in a given cell are mixed further by a PN-code of length 2^{15} chips [3].

Many digital cordless systems in the United States and Canada operate in the unlicensed bands reserved for Industrial, Scientific, and Medical (ISM), which include 902 to 928 MHz, and 2.4 to 2.4835 GHz. Cordless phones using direct-sequence (DS) or frequency-hopping (FH) spread-spectrum techniques (as will be discussed in Chapters 16 and 17), can operate in these ISM bands with up to 1-W transmit power. In some other applications they are limited to a field strength of 50 mV/m at a distance of 3 m [3].

Table 15.3 illustrates channel-bandwidth allocations for some wide area networks (WAN) and wireless local area networks (WLAN). The WAN/ CDPD system uses idle times to operate in a transparent mode on analog AMPS channels for transmission of packet data at 19.2 kb/s. ARDIS[4] and Mobitex operate with packet-data transmission, but over dedicated networks. WLAN networks, such as those in Table 15.3, are spread-spectrum

---

[4] ARDIS stands for *Advanced Radio Data Information Service.*

TABLE 15.2

DIGITAL MOBILE CELLULAR/CORDLESS TELEPHONE CHANNEL WIDTHS

Concept	System	Channel Width (kHz)	Number of Channels	Modulation
DIGITAL CELLULAR TELEPHONES	IS-54/-136 North American Digital Cellular (Dual Mode)	30	832 (3 Users/channel)	$\pi/4$ DQPSK
	IS-95 North American Digital Cellular	1250	20 (798 Users/channel)	QPSK/OQPSK
	European PCN/DCS 1800 Personal Communications Networks (PCN)	200	374 (8 Users/channel)	GMSK (0.3 Gaussian filter)
	GSM Global System For Mobile Communication	200	124 (8 Users/channel)	GMSK (0.3 Gaussian filter)
	Japan PDC Personal Digital Cellular	25	1600 (3 Users/channel)	$\pi/4$ DQPSK
DIGITAL CORDLESS TELEPHONES	Japan PHS Personal Handy Phone System	300	300 (4 Users/channel)	$\pi/4$ DQPSK
	DECT Digital European Cordless Telephone	1728	10 (12 Users/channel)	GFSK (0.5 Gaussian filter)
	CT2/CT2+	100	40	GFSK (0.5 Gaussian filter)

TABLE 15.3

CHANNEL WIDTHS FOR WIRELESS DATA (WAN/WLAN) SYSTEMS

Concept	System	Channel Width (kHz)	Number of Channels	Modulation
WIDE AREA NETWORK (WAN)	CDPD Cellular Digital Packet Data	30	832	GMSK (0.5 Gaussian filter)
	Ardis — RD — Lap	25	720	FSK (2- and 4-Level)
	RAM — Mobitex	12.5	480	GMSK (0.3 Gaussian filter)
WIRELESS LOCAL AREA NETWORK (WLAN)	IEEE — 802.11	FHSS: 1000 DSSS: 11,000	FHSS: 79 DSSS: 7	FHSS: GFSK (0.5 Gaussian filter) DSSS: DBPSK (1 Mb/s) DQPSK (2 Mb/s)

systems primarily intended for data high-speed transfer at high data transfer (i.e., $> \sim 1$ Mb/s) for indoor short-range communications.

Table 15.4 lists access schemes for North American Personal Communications Services (PCS), operating in the 1.9-GHz band (Rx: 1.930 to 1.990 GHz; Tx: 1.850 to 1.910 GHz). The various access schemes listed in Table 15.4 are derived from North American and European architectures, as will be evident in this chapter. The IF-filtering considerations here relate to the PCS-CDMA high-tier system, which is based on North American IS-95 digital cellular, with a channel width of 1.25 MHz. Wideband CDMA will be examined in the spread-spectrum coverage of Chapter 17.

15.1.2. SURFACE ACOUSTIC WAVE IF FILTERS IN MOBILE PHONES AND PAGERS

Surface acoustic wave IF filters are well suited to meet the stringent specifications for both narrow-channel and wide-channel signal processing in mobile phone and pager systems. For example, almost all GSM and DECT phones use a surface-wave filter in the first IF stages [4], [5]. Using sophisticated CAD methods, their frequency responses can be tailored to meet the performance expectations in terms of insertion loss, bandwidth, shape factor, and out-of-band suppression levels. To reduce effects of spurious electromagnetic coupling, increasing numbers of surface-wave filters for mobile/wireless circuitry are being fabricated with differential (balanced) outputs instead of conventional unbalanced outputs with one terminal

TABLE 15.4

NORTH AMERICAN PERSONAL COMMUNICATIONS SERVICES
(PCS)
(Rx: 1.930 to 1.990 GHz   Tx: 1.850 to 1.910 GHz)

Low-Tier	High-Tier
PACS (Based on PHS cordless)	PCS-TDMA (Based on IS-136 cellular)
	PCS-1900 (Based on GSM cellular)
DCT-U (Based on DECT cordless)	
	PCS-CDMA (Based on IS-95 cellular with 1.25-MHz channels)
Composite CDMA/TDMA	
	Wideband CDMA

grounded. They also have the miniaturization, ruggedness, and low cost demanded for mass production. For example, dimensions of surface-mount packages for IF filters in GSM systems are typically $5 \times 5 \times 2\,mm^3$ [4].

To meet stringent temperature-performance specifications, narrowband IF filters for 30-kHz AMPS channels are fabricated almost exclusively as Rayleigh-wave devices on temperature-stable ST-quartz substrates. In narrowband filtering applications, these structures can have superior out-of-band rejection performance over monolithic crystal filters (MCF), as illustrated in Fig. 15.4. On the other hand, wideband IF filters, such as for 200-kHz European DECT channels, have employed a variety of substrates, including: (i) 128° Y-X $LiNbO_3$, with temperature drift $\pm 1400\,ppm$ over 0 to $+40°C$; (ii) 112.2° Y-X $LiTaO_3$, with temperature drift $\pm 360\,ppm$ over 0 to $+40°C$; and (iii) 36° Y-X $LiTaO_3$, with temperature drift $\pm 600\,ppm$ over 0 to $+40°C$.

15.1.3. SCOPE OF THIS CHAPTER

This chapter examines the highlights of the design and performance of some low-loss and mid-loss surface-wave filters, for application to first IF filtering in mobile telephones and pagers. These include a variety of design examples for:

- Low-loss ($< \sim 3\,dB$) waveguide-coupled two- and four-pole resonator filters with fractional bandwidths $< \sim 0.1\%$. Depending on the filter

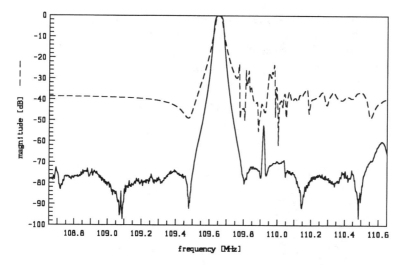

FIG. 15.4.   Comparison of 109.5-MHz waveguide-coupled narrowband SAW IF-filter response (solid line) vs. monolithic crystal filter (MCF) for channel selection in AMPS systems. SAW size: $9 \times 5\,mm^2$. MCF size: $6 \times 4\,mm^2$. Insertion loss: SAW = 3 dB, MCF = 2.5 dB; 3-dB bandwidth is 55 kHz. (Courtesy of Siemens Matsushita Components GmbH and Co. KG, Munich, Germany.)

centre frequency, these can range from narrowband IF filtering, as for the 30-kHz first IF channel for analog AMPS mobile-phone receivers, to ~ 360-kHz IF channels for combined European GSM/DCS1800 cellular receivers.

- Low-loss ($< \sim 3\,dB$) multiple-pole longitudinal-mode resonator-filters on ST-quartz, with fractional bandwidths ~0.15%, such as for IF-filtering at 250 MHz in the Japanese PHP digital cordless phones

- Low-loss SPUDT-based IF filters for European GSM and PCN phones

- Mid-loss Z-path filters [6], [7] with fractional bandwidths $< \sim 1\%$, for both reduced size and enhanced selectivity in IF stages of GSM mobile phones

- Multiple-track mid-loss SPUDT-based IF filters for 1.728-MHz channels in European DECT cordless phones [8], [9]

- CDMA IF filters for 1.25-MHz IF channels in North American Digital Cellular operating in the 800-MHz band

- IF filters for North American PCS-CDMA systems operating in the 1.9-GHz band

The basics of SPUDT filter operation were demonstrated in Chapter 12, as applied to a general design type with comb-filter or single-passband capability. Here, an efficient noncomb SPUDT filter geometry, which is routinely used in IF filter circuitry, will be highlighted.

15.2. Waveguide-Coupled SAW IF Resonator-Filters for Analog and Digital Phone Systems

15.2.1. GENERAL CONCEPTS

Before considering the basics of resonator-filter design, consider one significant difference between the amplitude/phase responses of a "standard" SAW transversal filter and those for a SAW resonator-filter. The bidirectional transversal filter structures considered in Chapter 4, as well as low-loss transversal ones, are inherently linear-phase devices, where the SAW designer has independent control over both the amplitude and the group-delay responses. This allows amplitude and/or group-delay equalization, or compensation, to be designed into the IDT structure. In contrast, SAW resonator-filters are not transversal filters. Their frequency responses have interdependent amplitude/phase characteristics that closely resemble those of lumped-element LC-resonators.

Figure 15.5 illustrates the geometry of a two-pole waveguide-coupled[5] resonator-filter. This consists of two (equal) one-pole SAW resonators transversely aligned and in close proximity to one another. Typical transverse separations are in the order of 2 to 3 λ_o. In this example the common ground strip between the resonators acts as the waveguide-coupling region. In practice, end reflection gratings usually employ shorted-metal strips, as these can give a higher grating reflectivity than with open grating strips. Optimum spacing (see Chapter 11) is normally employed between IDTs and adjacent end gratings in order to provide a symmetric amplitude response of the two-pole structure about centre frequency.

The IDT centre frequency and the grating stopband frequency are set at the same value if the film-thickness ratio h/λ is small enough for IDT finger reflections to be negligible. If IDT finger reflections are significant, however, the grating stopband frequency will need to be reduced by an amount equal to the downward-shift of the radiation-conductance peak in order to reduce insertion loss [10].

The acoustic Fresnel diffraction pattern, and its rapidly decaying sidelobes will dictate the degree of coupling between input and output

---

[5] Waveguide-coupled SAW resonator filters are also referred to as *transversely coupled* or *proximity-coupled* resonator-filters.

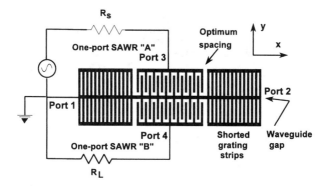

FIG. 15.5. Two-pole waveguide-coupled SAW resonator. Optimum spacing between reflection gratings and IDTs yields symmetric amplitude response. Shorted-metal grating strips generally give enhanced reflectivity over open ones. Waveguide gap is the metal ground-strip region between adjacent structures.

IDTs. From diffraction relationships given in Chapter 6, the Fresnel diffraction of the acoustic wave in a sample 83-MHz uniform IDT, with an acoustic aperture of 25 λ_o, is illustrated in Fig. 15.6 for the case of Rayleigh-wave propagation on ST-quartz. This particular Fresnel diffraction distribution was computed at a distance of 20 λ_o from the end of the IDT along the X-propagation axis (anisotropy and/or beam-steering were neglected). Note the distribution and rapid decay of acoustic amplitude in the transverse direction. At best, and even with typical separations in the order of 2 to 3 λ_o, the coupling between the two resonators will be weak. As a result, the maximum attainable bandwidth of this resonator-filter type will also be very small. Typically, this maximum fractional-bandwidth capability on quartz is $< \sim 0.1\%$, depending on the IDT acoustic aperture as well as on the gap size.

The waveguide-coupled resonator-filter has some very attractive physical and electrical features. Electrical merits include low insertion-loss capability (e.g., 1–2 dB) for a 2-pole filter, and good close-in rejection. The reason for enhanced close-in rejection capability is because the input and output transducers are not in line. As a result, there is virtually no coupling between the two for frequencies outside the reflection grating stopband.

These two-pole structures are routinely cascaded to give a four-pole response with insertion losses in the order of 3 to 5 dB, and close-in rejection in the order of 70 dB. Four-pole resonator filters have also been fabricated using four waveguide-coupled one-port resonators [11], [12]. They find application as first IF filters in analog-cellular phones such as the North American 800-MHz AMPS system with FDMA multiple access [13], or in

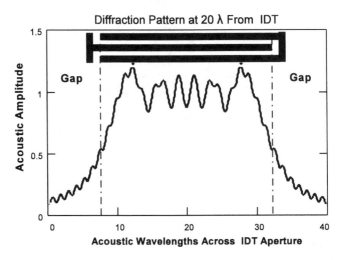

FIG. 15.6. Computation of SAW Fresnel diffraction from 86-MHz excited IDT as component of a two-pole waveguide-coupled resonator-filter on ST-quartz. IDT acoustic aperture is 25 λ_o. Wavefront is calculated at acoustic distance of 20 λ_o along propagation axis. Note "spillover" of wavefront into gap-coupling region. Anisotropy and/or beam steering are neglected here.

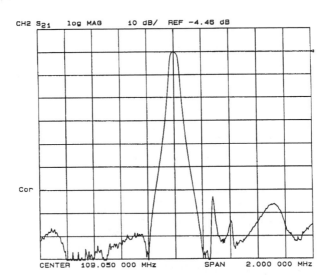

FIG. 15.7 Frequency response of 109.05-MHz four-pole waveguide-coupled SAW resonator-filter, with 4.45-dB insertion loss, and close-in rejection > ~ 70 dB. Network analyzer centre frequency: 109.05 MHz. Span: 2 MHz. (Courtesy of Sawtek Inc., Orlando, Florida.)

digital cellular networks such as the 1800-MHz European PCN/DCS network with TDMA/FDM multiple-access [14]. Figure 15.7 illustrates the frequency response of an 109.05-MHz four-pole waveguide-coupled resonator filter, with 4.45 dB insertion loss and close-in rejection greater than 70 dB. Figure 15.8 gives the frequency response of an 85.56-MHz balanced waveguide-coupled SAW IF filter, employing transverse and longitudinal coupling, for use in a North American IS-54 dual-mode cellular receiver. The out-of-band rejection is about 80 dB, without shielding of the SAW package [15]. Figure 15.9 shows the frequency response of the 459-MHz IF filter of Fig. 15.3, for application to a combined GSM/DCS1800 European cellular radio. The insertion loss marker is at 2.4 dB. Figure 15.10 shows its close-in amplitude and group delay response, with a bandwidth of 359.77 kHz, and group delay of 2.2 µs at centre frequency.

Because of their extremely small fractional bandwidth, IF resonator-filters for mobile phones normally are fabricated as Rayleigh-wave struc-

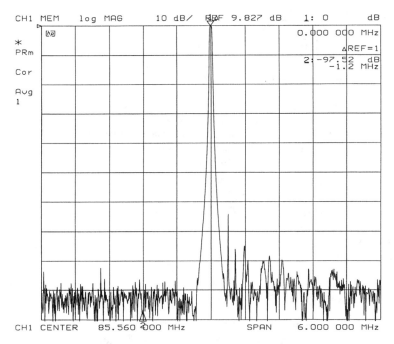

FIG. 15.8. Frequency response of 85.56-MHz four-pole waveguide-coupled SAW resonator-filter with combined longitudinal and transverse modes, and large out-of-band suppression. For application in North American IS-54 dual-cellular system. Horizontal scale: Centre frequency 85.560 MHz, with span of 6 MHz. Vertical scale: 10 dB/div. (Courtesy of NORTEL, Northern Telecom, Canada.)

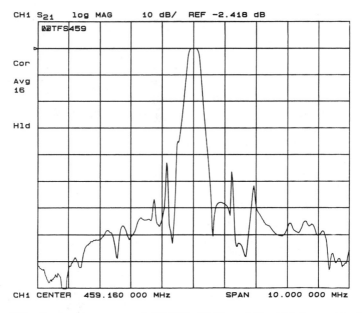

FIG. 15.9.   Frequency response of 459-MHz IF filter of Fig. 15.3 for combined GSM/
DCS1800(PCN) cellular radio. Horizontal scale: Center frequency 459.16 MHz, with span of
10 MHz. Vertical scale: 10 dB/div. Insertion loss marker at 2.418 dB. (Courtesy of VI
TELEFILTER, Teltow, Germany.)

tures on temperature-stable quartz substrates. As considered in Chapter 11,
the sharpness of the resonance response in each one-port structure will be
dictated by the effective separation between its reflection grating "mirrors."
A single, but weak, resonance response will result for each one-port structure if the IDT length is small. Multimode responses will be obtained,
however, if the IDT length is increased sufficiently[6]. Combinations of transverse and longitudinal mode coupling are used in some mobile base-station
designs to increase the number of resonator-filter poles and enhance the
overall bandwidth response [15].

15.2.2.   MODELLING THE FREQUENCY RESPONSE

Several methods can be employed to model the frequency response of
a two-pole waveguide-coupled SAW resonator-filter. In one method,
lumped-element equivalent circuits are used to model the *passband* re-

---

[6] The effective length between mirror gratings can also be increased by shifting the optimum position of the reflection gratings by an *additional* integral number of acoustic half-
wavelengths.

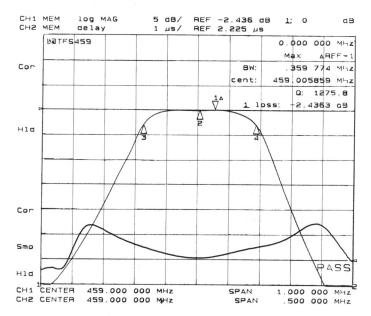

Fig. 15.10. Close-in insertion loss and group delay of Fig. 15.9 with BW = 359.77 kHz. Horizontal scales: Center frequency 459 MHz. Channel 1 (insertion loss) span is 1 MHz. Channel 2 (delay) span is 0.5 MHz. Vertical scales: 5 dB/div and 1 μs/div. Markers at 2.435 dB and 2.225 μs. (Courtesy of VI TELEFILTER, Teltow, Germany.)

sponse of each constituent one-port SAW resonator. The LCR lumped-equivalent parameters are derived from the SAW motional parameters for the constituent resonators [16]–[19]. A second method involves P-matrix computations, as introduced in Chapter 4. A third one employs coupling-of-modes (COM) analyses [20]–[23]. An advantage of the P-matrix or COM methods is that they can take into consideration IDT finger reflections, velocity perturbations, triple-transit interference (TTI), as well as the transverse modes discussed in Chapter 11 [24], [25].

The COM techniques may be classified as coupling-of-modes in space (COMS), since they involve the spatial mutual-coupling coefficient κ_{12} (with the units of m^{-1}), as presented in Chapter 9. An expanded technique [26], [27], as employed here, involves coupling-of-modes in time (COMT).

15.2.3. COUPLING-OF-MODES IN TIME (COMT)

To illustrate the COMT method, let us apply the LCR lumped-equivalent circuit of Fig. 15.11 to model the passband response of a waveguide-coupled SAW resonator. In the absence of any coupling between resonator A and

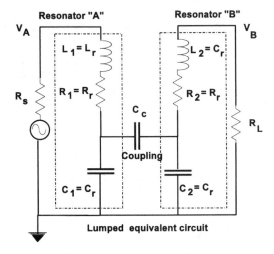

Fig. 15.11. Lumped-equivalent circuit model for waveguide-coupled resonator in Fig. 15.5. Identical one-port resonators are considered here.

resonator B, the unperturbed angular resonant frequencies and coupling parameters are [27]

$$\omega_1 = \frac{1}{\sqrt{L_1 C_1}} \tag{15.1a}$$

$$\omega_2 = \frac{1}{\sqrt{L_2 C_2}} \tag{15.1b}$$

$$k_{21} = \frac{j\omega_1 C_c}{2\sqrt{C_1 C_2}} \tag{15.2a}$$

$$k_{12} = \frac{j\omega_2 C_c}{2\sqrt{C_1 C_2}} \tag{15.2b}$$

This gives the mode-splitting frequencies as

$$\omega = \frac{\omega_1 + \omega_2}{2} \pm \left[\frac{(\omega_1 - \omega_2)^2}{2} + |k_{12}|^2 \right]^{1/2} \tag{15.3}$$

Now consider identical resonators in Fig. 15.11, with $L_1 = L_2 = L_r$, $C_1 = C_2 = C_r$, $R_1 = R_2 = R_r$, and where coupling capacitor $C_c << C_r$. For

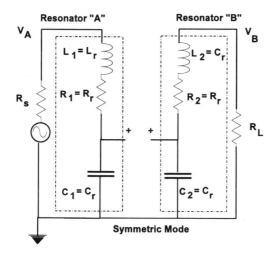

FIG. 15.12. Symmetric coupling-of-modes in time (COMT) mode for circuit of Fig. 15.5.

$k_{12} = k_{21} = k$ and uncoupled $\omega_1 = \omega_2 = \omega_o$, the frequency-splitting of the two coupled modes is $\Delta\omega = 2|k|$.

The frequency response of the coupled-resonator circuit of Fig. 15.11 can be considered in terms of the symmetric and antisymmetric modal contributions. For the symmetric mode of Fig. 15.12 the coupling capacitor C_c is unexcited, so that the resonance frequency ω_s of both resonators is

$$\omega_s = \omega_o = \frac{1}{\sqrt{L_r C_r}} \qquad (15.4)$$

In Fig. 15.12, input and output voltages V_A and V_B have the same polarity in this symmetric mode. For the antisymmetric mode of Fig. 15.13, however, coupling capacitor C_c is grounded and input and output voltages have opposite polarity. Thus, the resonant frequencies ω_a are now

$$\omega_a = \frac{1}{\sqrt{L_r \left(C_r + 2C_c \right)}} \qquad (15.5)$$

15.2.4. MODELLING THE COUPLING CAPACITANCE

The lumped equivalent coupling capacitance C_c will be proportional to series resonant capacitance C_r, with proportionality factors A_c and D_c. Thus,

$$C_c = A_c D_c C_r \qquad (15.6)$$

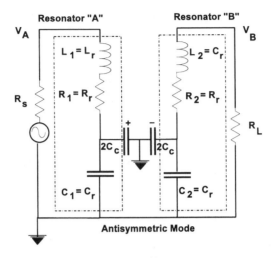

FIG. 15.13. Antisymmetric COMT-mode for circuit of Fig. 15.5.

In Eq. (15.6) factor A_c relates to the energy storage in the coupling gap, relative to that of the high-field resonator area. For a long IDT (and acoustic-field maxima in this region) we can approximate $A_c \approx C_g/C_T$, where C_g = effective coupling gap capacitance and C_T = IDT static capacitance. Diffraction parameter factor D_c is taken as an average value, over the coupling gap, for the relative intensity of the SAW field $S(y, x)$ at a point x from the IDT. Distance x is set at the plane of an effective grating mirror in the cavity where diffraction should be maximum. The Fresnel diffraction of the SAW IDT field, as illustrated in Fig. 15.6, can be computed using the parabolic approximation to the SAW velocity surface near a pure-mode axis, as given in Chapter 6, using standard Fresnel integrals [28].

15.2.5. Resonator-Lumped Equivalent-Circuit Parameters

In the symmetric mode representation in Fig. 15.12, the component values for the series-resonators are taken to correspond to those for a one-pole SAW resonator [16]–[19]. Assuming identical resonators, the equivalent series inductance L_r, resistance R_r and capacitance C_r are

$$R_r = \frac{1\left(1 - |\Gamma|\right)}{G_r\left(1 + |\Gamma|\right)} \qquad (15.7)$$

$$C_r = \frac{1}{\omega_s^2 L_r} \tag{15.9}$$

$$L_r = \frac{N_T G_r}{4 f_s} \tag{15.8}$$

where $|\Gamma|$ = grating reflection coefficient magnitude at the symmetric resonance frequency $f_s = \omega_s/2\pi$, and $N_T \lambda_o$ = total effective grating mirror separation (including grating penetration); G_r = peak radiation conductance at centre frequency. This can be for the $|(\sin X)/X|^2$ unperturbed radiation conduction function when IDT finger reflections are negligible, or for the peak radiation conductance as distorted by IDT finger reflections for large film-thickness ratio h/λ.

For the antisymmetric mode, we assume that R_r and L_r are unchanged to first order, while Eq. (15.9) is replaced by

$$C_r + 2C_c = \frac{1}{\omega_a^2 L_r} \tag{15.10}$$

where C_c = the equivalent coupling capacitance and $\omega_a = 2\pi f_a$ = resonance frequency for the antisymmetric mode. Frequency ω_a can be obtained once C_c is known. From Eqs. (15.9) and (15.10) the mode-splitting Δf is

$$\Delta f = \frac{(\omega_s - \omega_a)}{2\pi} \tag{15.11}$$

15.2.6. USING COUPLING-OF-MODES IN SPACE (COMS)

Now apply the COMS transmission-matrix analytical techniques of Chapter 11 to obtain the relationships for the waveguide-coupled SAW resonator-filter. Figure 15.14 identifies the matrices and matrix-notation involved. For input resonator A the overall 2×2 acoustic matrix $[M^a]$ in the absence of coupling to resonator B is

$$\left[M^a \right] = \begin{bmatrix} M_{11} & M_{12} \\ M_{21} & M_{22} \end{bmatrix}^a = \left[G_1^a \right] \left[G_2^a \right] \left[t_3^a \right] \left[D_4^a \right] \left[G_5^a \right] \tag{15.12}$$

where the 2×2 matrices for G and D apply to the reflection gratings and transmission-line gaps, respectively, while t_3 is the 2×2 acoustic submatrix of the 3×3 IDT matrix. In the absence of acoustic wave inputs from the left of grating $[G_1^a]$, and from the right of grating $[G_5^a]$, the output SAW components $w^\pm$ at these extremities are given by

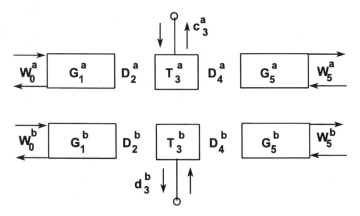

Fɪɢ. 15.14.   Transmission matrix notation used here for computing frequency response of two-pole waveguide-coupled SAW resonator filter.

$$\begin{bmatrix} 0 \\ w_o^{a-} \end{bmatrix} = \begin{bmatrix} M^a \end{bmatrix} \begin{bmatrix} w_5^{a+} \\ 0 \end{bmatrix} + c_3^a \begin{bmatrix} G_1^a \end{bmatrix} \begin{bmatrix} D_2^a \end{bmatrix} \begin{bmatrix} \tau_3^a \end{bmatrix} \qquad (15.13)$$

where c_3^a is the IDT input voltage and $[\tau_3^a]$ is the input electrical coupling vector component of IDT matrix T_3. This allows the outgoing acoustic wave component w_5^{a+} to be evaluated. Working backwards, this yields the acoustic vector $[W_2^a]$ at IDT T_3 as

$$\begin{bmatrix} W_2^a \end{bmatrix} = \begin{bmatrix} t_3^a \end{bmatrix} \begin{bmatrix} D_4^a \end{bmatrix} \begin{bmatrix} G_5^a \end{bmatrix} \begin{bmatrix} W_5^a \end{bmatrix} \qquad (15.14)$$

Now assume that the acoustic energy transfer between resonators is mainly between the IDT regions. The relative SAW amplitude at T_3 in resonator B is taken as a fraction F of the acoustic intensity $[W_2^a]$, so that for identical input and output resonators Eq. (15.13) may be approximated by

$$\begin{bmatrix} 0 \\ w_o^{b-} \end{bmatrix} = \begin{bmatrix} M^b \end{bmatrix} \begin{bmatrix} w_5^{b+} \\ 0 \end{bmatrix} + F \begin{bmatrix} G_1^a \end{bmatrix} \begin{bmatrix} D_1^a \end{bmatrix} \begin{bmatrix} W_2^a \end{bmatrix} \qquad (15.15)$$

from which w_5^{b+} can be evaluated. In addition, acoustic vector W_3^b at the IDT of resonator B is

$$\begin{bmatrix} W_3^b \end{bmatrix} = \begin{bmatrix} D_4^b \end{bmatrix} \begin{bmatrix} G_5^b \end{bmatrix} \begin{bmatrix} W_5^b \end{bmatrix} \qquad (15.16)$$

From the foregoing, the electrical output d_3^b from resonator B is obtained as the scalar product given by

$$d_3^b = \begin{bmatrix} \tau_3^{lb} \end{bmatrix} \bullet \begin{bmatrix} W_3^b \end{bmatrix} \qquad (15.17)$$

where τ_3^{lb} is the output electrical coupling vector component of the IDT matrix.

15.2.7. FREQUENCY-RESPONSE COMPUTATION

From the preceding transmission relationships, the frequency response $H_s(\omega_s)$ of the coupled resonators for the symmetric-mode case is obtained as a function of the resonance frequency $f_s = \omega_s/2\pi$ in Eq. (15.4). In a similar manner, the frequency response $H_a(\omega_a)$ for the antisymmetric mode is obtained using resonance frequency $f_a = \omega_a/2\pi$ in Eq. (15.5). In this example, acoustic factor F in Eq. (15.15) was taken as the average $F = 0.5$ for the two modal coupling conditions in Figs. 15.11 and 15.12. Recalling the polarity difference between the modes in the COMT evaluation, the overall frequency response $H(\omega)$ is

$$H(\omega) = \frac{H_s(\omega_s) - H_a(\omega_a)}{2} \tag{15.18}$$

Figure 15.15 illustrates a frequency response computation using the forementional method for a 100-MHz low-loss two-pole waveguide-coupled resonator filter on ST-quartz operating into a 100-Ω load, as well as for a four-pole cascaded structure. The IDTs had 401 solid electrodes with

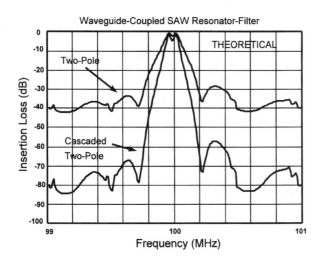

FIG. 15.15. Frequency response of illustrative two-pole waveguide-coupled 100-MHz resonator-filter on quartz using transmission-matrix modelling. Upper plot gives two-pole response. Lower plot is for four-pole cascaded filter. Horizontal scale: 99 to 101 MHz. Vertical scale: 10 dB/div.

$18\lambda_o$ acoustic aperture, while the reflection gratings had 600 shorted strips. The frequency separation between modes in this example was 50 kHz; transverse modes were not included in the computation. The general features of the computed frequency response may be compared with those of Fig. 15.7.

15.2.8.   P-MATRIX MODELLING TECHNIQUES

Waveguide-coupled SAW IF resonator-filters are also readily designed by using P-matrix analysis [25]. One method [14] employs 4×4 P-matrix relationships [25] for the electrical (I_3, I_4, V_3, V_4) responses at Ports 3 and 4 and acoustic responses (a_1, a_2, b_1, b_2) at Ports 1 and 2. An excitation cell is assigned to each waveguide mode. As applied to Port 3 this gives

$$\begin{pmatrix} b_1 \\ b_2 \\ I_3 \\ I_4 \end{pmatrix} = \begin{pmatrix} 0 & 1 & P_{13} & 0 \\ 1 & 0 & P_{23} & 0 \\ P_{31} & P_{32} & P_{33} & 0 \\ 0 & 0 & 0 & 0 \end{pmatrix} \begin{pmatrix} a_1 \\ a_2 \\ V_3 \\ V_4 \end{pmatrix} \tag{15.19}$$

In the same way, the excitation cell applied to Port 4 is

$$\begin{pmatrix} b_1 \\ b_2 \\ I_3 \\ I_4 \end{pmatrix} = \begin{pmatrix} 0 & 1 & 0 & P_{14} \\ 1 & 0 & 0 & P_{24} \\ 0 & 0 & 0 & 0 \\ P_{41} & P_{42} & 0 & P_{44} \end{pmatrix} \begin{pmatrix} a_1 \\ a_2 \\ V_3 \\ V_4 \end{pmatrix} \tag{15.20}$$

Similarly, the 4×4 P-matrix for a reflection cell is given by

$$\begin{pmatrix} b_1 \\ b_2 \\ I_3 \\ I_4 \end{pmatrix} = \begin{pmatrix} R_{11} & R_{12} & 0 & 0 \\ R_{21} & R_{22} & 0 & 0 \\ 0 & 0 & 0 & 0 \\ 0 & 0 & 0 & 0 \end{pmatrix} \begin{pmatrix} a_1 \\ a_2 \\ V_3 \\ V_4 \end{pmatrix} \tag{15.21}$$

in terms of reflection-coefficient parameter. Following the derivations of matrix parameters, elementary excitation cells for each waveguide mode are cascaded to give a composite 4×4 P-matrix. The "33," "34," "43," and "44" elements of this resultant P-matrix represent a two-port Y-matrix. The frequency response of the waveguide-coupled resonator filter is then obtained by summing the Y-matrices for each mode, adding the static capacitances to the "33" and "44" elements, and converting the resultant Y-matrix to an S-matrix [14].

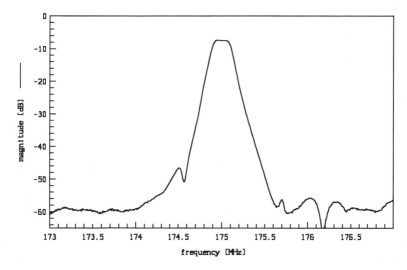

FIG. 15.16. Example of 175-MHz transverse- coupled resonator-filter for IF channel filtering in European PCN phone. Insertion loss is 7 dB, with 240-kHz 3-dB bandwidth. In a QCC10 ($9 \times 7\,\text{mm}^2$) package. (Courtesy of Siemens Matsushita Components GmbH and Co. KG, Munich, Germany.)

Using such design techniques, Fig. 15.16 illustrates the frequency response of a 175-MHz transverse-mode coupled resonator-filter, for channel-IF selection in 1.8-GHz European PCN phones. This was fabricated in a QCC10 package ($9 \times 7\,\text{mm}^2$), and exhibited an insertion loss of 7 dB, with a 3-dB bandwidth of 240 kHz.

15.3.   SAW In-Line IF Resonator-Filters

15.3.1.   COVERAGE OF THIS SECTION

Waveguide-coupled SAW resonator-filters have a small fractional-bandwidth capability that is typically less than about 0.1%. The actual obtainable bandwidth will, therefore, depend on the choice of centre frequency. For a given centre frequency, however, increased fractional-bandwidth capability to suit the preceding filtering can be obtained by employing the stronger coupling of longitudinal acoustic modes in a single-track structure. An example of this latter design method is given later for a 250-MHz IF filter applied to 300-kHz channels in the Japanese Cordless Personal Handy Phone (PHP) requiring a fractional bandwidth of about 0.2%, to account for channel width as well as fabrication and temperature tolerances. Before doing so, however, consider first some general concepts relating to SAW In-Line filter structures and limitations.

15.3.2. GENERAL CONCEPTS FOR SAW IN-LINE RESONATOR-FILTERS

For a given centre frequency, increased-bandwidth capability to suit the forementioned filtering can be obtained by employing the stronger coupling of longitudinal acoustic modes in a single-track structure. To date, several design techniques have been applied to achieve the filtering responses required. The general method is to add appropriately spaced resonance cavities within a two-port single-track IDT structure, These additional resonant cavities can be within or between IDTs. An example of the IDT cavity-type is shown in Fig. 15.17, which can be employed in one or both ports of a two-port SAW resonator-filter. By careful selection of the lengths of the two IDT-sections around the cavity—together with control of phases—selectively placed nulls in the frequency response can be created to suppress local oscillator and image signals in single-conversion receivers [29].

Figure 15.18 illustrates two in-line resonator-filter designs suitable for 200- to 300-kHz channel-bandwidth requirements. In the version of Fig. 15.18(a), a short *partially* reflecting grating G5 is placed between the input/output IDTs. This acts as a coupling element between the two cavity modes shown [23]. Except for its wider bandpass characteristic, the frequency response for this particular resonator example is similar to that of the waveguide-coupled response of Fig. 15.15. The version of Fig. 15.18(b) employs additional cavities for prescribing the response. The frequency response of both of these structures can be calculated using the transmission-matrix methods previously described, with suitable parameters for the

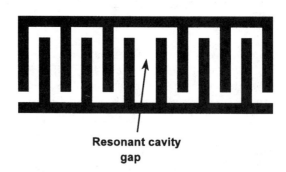

**Resonant cavity
gap**

FIG. 15.17.   IDT for version of an in-line SAW resonator-filter employing a resonant-cavity gap within IDT itself.

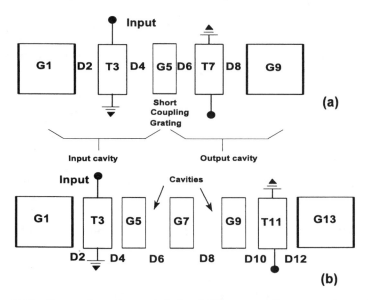

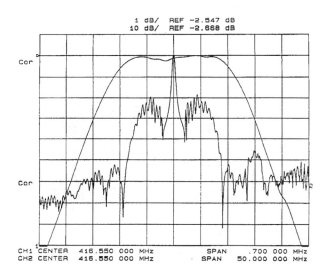

FIG. 15.18. Two versions of two-pole in-line SAW resonator filter geometries employing resonant-cavity gaps between IDTs.

FIG. 15.19. Frequency response of commercial two-pole in-line SAW resonator filter with insertion loss 2.547 dB, and 1-dB bandwidth ~ 300 kHz. Horizontal scale: Centre frequency 415.55 MHz, Span 1 is 0.7 MHz, Span 2 is 50 MHz. Vertical scale: 1 dB/div and 10 dB/div. (Courtesy of Sawtek Inc., Orlando, Florida.)

coupling gratings [23]. Figure 15.19 illustrates the frequency response of a commercial 415-MHz two-pole in-line resonator filter, with an insertion loss of 2.547 dB, 1-dB bandwidth of about 300 kHz, and close-in sidelobe rejection of about 20 dB.

15.3.3. A SAW In-Line IF Resonator-Filter for Japanese Cordless Personal Handy Phone (PHP)

Here, the frequency-response modelling of an illustrative 250-MHz SAW IF three-pole resonator-filter design with fractional bandwidth ~ 0.2% will be examined for application to PHP phone architecture, employing $\pi/4$ DQPSK modulation. This resonator-filter is fabricated on ST-quartz and, as sketched in Fig. 15.20(b), has the same two-port IDT structure as that for a conventional one-pole SAW resonator. In conventional one-pole SAW

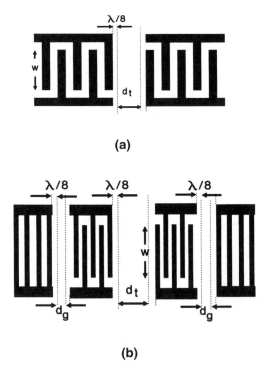

(a)

(b)

Fig. 15.20. (a) Distance parameters for Rayleigh-wave filter using long-pair IDTs with significant finger reflections ($h/\lambda = 2\%$) and inter-IDT resonant cavity. (b) Distance parameters for composite three-pole IF-channel filter, suitable for Japanese PHP Handy Phone receiver.

resonators, however, both the IDT electrode number and electrode metallization thickness ratio (h/λ) often are chosen to be small enough so that the effects of finger reflections do not corrupt the desired frequency response. In contrast, the three-pole SAW in-line resonator-filter of Fig. 15.20(b) uses long IDTs and large film-thickness ratio $h/\lambda = 2\%$, to achieve the desired response, in conjunction with a resonant cavity between IDTs [30]. Figure 15.20(a) illustrates the basic geometry before the addition of end reflection gratings in Fig. 15.20(b). Equal numbers of IDT fingers pairs $(N_p = 200)$ are employed in the long input/output IDTs. Resonance or near-resonance conditions are obtained with the structure of Fig. 15.20(a). The resonant peak is due to an inter-IDT resonance, governed by IDT separation and finger reflectivity. The near-resonant one is due to the peaking of the distorted radiation conductance of the IDT [30]. A third pole is provided by the end reflection gratings in Fig. 15.20(b). For optimum performance, the reflection grating stopband centre frequency is shifted downwards from the IDT one, to account for the radiation conductance distortion by finger reflections [10]. In addition, the grating spacing d_g is positioned to place the third pole for the desired bandwidth.

Using the modelling method given in the following, Fig. 15.21 illustrates the predicted two-pole response of the basic structure of Fig. 15.20(a) for a 250-MHz design on temperature-stable ST-quartz, using $N_p = 200$ finger pairs in each IDT, together with a film metallization thickness ratio $h/\lambda = 2\%$. The inter-IDT resonance peak occurs on the high-frequency side. Its position is determined by the inter-IDT separation d_t.

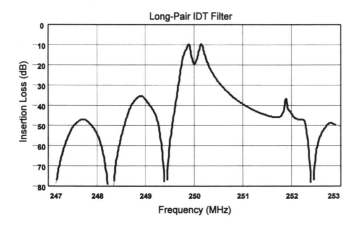

FIG. 15.21.   Frequency response of 250-MHz long-pair IDT filter section of Fig. 15.20(a), with $N_p = 200$ finger pairs in each IDT, and gap $d_t = 0.39\ \lambda_o$. Acoustic aperture is 20 λ_o, with load impedance $Z_L = 200\ \Omega$. Insertion loss ~ 7 dB.

15.3.4. Modelling Method

This three-pole SAW in-line IF resonator-filter is modelled here for a centre frequency of 250 MHz, with fractional bandwidth ~0.2% and insertion loss IL < 3 dB. As shown in Fig. 15.20, IDT and grating reference axes for computation were set $\lambda_o/8$ from the metal edges [23]. Self-coupling (k_{11}) and mutual -coupling ($k_{12}=k$) coefficients for ST-quartz can be derived from Table 9.1 as $k_{11} = 1985$ and $k_{12} = 1650$, respectively.

Using transmission-matrix relationships given in Chapter 11, the reflection coefficients for the end reflection gratings and the IDTs are obtained from the G_{21}/G_{11} ratios in their respective 2×2 grating matrices [G] using appropriate length parameters. In addition, the phase-angle reference parameter θ in both gratings is $\theta = \pi$ for ST-quartz. The SAW velocity v_m with metallization is approximated as $v_m = v_o(1 - k_{11}/\beta_{mid})$, where v_o = unperturbed velocity on ST-quartz, and θ_{mid} = phase constant at midband. To allow for the SAW velocity perturbation by the IDTs with $N_p = 200$ finger pairs in each, their unperturbed centre frequency f_o is chosen as $f_o = v_o/\lambda_o = 251.9$ MHz for a shifted frequency $f_{idt} = v_m/\lambda_o = 250.9$ MHz, with 2% metallization. For this example 311 grating strips are employed for each end grating, with a shifted centre frequency $f_{og} = 249.79$ MHz.

As demonstrated in Fig. 15.22, the distorted IDT radiation conductance G_d due to finger-edge reflections may be approximated as

$$G_D \approx \frac{C|P_{13}|^2}{\left[1 - |\rho_{idt}|^2\right]} \tag{15.22}$$

In Eq. (15.22) C = ohmic constant, ρ_{idt} = IDT reflection coefficient, while $P_{13} = P$-matrix parameter for an IDT, in the notation of Chapter 4. (Note that $P_{13} = S_{13}$, where S_{13} is the corresponding scattering-matrix term in Chapter 4.) P_{13} is approximated here from the method of Reference [31], using a transduction parameter α_t.

The (dimensionless) scattering-matrix {S} parameters for each IDT may be approximated as

$$S_{11} \approx \rho_{idt} + \rho_L \cdot \sqrt{1 - |\rho_{idt}|^2} \cdot \exp\left(-jN_{idt}\Lambda \cdot \delta_{idt}\right) \tag{15.23}$$

$$S_{31} \approx \frac{\sqrt{2\left[1 - |\rho_{idt}|^2\right]G_D \cdot Re(Y_L) \cdot \exp\left(-jN_{idt}\Lambda\delta_{idt}\right)}}{Y_T} \tag{15.24}$$

$$S_{21} \approx \sqrt{1 - |S_{11}|^2 - |S_{13}|^2} \cdot \exp\left(-jN_{idt}\Lambda\delta_{idt}\right) \tag{15.25}$$

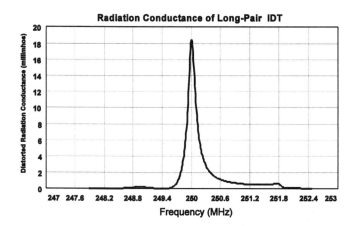

Fig. 15.22. IDT distorted radiation conductance due to IDT finger reflections, for filter response of Fig. 15.21.

In the foregoing S-matrix relations, ρ_{idt} = IDT reflection coefficient, $\rho_L = G_{D/}$ Y_L = input/load effective reflection coefficient, $N_{idt} = 2N_p$ = number of IDT electrodes, Λ = adjacent finger period = $\lambda_o/2$, $\delta_{idt} = [2\pi(f - f_o)/v_o] + k_{11}$ = IDT detuning parameter, Y_L = input/load admittance, $Y_T = Y_L + G_D +$ $j(2\pi f C_T + B_{eff})$. From Eq. (15.22), G_D = distorted radiation conductance of the IDT, B_{eff} = Hilbert transform of G_D, and C_T = IDT total capacitance.

The frequency response $A(f)$ of the basic SAW filter of Fig. 15.20(a) is approximated as

$$A(f) \approx 20 \cdot \log_{10} \left| \frac{d_t S_{13}^2}{1 - d_t^2 S_{11}^2} \right| \qquad (15.26)$$

where d_t = inter-IDT spacing between reference planes in Fig. 15.20(a).

The response of the three-pole IF filter in Fig. 15.20(b) may now be computed using {S}-to-{T} matrix transformations for the IDTs [32]. Specifically,

$$S_{11} = S_{22}; \quad S_{12} = S_{21}; \quad s = (-1)^{N_{idt}} \qquad (15.27)$$

$$T_{11} = \frac{1}{S_{12}}; \quad T_{12} = \frac{S_{11}}{S_{12}}; \quad T_{13} = \frac{S_{13}}{S_{12}}; \quad T_{21} = -T_{12} \qquad (15.28)$$

$$T_{22} = \frac{S_{12} - S_{11} \cdot S_{22}}{S_{12}}; \quad T_{23} = S_{13} - \frac{S_{11} \cdot S_{13}}{S_{12}} \qquad (15.29)$$

$$T_{31} = s T_{13}; \quad T_{32} = -s T_{23} \exp(-j N_{idt} \Lambda \delta_{idt}) \qquad (15.30)$$

$$T_{33} = \frac{Y_L - j\left(2\pi f C_T - B_D\right)}{Y_L + j\left(2\pi f C_T - B_D\right)} \tag{15.31}$$

Figure 15.23 illustrates a frequency-response computation for such a 250-MHz three-pole resonator-filter on ST-quartz. Parameters employed here included an unperturbed IDT centre frequency $f_o = 251.9\,$MHz, with a reflection-grating stopband frequency set at $f_{og} = 249.79\,$MHz. The IDTs had an acoustic aperture $w = 20\ \lambda_o$, and employed 200 electrode-pairs. The reflection gratings had 311 shorted-electrode strips. A film-thickness-ratio $h/\lambda = 2\%$ was used for IDTs and gratings. Distance parameters were $d_t = 0.39\ \lambda_o$, and $d_g = 0.2\ \lambda_o$. The load/source impedances were set at 200 Ω. The computed fractional bandwidth in Fig. 15.23 is about 0.18%, while the insertion loss is about 2 dB.

The computations leading to Fig. 15.23 did not include anisotropy or diffraction, which can be significant for the long-pair IDTs on nonautocollimating ST-quartz, as given in Table 6.2. These effects can be expected to "flatten" the passband response, as indeed indicated in the experimental response of Fig. 15.24, for a 250-MHz three-pole in-line filter on quartz. In Fig. 15.24, the reduced upper-sideband rejection can be attributed to the remnants of the distorted radiation conductance response in Fig. 15.22.

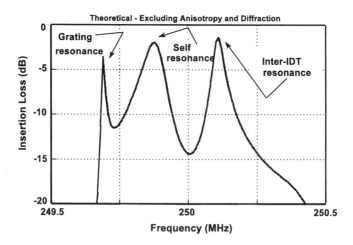

FIG. 15.23. Computed frequency response of 250-MHz three-pole in-line IF filter on ST-quartz, for long-IDT resonator-filter of Fig. 15.20(b). Anisotropy and diffraction not included. Horizontal scale: 249.5 to 250.5 MHz. Vertical scale: 0 to − 20 dB.

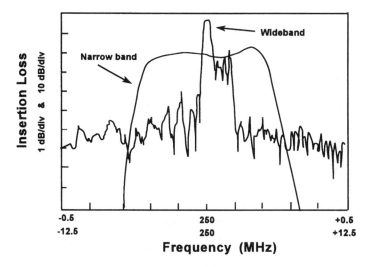

Fig. 15.24. Experimental in-band and out-of-band frequency responses of 250-MHz three-pole IF filter of Fig. 15.20(b). Horizontal scales: Centre frequency 250 MHz. Spans: 1 MHz and 25 MHz. Vertical scales: 1 dB/div and 10 dB/div. (After Reference [30].)

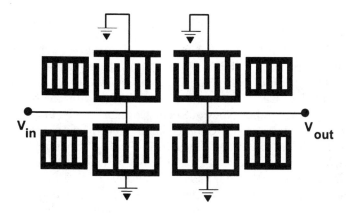

Fig. 15.25. Parallel-connected three-pole in-line IF filters for improved out-of-band rejection.

The out-of-band response can be improved by connecting two of these structures in parallel, as sketched in Fig. 15.25, as well as by selective withdrawal-weighting of IDT fingers [30]. In this way, Fig. 15.26 demonstrates the matched experimental response of the structure of Fig. 15.25 with insertion loss of 2.2 dB, fractional bandwidth of 0.15%, suitable for IF filtering in Japanese PHP phones [30].

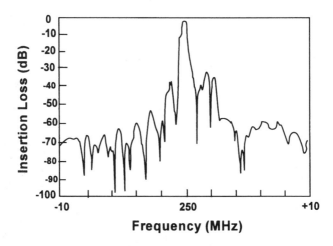

FIG. 15.26.   Experimental response of 250-MHz IF filter on quartz with structure of
Fig. 15.25. Horizontal scale: Centre frequency 250 MHz, and span of 20 MHz. Vertical scale:
0 to − 100 dB. (After Reference [30].)

15.4.   SPUDT-Based IF Filters for European GSM/PCN Phones

The principles of SPUDT-based filter operation have already been reviewed in Chapter 12. SPUDT filters have wide application for first-IF
channel filtering in European GSM/PCN systems. While the SPUDT structure illustrated in Chapter 12 was a general type with comb- or single-
response capability, other types of SPUDT designs can be employed. One
commonly used noncomb SPUDT geometry shown in Fig. 15.27 can be
considered to be modified split-finger IDTs [7]. This structure is known as
the electrode width-controlled SPUDT (EWC-SPUDT) [33]. Its interlaced
reflection strips are $\lambda/4$ wide, and withdrawal-weighting of reflectivity is
achieved by replacing the $\lambda/4$ electrodes by two grounded $\lambda/8$ electrodes
with $\lambda/8$ spacing. Relative positions between split-fingers and reflector strips
determine the shift between centres of transduction and centres of reflection. The reflector elements can be chosen so that reflection and regeneration effects cancel over a broad frequency range when the IDTs are
properly matched [7], [34]. In this latter structure, normally fabricated on
ST-X quartz, passband response shaping is achieved by selective withdrawal-weighting of electrodes. In designs for GSM/PCN IF filters, the
general aim is to minimize the passband distortions due to spurious signals
rather than to achieve very low insertion loss. Complex matching networks
may be required for enhanced operation [33].

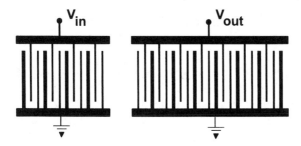

FIG. 15.27. SPUDT-based design for IF filter in European GSM/PCN phone systems using interlaced reflector strips for directivity. Withdrawal-weighting not shown here. (After Reference [7].)

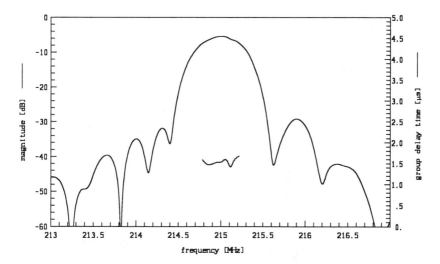

FIG. 15.28. Frequency response of illustrative 215-MHz SPUDT first IF filter for European GSM/PCN phone systems with insertion loss of 5- and 3-dB bandwidth of 500 kHz. Group delay ~ 1.5 μs. In a QCC12 (13×6.5 mm^2) package. (Courtesy of Siemens Matsushita Components GmbH and Co. KG, Munich, Germany.)

Figure 15.28 illustrates the frequency response of a sample 215-MHz SPUDT IF filter for GSM/PCN phone systems. This particular device had a 3-dB insertion loss of 5 dB, a 3-dB bandwidth of 500 kHz, and a group delay ~ 1.5 μs at centre frequency. This device was in a QCC12 (13×6.5 mm^2) package.

15.5. Z-Path IF Filters for GSM Phones

While SPUDT IF filters of the previous section give excellent performance, their acoustic lengths can be undesirably large for required IF-channel definition in some GSM phone systems. This is because they may be considered (to first order) to act as the transversal filters considered in Chapter 3. As such, they need lengthy IDT structures (e.g., 30-mm chip structures) to establish the large time-windows needed for GSM filter definition [4], [7].

The Z-path filter structure shown in Fig. 29 offers one solution to reducing the overall acoustic propagation path. As sketched, this structure incorporates two slightly inclined (~4°) reflection gratings to couple the surface wave from the input IDT to the output one. The triple reflection path within the interspace region reduces the overall size otherwise required, while providing required passband and out-of-band responses for 200-kHz GSM filter channels.

Since these structures are fabricated on quartz, only those acoustic waves travelling along the main propagation axis are subject to temperature stability. Provided that weakly inclined reflection gratings are employed, however, the Z-shaped acoustic propagation paths help to preserve the required temperature stability. This is because the distortions produced by the two opposite angles of reflection tend to cancel [6].

The Z-path IF filters are well suited for achieving high relative bandwidths on quartz. They are also suited for operation at fairly low IF frequencies, where small temperature tolerances as well as small fabrication tolerance can be maintained. Figure 15.30 illustrates the frequency response of a 71-MHz Z-path IF filter for GSM phone systems, contained in a DCC ($14 \times 8\,mm^2$) package with insertion loss of about 5 dB.

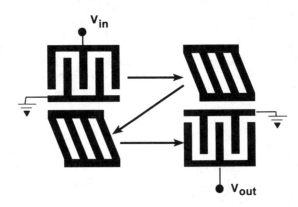

FIG. 15.29.   Basic Z-path filter.

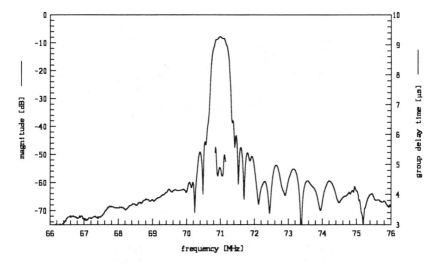

FIG. 15.30. Frequency response of a 71-MHz Z-path IF filter for GSM phone system. Insertion loss is 5 dB. In a DCC14 ($14 \times 8\,mm^2$) package. (Courtesy of Siemens Matsushita Components GmbH and Co. KG, Munich, Germany.)

15.6. IF Filters for Digital European Cordless Telephones (DECT)

The DECT system is designed for operation from 1880 to 1900 MHz, with GFSK modulation using a 0.5 Gaussian filter and employing 10 carriers with a channel spacing of 1.728 MHz. TDMA/FDM multiple access is employed with this system, with channel bit-rates of 1.152 Mb/s. With Gaussian filtering, the channel bandwidth is about 1.2 MHz. For typical IF frequencies for DECT systems in the 110-MHz range, this requires receiver first-IF filter bandwidths in the order of 1% [9].

IF filter designs for DECT phones can be based on "conventional" transversal-filter mid-loss designs or on SPUDT/reflector designs for reduction in both insertion loss and packaging size. One SPUDT/reflector design is illustrated in Fig. 15.31 [35]. As with the Z-path filter considered in the foregoing, the structures of Fig. 15.31 lead to a reduction in filter length.

In the structure of Fig. 15.31(a) the SPUDTs and the central reflection gratings are identical in the two parallel tracks. In the lower track, however, the separations between SPUDTs and gratings are increased by $\lambda/4$ compared with those in the upper tracks.

There are three main paths that acoustic waves can travel. In the first of these, the waves that travel directly from input SPUDTs to output SPUDTs in each track arrive 180° out of phase with one another and cancel at the

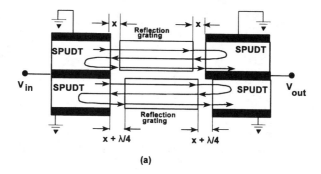

(a)

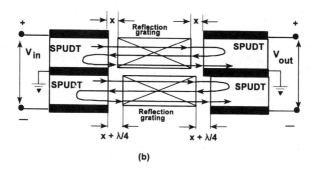

(b)

FIG. 15.31. (a) Basic SPUDT/grating dual-track SAW IF filter. (b) Modified design for balance input-output active drive circuits. (After Reference [35].)

output port. For the second main path, waves from input SPUDTs can be reflected by output SPUDTs and reflection gratings and arrive in-phase at the output summing port. Along the third main path, waves from the output SPUDTs can be reflected by the centre gratings and the input SPUDTs and again arrive in-phase at the output SPUDTs. Waves along Paths 2 and 3 combine to give an output signal [35]. The geometry of Fig. 15.31(b) is particularly suitable for IF-filtering in the GSM/DCS-1880 digital cordless phone systems operating in the 1800-MHz band, in active circuits that require differential-balanced input and output drives. Balanced drives in active circuits can reduce power-consumption requirements for a given signal level while causing improved amplitude- and phase-compression over single-ended circuits. In some instances, the use of a balanced active circuit can lead to elimination of even-order intermodulation distortion and reduced odd-order intermodulation distortion.

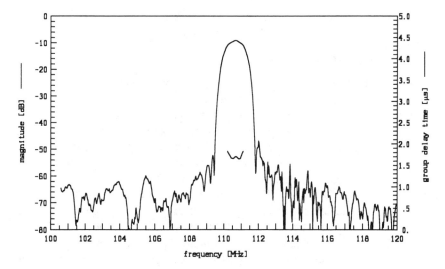

FIG. 15.32. SAW IF filter for DECT cordless telephones using SPUDT/ reflector design. Insertion loss is 8 dB with 3-dB bandwidth of 1.1 MHz. In a ceramic QCC10 (9×7 mm^2) package. (Courtesy of Siemens Matsushita Components GmbH and Co. KG, Munich, Germany.)

As a variation of the geometry of Fig. 15.31(a), a design for a DECT IF filter starts with identical central reflection gratings, and then removes one reflector strip from the lower Track-2 grating (with a grating-pitch of $\lambda/2$). In addition, the polarity of the input SPUDT in Track 1 is reversed. As a result, the phase difference of 180° between tracks is not a frequency-dependent one, so that the direct-path signal is cancelled for all frequencies. To first order, the filter response function $H(f)$ for this structure is given as [35]

$$H(f) \approx 2\left[T_{sp}(f)\right]^2 T(f) R(f) R_{sp}(f) \tag{15.32}$$

where $T_{sp}(f)$ and $R_{sp}(f)$ relate the forward transfer function and the reflection function of the SPUDTs, respectively, while $T(f)$ gives the transmission through a centre grating, and $R(f)$ is the reflection of a centre grating. These structures have been fabricated on 112.2° *Y-X* LiTaO$_3$, with high K^2 and moderate temperature coefficient.

Figure 15.32 illustrates the frequency response of a 110.5-MHz SPUDT/ reflector-type IF filter for the receiver channel in a DECT cordless telephone. This particular device has an insertion loss of 8 dB, with a 3-dB bandwidth of 1.1 MHz, and is in a ceramic QCC10 (9×7 mm^2) package.

15.7. IF Filters for North American Digital Cellular IS-95 and PCS Phones

15.7.1. RATIONALE FOR USING CDMA

As indicated in Tables 10.3 and 15.2, the access scheme for the North American IS-95 phone system employs Code Division Multiple Access (CDMA), with Frequency Division Multiplexing (FDM). As will be presented later in Chapters 17 and 18, CDMA is a spread-spectrum system. With CDMA, the spectral energy of the message signal is "spread" (i.e., expanded) by further encoding each digital message bit with a high-speed digital chip sequence. In this way, a frequency channel can be used simultaneously by a number of mobile units within a given cell, as the message signal of each subscriber can spread with different chip codes for each unit. As with TDMA architecture, the CDMA system can support more users per base station per megahertz of frequency spectrum. The CDMA and TDMA architectures allow flexibility for voice/data communications, as well as reduced transmit power requirements for the mobile phone—thus increasing battery-charge life.

15.7.2. SOME FEATURES OF NORTH AMERICAN DIGITAL CELLULAR IS-95

In the North American IS-95 system, the basic user-channel message bit-rate is 9.6 kb/s. The spectrum for this transmission is spread by a factor of 128, using a channel chip-rate of 1.2288 Mchip/s. To reduce interference between mobile units in the same cell that use the same chip-spreading code, all signals in a particular cell are further scrambled using a pseudorandom (PN) sequence of 2^{15} chips. Similar to IS-54, the IS-95 system is a dual-mode cellular one that can operate in the CDMA mode or the AMPS mode [3]. As indicated in Fig. 10.2, the channel width for IS-95 is 1.25 MHz, with 45-MHz spacing between transmit and receive bands. This system has 20 channels with 798 users/channel.

15.7.3. ASPECTS OF NORTH AMERICAN PCS

In the United States, a spectral amount of 140 MHz, around 1.9 GHz, has been allocated for Personal Communications Services (PCS). The transmit band covers 1.850 to 1.910 GHz, while the receive band is 1.930 to 1.990 GHz. For high-tier standards, this band is available for PCS-TDMA (based on IS-136 cellular), PCS-CDMA (based on IS-95 cellular), PCS-1900 (based on GSM cellular), and Wideband-CDMA. This chapter highlights some IF filters for IS-95 and PCS-CDMA. SAW devices for Wideband-CDMA spread-spectrum systems are examined in Chapter 18.

15.7.4.   SAW IF FILTERS FOR IS-95 AND PCS-CDMA

To avoid message-signal degradation, IF filters for both receiver and trans-mitter base-station applications for IS-95 or PCS-CDMA require flat pass-band response, and excellent in-band phase linearity (i.e., flat group-delay response), as well as good out-of-band rejection. These can be readily achieved using SAW transversal-filter techniques.

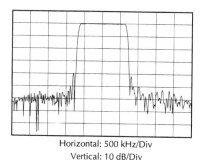

Horizontal: 500 kHz/Div
Vertical: 10 dB/Div

(a)

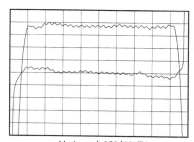

Horizontal: 150 kHz/Div
Vertical: 1 dB/Div
Vertical: 5 deg/Div

(b)

FIG. 15.33.   Example of a 69.99-MHz CDMA base-station receiver with ultraflat passband, and minimal phase deviation from linearity. Horizontal scale: (a) 500 kHz/div; (b) 150 kHz/div. Vertical scale: (a) 10 dB/div; (b) 1 dB/div, and phase scale set for 5°/div deviation from linearity. (Courtesy of Sawtek Inc., Orlando, Florida.)

Figure 15.33(a) gives the frequency response of a commercial 69.99-MHz CDMA base-station IF filter fabricated on quartz. The insertion loss for this filter is 22 dB (max.), while the passband width is 1.26 MHz (min.), and the out-of-band rejection level is 50 dB (min.). Fig. 15.33(b) shows the in-band response, with a peak-to-peak amplitude ripple of 0.7 dB (max.), and a phase deviation from linearity of 5° (max.).

15.8. "Standard" IF Frequencies For Mobile Phones

As mobile phone systems continue to develop, "standard" IF frequencies and packages are becoming increasingly available. For IF filters in very narrowband systems such as analog AMPS, standard "off-the-shelf" frequencies for SAW filters include those at 82.2, 83.16, 85.05, 86.85 and 90 MHz.

Standard IF frequencies are also appearing for digital phone systems. For GSM phone systems, IF frequencies of 71, 78, 163, 246, and 259 MHz are commonly used. Other standard IF frequencies include 110.593 MHz for DECT phones, 130 MHz for PDC, and 248.45 MHz for PHS [34].

15.9. Summary

This chapter has presented a variety of SAW IF filter designs such as are employed for channel-filtering in mobile/wireless receivers. These included designs for channel filtering in North American, European and Japanese cellular and cordless phone systems. The reader should note, however, that devices selected for demonstration purposes are not unique to a given cellular/cordless phone system, but were given here to illustrate the global nature of SAW filter applications.

15.10. REFERENCES

1. P. Vizmuller, *RF Design Guide*, Artech House, Boston, ch. 1, 1995.
2. K. Feher, *Digital Communications*, Prentice-Hall, Englewood Cliffs, ch. 4, 1983.
3. J. E. Padgett, C. G. Günther and T. Hattori, "Overview of wireless personal communications," *IEEE Communications Mag.*, pp. 28–41, Jan. 1995.
4. J. Machui, J. Bauregger, G. Riha and I. Schropp, "SAW devices in cellular and cordless phones," *Proc. 1995 IEEE Ultrason. Symp.*, vol. 1, pp. 121–130, 1995.
5. G. Riha, K. Anemogiannis, R. Dill and C. Kappacher, "SAW low-loss filters for mobile communications systems," *Proc. International Symp. on Surface Acoustic Wave Devices for Mobile Communication*, Sendai, Japan, pp. 113–120, 1992.
6. J. Machui and W. Ruile, "Z-path IF-filters for mobile phones," *Proc. 1992 IEEE Ultrason. Symp.*, vol. 1, pp. 147–150, 1992.
7. C. C. W. Ruppel *et al.*, "SAW devices for consumer communication applications," *IEEE Trans. Ultrason. Ferroelec. and Freq. Control*, vol. 40, pp. 438–452, Sept. 1993.

8. M. Solal, P. Dufilie and P. Ventura, "Innovative SPUDT based structures for mobile radio applications," *Proc. 1994 IEEE Ultrason. Symp.*, vol. 1, pp. 17–22, 1994.
9. R. Dill, J. Machui and G. Müller, "A novel SAW filter for IF-filtering in DECT systems," *Proc. 1995 IEEE Ultrason. Symp.*, vol. 1, pp. 51–54, 1995.
10. T. Uno and H. Jumonji, "Optimization of quartz SAW resonator structure with groove gratings," *IEEE Trans. Sonics Ultrason,* vol. SU-30, pp. 299–310, Nov. 1982.
11. G. Martin, B. Wall and M. Weihnacht, "Single mode waveguide resonator filters with weak coupling," *Proc. 1994 IEEE Ultrasonics Symposium*, vol. 1, pp. 125–128, 1994.
12. B. Wall and A du Hamél, "Balanced driven transversely coupled waveguide resonator filters," *Proc. 1996 IEEE Ultrasonics Symposium*, vol. 1, pp. 47–51, 1996.
13. S. Broderick, "SAW double-mode resonators for first IF in cellular radio," *RF Design*, pp. 71–73, March 1991.
14. G. Scholl, W. Ruile and P. H. Russer, "P-matrix modeling of transverse-mode coupled resonator filters," *Proc. 1993 IEEE Ultrason. Symp.*, vol. 1, pp. 41–46, 1993.
15. J. Saw, M. Suthers, Y. Xu, R. Leroux, J. Nisbet, G. Rabjohn and Z. Chen, "SAW technology in RF multichip modules for cellular systems," *Proc. 1995 IEEE Ultrason. Symp.*, vol. 1, pp. 171–175, 1995.
16. E. J. Staples, J. S. Schoenwald, R. C. Rosenfeld and C. S. Hartmann, "UHF surface acoustic wave resonators," *Proc. 1974 IEEE Ultrason. Symp.*, pp. 245–252, 1974.
17. W. R. Shreve and P. S. Cross, "Surface acoustic waves and resonators," in E. A. Gerber and A. Ballato (eds), *Precision Frequency Control*, vol. 1, Academic Press, Orlando, pp. 119–145, 1985.
18. E. A. Ash, "Fundamentals of signal processing devices," in A. A. Oliner (ed.), *Acoustic Surface Waves*, Springer-Verlag, Berlin, pp. 107–116, 1978.
19. S. Datta, *Surface Acoustic Wave Devices*, Prentice-Hall, New Jersey, pp. 225–239, 1986.
20. H. A. Haus and P. V. Wright, "The analysis of grating structures by coupling-of-modes theory," *Proc. 1980 IEEE Ultrason. Symp.*, vol. 1, pp. 277–281, 1980.
21. P. V. Wright, "A coupling-of-modes analysis of SAW grating structures," Ph.D. thesis in electrical engineering, Massachusetts Institute of Technology, Cambridge, Mass., 1981.
22. P. V. Wright, "Analysis and design of low-loss SAW devices using coupling-of-modes theory," *Proc. 1989 IEEE Ultrason. Symp.*, vol. 1, pp. 149–152, 1989.
23. P. S. Cross and R. V. Schmidt, "Coupled surface-acoustic-wave resonators," *Bell Syst. Tech. J.*, vol. 56, pp. 1447–1481, 1977.
24. C. K. Campbell, "Modelling the transverse-mode response of a two-port SAW resonator," *IEEE Trans. Ultrason. Ferroelec. and Freq. Control,* vol. 38, pp. 237–242, May 1991.
25. C. C. W. Ruppel, W. Ruile, G. Scholl, K. C. Wagner and O. Männer, "Review of models for low-loss filter design and application," *Proc. 1994 IEEE Ultrasonics Symposium*, vol. 1, pp. 313–324, 1994. (With 108 references).
26. H. A. Haus and W. Huang, "Coupled-mode theory," *Proc. IEEE*, vol. 79, pp. 1505–1518, Oct. 1991.
27. H. A. Haus, *Waves and Fields in Optoelectronics*, Prentice-Hall, New Jersey, 1984.
28. J. M. Stone, *Radiation and Optics*, McGraw-Hill, New York, ch. 10, 1963.
29. L. W. Heep, "Selective null placement in SAW coupled resonator filters," *Proc. 1991 IEEE Ultrason. Symp.*, vol. 1, pp. 185–188, 1991.
30. Y. Yamamoto and R. Kajihara, "SAW composite longitudinal mode resonator (CLMR) filters and their application to new synthesized resonator filters," *Proc. 1993 IEEE Ultrason. Symp.*, vol. 1, pp. 47–51, 1993.
31. C. S. Hartmann, D. P. Chen and J. Heighway, "Modeling of SAW transversely coupled resonators using coupling-of-modes modeling technique," *Proc. 1992 IEEE Ultrason. Symp.*, vol. 1, pp. 39–43, 1992.

32. C. B. Saw, "Single-Phase Unidirectional Transducers for Low-Loss Surface Acoustic Wave Devices," Ph.D. Thesis in Electrical Engineering, McMaster University, Hamilton, Ontario, Canada, 1988.
33. C. S. Hartmann and B. P. Abbott, "Overview of design challenges for single phase unidirectional SAW filters," *Proc. 1989 IEEE Ultrason. Symp.*, vol. 1, pp. 79–89, 1989.
34. C. S. Lam, D-P Chen, B. Potter, V. Narayanan and A. Vishwanathan, "A review of the applications of SAW filters in wireless communications," *International Workshop on Ultrasonics Application*, Nanjing, China, Sept. 1996.
35. M. Solal and J. M. Hode, "A new compact SAW low loss filter for mobile radio," *Proc. 1993 IEEE Ultrason. Symp.*, vol. 1, pp. 105–109, 1993.

—16—

Fixed-Code SAW IDTs for Spread-Spectrum Communications

16.1. Introduction

16.1.1. SPECTRAL EFFICIENCY OF CDMA IN MOBILE CELLULAR COMMUNICATIONS

With the current revolutionary expansion of radio frequency (RF) consumer communications technology into such fields as local-area networks (LAN), mobile telephones, indoor communications, and home and office communications, the demands imposed on circuitry for achieving efficient, interference-free data transmission and reception are becoming increasingly stringent. One of the key technologies to relieve such signal congestion is spectrum-spectrum technology and its code-division multiple access (CDMA) signalling method, which is more spectrally efficient than either frequency-division multiple access (FDMA), or time-division multiple access (TDMA) [1].

The FDMA technique is implemented by dividing an allocation RF spectrum into different radio channels. The TDMA scheme divides a radio channel into many time slots, where each carries a traffic channel. On the other hand, the CDMA signalling, which uses code sequences as traffic channels in a common radio channel, permits many users to utilize the same frequency simultaneously.

To illustrate their relative efficiencies, consider an example of an FDMA cellular link with a spectral bandwidth of 1.5 MHz, which can be subdivided into 150 radio channels, each with a channel bandwidth of 10 kHz. With a TDMA link, the overall bandwidth could be divided into 50 radio channels, where, for example, each has a radio-channel bandwidth of 30 kHz and each channel carries three time-slots. This TDMA link would have the same channel-number capability (150) as the FDMA one. In contrast, the CDMA link would employ the entire spectral bandwidth of 1.5 MHz for one radio channel, while providing 50 simultaneous code-sequence traffic channels for each section of a communications "cell", where each cell can be divided into three "sectors." As far as *channel efficiency* is concerned, this CDMA scheme has the same 150-channel capability as the FDMA and TDMA examples. For the FDMA and TDMA systems, however, we have to consider their *frequency re-use factor K* [1]. Consider that a frequency re-use factor $K = 5$ may be required to maintain a carrier-to-interference ratio (C/I) of $C/I > 18\,dB$. This means that we will only be able to use $150/5 = 30$ channels per cell with FDMA or TDMA. With the CDMA system, however, every cell can have the same 150 channels—demonstrating that cellular CDMA has a greater spectral efficiency than either cellular FDMA or cellular TDMA. Moreover, the added advantage of using spread-spectrum techniques is that they can provide high levels of interference rejection while operating at low powers.

Unlicensed mobile spread-spectrum radio links operate at low RF power levels, with restrictions on them can often be severe, depending on the region involved. In Japan, for example, regulations for spread-spectrum mobile radios operating at frequencies of less than $322\,MHz$ require their field strengths to be less than $500\,\mu V$/meter at a distance of 3 m from the unit [2]. These kinds of restrictions have prompted developments of direct-sequence (DS) spreading codes[1] that will yield demodulation sequences with low autocorrelation sidelobes for enhanced receiver sensitivities [2].

16.1.2. SCOPE OF THIS CHAPTER

This chapter highlights SAW transducer techniques for use in spread-spectrum wireless communications that involve detection and extraction of digital message signals, where in the bandwidth has been "spread" prior to transmission. As will be demonstrated, this encoding and bandwidth expansion can reduce interference-corruption of the received signal, while also providing security. Individual coding can enable the CDMA to handle many simultaneous transmissions.

The SAW transducer structures considered in this chapter relate to the matched-filter detection of digital message waveforms that have been encoded with fixed codes prior to RF transmission. Detection involves extracting the original message-data extraction by correlation with *fixed* short codes, (e.g., up to about 1024 chips), which have been "built into" the SAW IDT finger-pattern, to act as a matched filter. One attraction of using fixed-code receiver structures is that they do not require synchronization between

---

[1] The modulation of a carrier by a digital code sequence whose bit-rate is much higher than the signal bandwidth is usually referred to as direct sequence (DS) modulation [3].

transmitter and receiver for message extraction. The added attraction of using linear SAW devices for this task is that they can have moderate processing gains (e.g., 20 to 25 dB), as well as simplistic features not readily implemented with digital matched-filter circuitry.

Coding techniques involving very long, or changing, codes—which require synchronization between transmitter and receiver—require nonlinear convolvers, as will be considered in Chapter 17. Due to the wealth of coding techniques that have evolved over the years for digital signal processing, only a brief encounter with some of these is possible within the framework of this book.

The design of SAW devices for code detection requires an appreciation of matched-filtering and correlation concepts. As a consequence, this chapter begins with a necessary review of the topic. In addition, the rationale for using spread-spectrum techniques is presented for the benefit of those readers unfamiliar with this field.

The versatility offered to coding applications by linear SAW devices is demonstrated here for four fixed-coding methods. These include SAW transducer matched-filter configurations for: a) binary phase-shift keying (BPSK); b) CDMA sequences using differential phase-shift keying; c) quadrature phase-shift keying (QPSK); and d) continuous phase-shift modulation (CPSM). (Continuous phase shift modulation is also known as minimum shift keying (MSK), and this notation will be used interchangeably here.)

For temperature stability, these structures are normally fabricated on ST-quartz and employ Rayleigh-wave propagation. The use of quartz may not be desirable in some applications, however, where insertion-loss levels are to be minimized and broadband matching to $50\,\Omega$ is to be maximized. Because of this, some CDMA indoor and mobile radio links with fixed-code IDTs use X112.2° rotated-Y lithium tantalate instead [4]. Because undesirable acoustic reflections from IDT finger discontinuities will degrade the correlation performance of these devices (especially with long-coded IDTs), some techniques for minimizing these are examined in the chapter.

16.2. Matched-Filter Concepts

The main aim in the efficient detection of a message over a mobile radio channel is to maximize its coherent extraction in the presence of corrupting interference or other degradations. These can be caused by a variety of factors such as: a) multipath fading due to multiple RF signal reflections from buildings; b) excessive path-loss caused by interference between directly- transmitted radio waves and ground-reflected waves; c) large levels of industrial electrical noise that will raise the receiver noise floor; and d) signal dispersion through indoor structures, resulting in multiple-signal reception at the mobile unit, with degradation of data-channel transmissions.

To enhance or optimize the strength of the received RF signal relative to the background noise environment, some type of predetection filtering is required. When the interference is due mainly to white noise of constant spectral density over the band of interest, the signal-to-noise ratio is optimized by the use of a *matched filter*. This capability is particularly useful in the detection of coded waveforms, where it is necessary merely to establish the presence or absence of a pulse in a specific bit-sequence interval. Synchronization between transmission and detection is not required.

It may be shown [5] that the optimum transfer function $H_{opt}(f)$ of a matched filter is given by

$$H_{opt}(f) = S^{\star}(f)e^{-j2\pi fT}, \tag{16.1}$$

where T is the time at which the amplitude of the received output signal is maximized and the output noise minimized. In Eq. (16.1), $S^{\star}(f)$ is the complex conjugate of the Fourier transform of the input time-domain signal $s(t)$. In simple terms Eq. (16.1) states that, except for a time delay factor $\exp(-j2\pi fT)$, the transfer function of the matched filter is the same as the complex conjugate of the spectrum of the input signal.

The significance of this last statement becomes more understandable in terms of SAW device implementation, if the matched filter is now considered in terms of its time-domain impulse response. This is obtained by applying an inverse Fourier transform to Eq. (16.1) to give the optimum impulse response

$$h_{opt} = \int_{-\infty}^{+\infty} S^{\star}(f)e^{-j2\pi f(T-t)}df \tag{16.2}$$

Because the spectral density $S^{\star}(f) = S(-f)$ for a real-valued signal, Eq. (16.2) reduces to

$$h_{opt} = \int_{-\infty}^{+\infty} S(-f)e^{-j2\pi f(T-t)}df \tag{16.3}$$

which corresponds to

$$h_{opt}(t) = s(T-t) \tag{16.4}$$

Equation (16.4) demonstrates that the impulse response of the optimum matched filter is just a time-reversed and delayed replica of the input signal $s(t)$, as sketched in Fig. 16.1.

In understanding the physical significance of Eq. (16.4) it is important to recognize that all of the input signal $s(t)$ must have entered the filter by time $t = T$, in order to obtain a maximum value for the output signal-to-noise

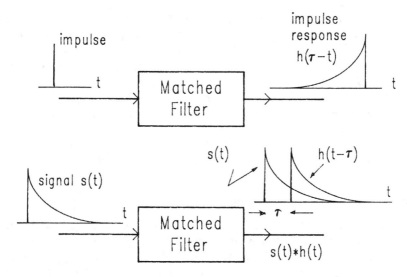

FIG. 16.1.   Impulse response of an optimum matched filter.

ratio. An alternative approach to this optimum filtering problem—which yields the same result—is through the use of a correlation receiver, which correlates the received signal with a stored replica of the transmitted one.

16.3.   Rationale for Using Spread Spectrum

16.3.1.   CHANNEL CAPACITY

In a spread-spectrum channel, the transmission bandwidth is extended well beyond that needed to transmit a signal message at a given rate. The technology may be considered to emanate from the basic studies of Shannon [6] on information theory. In mathematical terms, the channel capacity C of a transmission system is given as

$$c = W \log_2\left(1 + \frac{S}{N}\right) \tag{16.5}$$

where C = channel capacity (b/s), W = bandwidth (Hz), S = signal power, and N = noise power. Depending on the particular application, the system may be required to operate in an environment where the signal-to-noise ratio (S/N) is considerably less than unity.

Example 16.1 Required Bandwidth of Spread-Spectrum Communications Link.   A spread-spectrum wireless link is required to operate in an environment where interference at the mobile receiver due to a narrowband source is 100 times stronger than the transmitted signal power. If the message signal is to be transmitted at a data rate of 15000 bits per second, over what bandwidth W must the information be transmitted? ■

Solution. For low signal-to-noise ratios, as in this example, Eq. (16.5) may be approximated as $C = 1.4 W(S/N)$, where $(S/N) = (1/100)$, or $(S/N) = 10 \log_{10} (1/100) = -20 \, \mathrm{dB}$. Thus the required bandwidth is $W = CN/ (1.4S) = 16,000 \times 100/1.4 = 1.14 \, \mathrm{MHz}$. The message spectrum must be artificially broadened in order to occupy this 1.14-MHz bandwidth. ■

16.3.2. BANDWIDTH EXPANSION

By itself, bandwidth expansion is a necessary but not sufficient condition for defining the spread-spectrum condition in the context in which it is normally used. Conventional frequency-modulation (FM) and pulse-code-modulation (PCM) represent examples of circuits involving bandwidth expansion, but which are not classed as spread-spectrum systems. The difference lies in the fact that in circuits such as for FM and PCM, the bandwidth expansion that ensues is uniquely related to the modulating message signal itself. In the spread-spectrum context considered in this chapter, however, bandwidth expansion is not uniquely related to the message signal, but is achieved through a quite separate encoding operation. While bandwidth expansion in a spread-spectrum system does not serve to combat white noise interference in the way it does for FM and PCM communications, it has many valuable applications in mobile/wireless analog and digital communications links [5], [7], [8]. These include: 1) reducing corruption of the received signal due to narrow-band interference; 2) secure communications employing very low spectral energy; and 3) improved performance against multipath interference.

Spread-spectrum communications normally involve the transmission of *digitally-encoded* message modulation (rather than analog modulation) in systems requiring privacy and security in information transfer. This is because amplitude-modulated message signals, whether encoded or not, can be readily detected with the use of an envelope detector. Likewise, conventional FM signals can be demodulated in a straightforward manner with the aid of a squaring circuit and a discriminator.

16.3.3. SPREAD-SPECTRUM TECHNIQUES

Spread-spectrum operation can be achieved by a variety of techniques; these include:

1. Pseudo-noise (PN) coding, also known as direct-sequence (DS)
2. Frequency hopping (FH)
3. Hybrid operation using frequency hopping with PN-coding (FH/PN)

In the fixed-code detection processes considered in this chapter, spectral energy spreading is achieved by artificially increasing the message modulation rate. One conventional method is sketched in Fig. 16.2. As shown, message bits of period T_b are further coded by a high-speed code sequence of chip period T_c. The power spectrum of such a code will be noise-like (i.e., with a $|(\sin X)/X|^2$ distribution) when the coding sequence is long enough to approximate a random sequence, in which case it is termed as pseudo-noise (PN) or direct sequence (DS) coding.

16.3.4.   SYSTEM REQUIREMENTS FOR SPREAD SPECTRUM

Despite the variety of spread-spectrum techniques that exist, they are all bound by four common features:

1. The message signal must be encoded in some way prior to transmission.
2. The receiver must "remember" the exact coding sequence.
3. The receiver code must be used in the proper sequence if the message signal is to be faithfully extracted.

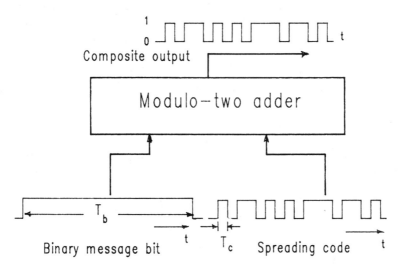

FIG. 16.2.   Use of a spreading code to increase the message rate artificially.

4. The process must result in an enhancement of the received message signal over intentional or unintentional noise interference.

16.4. Processing Gain with Binary Phase-Coded SAW IDTs

16.4.1. CORRELATION AND CONVOLUTION

A paramount requirement of a spread-spectrum mobile/wireless communications link is that, within the limits of its design, it should be able to achieve meaningful detection of a digitally encoded message signal, despite its corruption by narrowband noise in the transmission channel. To achieve this, it must sufficiently increase the signal-to-noise ratio of the signal in its passage through the receiver. In accordance with Shannon's principles, this involves encoding the transmission in some way, so that the spectrum of the transmitted signal is artificially broadened. A reference copy of the coding scheme used for such broadening must be contained within the receiver, for auto- or cross-correlation with the received signal.

To distinguish between convolution and correlation, recall the convolution relation given in Eq. (3.9) of Chapter 3 as

$$\left(Convolution\right) \quad f(t) = \int\limits_{-\infty}^{+\infty} f_1(\tau)f_2(t-\tau)d\tau = f_1(t) \star f_2(t) \qquad (3.9)$$

for the convolution of two time signals $f_1(t)$ and $f_2(t)$, where the right-hand expression is a "shorthand" notation for convolution. When applied to the response of a linear filter subjected to an input stimulus $f_1(t) = s(t)$, the output response $f(t)$ is given by the convolution of $s(t)$ with the impulse response $h(t)$ of the filter. Substitution of $h(t) = f_2(t)$ in Eq. (3.9) yields

$$\left(Convolution\right) \quad f(t) = s(t) \star h(t) = \int\limits_{-\infty}^{+\infty} s(\tau)h(t-\tau)d\tau \qquad (16.6)$$

for real-valued functions.

Next, the autocorrelation function $A(t)$ may be defined as the convolution of the input signal $s(t)$ with a time-reversed replica $s(-t)$. Substitution in Eq. (3.9) gives

$$\left(Autocorrelation\right) \quad A(t) = \int\limits_{-\infty}^{+\infty} s(\tau)s(\tau+t)d\tau \qquad (16.7)$$

If input signal $s(t)$ is itself an impulse response $s(t) = h(t)$, the autocorrelation becomes

$$\left(Autocorrelation\right)\quad A(t)=\int_{-\infty}^{+\infty} h(\tau)h(\tau+t)d\tau \tag{16.8}$$

for real-valued functions.

In applying these relations to matched filtering using a SAW filter in the receiver, the use of time-reversed terms in Eqs. (16.6) and (16.8) simply corresponds to the use of a receiver SAW transducer whose IDT pattern is a *spatially reversed sampled replica* of the impulse response $h(t)$ of the input signal $s(t)$. This is also in accord with the optimum filtering requirement in Eq. (16.4). By using a change of variables in Eq. (16.7) the autocorrelation may be given in terms of $A(\tau)$, where τ is a convenient variable, so that

$$\left(Autocorrelation\right)\quad A(\tau)=\int_{-\infty}^{+\infty} s(t)s(t+\tau)dt \tag{16.9}$$

as will be employed in Example 16.2.

Cross-correlation, on the other hand, relates to a comparison of two separate signals $f_1(t)$ and $f_2(t)$ for which the cross-correlation integral $C(\tau)$ is

$$\left(Cross\text{-}correlation\right)\quad C(\tau)=\int_{-\infty}^{+\infty} f_1(t)f_2(t+\tau)dt \tag{16.10}$$

for real-valued functions. As shown in Chapter 17, the cross-correlation relation is applicable to signal processing with nonlinear SAW convolvers employing two separate signal inputs.

Example 16.2 Autocorrelation Function for a Digital Sequence.   For the digital impulse sequence sketched in Example 16.2(a), sketch the form of the autocorrelation function over a relative period $-4 \le \tau \le +4\,\mu s$. ∎

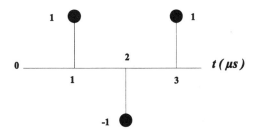

EXAMPLE 16.2(a).   A digital impulse sequence.

Solution. AS we are dealing just with pulses in the foregoing, the integration in Eq. (16.9) reduces to multiplication and addition. The autocorrelation of the sequence may then be evaluated one step at a time, in a "longhand" calculation. In this way, the individual contributions are as sketched in Example 16.2(b). This multiply-and-add procedure may be continued to cover all other τ values between -4 and $+4\,\mu s$. In this way the time dependence of the correlation pattern is shown in Example 16.2(c). ■

For $\tau = 0$

times

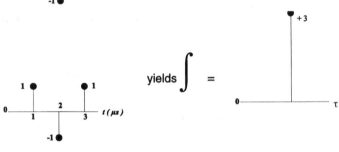

For $\tau = -1$

times

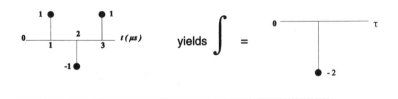

EXAMPLE 16.2(b). Individual correlation contributions for Example 16.2(a).

For $\tau = -2$

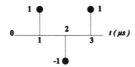

times

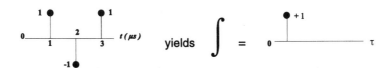

EXAMPLE 16.2(b). *Continued.*

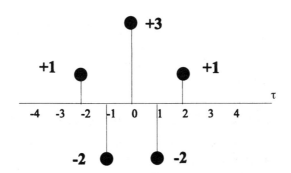

EXAMPLE 16.2(c). Resultant time-dependence of autocorrelation function.

As sketched in Fig. 16.3, an essential feature of a correlation processor is that the output signal bandwidth is "de-spread" (i.e., reduced) to the original message bandwidth B, while the spectrum of any narrow-band interference is simultaneously increased to at least the spreading bandwidth W. The narrowband IF filter that is included in or follows, the processor will remove most of the interference, thereby causing the processing gain. Following

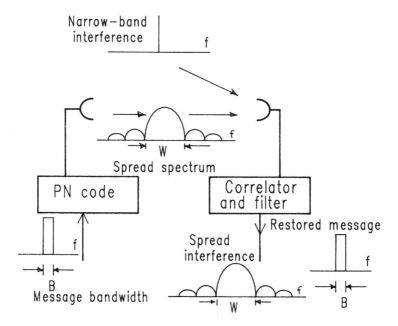

Fɪɢ. 16.3.   Effect of correlation process on both the message spectrum and the narrowband interference.

correlation, the "de-spread" message signal can be demodulated to restore the original baseband message signal.

16.4.2.   Pʀᴏᴄᴇssɪɴɢ Gᴀɪɴ Wɪᴛʜ Bɪɴᴀʀʏ Pʜᴀsᴇ-Sʜɪꜰᴛ Kᴇʏɪɴɢ

Processing gain (in dB) is just the difference in signal-to-noise ratios at the output and input of a correlation processor. It may be shown that this also corresponds to the time-bandwidth product in the linear FM filters considered in Chapter 8. For the correlation detection of PN-encoded transmissions, the processing gain (PG) in dB is [3]

$$PG = 10\log_{10}\frac{T_b}{T_c} \quad (in\ dB) \tag{16.11}$$

where T_b = message bit period and T_c = chip period within the PN coding sequence. Equation (16.11) may alternatively be expressed in terms of the number of chips N_c within a message bit as

$$PG = 10\log_{10} N_c \quad (in\ dB) \tag{16.12}$$

16.5.   Fixed-Code SAW Transducers for Binary Phase-Shift Keying

16.5.1.   GENERATION AND DETECTION OF BPSK FIXED-CODE WAVEFORMS

Both code generators and matched filters for BPSK modulation may be implemented by linear phase SAW transducers for fixed code lengths within the constraints imposed by substrate and IDT lengths. As this is a fairly straightforward circuit implementation, digital waveform generators are normally used to implement the spreading code in the transmitter sections. Such phase-coded sequences can also be readily generated using SAW techniques, simply by applying an impulse to a coded SAW transducer as depicted in Fig. 16.4. The conjugate transducer in the receiver implements *asynchronous* matched filtering, without the need for any synchronization clock circuitry. Temperature-stable substrates such as ST-X quartz are required when lengthy codes are employed.

Code times of $50\,\mu$s are possible with this technique, with filter fractional bandwidths of about 15% [9]. Operation at 200 Mb/s chip rates on phased coded IDTs has been demonstrated at operating frequencies of 1 GHz with peak-to-sidelobe ratios of 13 dB in the time response of the correlated output [10].

Figure 16.5 illustrates a conventional (non-SAW) technique for generating BPSK waveforms, where a digital sequence is mixed with a CW local oscillator to obtain phase reversals of the output waveform appropriate to the sequence bits. Illustrative experimental waveforms are shown in Fig. 16.6(a) for the generation of a 13-chip Barker digital sequence using active circuitry. Figure 16.6(b) shows the compressed pulse obtained using a SAW filter with conjugate IDT pattern in the receiver.

Figure 16.7 illustrates the form of one type of IDT for correlation of fixed BPSK codes. This configuration, illustrated here for correlation of a (0101) sequence, employs a wideband uniform input IDT with but few finger pairs, together with a narrowband tapped delay line output IDT. An alternative

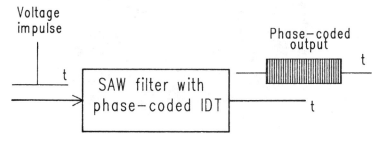

FIG. 16.4.   A technique for generating a phase-coded sequence, using impulse the excitation of a coded SAW transducer.

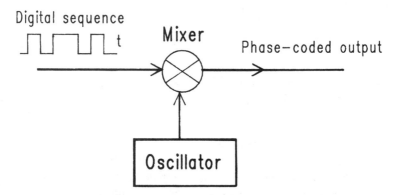

Fɪɢ. 16.5.   A conventional technique for generating a phase-coded RF sequence.

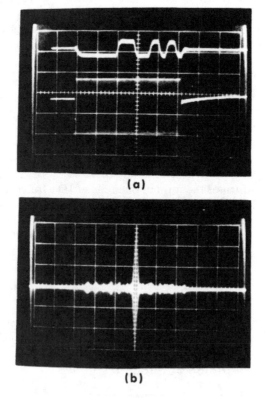

Fɪɢ. 16.6.   (a) Upper trace shows 13-chip Barker code used to generate phase-coded RF sequence in lower trace, using method of Fig. 16.5. Horizontal scale: 0.5 μs/div. (b) Compressed pulse with 21 dB sidelobe suppression, using SAW matched filter. Horizontal scale: 1 μs/div. (Reprinted with permission from Reference [14]. © IEEE, 1974.)

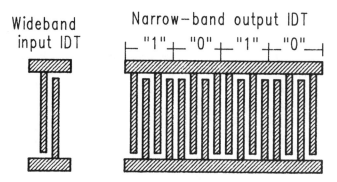

Fig. 16.7.   SAW filter for correlation of a (0101) sequence, using a wideband input IDT, and narrow-band coded output IDT.

version shown in Fig. 16.8 employs a narrowband input IDT and a wideband tapped delay line output IDT. In Fig. 16.8, the period between the tapped sections of the output IDT corresponds to the width of the input IDT. Unlike tapped-delay-line comb filters, however, the finger rungs in the output IDT of Fig. 16.8 are not identical, so that this structure does *not* have a comb frequency response. In both Figs. 16.7 and 16.8, the role of input and output transducers can be switched without detriment to the *principles* of operation because the devices are passive linear networks. The differing input-output impedance levels may, however, result in some performance variations.

When using phase-coded or other matched filter structures in the receiver, maximum correlation occurs when the input signal pattern matches exactly that of the conjugate SAW transducer. Maximum voltage amplitude is then obtained from the excited IDT, as all of the induced voltage components at each finger pair are in phase. The peak-to-sidelobe ratios of the output signal excursions for this maximum correlation condition will depend on the code employed. Correlation will be reduced at other times (or for other signal frequencies) because the voltage phasors for the various finger pairs will no longer add constructively. The rate of fall-off will again depend on the type and length of code employed. For very lengthy PN codes it can be shown that the output signal reduces to zero when the BPSK waveform and the correlation receiver code are displaced in time by one chip event.

16.5.2.   BARKER CODES

Figure 16.9 shows the structure of a SAW transducer with constant acoustic aperture, for correlation of a 5-chip Barker code with (11101) sequence, together with the distribution of surface wave for maximum correlation.

Narrow−band
input IDT

Wideband output IDT
"1"   "0"   "1"   "0"

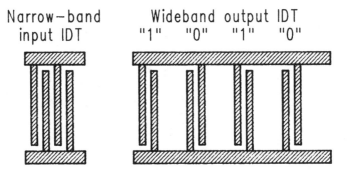

FIG. 16.8. Alternative version of correlation receiver of Fig. 16.7, using a narrowband input IDT and a wideband tapped delay-line output IDT. Separation between output electrode taps corresponds to one chip-period sampled by the input IDT.

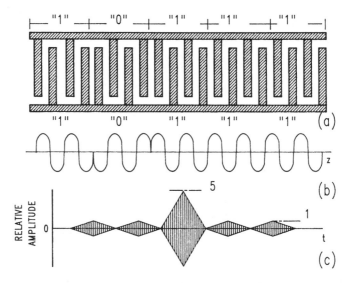

FIG. 16.9. (a) Narrowband IDT for correlation of (11101) 5-chip Barker sequence. (b) Underlying SAW waveform for maximum correlation. (c) Time-domain response.

Barker sequences form a class of binary codes that have correlation functions with associated peak-to-sidelobe voltage ratios of $20 \log_{10} N_c$. True Barker sequences only exist up to length 13, as given in Table 16.1. The 13-chip Barker code depicted in Fig. 16.6 represents the highest perfect code

TABLE 16.1

TRUE BARKER CODE SEQUENCES

Sequence Length N_c	Relative Polarity of Chip Sequence
2	+ + or + −
3	+ + −
4	+ + − + or + + + −
5	+ + + − +
7	+ + + − − + −
11	+ + + − − − + − − + −
13	+ + + + + − − + + − + − +

known[2], with equal amplitude sidelobes [11]. Longer sequences may be generated by further encoding each chip by another Barker sequence [12], [13]. Thus if a Barker sequence of N_c chips is further encoded with another (and faster) Barker sequence of length B_c, the processing power gain will be PG = 10 $\log_{10} (B_c N_c)$ while the peak-to-sidelobe (voltage) ratio will remain at $20 \log_{10} N_c$. The peak-to-sidelobe ratio is 20 $\log_{10} (N)^{1/2}$ for maximal sequence PN codes of length $N = 2^n - 1$, generated by a shift register with n stages and modulo-2 feedback [3].

A SAW tapped delay line with biphase encoding may contain more than one waveform cycle per chip. For the Barker sequence example of Fig. 16.9 employing two cycles per chip, the processing gain corresponds to the chip rate N_c, while the number of cycles per chip determines the center frequency f_o of the input and output SAW IDTs. The cycle number will also affect the correlation loss away from transducer center frequency f_o.

Example 16.3 Design of a Biphase-Coded IDT. A SAW-based spread-spectrum communications link employs biphase encoding of a digital message signal. The receiver uses a fixed-code SAW filter of the type shown in Fig. 16.7, with BPSK finger encoding of the narrowband output IDT. The biphase code length is $N_c = 127$ chips per message bit, at a chip rate $R_c = 90$ M-chip/s. Standard (non-SAW) transmission encoding is employed, with $C = 2$ cycles/chip. Determine (a) the number of finger pairs in the receiver coded IDT, (b) the center frequency of this SAW receiver filter and (c) the processing gain. ∎

---

[2] Recently, selected coded sequences for direct sequence (DS) spreading of DPSK signals have been reported that give sequence-reduced output autocorrelation sidelobes that are 6 dB lower than for the 13-chip Barker sequence [2].

Solution.   (a) The number of coded finger pairs is $N_p = CN_c = 2 \times 127 = 254$ finger pairs. (b) The center frequency f_o of the SAW receiver filter is given by $f_o = CR_c = 2 \times 90\,\text{MHz} = 180\,\text{MHz}$. (c) The processing gain is $\text{PG} = 10\log_{10}N_c = 10\log_{10}(127) = 21\,\text{dB}$. ∎

Example 16.4 Correlation of a Fixed BPSK Code using 127 Chips/ Message Bit.   A SAW receiver filter of the type shown in Fig. 16.8 is fabricated on ST-quartz for the correlation of a fixed BPSK code with 127 chips per message bit. The chip rate is $R_c = 25\,\text{MHz}$, with $C = 4$ cycles per chip in the output IDT. Determine: (a) the center frequency f_o of the SAW filter; (b) the number of finger pairs in the input IDT; (c) the separation between adjacent tapped sections in the output IDT; and (d) the fractional bandwidth of the input IDT section of the receiver filter. ∎

Solution.   The center frequency is $f_o = CR_c = 4 \times 25\,\text{MHz} = 100\,\text{MHz}$. (b) The receiver input IDT must accommodate one chip of code at a time, or four acoustic wavelengths. The number of input IDT finger pairs is $N_p = 4$. (c) At 100 MHz, the acoustic wavelength is $\lambda_o = v/f_o = 3158/ (100 \times 10^6) = 31.58 \times 10^{-6}\,\text{m} = 31.58\,\mu\text{m}$. The length of the receiver input IDT is $L = C\lambda_o = 4 \times 31.58 = 126.32\,\mu\text{m}$ $(1\,\mu\text{m} = 10^{-4}\,\text{cm})$. This is also the period between adjacent coded taps in the output IDT of the receiver filter. (d) The fractional bandwidth of the input section, to the 4-dB points, is BW_4 $\% = 100/N_p = 100/4 = 25\%$. ∎

16.6.   Second-Order Effects in SAW Tapped Delay Lines

16.6.1.   Sources of Phase Errors

The correlation performance of a SAW tapped delay line will be degraded by spurious finger reflections, which can increase in severity with the number of fingers or taps. This can be offset to some extent by the use of split-electrode geometry, as sketched for the wideband biphase-coded IDT in Fig. 16.10. As shown in this structure, dummy split-electrodes may be required. The reason for this is that although the IDT is not a comb filter, there will be periodic stopbands due to finger reflections, at harmonics of half the chip rate. Filling in with dummy electrodes removes this stopband periodicity. Figure 16.11 shows experimental pulse expansion waveforms for a 127-chip biphase encoded SAW IDT before and after such dummy fingers are included [14].

Phase errors also cause a degradation of the correlation response, in matching the incoming phase-modulated waveform with the conjugate SAW IDT. These errors can be caused by imperfect lithography of the coded IDT pattern, frequency shifts, improper crystal axis orientation, and

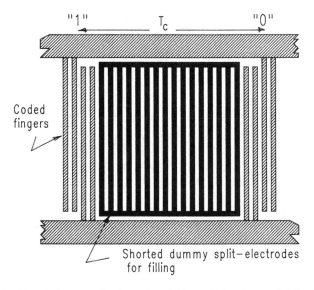

FIG. 16.10. Use of dummy split-electrodes within a wideband tapped delay-line IDT, to move stopband due to finger reflections up to the second harmonic of the passband center frequency. (After Reference [14].)

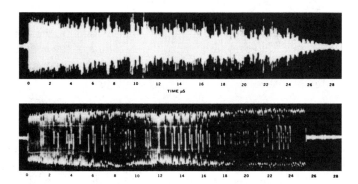

FIG. 16.11. Upper trace shows expansion of a 127-chip biphase-encoded waveform, using a wideband tapped delay-line SAW transducer without dummy fingers between taps. Lower trace is response after including the dummy fingers of Fig. 16.10. Waveform duration about 26 μs. (Reprinted with permission from Reference [14]. © IEEE, 1974.)

substrate temperature changes. A deviation in the orientation of the SAW IDT from the correct axis will affect the SAW velocity, and thus the SAW phase shift through the device. Temperature changes will also affect the SAW velocity as well as the substrate size. These can be avoided to some extent by fabricating the coded device on ST-X quartz; the drawback is increased conversion loss. In some designs where such insertion losses on ST-quartz may be unacceptably high, the piezoelectric 112.2° rotated-Y lithium tantalate (LiTaO$_3$) is employed as an alternative substrate. [4].

16.6.2. THE AMBIGUITY FUNCTION

The ambiguity function $\chi(\tau, \phi)$ serves as a useful tool in depicting the degradation from optimum correlation with a fixed-code IDT due to phase errors, or frequency errors due to Doppler frequency shifts. Such ambiguity functions are used in pulsed radar signal processing to obtain two- or three-dimensional plots of correlation amplitude loss due to combined effects of delay τ and Doppler frequency shift f_d [11]. Within the context used here for BPSK fixed-code SAW transducers, the ambiguity function is given as [13]

$$\chi(\tau, \ \phi) = \int_{-\infty}^{\infty} s(t)s(t-\tau)*e^{-j\phi t}dt, \tag{16.13}$$

where τ is the time shift from optimum correlation (see Example 16.2), ϕ is the angular frequency shift between two otherwise identical signals to be correlated, $s(t)$ is the signal to be correlated and $s(t-\tau)*$ is the reference signal associated with the coded IDT in the receiver.

Equation (16.13) assumes that the angular frequency shift ϕ is quite small. It may be conveniently evaluated to give degradation of the correlation peak at $\tau = 0$. For the special case where the envelope of the coded signal is a rectangle of time duration T, the ambiguity function at correlation time $\tau = 0$ takes the approximate sinc function form [10]

$$\chi(0, \ \phi) = \frac{\sin(\pi\phi T)}{\pi\phi T}. \tag{16.14}$$

Figure 16.12 is a two-dimensional plot of Eq. (16.14) applied to a fixed-code SAW transducer, which chip length N_c is proportional to impulse time T in Eq. (16.14). The loss in correlation peak is plotted over two horizontal and two vertical scales as a function of phase or frequency error ϕ and chip sequence length N_c [14]. Three illustrative device specifications are indicated by the dotted lines in this figure, for allowable phase or frequency errors ranging from 27 ppm for a biphase coded IDT with chip sequence length N = 10,000 to 275 ppm for a sequence of $N_c = 1600$ chips.

Example 16.5 Correlation Receiver Using 13-Chip Barker Code IDT. A correlation receiver employs a fixed-code narrow-band SAW transducer with a 13-chip Barker code sequence where $C = 6$ cycles per chip. The chip rate is $R_c = 30$ Mchip/s. Determine (a) the center frequency f_o of

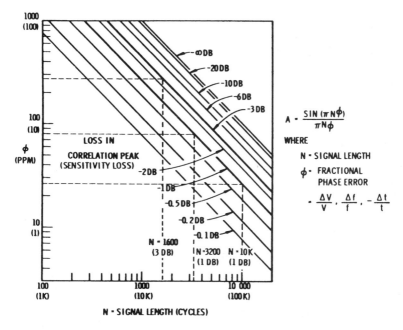

FIG. 16.12. Effects of phase or frequency errors on loss in correlation response peak. Dotted lines show three illustrative cases. Note that there are dual scales for vertical and horizontal axes. (Reprinted with permission from Reference [14]. © IEEE, 1974.)

this coded IDT and (b) the frequencies of the first minima on either side of the compressed pulse. ∎

Solution. (a) The center frequency is $f_o = CR_c = 6 \times 30 = 180\,\text{MHz}$. (b) For a BPSK-coded IDT, it is found that the variation of correlated pulse amplitude with frequency closely follows that for a conventional linear phase filter with the same number of finger pairs. From Chapter 3 this condition is given by $\sin(N_p \pi (f - f_o)/f_o) = 0$, corresponding to $f = f_o\,(1 \pm 1/N_p)$. The equivalent number of finger pairs is $N_p = 6 \times 13 = 78$. The correlation pulse minima are at frequencies $f = 80 \times (1 \pm 1/78)\,\text{MHz} = 81.025\,\text{MHz}$ and $78.974\,\text{MHz}$. ∎

16.7. Spread-Spectrum IDT Coding for Differential Phase Shift Keying (DPSK)

16.7.1. Use of DPSK in Mobile/Wireless Communications

In wireless spread-spectrum communications, the bandwidth required for CDMA signalling is obtained by spreading the information contained in each bit of a digitally encoded message signal. This is done through mixing with a fast pseudo-noise (PN) code sequence, which modulates each *bit* of the message-waveform sequence. Several CDMA users may simultaneously access the same radio channel, with each using a different coding.

One technique involves direct-sequence spread-spectrum (DSSS) modulation with differential phase shift keying (DPSK). With DPSK, the status of the binary input information is contained in the *difference* between two successive message bits, rather than through their absolute phase. An advantage of this technique is that it simplifies the reception circuitry, in that it is not necessary to recover a phase-coherent carrier. One disadvantage is that it requires an increase of about 1–3 dB in signal-to-noise (S/N) performance (more than is the case for absolute BPSK) in order to achieve the same bit-error rate (BER). [15].

16.7.2. DPSK MODULATION AND DEMODULATION PRINCIPLES

This section illustrates the principles of DPSK modulation of binary-coded message waveforms[3], with reference to the block diagram of Fig. 16.13(a). The DPSK examples given here relate to the modulation of BPSK signals considered in the previous section. As shown, one technique for implementing this involves an exclusive-NOR (XNOR) digital logic gate [15]. The binary message waveform is applied to one input of the XNOR-gate, while the second input is derived from a feedback stage—where the XNOR output is returned to its input through a 1-bit delay. The composite signal is applied to the input of a balanced-modulator (i.e., balanced-mixer), to modulate the carrier signal. The example of Fig. 16.13(b) illustrates this modulation technique. Whereas the first message-data bit has no preceding bit for reference comparison, an initial reference-bit is assumed. As shown here, a logical "0" value is assumed for the initial reference. If the initial reference bit is a logical "1," the output from the XNOR gate is just the complement of that shown in Fig. 16.13(b). Figure 16.14 illustrates the basic receiver-stage technique for recovering the initial binary-message data. The received signal is delayed by one bit and then compared with the next signal-bit entering the balanced modulator. If they are the same, a logical "1" (voltage) output is obtained. If they are different, a logical "0" (voltage) output results. If the reference phase is incorrectly assumed, its sole effect is that only the first demodulated bit will be in error. (A BGS wave resonator may be used for the one-bit delay, as demonstrated in Fig. 11.30 [16].)

---

[3] The DPSK examples given here relate to the modulation of simple *binary-coded* message signals. Differential encoding can be achieved by modulating with other M-ary codes, using more complicated coding algorithms.

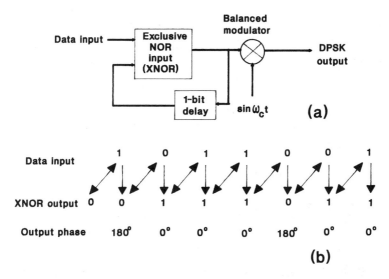

FIG. 16.13.   (a) DPSK modulator using exclusive NOR (XNOR) logic gate and 1-bit delay line. (b) Example illustrating output phase response, for given binary-data input signal. (After Reference [15].)

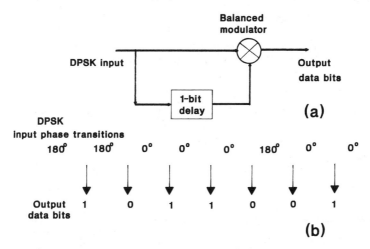

FIG. 16.14.   DPSK demodulator stage using 1-bit delay line, showing message bit restoration of input signal in Fig. 16.13. (After Reference [15].)

16.7.3. DSSS/DPSK Receivers Employing SAW Transducers

The preceding section has outlined the principles of DPSK modulation and demodulation of binary message signals. For spread-spectrum communications, however, we still have to consider how to incorporate the fast spreading codes required for spread-spread signalling, and how, and where, SAW devices can be included in this implementation. The coding techniques considered in this chapter apply to *fixed-length* codes, which can be incorporated into a SAW transducer to provide a matched filtering capability. (If such fixed-length codes are lengthy enough they can, of course, be made to approximate a pseudo-noise (PN) random sequence for additional signalling security.)

In this section we now consider the employment of fixed-length codes to DSSS circuitry. For the most part, spreading techniques can be readily attained in the transmitter circuitry, by voltage modulation of the binary message bits by data generators, prior to RF carrier modulation. Demodulation in the receiver circuitry is not straight-forward, however, as it requires restoration of the incoming analog signal carrying the digital message as a modulation of the RF carrier.

To illustrate these points, the block diagram of Fig. 16.15 depicts one example of a DSSS/DPSK transmitter stage. This merely extends the basic circuit of Fig. 16.13 by incorporating an additional mixing stage for modulation of each message bit by a fast fixed code. (Code lengths of 64 chips or more can be made to reasonably approximate a pseudo-noise (PN) sequence.) As indicated, this transmitter circuitry does not require SAW technology.

The corresponding DSSS/DPSK receiver stage of Fig. 16.16 can, however, be readily implemented using SAW filter technology. The one-bit delay circuit can be a SAW delay line, (or BGS wave resonator), which delay time corresponds to the bit duration in the message signal. Moreover, the matched filtering is achieved by incorporating SAW matched filters in each arm of the balanced demodulator, where each SAW filter has a coded

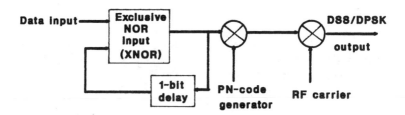

Fig. 16.15. Illustrative block diagram of spread-spectrum modulator for DSSS/DPSK transmission.

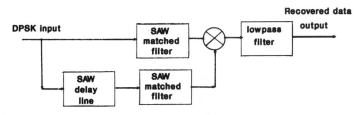

FIG. 16.16. SAW-based demodulator for DSSS/DPSK spread-spectrum signals, using SAW matched filters and 1-bit SAW delay line. (After Reference [17].)

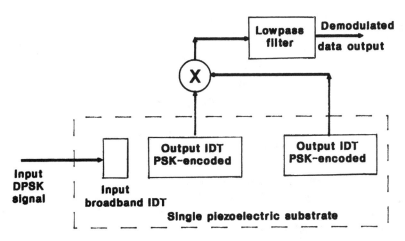

FIG. 16.17. Improved SAW -based demodulator for DSSS/DPSK spread-spectrum signals incorporates 1-bit delay and two SAW matched filters on a single piezoelectric substrate. (After Reference [17].)

IDT corresponding to the conjugate of the fixed-code spreading sequence employed in the transmitter [17]. One design problem for the SAW-based receiver demodulator of Fig. 16.16 is that the combined insertion loss can be large for the serially connected SAW delay line and matched-filter. Degradation of the differential decoding processes can arise from temperature differences between the SAW delay line and the SAW matched filters. This problem has been overcome in recent DPSK receiver designs, however, by fabricating the structure on a single piezoelectric substrate, and incorporating the required one-bit delay into the design of the two SAW matched filters, as sketched in Fig. 16.17 [17].

16.8. SAW Transducers for Quadraphase Codes

16.8.1. REASONS FOR USING QUADRAPHASE CODES

Codes employing quadraphase phase-shift keying (QPSK)[4] also find employment in direct sequence spread-spectrum links. Despite its added complexity, the main reasons for using QPSK are that: (1) it is not as seriously degraded as biphase signals when passing through a nonlinearity simultaneously with interference; and (2) the RF bandwidth for a quadraphase code is only one-half of that required for a BPSK signal. (With the corollary that twice as much data can be transmitted in the same bandwidth as for a biphase signal). The power envelope for a quadraphase signal has the same $|(\sin X)/(X)|^2$ (i.e., sinc-function^2) response as for a biphase-modulated signal. However, its main-lobe bandwidth is only one-half of that for a biphase signal at the same coding rate.

There may be situations or environmental conditions in which the use of biphase carrier modulation is undesirable. With pulsed-carrier transmissions, for example, the output spectrum is dependent on the modulation envelope. An undesirable amount of transmitted power may be contained in the sidebands. For the case of single pulse of a carrier of angular frequency ω with rectangular modulation envelope of duration T, the transmission power spectrum is proportional to $|(\sin \omega T/2)/(\omega T/2)|^2$ for $-T/2 \leq t \leq +T/2$, with adjacent sidelobes at about 13 dB and a power roll-off at 6 dB/octave. The null-to-null bandwidth of the main lobe is $2/T$. For continuous coded transmissions, code or carrier unbalance in the transmitter can also lead to narrowband interference in the receiver, while the signal-to-noise performance with biphase modulation may be degraded by nonlinearities in the receiver circuit stages [15].

The QPSK signalling features subpulses with a half-cosine shape, and represents one method for improving bandwidth efficiency and reducing spectral splatter [18]. While the conventional circuit implementation of QPSK modulation is more complex than for BPSK, fixed QPSK codes of short length may be generated readily using SAW-based techniques, as will be demonstrated in the next section.

Desirable features of a quadraphase coded signal are as follows [18]:

1. It does not rise from zero instantaneously.
2. During a bit modulation period, the phase of the carrier signal shifts linearly over a 90° range.
3. The direction of this phase shift can be ±90°, depending on the chosen biphase code sequence.

---

[4] Quadraphase phase shift keying is also known as quadriphase phase shift keying.

4. The digitally modulated carrier frequency f_o should be limited to two frequency excursions $(f_o \pm \Delta f)$.

16.8.2. CONVERSION FROM BARKER CODES TO QUADRAPHASE CODES

The coding examples in the previous sections of this chapter dealt with Barker codes to some extent so we will demonstrate how a Barker coding scheme can be converted to a quadraphase one. The illustration in Table 16.2 is applied to the 13-chip Barker sequence.

To convert a Barker code to the corresponding quadrature (4-phase) code, simply multiply each kth bit by $j^{(k-1)}$, where $j = 90°$ [18]. Table 16.2 shows the quadrature code sequences derived in this way from the 13-chip Barker sequence. From Table 16.2 the phase excursions over the time duration of the 13-chip quadrature phase code sequence are as shown by the solid line in Fig. 16.18(a). The dotted line in this illustration shows the desired phase shift that is reasonably approximated by this code form. From the frequency-phase relationship $\Delta f = \Delta\phi/\Delta t$, it is seen that a shifting between two frequencies is involved.

16.8.3. APPLICATION TO A SAW IDT

The finger spacing and structure of an IDT that will accommodate those phase-frequency excursions already mentioned here may be derived readily. For a "standard" IDT in a linear phase design with finger metallization ratio of $\eta = 0.5$, the finger widths and spacings are $\lambda_o/4 = v/4f_o$ at center frequency f_o. For the quadrature-code IDT, the finger widths and spacings are given by either $v/4(f_o + \Delta f)$ or $v/4(f_o - \Delta f)$, in terms of SAW velocity v and the two frequency excursions $\pm\Delta f$ in Fig. 16.18(b). In this example, finger widths and finger spacings are equal during the first four chips of the encoded pulse, so that the amplitude of an impulse-induced surface wave passing under this section will vary sinusoidally at frequency $(f_o + \Delta f)$. For the fifth code chip, however, the IDT finger width and spacing

TABLE 16.2

QUADRAPHASE CODE SEQUENCE DERIVED FROM 13-CHIP BARKER CODE

Barker bit #:	1	2	3	4	5	6	7	8	9	10	11	12	13
Code Polarity:	+	+	+	+	+	−	−	+	+	−	+	−	+
QPSK Code:													
Quadrature bit:	1	2	3	4	5	6	7	8	9	10	11	12	13
Relative value:	1	j	−1	−j	1	−j	1	−j	1	−j	−1	j	1
Phase (degrees):	0	90	180	270	0	270	0	270	0	270	180	90	0

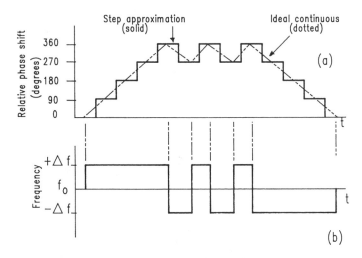

FIG. 16.18. (a) Solid line gives quadraphase code derived from 13-chip Barker sequence, while dotted-line gives continuous-phase approximation. (b) Associated frequency transitions from $\Delta f = \Delta \phi / \Delta t$. (After Reference [18].)

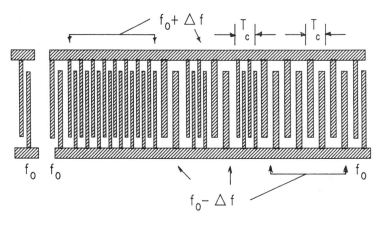

FIG. 16.19. SAW IDT implementation of quadraphase-coding of 13-chip Barker sequence in Fig. 16.18.

is given by $v/4(f_o - \Delta f)$ etc. While the metallization ratio $\eta = 0.5$ is preserved in the QPSK-coded IDT, the *actual* finger widths and spaces will vary as sketched in Fig. 16.19 for this example.

Application of a voltage impulse stimulus to the SAW transducer of Fig. 16.19 will cause it to modulate the surface wave with the QPSK code.

In the receiver, correlation of this coded waveform is achieved using a conjugate SAW transducer.

16.9. SAW Filters for Continuous Phase-Shift Modulation (CPSM)

16.9.1. HIGHLIGHTS OF CPSM

In spread-spectrum systems employing continuous phase-shift modulation (CPSM), also known as minimum-shift keying (MSK), transmissions are in the form of sequences of contiguous pulses at one of two frequencies. As shown in Fig. 16.20, the power spectrum of CPSM transmissions has higher suppression of close-in sidelobes (23 dB) than that for biphase modulation (13 dB), thereby causing less interference or cross-modulation. The null-to-null bandwidth ($1.5/T$) of the major lobe of the CPSM power spectrum is also less than that for BPSK modulation ($2/T$). Also as sketched in Fig. 16.20, a most important feature of a CPSM waveform is that it has continuous phase excursions in going from frequency f_1 to f_2.

The MSK condition represents the minimum frequency difference that will enable the "1" and "0" modulation excursions of the coded transmission to be separately correlated and detected. It may be shown that these two frequencies are

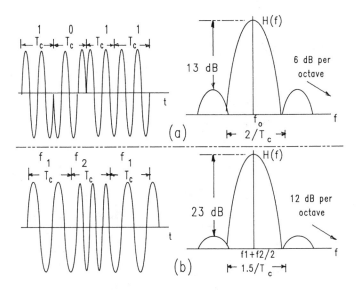

FIG. 16.20. (a) Power spectrum of biphase-encoded PN sequence with 13-dB close-in sidelobe suppression, and fall-off at 6 dB per octave. (b) With CPSM, the close-in sidelobe suppression is increased to 23 dB, with a fall-off of 12 dB per octave.

$$f_{1,2} = f_o \pm \frac{N_c R_b}{4} = f_o \pm \frac{1}{4T_c} \qquad (16.15)$$

where f_o = centre (IF) frequency, N_c = number of chips per message bit, R_b = digital message bit rate and T_c = chip time. When $2\pi (f_1 - f_2)T = \pi$ after chip time T_c, the two signals will 180° out of phase with each other. The correlation between the two signals will then be zero.

16.9.2. SAW IDT IMPLEMENTATION OF CPSM

The SAW-based generation of CPSM may be achieved in one of two ways, with the same end result. In one method, the baseband PN-coded waveform is used to generate a biphase coded signal [19]. As sketched in Fig. 16.21(a) this biphase signal is applied to an appropriately designed IDT for conversion to CPSM coding.

In the alternative method, a succession of impulses is directly applied to the SAW filter in question to obtain the CPSM output [19]. As the principles behind both designs are inherently equivalent and yield the same result, only the second technique will be reviewed here.

With this in mind, consider how the SAW filter of Fig. 16.22 is employed to generate CPSM [20], [21]. As shown, its input IDT has weighted fingers,

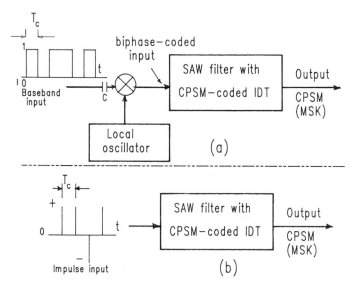

FIG. 16.21. (a) Generation of CPSM sequence using coded SAW IDT and baseband binary-message signal. (b) Alternative method for obtaining CPSM sequence using coded SAW IDT and impulse input excitation.

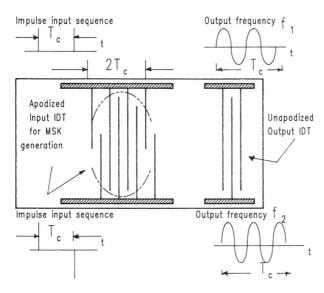

FIG. 16.22. SAW transducer geometry for generating CPSM (MSK) sequence, using impulse excitation method of Fig. 16.21(b).

while the output one is merely a broadband receiver. Three essential features of the input IDT are that: a) its fingers have a sine function (not sinc function!) apodization; b) the overall length is $2T_c$, where T_c is the input chip period; and c) it must be an odd acoustic number of half-cycles in length (i.e., an even number of fingers N). Its input is driven by a sequence of voltage impulses of "1" or "0" relative polarity that are derived from the PN chip-stream. The frequency of the output signal will be f_1 or f_2, depending on whether consecutive input impulses have the same or opposite voltage polarity.

16.9.3. Simplified Mathematical Explanation for CPSM IDT Design

The generation of the CPSM waveform with the SAW filter design of Fig. 16.22 may be explained in simple fashion by recourse to the trigonometric identity,

$$\sin(A)\sin(B) = 0.5\left[\cos(A - B) - \cos(A + B)\right] \qquad (16.16)$$

How do we relate Eq. (16.16) to the desired impulse response $h(t)$ of the CPSM filter of Fig. 16.22? To understand this, consider the form that the impulse response $h(t)$ is required to take. For convenience, the desired form of $h(t)$ for a normalized impulse response is

$$h(t) = \sin(A)\sin(B) = \sin(2\pi f_o t)\sin\left(\frac{\pi t}{2T_c}\right) \quad for \ 0 < t < T_c$$

$$= 0 \quad otherwise \tag{16.17}$$

where T_c = chip interval and f_o = IDT synchronous frequency. The first sine function in Eq. (16.17) gives the synchronous frequency and IDT finger period, the second one relates the apodization pattern of the input IDT. With recourse to Eq. (16.16), an expansion of Eq. (16.17) yields,

$$h(t) = 0.5\left[\cos(2\pi f_1 t) - \cos(2\pi f_2 t)\right] \quad for \ 0 < t < T_c$$

$$= 0 \quad otherwise \tag{16.18}$$

where the two frequency components f_1 and f_2 present are

$$f_1 = f_o - \frac{1}{4T_c} \tag{16.19a}$$

$$f_2 = f_o + \frac{1}{4T_c} \tag{16.19b}$$

These are the same frequency relationships as given in Eq. (16.15).

Equation (16.17) initially appears to be troublesome because it contains both of these frequency terms instead of one or the other. This dilemma is automatically overcome, however, by selecting the IDT length to be an odd number of half cycles in length at centre frequency.

To understand this, recall that the input voltage impulse sequence is only at chip period T_c. At any instant, therefore, the response of this SAW filter will be the resultant of *two* consecutive impulse stimuli. This leads to two possible solutions. For consecutive and overlapping impulse responses $h_n(t)$ and $h_{n+1}(t)$, the output voltage V_{out} from this SAW filter will be

$$V_{out} = h_n(t) + h_{n+1}(t) \quad for \ nT_c < t < (n+1)T_c \tag{16.20}$$

By choosing an appropriate reference time, the $h_n(t)$ term in Eq. (16.20) can be expanded in the cosine difference form of Eq. (16.18). The time difference of T_c for $h_{n+1}(t)$ in Eq. (16.20) corresponds to a phase shift of 90° in both the sin(A) and sin(B) terms in Eq. (16.17), allowing them to be written in cosine form for this second normalized impulse response term, so that

$$h_{n+1}(t) = \cos(A)\cos(B) = 0.5\left[\cos(A - B) + \cos(A + B)\right] \tag{16.21}$$

where $A = 2\pi f_o t$ and $B = [\pi t /(2T_c)]$ from Eq. (16.17).

The output voltage and frequency from the CPSM (MSK) modulator are readily obtained by considering whether consecutive input voltage stimuli have the same or opposite voltage polarity. This gives:

Case 1.   If two successive excitation impulses have the *same* polarity, the output voltage obtained as the sum of Eqs. (16.18) and (16.21) reduces to a single $(A-B)$ term, giving

$$V_{out} \propto \cos(2\pi f_1 t) \quad for \, nT_c < t < (n+1)T_c, \qquad (16.22)$$

with a single frequency component $f_1 = f_o - 1/(4T_c)$.

Case 2.   On the other hand, if two successive excitation impulses have *opposite* polarity, the output voltage is obtained as the difference between Eq. (16.18) and Eq. (16.21), giving a single $(A+B)$ term so that

$$V_{out} \propto \cos(2\pi f_2 t) \quad for \, nT_c < t < (n+1)T_c \qquad (16.22)$$

with a single frequency component $f_2 = f_o + 1/(4f_o)$.

It can also be deduced that in switching from one frequency to the other, transitions will occur at instants when the cosine function is a maximum. As a result the MSK waveform will have a continuous phase shift.

Operation of the SAW device discussed here was demonstrated for the generation of CPSM waveforms. A spatially reversed IDT pattern can also be employed in a receiver, however, for matched filter operation.

Example 16.6 CPSM Encoding with PN Chip Sequence for Spread-Spectrum Signal.   The SAW-based CPSM (MSK) waveform generator of Fig. 16.22 is required to generate a spread-spectrum signal for a digital message signal with bit times of $11\,\mu s$. Each message bit is further encoded with a 1024 PN chip sequence. If the synchronous frequency f_o of this SAW filter is $f_o = 300\,\text{MHz}$, determine: (a) the chip-encoding rate R_c; (b) the chip period T_c; (c) the two MSK frequencies f_1 and f_2 generated by the device; and (d) the frequency difference f_d between the two states. ∎

Solution.   (a) The chip encoding rate is $R_c = N_c R_b$ where $N_c = 1024$ chips per message bit and $R_b =$ bit rate $= 1/(11\,\mu s) = 90.91\,\text{kb/s}$. The PN chip rate is $R_c = 1024 \times 90.91 \times 10^3 = 93.09\,\text{Mchip/s}$. (b) The chip period $T_c = 1/R_c = 1/(93.09 \times 10^6) = 10.7\,\text{ns}$. (c) From Eq. (16.19) the MSK frequencies are $f_2 = f_o + 1/(4T_c) = (300 \times 10^6) + 1/(4 \times 10.7 \times 10^{-9}) = 323.3\,\text{MHz}$ and $f_1 = f_o - 1/(4T_c) = (300 \times 10^6) - 1/(4 \times 10.7 \times 10^{-9}) = 276.7\,\text{MHz}$. (d) Difference frequency $f_d = 1/(2T_c) = 46.6\,\text{MHz}$. ∎

Example 16.7 SAW MSK Modulator on ST-X Quartz.   The SAW MSK modulator of Example 16.6 is to be fabricated on ST-X quartz. Determine: (a) its impulse response time; (b) the number of electrode pairs N_p in the apodized IDT; (c) the number of acoustic half-cycles in the coded IDT; and (d) the number of IDT fingers. Neglect the broadband response of the output IDT. ∎

Solution. (a)   From   Example   16.6   the   impulse   response
time $= 2T_c = 21.4$ ns. (b) At 300 MHz IF center frequency, the acoustic wavelength $\lambda_o = v/f_o = 3158/(300 \times 10^6) = 10.5\,\mu$m. The impulse time of 21.4 ns
yields a propagation length $d = vt = 3158 \times 21.4 \times 10^{-9} = 67.6\,\mu$m in the
apodized IDT. The number of finger pairs (periods) in the apodized IDT is
$N_p = 67/10.5 \approx 6.5$. (c) the number of acoustic half-cycles in the coded IDT is
$2N_p = 13$. (d) the number of IDT fingers is $N = 2N_p + 1 = 14$. ∎

16.9.4. SAW-BASED MSK RECEIVER FOR A 2.4-GHz SPREAD-SPECTRUM INDOOR RADIO LINK

As an example of the application of the SAW-based CPSM (MSK) techniques described in the preceding material, it is instructive to give some
details of a system employing such techniques. The system in question was
a hybrid CDMA/TDMA spread spectrum 2.4-GHz indoor radio link for
processing 128-chip Gold codes[5] with an integration time of $3\,\mu$s, and a
bandwidth greater than 50 MHz. [4]. The CDMA receiver section tapped-
delay line (TDL) system operated around an IF frequency of
$f_o = 266.66$ MHz, with broadband matching to 50-Ω impedance levels. Sine-
function weighting of the IDT fingers was employed in accordance with the
layout of Fig. 16.22. The SAW structure was fabricated on X112.2° rotated-
Y lithium tantalate so as to circumvent the insertion loss of 50 dB otherwise
obtained with the use of ST-quartz. The required bit rate of 333.3 kb/s
yielded a chip time of 23.44 ns. The frequency excursions about the IF
center frequency were $f_1 = 256.00$ MHz and $f_2 = 277.33$ MHz. To minimize
finger reflections, the IDTs employed split electrodes with $\eta = 0.5$, for a
width of $\lambda_o/8$. This design yielded a matched insertion loss of 18 dB, with an
amplitude ripple of less than 2 dB and a close-in selectivity of 23 dB [4].

16.10. Summary

This chapter first reviewed the principles of autocorrelation and matched
filtering using fixed length codes. This was followed by a presentation of the
principles of operation of some coding techniques for spread-spectrum
communications employing SAW IDTs with fixed code lengths as matched
filters in the receiver stages. These included fixed-code designs for: 1)
binary-phase coding (BPSK), including Barker codes; 2) differential phase
shift keying (DPSK); 3) quadraphase shift keying (QPSK); and 4)

---

[5] Gold codes are nonmaximal composite codes generated from maximal code sequences.
These allow construction of families of $(2^n - 1)$ codes from pairs of *n*-stage shift registers, in
which all codes have well defined correlation characteristics [3].

minimum-shift keying (MSK)—also known as continuous phase-shift modulation (CPSM). A number of actual examples were given so as to clarify chip and bit specifications for differing designs. The chapter concluded by highlighting an example of an MSK fixed-code receiver for an indoor 2.4-GHz spread-spectrum radio link.

16.11. REFERENCES

1. W. C. Y. Lee, "Mobile radio and cellular communications," in R. C. Dorf, (Ed.), *The Electrical Engineering Handbook*, CRC Press, Ann Arbor, Michigan, ch. 69, pp. 1546–1553, 1993.
2. T. Shiba, A. Yuhara, M. Moteki, Y. Ota, K. Oda and K. Tsubouchi, "Low loss SAW matched filters with low sidelobe sequences for spread spectrum communications," *Proc. 1995 IEEE Ultrason. Symp.*, vol. 1, pp. 107–112, 1995.
3. R. C. Dixon, *Spread Spectrum Systems*, Second Edition, John Wiley and Sons, New York, 1984.
4. R. Weigel, C. Knorr, K. C. Wagner, L. Reindl and F. Seifert, "MSK SAW tapped delay lines on LiTaO3 with moderate processing gains for CDMA indoor and mobile radio applications," *Proc. 1995 IEEE Ultrason. Symp.*, vol. 1, pp. 167–170, 1995.
5. S. Haykin, *Communications Systems*, Second Edition, John Wiley and Sons, New York, 1983.
6. C. E. Shannon, "A mathematical theory of communication," *Bell System Technical Journal*, vol. 27, pp. 379–423 and pp. 623–656, 1948.
7. G. R. Cooper and C. D. McGillem, *Modern Communications and Spread Spectrum*, McGraw-Hill Book Company, New York, 1986.
8. D. J. Torrieri, *Principles of Secure Communications Systems*, Artech House Inc, Dedham, 1985.
9. R. D. Colvin, "Correlators and convolvers used in spread spectrum systems," *Proc. 1980 IEEE National Telecommunications Conference*, vol. 1, pp. 22.41–22.44, 1980.
10. P. H. Carr, R. D. Colvin and J. H. Silva, "Encoding and decoding at 1 GHz with SAW tapped delay lines," *Proc. 1979 IEEE Ultrason. Symp.*, pp. 757–760, 1979.
11. A. I. Sinsky, "Waveform Selection and Processing," in E. Brookner (ed), *Radar Technology*, Artech House, Dedham, Chapter 7, 1977.
12. D. T. Bell Jr. and L. T. Claiborne, "Phase Code Generators and Correlators," in H. Matthews (Ed.): *Surface Wave Filters*, John Wiley and Sons, New York, Chapter 7, 1977.
13. J. L. Eaves and E. K. Reedy (eds), *Principles of Modern Radar*, Van Nostrand Reinhold, New York, 1987.
14. M. G. Holland and L. T. Claiborne, "Practical surface acoustic wave devices," *Proceedings of the IEEE*, vol. 62, pp. 582–611, May 1974.
15. W. Tomasi, *Advanced Electronic Communications Systems*, Prentice-Hall, New Jersey, 1987.
16. K. Morozumi, M. Kadota and S. Hayashi, "Characteristics of BGS wave resonators using ceramic substrates and their applications," *Proc. 1996 IEEE Ultrasonics Symp.*, vol. 1, pp. 81–86, 1996.
17. F. Moeller, A. Rabah, S. M. Ritchie, M. A. Belkerdid and D. C. Malocha, "Differential phase shift keying direct sequence spread spectrum single SAW based correlation receiver," *Proc. 1994 IEEE Ultrason. Symp.*, vol. 1, pp. 189–193, 1994.
18. C. R. Vale, "SAW quadraphase code generator," *IEEE Trans. Sonics and Ultrasonics*, vol. SU-28, pp. 132–136, May 1981.

19. W. R. Smith, "SAW filters for CPSM spread spectrum waveforms", *Proc. 1980 IEEE National Telecommunications Conference*, vol. 1, pp. 22.1.1–22.1.6, 1980.
20. J. H. Goll and D. C. Malocha, "An application of SAW convolvers to high bandwidth spread spectrum communications," *IEEE Trans. Sonics and Ultrasonics*, vol. SU-28, pp. 195–205, May 1981.
21. D. C. Malocha, J. H. Goll and M. A. Heard, "Design of a compensated SAW filter used in wide spread spectrum signals," *Proc. 1979 IEEE Ultrason. Symp.*, pp. 518–521, 1979.

Real-Time SAW Convolvers for Voice and Data Spread-Spectrum Communications

17.1. Introduction

17.1.1. MULTIPATH PROBLEMS IN INDOOR ENVIRONMENTS

Due to the effects of multipath fading and shadowing the electromagnetic environment for indoor mobile/wireless communications can be very harsh. Multipath fading arises from the delay spread between received signals, while shadowing is generally caused by walls, floors, and other building discontinuities [1], [2].

Consider the simple illustration of Fig. 17.1, as applied to a spread-spectrum direct-sequence receiver[1], where the direct and delayed signals have the same amplitude. The delay spread Δ is given by the time delay between paths, so that $\Delta = $ Delay 1 - Delay 2. To minimize the corrupting effect in the receiver due to multipath fading, we require the *coherence bandwidth* (B_C) to be less than the transceiver RF bandwidth (B_{RF}). The coherence bandwidth B_C is defined as the bandwidth within which the amplitudes or phases of two received signals have a high degree of similarity. The coherence bandwidth for the fading bandwidths of two received signals is given as

$$B_c = \frac{1}{\alpha \Delta} \tag{17.1}$$

where $2\pi \leq \alpha \leq 4\pi$, for various communications environments [3]–[5]. For the case of a CDMA communications system, with $B_{RF} = 1.23 \text{MHz}$, this

---

[1] Direct *sequence modulation* is more exactly defined as *directly carrier-modulated code sequence modulation*. Here, we consider direct sequence spectral-spreading to be implemented by binary-code chip modulation of each message data bit.

Delay Spread = Δ = Delay 1 - Delay 2
Coherence Bandwidth = constant

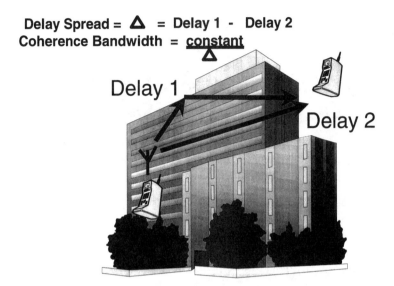

FIG. 17.1.   Illustrating delay spread Δ and coherence bandwidth B_c.

requires the delay spread $\Delta \approx 1/B_C$ to be greater than about 100ns, for minimal multipath interference. Indoor use of CDMA communications may thus require additional time diversity to be artificially implemented by the use of distributed antennas or other techniques [2].

Let us re-examine the multipath problem in terms of the direct-sequence data waveforms considered here. If the delayed signal exceeds the direct-path signal by more than one chip, it will be treated exactly as any other uncorrelated input signal and suppressed accordingly [6]. The corrupting effect of multipath reception, as functions of both time and distance, will be reduced as the coding chip-rate is increased—with commensurate increase in the RF bandwidth. With a chip rate of 5 Mb/s, for example, the difference between delayed and direct paths must be less than about 60m ($\Delta \sim 200$ns) before there is any significant corruption of the message signal. If the chip rate is increased to 20 Mb/s the path length difference of onset of coherent interference is reduced to about 15m ($\Delta \sim 50$ns).

In typical indoor radio environments, the rms delay spread due to multipath reception is less than 100ns [7]. From indoor radio-channel measurements of delay spread it has been determined that for line-of-sight transmission echoes arriving closely spaced in time and with different powers, the coherence bandwidth is in the range $B_c \approx 6$–30 MHz, where the higher values of B_c are associated with waveguiding wall geometries [5].

With their advantages of broad bandwidth, large processing gain, compactness, and relatively simple construction, broadband monolithic SAW convolvers with RF bandwidths greater than the coherence bandwidths are well suited for indoor spread-spectrum communications, especially in buildings with highly reflecting walls and corridors. In addition to improved performance against multipath interference, these systems can also offer protection against jamming if pseudo-noise spreading codes are employed [8]. They are well suited to circuitry for data reception with differential phase shift keying (DPSK). They have also been designed for operation in packet-data communication modems employing code-shift keying (CSK), in conjunction with binary phase-shift keying (BPSK) and frequency hopping. Packet-voice SAW-convolver spread-spectrum systems have also been implemented, where voice packets from several systems may be multiplexed in time. It has also been demonstrated that Kasami chip-codes can readily be applied, for PN-encoding in the transmission modems for such packet-data and packet-voice spread-spectrum systems, as well as for time-reversed comparison sequences in the receiver modules [8], [9].

17.1.2. A NOTE ON MATCHED FILTERING AND CORRELATION

While the terms "matched filtering" and "correlation" are often used interchangeably, as in this text, there are some subtle differences between the two processes [10]. Matched filtering of a data bit provides a completely asynchronous method of demodulation, while correlation against an actively generated reference provides a synchronous method of demodulation. Matched filtering provides a continuous output in time that represents all time relations between signal and reference. Correlation creates only one output sample at a time — as one value of the evolving relationship between signal and reference.

As the nonlinear convolution process compresses the output signal by a factor of two relative to the input, the output signal bandwidth will have to be doubled accordingly. With the linear matched filter process on the other hand, the same filter bandwidth can be employed in output and input stages because there is no time compression.

17.1.3. SCOPE OF THIS CHAPTER

While differing real-time SAW convolver types have been fabricated over the years, principally for sonar and radar signal processing [11]–[24], this chapter will focus on the compact monolithic type known as the elastic SAW convolver[2], for application to spread-spectrum indoor communications. This is a three-port *nonlinear* device, well suited to system-bandwidths of about 100 MHz or less.

The coverage begins with a presentation of the signal relations applicable to nonlinear convolution in the elastic convolver, including trade-offs between bandwidth and convolution. With this background in hand, the chapter proceeds to illustrate some design features of three spread-spectrum transceivers for indoor mobile communications. Two of these relate to SAW convolvers operating under *synchronous* recovery conditions. Of these, one relates to packet-data transmissions and the other is for packet-voice. Both designs highlighted here employ dual elastic SAW convolvers in conjunction with processors accommodating Kasami chip-codes for direct-sequence spectral spreading.

The third spread-spectrum convolver-system illustrated here is for *asynchronous* data recovery. Versions of this third type have been fabricated for a) license-free operation in the U.S. (902 to 928 MHz, at 1 W), b) full-duplex transmission in another U.S. spread-spectrum band (2.4000 to 2.4835 GHz, at 1 W), as well as for c) low-power license-free operation in Japan, where the allowed transmitter field-strength is less than $500\,\mu$V/m at a distance of 3 m, in the frequency range below 322 MHz—corresponding to a transmitted power of less than $0.05\,\mu$W/120-kHz.

17.2. Operation of SAW Devices under Nonlinear Conditions

The SAW devices examined so far in the text have all been those with a linear response between the input electrical signal and the output one at the same frequency. A tacit underlying assumption for such linear operation is that the SAW power densities at the piezoelectric substrate surface are low enough to ensure that the accompanying mechanical deformations of the piezoelectric are elastic ones. Nonlinear operation will result, however, for sufficiently high levels of SAW power density—or when a semiconductor nonlinearity is involved. Because the surface waves have a limited penetration depth, nonlinear piezoelectric operation of SAW devices can normally be realized with much lower electrical power levels than for bulk-wave devices. To date, convolver designs have been based on Rayleigh-wave propagation, as these have higher acoustic-energy densities at the piezolelectric surface than either LSAW or SSBW under comparable excitations, and can be more easily operated in nonlinear fashion.

---

[2] The monolithic *elastic* SAW convolver is also known as the *waveguide* SAW convolver, for designs where the convolving plate is reduced to a narrow waveguide-type strip.

For Rayleigh-wave propagation, it has been shown that a maximum power level of 10 mW/mm of acoustic aperture is the upper limit for linear operation on *YZ*-lithium niobate at 1 GHz [25]. By reducing the acoustic aperture of the IDT, nonlinear piezoelectric behavior can be achieved with signal levels in the order of a few milliwatts. Despite the fact that this piezoelectric nonlinearity is a very weak one, there are signal processing applications that can be exploited by operating in this fashion, as realized early in the development of SAW devices [26]. These relate to the use of SAW convolvers as nonlinear SAW devices with large time-bandwidth products, for programmable matched filtering of lengthy coded sequences. In such operation the convolvers are configured to act as correlators for the matched filtering requirements.

17.2.1. Nonlinear Piezoelectric Behavior of SAW Devices

As already noted, all of the SAW devices examined in previous chapters have been assumed to operate in linear fashion by use of sufficiently low levels of SAW power density at the piezoelectric surface. In this situation there is elastic deformation of the piezoelectric surface by SAW waves emanating from an excited IDT. This elastic deformation is that related by Hooke's Law in Chapter 2. For nonpiezoelectric elastic solids this was expressed in the tensor form of Eq. (2.1) as:

$$[T] = [c] : [S] \qquad (2.1)$$

where $[c]$ = elastic stiffness tensor, $[T]$ = stress tensor and $[S]$ = strain tensor. For this situation, the elastic potential energy Φ is a quadratic function of the strains [27]

$$\Phi = \frac{1}{2} c_{ijkl} S_{ij} S_{kl} \qquad (17.2)$$

Consideration of nonlinear behavior of the piezoelectric requires addition of a cubic term to Eq. (17.2). This nonlinearity is also incorporated in quadratic terms of the two constitutive relationships for electrical displacement and mechanical stress T given in reduced matrix form in Eqs. (2.8) and (2.9) of Chapter 2. For the electrical relationship we had

$$D = [e][S] + [\varepsilon] E \qquad (2.8)$$

where electrical displacement density D (C/m^2) is a three-dimensional parameter in the rectangular x, y, z coordinate system, $[e]$ is a 3×6 piezoelectric matrix with 18 elements, $[S]$ is the strain matrix with 6 components.

Additionally, $[\varepsilon]$ is the electric permittivity matrix with 6 terms, while E is the electric field intensity (V/m).

The mechanical relation was given as

$$[T] = [c][S] - [e^t]E, \qquad (2.9)$$

in terms of stress matrix $[T]$, where $[c]$ is an elastic stiffness matrix and $[e^t]$ is a 3×6 matrix, which is the transpose of the piezoelectric constant matrix $[e]$ in Eq. (2.8) (i.e. matrix rows and columns are interchanged).

The source of piezoelectric nonlinearity in Eqs. (2.8) and (2.9) is in their quadratic terms. For a one-dimensional illustration of these for YZ-lithium niobate, Eq. (2.8) reduces to

$$D = (e_l S + \varepsilon_l E) + (0.5\varepsilon_n E^2 + p_n ES + 0.5q_n S^2) \qquad (17.3)$$

while Eq. (2.9) reduces to

$$T = (C_l S - e_l E) + (0.5C_n S^2 - 0.5p_n E^2 - q_n ES) \qquad (17.4)$$

for this one-dimensional case, where the subscript l relates to linear coefficients and the subscript n relates to nonlinear ones. Thus the coefficient p_n is an electrostrictive one, while q_n relates to an electrooptic coefficient [28].

These second-order coefficients are very small at best so the nonlinear responses obtained from SAW convolvers will also be very small. This leads us to seek fabricational techniques that will optimize device operation.

17.3. Convolution Relations for the Elastic SAW Convolver

17.3.1. Signal, Reference and Convolution Relationships

The convolver type considered in this chapter for application to indoor spread-spectrum communications is that of the three-port elastic SAW convolver. To understand its operation, consider the simplified illustrative structure of Fig. 17.2. This consists of two input IDTs on a piezoelectric substrate, with a thin metal surface film of length L between them. The signal under scrutiny at IF carrier frequency f_o is applied to one input port (designated here as Port 1), while the local reference waveform—also at f_o—is simultaneously applied at the second input port (Port 2). Under nonlinear operation, the output signal generated between the metal surface plate of length L and a bottom ground electrode film is the convolution result (or the correlation one depending on the time-ordering of the reference sequence). The convolver's time-bandwidth TB product is dictated by

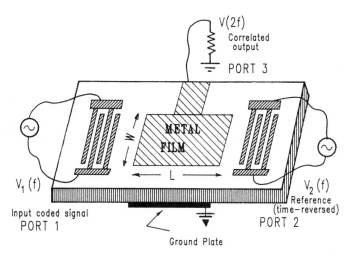

FIG. 17.2. Primitive form of monolithic elastic SAW convolver.

the length L of the metal plate giving interaction time $T = L/v$. The bandwidth is approximately that of the SAW delay line between Port 1 and Port 2. For wideband operation, the IDTs at these ports should have as few fingers as possible.

Further consider that an IF signal voltage $V_1 = |V_1(t)|\exp(j\omega_o t)$ is applied to Port 1 at angular frequency $\omega_o = 2\pi f_o$, while a reference voltage $V_2 = |V_2(t)|\exp(j\omega_o t)$ is applied to Port 2. These will generate contradirective surface acoustic travelling waves in the region between the two IDTs. (Symbols S, s for *signal* and R, r for *reference* are used here as a visual aid). If the system is operating nonlinearly, the mixing of the contra-propagation of surface waves under the convolver plate will result in an output signal with a *second harmonic* component, relating to the correlation of the two signals.

At any subsequent time t along the z reference axis between the two IDTs, the surface wave potential $s(t,z)$ due to the input signal at Port 1 can be expressed in transmission line form as

$$\left(\textit{From Port 1}\right) \quad s(t, z) = S(t - z/v)e^{j(\omega_o t - \beta z)} \qquad (17.5)$$

where the signal modulation envelope $S(t - z/v)$ is a function of SAW velocity $v = f\lambda$ and phase constant $\beta = 2\pi/\lambda$. The carrier phase is contained in the exponential term. In the same way the potential of the contradirective SAW wave emanating from reference Port 2 can be expressed as

$$(\textit{From Port 2}) \quad r(t,z) = R(t + z/v)e^{j(\omega_o t - \beta z)} \tag{17.6}$$

where the change in the sign of z in Eq. (17.6) relates to propagation along the $-z$ axis.

Under nonlinear deformation of the piezoelectric surface, the interaction between these two surface waves will give rise to second-order potential terms $\Phi(z,t)$ at the surface of the piezoelectric with angular frequencies ($\omega_1 \pm \omega_2$), $2\omega_1$, $2\omega_2$ as well as dc potentials proportional to the squares of the input and reference signals [29]. (In effect, the result is similar to that obtained using a conventional diode mixer operating under nonlinear excitation.) In the case of the elastic SAW convolver with $\omega_1 = \omega_2 = \omega_o$, the most significant of these second-order surface potential terms is Φ_2 (z,t) given by [30],

$$\Phi_2(t,z) = 2Ds(t - z/v)r(t + z/v)e^{2j\omega_o t}, \tag{17.7}$$

where D = a nonlinearity constant.

There are two notable features in the exponential term of Eq. (17.7). The first is that the angular frequency is $2\omega_o$. This change of frequency is most useful in that it allows the output to be filtered from the input signal f_o and any spurious responses accompanying it, such as due to timing pulses. The second significant feature is that it is not a travelling wave term; it is only a time-dependent function. This means that the phase constant $\beta = 0$, so that $\Phi_2(z,t)$ is spatially uniform over the interaction area. This allows the geometry of the metal film overlay to be in the simple form of a rectangular plate, so that the second-harmonic voltage $C(t)$ extracted at Port 3 is the convolution result,

$$C(t) = P \int_{-L/2}^{+L/2} S(t - z/v)R(t + z/v)dz \, e^{j2\omega_o t} \tag{17.8}$$

over the length L of the thin-film metal plate, where P is a constant dependent on the strength of the nonlinear interaction. If the two signals are totally contained within the region of the metal plate, the $\pm L/2$ limits of the integral can be extended to $\pm\infty$ without affecting the value of $C(t)$.

Using a change of variable $\tau = (t - z/v)$, Eq. (17.8) may be reformulated as

$$C(t) = Pve^{2j\omega_o t} \int_{t-L/v}^{t+L/v} S(\tau)R(2t - \tau)d\tau \tag{17.9a}$$

or

$$C(t) = Mve^{2j\omega_o t} \int_{-\infty}^{+\infty} S(\tau)R(2t - \tau)d\tau \qquad (17.9b)$$

if the interacting travelling waves are contained within the metal plate, where M is a constant dependent on the strength of the nonlinear interaction. Equation (17.9) can be extended to include a time delay term τ_d associated with the distance $d = v.\tau_d$ between input and output IDTs so that

$$C(t) = Mve^{2j\omega_o t} \int_{-\infty}^{+\infty} S(\tau)R(2t - \tau - \tau_d)d\tau \qquad (17.10)$$

Equation (17.10) has the form of a convolution integral. If we compare it with the mathematical relation for convolution in Eq. (16.6) of Chapter 16, repeated here:

$$(Convolution) \quad f(t) = s(t) \star h(t) = \int_{-\infty}^{+\infty} s(\tau)h(t - \tau)d\tau \qquad (16.6)$$

the only significant difference is that the argument of R differs by a factor of two from that in f_2, so that the output signal in Eq. (17.10) is compressed in time by this factor. The simple physical reason for this is due to the contradirective SAW waves in the region between the two IDTs in the elastic convolver. As these two surface waves of velocity v are travelling towards one another, their relative velocity is $2v$, and the interaction is over in half the time. (As a result, the output signal bandwidth will have to be doubled accordingly!)

17.3.2. CONVOLUTION EFFICIENCY OF THE ELASTIC CONVOLVER

The convolution efficiency of a SAW convolver η_c may be defined as

$$\eta_c = 10 \log_{10} \frac{P_{out}}{P_{sa}P_{ra}} \quad (dBm^{-1}) \qquad (17.11)$$

where P_{out} = output power at Port 3, while P_{sa} and P_{ra} are the signal and reference powers entering the interaction area due to input powers P_a and P_s at Port 1 and Port 2, respectively. The trend is for η_c to be evaluated when P_{sa} and P_{ra} both have power levels of $0\,dBm$ ($0\,dBm = 1\,mW$). Convolution efficiencies of elastic SAW convolvers—small at best—would be very low indeed for the simple structure of Fig. 17.2, due to the low SAW power density that would be involved in this case. Its surface plate geometry needs to be optimized in order to boost η_c as much as possible. This is the rationale for the design of the waveguide-type of elastic SAW convolver, with convolution efficiencies of up to about $-55\,\mathrm{dBm}^{-1}$.

Example 17.1 Output Power and Dynamic Range of Elastic SAW Convolver. An elastic SAW convolver has a convolution efficiency $\eta_c = -70\,\mathrm{dBm}^{-1}$. Input and output IDTs are matched at operating frequency f_o for signal and reference inputs. (a) Determine the output power at $2f_o$ when the signal and reference powers at the input ports are $P_s = P_r = 20\,\mathrm{dBm}$ (100 mW). (b) If the output noise floor level is $-80\,\mathrm{dBm}$, determine the dynamic range of the device, given by the value of the output-signal/thermal-noise ratio. ■

Solution. (a) Expressed in dBm units, Eq. (17.11) becomes $P_{out} = \eta_c + P_{sa} + P_{ra}$. Because the signal and reference inputs are applied to matched IDTs, half of the acoustic power from each is lost through bidirectionality. Thus $P_{sa} = P_{sa} = (20 - 3) = 17\,\mathrm{dBm}$. This gives $P_{out} = (-70) + (17) + (17) = -36\,\mathrm{dBm} = 0.25$ microwatt.

(b) The output-signal/thermal-noise ratio expressed in dB is $\mathrm{S/N} = -36 - (-80) = 44\,\mathrm{dB}$. ■

17.3.3. FIGURE OF MERIT F AND CONVOLUTION EFFICIENCY η_c

To optimize the inherently low value of convolution efficiency η_c of a monolithic SAW convolver, we need to re-examine the "basic" structure of Fig. 17.2. For given input signal and reference power levels, this necessitates boosting the nonlinear effect and the convolution efficiency by increasing the SAW power density in the SAW interaction area. The open-circuit rms output voltage V_{out} from convolution Port 3 is proportional to the power densities due to the signal and reference inputs [18] so that

$$V_{out}\left(2f_o\right) = \frac{F\left(P_{sa}P_{ra}\right)^{1/2}}{W}, \qquad (17.12)$$

where F = a figure of merit for the nonlinear behavior of the piezoelectric, and is independent of frequency. P_{sa} and P_{ra} are as given for Eq. (17.11), while W = acoustic beamwidth of the contradirective surface waves. The open-circuit output voltage is proportional to f_o and inversely proportional to W. Note that the figure of merit F (with units of volt-meter/watt), cannot be increased without limit by increasing the input power, as saturation of the nonlinear mixing process will eventually occur [11].

To optimize the output power and voltage V_{out}, it is further necessary to examine the output impedance at convolution Port 3, looking back into the surface metal plate. The resistive portion R_{out} of this complex output may be written as

$$R_{out} = \frac{r_a}{WL} \tag{17.13}$$

where $r_a = R_{out} WL$ in terms of acoustic aperture W and interaction length L. From the maximum power transfer theorem, for which $P_{max} = V^2 / 4R_{out}$, the convolution efficiency η_c in Eq. (17.11) and the figure of merit F are related by [18]

$$\eta_c = \frac{F^2}{4r_a} \frac{L}{W} \tag{17.14}$$

This important result shows that the convolution efficiency increases as the acoustic aperture W decreases.

17.4. Using the Elastic SAW Convolver as a Correlator

17.4.1. ANOTHER LOOK AT CONVOLUTION AND AUTOCORRELATION

SAW convolvers are so frequently used as correlators in radar and spread spectrum applications that it is necessary to review some of these aspects before examining the waveguide SAW convolver in more detail. Recall the pertinent relations given for convolution and autocorrelation in Chapter 16. For the convolution process we had

$$\left(Convolution\right) \quad f(t) = \int_{-\infty}^{+\infty} f_1(t) f_2(t-\tau) d\tau = f_1(t) \star f_2(t) \tag{16.6}$$

where the right-hand expression is a "shorthand" notation for convolution. For the autocorrelation and matched filtering process we also had

$$\left(Autocorrelation\right) \quad A(\tau) = \int_{-\infty}^{+\infty} s(t) s(t+\tau) dt \tag{16.9}$$

As depicted graphically in Fig. 17.3 for the special case $f_1(t) = f_2(t)$, the difference between the two processes involves a time-reversal of the signal waveform.

17.4.2. BRINGING IN CROSS-CORRELATION

While the use of autocorrelation was applicable to the fixed transversal filters of Chapter 16, in the case of the SAW convolver it is really cross-correlation between the input and reference signals that must be

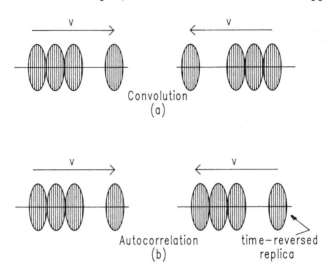

FIG. 17.3. (a) Illustrating the convolution of two surface waves in a SAW convolver. (b) Autocorrelation is obtained using time-reversal of one of the waveforms.

considered. Recall the relation given in Chapter 16 for cross-correlation between the two signals as

$$\left(Cross\text{-}correlation\right) \quad C(\tau) = \int_{-\infty}^{+\infty} f_1(t)f_2(t+\tau)dt \tag{16.10}$$

for real-valued functions. For the elastic SAW convolver, $f_1(t)$ may be associated with the signal input at Port 1 and $f_2(t)$ may be compared with the reference input at Port 2, which is applied simultaneously as a *time-reversed* sequence as sketched in Fig. 17.4.

17.4.3. EXTRACTING THE MESSAGE MODULATION IN THE SAW CONVOLVER

In spread-spectrum communications applications involving the correlation of PN sequences, the reference signal applied at Port 2 is not necessarily an exact time-reversed replica of the input signal. The reason for this is that, as demonstrated in Fig. 17.2, the modulation envelope of the spread spectrum transmission is obtained from a composite digital sequence due to the PN spreading signal and the digital message signal. If the original digital message sequence is to be recovered by the cross-correlation process, it will be appreciated that the reference signal at Port 2 must be only that associated

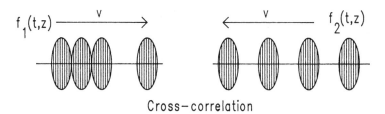

Cross−correlation

FIG. 17.4.   Cross-correlation of two surface-wave sequences.

with the PN spreading code. Otherwise the phase reversals accompanying the "1", "0" message excursions will be lost.

Example 17.2   Cross-Correlation of Two Digital Impulse Sequences.   (a) Sketch the form of the cross-correlation response of the two digital impulse sequences given here, over a relative period $-4 \leq \tau \leq +4 \, \mu s$. (b) Compare this result with that for the autocorrelation Example 16.2 in Chapter 16. ■

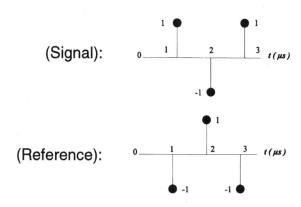

EXAMPLE 17.2.   Illustrations for question section.

Solution.   (a) Following Example 16.2, the cross-correlation in Eq. (16.10) reduces to multiplication and addition , which can be evaluated one step at a time in a "longhand" calculation. In this way, the individual contributions are obtained as

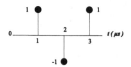

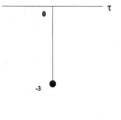

For τ = 0

times 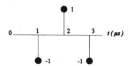 yields ∫ =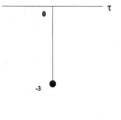

- -

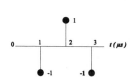

For τ = -1

times 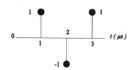 yields ∫ =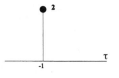

EXAMPLE 17.2a.   Illustration for first part of solution.

This multiply-and-add procedure may be continued to cover all other τ values between –4 and +4 μs. In this way the time dependence of the cross-correlation is obtained as

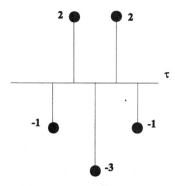

For -4 $\leq \tau \leq$ 4 μs

EXAMPLE 17.2b.   Illustration for second part of solution.

(b) The resultant sequence is just a phase-inverted version of that in Example 16.2(c). ■

17.5.   Monolithic Single-Track Waveguide Type of Elastic SAW Convolver

17.5.1. DESIGN FEATURES

Figure 17.5 depicts the geometric form of a single-track waveguide type of elastic SAW convolver. In accord with the preceding discussion on optimizing convolution efficiency, the contradirective SAW waves from IDTs at input and reference ports undergo a beam compression so that they are focused into a narrow central region of length L and acoustic width $W \geq \approx 2\lambda_o$. Techniques for doing this include the use of parabolic horn compressors, prism couplers, curved IDTs, and multistrip coupler compressors. This allows IDTs of "normal" (50 - 100 λ_o) width to be employed at the two input ports. One advantage of using wider IDTs is that they have lower input impedances than their narrow-beam counterparts, so that they are generally easier to tune and match. Another advantage is that the power-handling capability of the IDT increases with width, for a given mismatch ratio.

Another facet of the convolver plate structure in Fig. 17.5 is that two ground-return metal planes are deposited on either side of the central convolution area. This is superior to the construction as shown in Fig. 17.2,

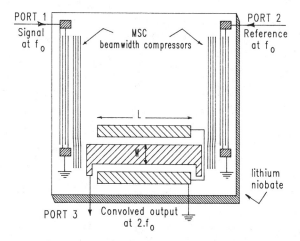

Fɪɢ. 17.5. Waveguide type of elastic SAW convolver, using multistrip coupler for beam compression. (After Reference [18].)

where the ground return metallization was on the bottom surface of the piezoelectric. A mechanical advantage is that the entire metallization can be carried out in one process step. An electric advantage is that the output capacity at Port 3 is independent of substrate thickness and any attendant variations in manufacturing. In addition, the plate capacity C_p is independent of the plate width W. The Q of the output circuit is proportional to $WL/C_p\omega_o$, and can be kept low enough to retain the large bandwidth capability of the device. An early design of such a waveguide elastic convolver on lithium niobate [18] attained a time-bandwidth product $TB = 600$ at IF center frequency $f_o = 156\,\text{MHz}$, with bandwidth $B = 50\,\text{MHz}$, interaction time $T = 1.2\,\mu s$, and convolution efficiency $\eta_c = -54.7\,\text{dBm}^{-1}$. In more recent designs [11], this performance has been extended to time-bandwidths $TB = 2000$, with bandwidth $B = 100\,\text{MHz}$, interaction time $T = 20\,\mu s$ and a convolution efficiency $\eta_c = -70\,\text{dBm}^{-1}$.

Example 17.3 Single-Track Waveguide Monolithic SAW Convolver. A single-track waveguide type of monolithic elastic SAW convolver is to be designed for operation at an IF frequency $f_o = 170\,\text{MHz}$, with a time bandwidth product $TB = 600$ and a 3-dB bandwidth of 50 MHz. Estimate the required number of fingers in each IDT. Neglect the effect of the convolver plate on bandwidth degradation. ∎

Solution.   Assume that the 50-MHz bandwidth specification is about the same when the device is operated either as a two-port SAW delay line or as a three-port convolver. As a SAW delay line the amplitude roll-off to the 3-dB level is a $|(\sin X)/X|^2$ one for input and output IDTs in cascade. From sinc function tables, [31] it can be deduced that the 3-dB bandwidth of the individual IDTs is a factor of about 1.3 higher than for their product. The individual 3-dB bandwidths are $BW_3 \approx 1.3 \times 50 = 65\,MHz$, while the 4-dB bandwidth is about another factor of 1.1 higher, giving $BW_4 \approx 1.1 \times 65 = 71\,MHz$. From Chapter 3 the number of IDT finger pairs N_p is given approximately by $N_p \approx f_o/BW_4 = (170 \times 10^6)/(71 \times 10^6) \approx 2.5$ finger pairs, or 5 single electrodes. ∎

17.5.2. Trade-off between Bandwidth and Convolution Efficiency

In the structure of Fig. 17.5, the narrow metal film covering the acoustic interaction area reduces the velocity of the SAW under it, so as to confine the acoustic beam. In this way it acts as an acoustic waveguide—hence the name given to this particular convolver structure. As the width of this acoustic waveguide is reduced, however, an increasing proportion of the SAW energy is propagated outside the metallization. This results in undesirable dispersion, where the SAW velocity varies rapidly with frequency. In the devices tested [18], this minimum critical width was found to be in the order of $W_{min} \approx 2\lambda_o$.

Dispersion will place a limit on the fractional bandwidth over which linear phase operation can be preserved. To accommodate large fractional bandwidths, as in Example 17.3, the acoustic aperture must be increased from the value of W_{min}. Because convolution efficiency in Eq.(17.14) decreases with increasing aperture width, however, this will necessitate a trade-off against fractional bandwidth requirements.

17.5.3. Using a Multistrip Coupler for Beam Compression

Figure 17.6(a) illustrates the structure of a multistrip coupler (MSC) with N_{bwc} metal strips, for SAW beamwidth compression in the waveguide elastic SAW convolver of Fig. 17.5, while Fig. 17.6(b) shows a "standard" MSC with N_M metal strips as examined in Chapter 6. The input and output portions of the standard MSC have equal acoustic apertures and the same finger periodicity P. The beamwidth compressor (BWC) on the other hand has unequal input and output apertures and a slightly different periodicity of finger spacing in each section, with the same metallization ratio η along each strip. From an analysis of the MSC beamwidth compressor it may be shown [32], [33] that the required number of strips N_{bwc} is

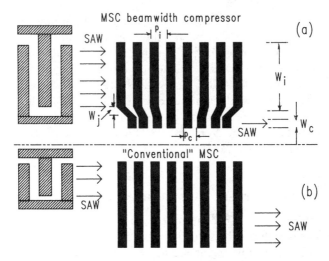

Fig. 17.6. (a) Multistrip coupler used as a SAW beamwidth compressor. (b) "Conventional" multistrip coupler.

$$N_{bwc} = N_M \frac{\left(W_i + W_j + W_c\right)}{2\sqrt{\left(W_i W_c\right)}} \tag{17.15}$$

where W_i = input acoustic aperture, W_c = acoustic aperture of the compressed SAW beam, W_j = acoustic aperture of the strips joining the two regions and N_M = number of strips in an equivalent standard MSC. Note that $N_{bwc} \geq N_M$ in Eq.(17.15).

In addition, the difference between the input strip periodicity P_i and the periodicity P_c of the strips in the compression region is

$$P_i - P_c = \frac{\pi}{\beta_o N_M} \frac{\left(W_i - W_c\right)}{\left(W_i + W_j + W_c\right)}, \tag{17.16}$$

where $\beta_o = 2\pi/\lambda_o$ = phase constant at center frequency $f_o = v/\lambda_o$. Beamwidth compression using this asymmetric MSC structure has been demonstrated for compression ratios of up to 15:1, with losses of less than 1.5 dB in each beamwidth compressor, plus an additional loss of about 1 dB propagation under the convolution plate [18]. The difference in strip periodicities is typically not more than 1%.

Example 17.4 Design of SAW Beamwidth Compressor. A multistrip coupler with metallization ratio $\eta = 0.5$ is to be used as a beamwidth compressor in a single-track waveguide SAW elastic convolver on lithium niobate, operating at a center frequency $f_o = 100$ MHz. The input IDTs are required to have an acoustic aperture of 75 λ_o in conjunction with a convolving plate with an acoustic aperture of 5λ_o for a compression ratio of 15:1 and large bandwidth capability. Determine (a) the number of strips N_{bwc} in the beamwidth compressor and (b) the difference in periodicities in the input and compression tracks. Assume that aperture W_j of the linking region is negligible. ■

Solution. (a) From Example 6.1 of Chapter 6 the number of strips in a standard MSC at $f_o = 100$ MHz was $N_M \approx 98$. Expressing W_i and W_c in acoustic wavelengths in Eq.(17.15) yields $N_{bwc} = 98 \times (75 + 5)/2\sqrt{(75 \times 5)} = 202$ strips. (b) Again from Example 6.1 (and using its notation) the finger period of the input section is $P_i = b = 0.375\lambda_o = 0.375v/f_o = 0.375 \times 3488/(100 \times 10^6) = 13 \times 10^{-6}$ m $= 13\,\mu$m. The phase constant $\beta_o = 2\pi/\lambda_o = 2\pi f_o/v = (2 \times \pi \times 100 \times 10^6)/3488 \approx 180,000$ rad/m. Substituting in Eq.(17.16) yields the period difference $P_i - P_c = \{\pi/(1.8 \times 10^5 \times 98)\} \times \{(75 - 5)/(75 + 5)\} = 1.5 \times 10^{-7}$ m $= 0.15\,\mu$m. The period difference ratio is 0.15/13 or about 1.1%. ■

17.6. Dual-Track Waveguide Type of Elastic SAW Convolver

17.6.1. DESIGN HIGHLIGHTS

One of the problems that may be encountered with single-track waveguide elastic SAW convolvers, as shown in Fig. 17.5, relates to spurious responses due to self-convolution of the signal and reference waveforms, caused by reflections from the transducers at each end. One technique for overcoming this involves the design of a dual-track waveguide convolver as sketched in the elementary representation of Fig. 17.7(a). The finger polarities of the IDTs are arranged so that the SAWs excited by the two transducers in the signal input are in phase (symmetric), while the SAWs excited in the reference IDTs are 180° out of phase (antisymmetric). In this way the desired convolution outputs of the two waveguide tracks are 180° out of phase with one another, and can be combined in a difference structure. When the SAWs with symmetric wavefronts excited by the signal IDTs reach the reference transducers, the latter will not regenerate a signal because of their antisymmetric polarity. However, there will be SAW reflections due to the velocity mismatch under the waveguide tracks. But as these will be in phase, they will cancel out in the output circuitry. With this technique, a self-convolution suppression of greater than 40 dB has been achieved in dual-track convolvers with interaction times of as high as 22 μs, even with solid-electrode IDT fingers [34].

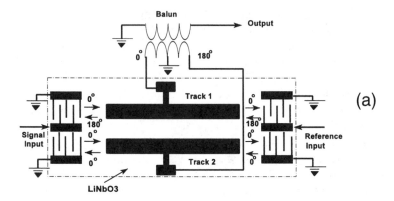

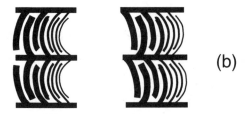

Fɪɢ. 17.7. (a) Dual-track waveguide SAW convolver, for suppression of spurious self-convolution. Beam compression not shown here. Output transformer can be replaced by active differential summer. (b) linear IDTs in (a) can be replaced by focusing linear FM chirp transducers. (After Reference [35].)

For pictorial simplicity, Fig. 17.7(a) did not include any SAW beam compression for increasing the device nonlinearity. In practice, beam compression could be carried out using the multistrip techniques shown in Fig. 17.6(a). Another technique for beamwidth compression is that using focused chirp transducers in Fig. 17.7(b). The design technique involves fitting the IDT structure to the diffraction pattern of the fundamental waveguide mode, as it diffracts on leaving the ends, rather than trying to make the converging SAW distribution fit with the waveguide mode at its entrance. The advantage of the first technique is that the diffraction pattern of the waveguide mode is a well-behaved one, and must be determined once [8], [35]. Transducer tap-positions for the focusing chirp transducers along the

z-axis of the YZ-LiNbO$_3$ substrate are determined, for linear chirps, from the relationship

$$z_n = v \cdot \frac{T}{B} f_1 \left[-1 + \sqrt{\frac{nB}{Tf_1^2} + 1} \right] \tag{17.17}$$

where f_o = center frequency, B = nominal bandwidth, $f_1 = f_o - B/2$, v = SAW velocity and T = chirp duration. The tap number n varies between 0 and $2Tf_o$. The chirp duration must be chosen so as to adjust the impedance to the desired level. For a 50-Ω system, and as the input transducers in each track are in parallel, the real part of the transducer impedance should be slightly below 100 Ω for a broadband match, in conjunction with series-inductance tuning [35].

Waveguide widths in the convolver of Fig. 17.7(a) are about the same as for the single-track convolver (e.g., 2 to 3 λ). To handle the resulting high impedance of the waveguide strips, their outputs are usually taken from multiple taps along the guide, rather than the single ones shown in the simplified illustration. The passive balanced-transformer differential summer shown here would normally be replaced by an active differential summer.

17.6.2. EXAMPLE OF MINIATURE SAW CONVOLVER USED FOR INDOOR COMMUNICATIONS

A miniaturized dual-track SAW convolver has been reported for indoor communications in order to meet demands for low-cost, compact size, and convolution efficiency [8]. This structure employed the fore-mentioned focused chirp transducers on YZ-LiNbO$_3$, and was fabricated within an SMD package of size $18 \times 8.2 \times 2.4$ mm^3. It operated at a center frequency $f_o = 350$ MHz, with an integration time $T = 3\,\mu$s, and bandwidth $B = 50$ MHz, for a time-bandwidth product $TB = 150$. Its convolution efficiency was $\eta_c = -70$ dBm, while the deviation from linear phase was less than $5°$.

17.7. Example of a Packet-Data System for Indoor Communications

17.7.1. RATIONALE AND USE OF KASAMI CODE SEQUENCES

Before commencing to examine an example of an indoor packet-data spread-spectrum receiver employing SAW convolvers, let us first examine some coding techniques for direct-sequence binary-coding of the message-data bits. In this coding method it is desirable to employing pseudo-noise (PN) binary-chip sequences that satisfy the following requirements [9]:

- The PN code chip sequences used for direct-sequence spectral spreading should not be deciphered by unintended parties.

- As the convolver reference-input codes have to be in time-reversed order, the original PN chip sequences should be easily time reversed.

- The chip sequences should have low autocorrelation sidelobes in order to obtain a high probability of correct acquisition of the preamble code sequence.

- For systems involving different PN coding of "1" and "0" message data bits (i.e., for code-shift keying (CSK)), there should be a low cross-correlation between PN code patterns assigned to "1" and "0" message data bits.

A code set that satisfies the foregoing requirements is that relating to Kasami sequences [9]. These Kasami codes can be generated by the modulo-2 addition of the output sequences of three configured linear-feedback shift registers (LFSRs), as depicted in Fig. 17.8. In the shift-register circuitry of Fig. 17.8, two of the shift registers have length $n = 8$, while the third has length $n = 4$. Each of these has code length of 2^{n-1} for even n. With the feedback connections shown, and by varying the initial shift-register states, the circuit of Fig. 17.8 can yield 4111 different sequences for the 255-chip codes employed here.

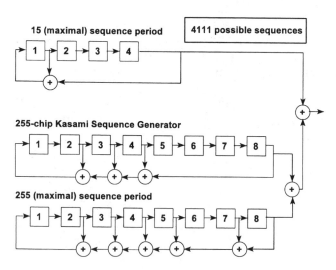

Fig. 17.8. A 255-chip Kasami sequence generator, with 4111 possible sequences, using linear-feedback shift registers. (After Reference [9].)

17.7.2. DATA-BURST TRANSMITTER FOR THE SPREAD-SPECTRUM
TRANSCEIVER

The transmitter section of an illustrative spread-spectrum packet-data system is outlined in Fig. 17.9 [9]. This is a hybrid one involving both DS and FH modulations, where the hopping rate is equal to the data rate. This hybrid BPSK/FH combination enables a large spectral-spreading to be attained, while using considerably lower code-chip rates than with nonhybrid systems. The result is to reduce problems that could otherwise occur with the sole use of higher-speed logic circuits. The total RF bandwidth for this hybrid system was 340 MHz, with orthogonal code-shift keying (CSK) of the "1", "0" message data bits, using 255-chip Kasami code sequences. In this packet data system [9], each block of 1000 message data bits is preceded by a 7-bit preamble for synchronization, and transmitted at an 84 kb/s rate. The receiver circuit employs two elastic SAW convolvers in parallel, with a time-bandwidth product TB ~480, for extraction of the "1" and "0" message data bits. Frequency-hopping was over 16 frequencies, with a frequency spacing of 20 MHz. The processing gain of the system was 30 db, with a dynamic range of about 55 dB.

Figure 17.10 illustrates the data-code generator portion of the transmitter of Fig. 17.9. This uses an expanded version of the Kasami sequence generator of Fig. 17.8, with a chip rate of 21.4 MHz. Two four-bit shift registers are included for code-shift keying (CSK) of the binary data bits, instead of the

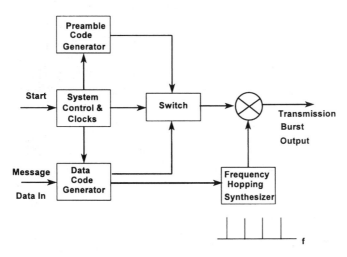

FIG. 17.9. Block diagram of transmitter section for packet-data burst-generation in illustrative hybrid direct-sequence and frequency-hopping spread spectrum communications systems. (After Reference [9].)

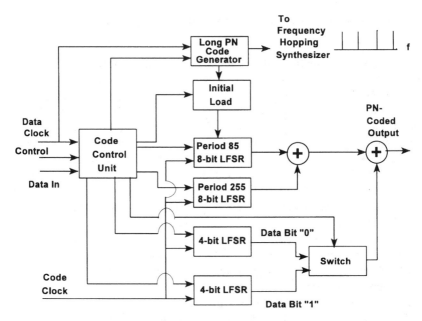

Fɪɢ. 17.10. Block diagram of Kasami-code sequence generator in Fig. 17.9. (After Reference [9].)

single one shown in Fig. 17.8. In this way, orthogonal Kasami-code sequences can be applied to chip-modulation of the "1" and "0" message-data bits. This is achieved by assigning 7 of the possible 15 initial states to message "1" bits for direct sequence encoding, together with 7 state assignments for message '0" bits.

Attainment of the large sequence set is achieved by varying the initial state of the non-maximal 8-bit shift register with period 85. This is controlled by an initial loading block. The initial loading block is controlled in turn by a 20-bit PN-coded maximal-sequence (i.e., *m*-sequence) shift-register[3], run at the data clock speed. The states of the two 4-bit shift registers were also changed after about every 150 message data bits, so as to generate two orthogonal chip-code sets, each containing 1764 different sequences. The 7-bit preamble code, for synchronization, was implemented using the set of 255 Kasami codes, from the sequences of the two 8-bit shift registers in Fig. 17.8.

---

[3] A maximal code is, by definition, the longest code that can be generated by a shift register of given length *n*. For the binary shift register sequences generators considered here, the maximal sequence (*m*-sequence) length is $2^n - 1$ chips.

17.7.3. DUAL-CONVOLVER RECEIVER FOR PACKET DATA SPREAD-SPECTRUM EXAMPLE

Figure 17.11 gives a block diagram sketch of the dual-convolver receiver for the packet-data communication example highlighted here [9]. Each convolver was of the dual-track waveguide-type illustrated in Fig. 17.7. These convolvers were designed for an interaction time $T = 11.9\,\mu s$, and input 3-dB bandwidth $B = 40\,MHz$, for a time-bandwidth product TB = ~480, where the interaction time T corresponds to the duration of one message data bit. Their convolution efficiency was $\eta_c = \sim\!-69\,dBm$, with a self-convolution suppression of 40 dB.

Time-reversed Kasami sequences are applied to the dual monolithic elastic SAW convolvers. For message detection and extraction, the receiver data code generator has to apply both the time reversed "1"-codes and "0"-codes simultaneously, so that a message "1" or "0" bit can be extracted from either convolver, through envelope detectors.

For synchronization with an incoming message burst, the first two preamble codes are continuously circulated in both convolver channels. The delay time for successful correlation detection of the preamble code is used

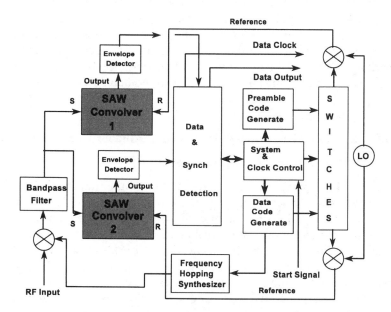

FIG. 17.11.   Receiver for hybrid direct-sequence and frequency hopping spread spectrum system, using parallel monolithic SAW convolvers and time-reversed orthogonal Kasami sequences in reference ports for "1", "0" data-bit extraction. (After Reference [9].)

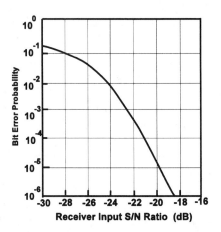

Fɪɢ. 17.12.   Theoretical bit-error probability for the packet-data spread-spectrum receiver example. (After Reference [9].)

to correct the data code timing, with a residual uncertainty of one-chip code time. Only 5 preamble bits are processed in the receiver. The remaining two code times allow for the transient response of the system.

Figure 17.12 shows the theoretical response of the bit-error probability in the presence of additive Gaussian white noise, as averaged over a large number of data bursts. The difference of about 2 dB between theory and experiment was attributed to a number of possible effects, such as amplifier and mixer noise, nonideal switching, and nonideal receiver bandpass filters [9].

17.8.   Example of a Packet-Voice System for Indoor Communications

17.8.1.   Hɪɢʜʟɪɢʜᴛꜱ ᴏꜰ Tʀᴀɴꜱᴍɪꜱꜱɪᴏɴ Sʏꜱᴛᴇᴍ

Figure 17.13 is a block diagram representation of the transmission portion of an illustrative spread-spectrum packet-voice system, suitable for indoor communications [36]. This system employs only direct-sequence spectral spreading. In the transmission portion of this packet-voice system, the incoming analog speech signals are encoded at a 16 kb/s rate, using a delta-modulator [37]. The encoded bit stream is stored in a random access memory (RAM) in blocks of 1024 bits, as indicated in Fig. 17.14(a). A preamble code is generated when a complete block has been read into the RAM, and a start pulse is sent to the spread-spectrum transmitter. A 14-bit header, together with the 1024-bit message data, and 12-bit control sequences—for a total of 1050 bits in Fig. 17.14(b)—are then transferred to

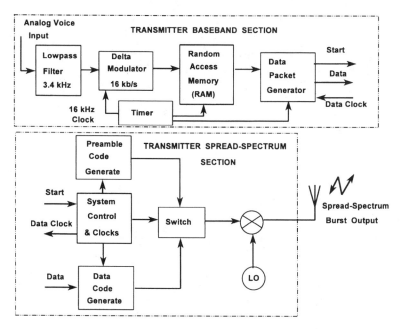

FIG. 17.13. Baseband and transmitter sections of an illustrative packet-voice spread-spectrum system suitable for indoor communications. (After Reference [36].)

the spread-spectrum unit at a higher rate of 84 kb/s, which allows voice sources to be multiplexed in time. An 11-bit preamble sequence for synchronization is added to the 1050-bit packet, as shown in Fig. 17.14(b). Each bit of the preamble and message block is direct-sequence modulated by the 255-chip Kasami sequences as used in the previously discussed packet-data system, again with orthogonal code-shift keying (CSK) chip-modulation of the "1" and "0" data bits. This chip stream is applied to BPSK modulation of the carrier at the chip rate of 21.4 MHz.

17.8.2. RECEIVER DETECTION AND SYNCHRONIZATION

Figure 17.15(b) is a block diagram of the spread-spectrum receiver for this system. This has basically the same architecture as for the packet-voice example, except that there is no frequency-hopping. For dual SAW convolvers the incoming transmission bursts are down-converted to the convolver IF frequency of 155 MHz.

For synchronization acquisition, the first preamble code 1 in Fig. 17.15(c) is periodically repeated in the parallel SAW-convolver reference channels

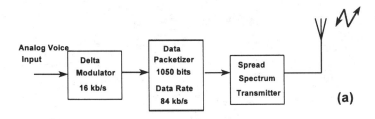

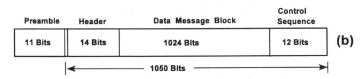

Fig. 17.14.   (a) Delta-modulator and data-packetizer for transmitter of Fig. 17.13. (b) Bit assignments for preamble, header, data message block, and control sequences. (After Reference [36].)

until correlation is achieved. Preamble code bits 2 and 3 are used as dead spaces to allow for system transient response. The acquisition timing information obtained for code bit 1 enables correct timing determination of preamble reference codes in bits 4 through 11, for data decision.

After synchronization is attained, the receiver code generator simultaneously applies the time-reversed Kasami sequences for "1" and "0" codes to the reference ports of both SAW convolvers. Outputs are obtained through envelope detectors as shown.

As the recovered data stream enters the baseband section, the header and control data information is stripped away. The remaining message data is stored in RAM, for subsequent application in blocks to a delta demodulator, with analog voice restoration after passage through a lowpass filter in Fig. 17.15(a).

While the bit-error probability *within* a burst follows the same fall off with receiver S/N ratio in Fig. 17.12, it also useful to obtain the burst loss probability itself, as a function of false-alarm probability. This burst-loss probability P_L may be derived from the relationship [36]

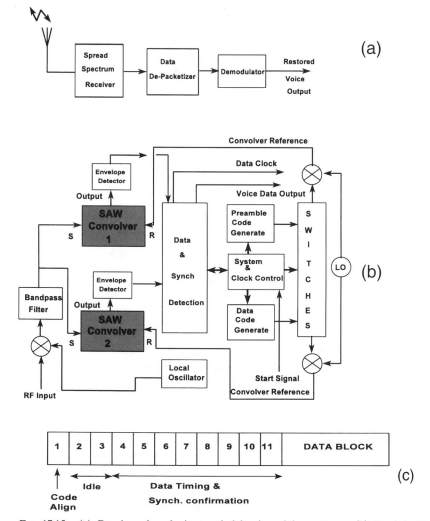

Fig. 17.15.   (a) Receiver depacketizer and delta-demodulator stages. (b) Dual SAW convolver receiver for direct sequence packet voice (b) Dual-convolver receiver with time-reversed Kasami-code reference sequences. (c) Assignment of 11-bit preamble code for data timing and sync acquisition. (After Reference [36].)

$$P_L = 1 - \left(1 - P_B\right) P_D P_S \qquad (17.18)$$

where P_B is the probability that the receiver is already blocked when a data burst arrives, P_D is the probability of detection of the first preamble bit 1, and P_S is the probability of synchronization confirmation at the end of the

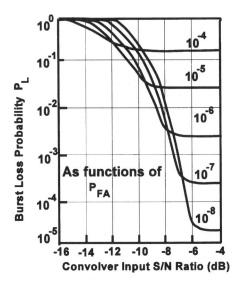

FIG. 17.16. Theoretical plots of burst-error probability, as a function of false-alarm probability for dual-convolver packet voice spread-spectrum receiver of Fig. 17.15. (After Reference [36].)

preamble processing. Figure 17.16 shows theoretical plots of such burst-loss probability, as functions of false alarm probability [36].

Example 17.5 Interaction Time Under SAW Convolver Plate. A spread-spectrum system employs a three-port SAW convolver system for correlation of m[7,1] 127-chip maximal-length PN-code sequences, running at a chip-rate of 14 Mb/s. What is the required time duration of each sequence under the convolving plate for correlation to be achieved? ■

Solution. The duration of each chip in the 14-Mb/s sequence is $1/(14 \times 10^6) = 7.14 \times 10^{-8}$ s. However, all 127 chips must be under the convolver plate for full correlation. Its time-length T_p is therefore required to be $127 \times 7.14 \times 10^{-6} = 9.07 \, \mu$s. The actual physical length L of the convolver plate would need to be $L = v \cdot T_p$, where v = surface wave velocity appropriate to the convolver structure. ■

17.9. An Asynchronous Spread-Spectrum SAW-Convolver System

17.9.1. Some Concepts for Asynchronous Operation

The two SAW convolver-systems considered here operate as synchronous demodulators. Following synchronization acquisition in these systems, a time-reversed reference code bit is applied in exact step with the input data bit. With lengthy PN sequences, there will be a loss of correlation gain if the two sequences are out of step by one chip. The attendant problems of code search and acquisition can be serious if synchronization is not exact and the time of the signal arrival is unknown.

Asynchronous matched filtering of long waveforms can be attained with SAW convolver technology in power-limited spread spectrum applications. In one scheme [23], [24] the signal and reference waveforms are each required to have a duration of $2T$ for application to a SAW convolver with interaction time T. No gaps are allowed in application of the reference coding samples. This requires the use of two convolvers to provide a continuous interaction. Detection of a signal of unknown arrival time is immediate once the signal passes completely though the convolution interaction area. Such techniques have been applied to wideband packet radio networks — intended to provide data communications to users over a broad geographic area [24].

17.9.2. A Layered Surface-Wave Convolver for Spread Spectrum

As will be highlighted in what follows, another asynchronous SAW convolver system for spread-spectrum communications operates with a trade-off between circuit simplicity and PN-coding diversity. The receiver system employs a single highly efficient SAW convolver as sketched in Fig. 17.17 [38]. This involves a layered structure of the $ZnO/SiO_2/Si$ type,

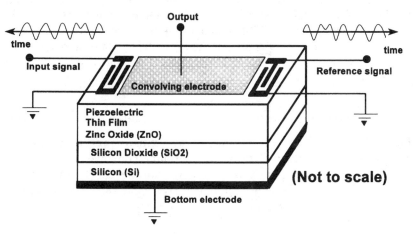

Fig. 17.17. A 215-MHz layered $ZnO/SiO_2/Si$ three-port convolver employing Sezawa surface waves, with convolution efficiency $\eta_c = -42\,dBm$, for use in an asynchronous spread-spectrum receiver. (After Reference [38].)

incorporating piezoelectric zinc oxide (ZnO), insulating silicon dioxide (SiO$_2$) and a silicon semiconductor (Si). The SiO$_2$ insulator is used here to isolate the silicon and allow field penetration in it. The piezoelectric ZnO layer here may be considered to be crudely analogous to a microwave waveguide structure, in that it can support propagation of a fundamental Rayleigh wave and higher-order Rayleigh waves, depending on its thickness and the structure on which it is deposited. These higher-order Rayleigh waves — known as Sezawa waves — have higher velocities than the fundamental Rayleigh wave. In some layered structures, such as on diamond[4], the first-higher Sezawa wave can have a velocity which is about twice that for the corresponding Rayleigh wave [39]. (The rationale for using higher-velocity materials is, of course, to allow the fabrication of higher-frequency surface-wave devices, within the constraints of the available microlithographic facility.)

In operation, the normal component of E_N of the electric field due to the surface wave in the piezoelectric penetrates into the epitaxial silicon layer, and gives rise to a depletion region. The voltage generated in this region is proportional to E_N^2. When signals that are frequency f_o are applied to the input and reference ports of the convolver, two electric fields E_{Ni} and E_{Nr} are produced by the contradirective surface waves. The nonlinear mixing that ensues produces a second harmonic voltage and dc terms across the depletion layer that are proportional to $(E_{Ni}^2 + E_{Nr}^2)$. This effect, known as the "transverse acoustoelectric effect," serves to increase convolver efficiency [19]. In this instance, the convolution efficiency of the layered convolver in Fig. 17.17 was $\eta_c = -42\,\text{dBm}$, which is a factor of about 30 dBm better than for the "conventional" monolithic convolver. Other parameters for the 315-MHz convolver of Fig. 17.17 included a time-bandwidth product $TB = 207$, an interaction time of 9 μs under the convolving plate, and bandwidth $B = 23\,\text{MHz}$, as well as a dynamic range of 50 dB and a self-convolution suppression of 35 dB.

17.9.3. Operation of Illustrative Asynchronous Convolver System

Figure 17.18 illustrates the pertinent waveforms for the transmitter and receiver sections of the illustrative asynchronous convolver. As shown, this particular system can operate in a frequency-shift-keying (FSK) mode, with direct-sequence (DS) spectral spreading (i.e., DS/FSK) or in a DS-mode

---

[4] Diamond has the highest value of Young's Modulus among all substrate materials, so that the highest acoustic velocity can be realized with this material (e.g., up to 10,000 m/s) in layered structures of ZnO/bulk diamond (001) and ZnO/thin-film diamond/Si [39].

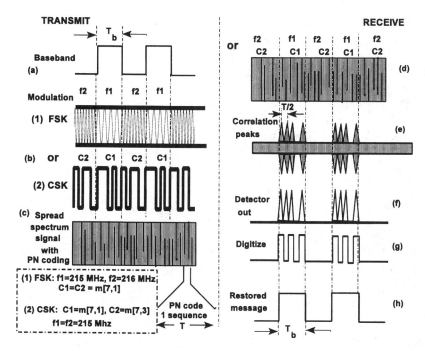

FIG. 17.18. Outline of waveforms for illustrative asynchronous spread-spectrum system, using convolver of Fig. 17.17 in receiver. System can employ DS/FSK or DS/CSK modulation. (After Reference [38].)

with code-shift-keying (CSK) (i.e., DS/CSK). The reason that asynchronous operation is possible is that only two PN 127-chip maximal sequence generators are used in the transmitter modulator, while only one of these is used as the reference code for the convolver in the receiver modem.

As indicated in Fig. 17.18(b), the two 127-chip m-sequences are $C_1[7,1]$ and $C_2[7,3]$, with low cross-correlation magnitudes, while the two frequencies are $f_1 = 215\,\text{MHz}$ and $f_2 = 216\,\text{MHz}$. For the DS/FSK mode the primary modulations are $f_1 = 215\,\text{MHz}$, $f_2 = 216\,\text{MHz}$, and $C_1 = C_2 = \text{m}[7,1]$. For the alternate DS/CSK mode, the modulations are for $f_1 = f_2 = 215\,\text{MHz}$, $C_1 = \text{m}[7,1]$, and $C_2 = \text{m}[7,3]$. Although the convolution time T under the convolver plate is $T = 9\,\mu\text{s}$, the reader may note in Fig. 17.18(e) that the correlation peaks for each of the repetitive chip sequences are separated by only $4.5\,\mu\text{s}$. The reason for this is that the signal and reference waveforms are moving in opposite directions under the convolver plate, so that the correlation is a maximum in half of the traverse-time for one wave.

The occurrence of convolution peaks corresponds to a binary "1" state and the absence of convolution peaks corresponds to a binary "0." Only *one* m-sequence code is required. Two codes are employed so that the automatic gain control (AGC) in the modem v/ill not suffer from data-dependent fluctuations [40].

The PN codes C_1 and C_2 are continuously refreshed during each bit-length in the 1200 b/s data signal. The signal applied to the reference input of the convolver is limited to frequency $f_1 = 215$ MHz and the time-inverted reference for PN-code $C_1 = m[7,1]$, which is continually refreshed. This allows asynchronous correlation to take place.

17.9.4. SPREAD-SPECTRUM IMPLEMENTATION ASYNCHRONOUS SAW CONVOLVER

The asynchronous spread-spectrum system illustrated in the preceding discussion has been evaluated for three applications, as follows:

1. For operation in the USA license-free 900-MHz spread-spectrum band (902 to 928 MHz, 1 W), using the DS/CSK mode. With this system, BER $< 10^{-6}$ were obtained at a distance of 16 km, for an information rate of 32 kb/s.

2. For duplex operation in Japan in the very low-power license-free spread-spectrum band, where the allowed transmitter field-strength is less than 500 μV/m at a distance of 3 m, in the frequency range below 322 MHz—corresponding to a transmitted power of less than 0.05 μW/ 120-kHz. In this operational mode, error-free operation was maintained up to a distance of 100 m.

3. For full-duplex operation at 2.45 GHz, in the 2–GHz USA spread-spectrum band (2.40000 to 2.4835 GHz, 1 W). A block diagram of this full-duplexer is shown in Fig. 17.19, which used DS/CSK modulation. The center frequency of the transmitter modem was 2.45 GHz; that of the receive modem was 2.450941 GHz. The RF bandwidth of the transmitted signal was 28 MHz, for PN-code chip-rates of 14 Mb/s. PN-codes used in the transmit and receive sections were m[7,1], m[7,3], m[7,1,2,3] and m[7,1,4,6]. In this system, the RF carrier was down-converted to a 216-MHz IF at the center frequency of the convolver. Self-jamming at the common antenna input/output was reduced by the use of an isolator, in addition to that provided by the processing gain in the convolver. Evaluation of this system was made under conditions of bit-error rate as a function of desired-to-undesired (D/U) power-ratio at the antenna port. No bit-error detection was measured for D/U ratios as low as −78.3 dB, with operation up to 9600 b/s [40].

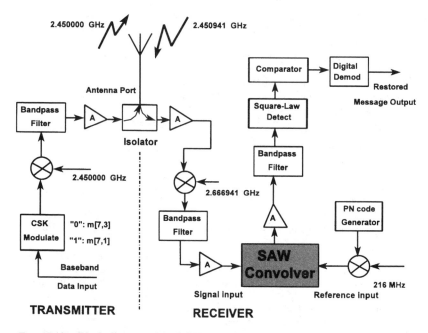

FIG. 17.19. Block diagram of 2.45-GHz asynchronous spread-spectrum system, for full-duplex 1-W operation, using convolver structure of Fig. 17.17. (After Reference [38].)

17.10. Summary

This chapter first examined multipath problems facing mobile/wireless communications system performance in indoor environments. This was followed by an examination of the principles of operation of a monolithic surface-wave convolver. Two systems developed for packet-data and packet voice spread-spectrum indoor communications were then highlighted, using synchronous correlation circuitry. The chapter concluded with a description of an asynchronous convolver circuit for spread-spectrum communications employing a layered high-efficiency ZnO/SiO$_2$/Si convolver for differing applications, including that for full-duplexer operation in the 2.4-GHz USA spread-spectrum band.

17.11. REFERENCES

1. D. M. J. Devasirvatham, "Time delay spread and signal level measurement of 850 MHz radio waves in the building environment," *IEEE Trans. Antennas Prop.*, vol. AP-34, pp. 1300–1305, Nov. 1986.

2. S. Gopani, J. H. Thompson and R. Dean, "GHz SAW delay line for direct sequence spread spectrum, CDMA in-door communications system," *Proc. 1993 IEEE Ultrasonics Symp.*, vol. 1, pp. 89–93, 1993.

3. W. C. Y. Lee, *Mobile Cellular Telecommunications Systems*, McGraw-Hill, New York, p. 22, 1989.

4. W. Pietsch and F. Seifert, "Measurement of radio channels using an elastic convolver and spread spectrum modulation,: Part 1—Implementation," *IEEE Trans, Instrumentation Meas.*, vol. 43, pp. 689–694, Oct. 1994.

5. W. Pietsch, "Measurement of radio channels using an elastic convolver and spread spectrum modulation,: Part 2—Results," *IEEE Trans, Instrumentation Meas*, vol. 43, pp. 695–699, Oct. 1994.

6. R. C. Dixon, *Spread Spectrum Systems*, Second Edition, John Wiley and Sons, New York, 1984.

7. K. Pahlavan and A. H. Levesque, "Wireless data communications," *Proc. IEEE,* vol. 82, pp. 1398–1430, Sept. 1994.

8. A. Fauter, L. Reindl, R. Weigel, P. Russer and F. Seifert, "Miniaturized SAW convolver for indoor mobile communication," *Proc. 1993 IEEE Ultrasonics Symp.*, vol. 1, pp. 73–77, 1993

9. M. Kowatsch, J. Lafferl and F. J. Seifert, "Burst-communication modem based on SAW elastic convolvers," *Proc. 1982 IEEE Ultrasonics Symp.*, vol. 1, pp. 262–267, 1982

10. J. H. Cafarella, "Application of SAW convolvers to spread spectrum communications," *Proc. 1984 IEEE Ultrasonics Symp.*, vol. 1, pp. 121–126, 1984.

11. H. Gautier and C. Maerfeld, "Wideband elastic convolvers,"*Proc. 1980 IEEE Ultrasonics Symp*, Boston, Massachusetts, 5–7 November 1980, vol. 1, pp. 30–36.

12. See the Special Issue on SAW Convolvers and Correlators, *IEEE Trans. on Sonics and Ultrasonics*, vol. SU-32, September 1985.

13. K. A. Ingebrigtsen, "The Schottky diode acoustoelectric memory and correlator—a novel programmable signal processor,"*Proc. IEEE,* vol. 64, pp. 764–771, May 1976.

14. R. S. Wagers, "Principles of strip-coupled SAW memory correlators," *IEEE Transactions on Sonics and Ultrasonics,* vol. SU-32, pp. 716–727, September 1985.

15. G. S. Kino, S. Ludvik, H. J. Shaw, W. R. Shreve, J. M. White and D. K. Winslow, "Signal processing by parametric interactions in delay line devices,"*IEEE Trans. Microwave Theory and Techniques,* vol. MTT-21, pp. 244–255, April 1973.

16. M. E. Motamedi, M. K. Kilcoyne and R. K. Asatourian, "Large-scale monolithic SAW convolver/correlator on silicon,"*IEEE Trans. on Sonics and Ultrasonics,* vol. SU-32, pp. 663–669, September 1985.

17. M. F. Lewis, C. L. West, J. M. Deacon and R. F. Humphryes, "Recent developments in SAW devices," *IEE Proceedings (Great Britain)*, vol. 131, Part A, No. 4, pp. 186–213, June 1984.

18. P. Defranould and C. Maerfeld, "A SAW planar piezoelectric convolver," *Proceedings of the IEEE,* vol. 64, pp. 748–753, May 1976.

19. J. M. Smith, E. Stern, A. Bers and J. Cafarella, "Surface acoustoelectric convolvers," *Proc. 1973 IEEE Ultrasonics Symp.*, pp. 142–144, 1973.

20. S. A. Reible, J. H. Cafarella, R. W. Ralston and E. Stern, "Convolvers for DPSK demodulation of spread spectrum signals,"*Proc. 1976 IEEE Ultrasonics Symp.*, pp. 451–455, 1976.

21. S. A. Reible, "Acoustoelectric convolver technology for spread-spectrum communications," *IEEE Trans. Sonics and Ultrasonics,* vol. SU-28, pp. 185–195, May 1981.

22. J. H. Goll and D. C. Malocha, "An application of SAW convolvers to high bandwidth spread spectrum communications," *IEEE Trans. Sonics and Ultrasonics,* vol. SU-28, pp. 195–205, May 1981.

23. I. Yao and J. H. Cafarella, "Applications of SAW convolvers to spread-spectrum communication and wideband radar," *IEEE Trans. Sonics and Ultrasonics,* vol. SU-32, pp. 760–770, September 1985.

24. J. H. Fischer, J. Cafarella, D. R. Arsenault, G. T. Flynn and C. B. Bouman, "Wide-band packet radio technology," *Proceedings of the IEEE,* vol. 75, pp. 100–115, January 1987.

25. A. J. Slobodnik, Jr., "Materials and Their Influence on Performance," in A. A. Oliner (ed.), *Acoustic Surface Waves,* Springer-Verlag, New York, Chapter 6, 1978.

26. C. F. Quate and R. B. Thompson, "Convolution and correlation in real time with nonlinear acoustics," *Applied Physics Letters,* vol. 16, pp. 494–496, 1970.

27. E. Dieulesaint and D. Royer, *Elastic Waves in Solids,* John Wiley and Sons, New York, 1980.

28. G. W. Farnell and E. A. Adler, *An Overview of Acoustic Surface-Wave Technology,* Final Report to Communications Research Centre, Ottawa, Canada, 12 August 1974, on DSS Contract 36001-3-4406.

29. G. S. Kino, "Acoustoelectric interactions in acoustic-surface-wave devices," *Proceedings of the IEEE,* vol. 64, pp. 724–748, May 1976.

30. A. Chatterjee, P. K. Das and L. B. Milstein, "The use of SAW convolvers in spread-spectrum and other signal processing applications," *IEEE Trans. on Sonics and Ultrasonics,* vol. SU-32, pp. 745–759, September 1985.

31. See for example, R. E. Ziemer, W. H. Tranter and D. R. Fannin, *Signals and Systems,* Macmillan Publishing Co, New York, Appendix D, 1983.

32. C. Maerfeld and G. W. Farnell, "Nonsymmetrical multistrip coupler as a surface wave beam compressor for large bandwidths,"*Electronics Letters*, vol. 10, p 209, 1974.

33. D. P. Morgan, *Surface-Wave Devices for Signal Processing,* Elsevier, New York 1985.

34. I. Yao, "High performance elastic convolver with extended time-bandwidth product," *Proc. 1981 IEEE Ultrasonics Symp.,* vol. 1, pp. 181–185, 1981.

35. H. P. Grassl and H. Engan, "Small-aperture focusing chirp transducers vs. diffraction-compensated beam compressors in elastic SAW convolvers," *IEEE Trans. on Sonics and Ultrasonics,* vol. SU-32, pp. 675–684, September 1985.

36. M. Kowatsch, "Design of a convolver-based packet voice spread-spectrum system," *Proc. 1984 IEEE Ultrasonics Symp.,* vol. 1, pp. 127–131, 1984.

37. See, for example, L. W. Couch II, *Digital and Analog Communications Systems,* Third Edition, Macmillan Publishing Co., New York, ch. 3, 1990.

38. K. Tsubouchi, "An asynchronous spread-spectrum wireless modem using SAW convolver," *Proc. International Symp. on Surface Acoustic Wave Devices for Mobile Communications,* Sendai, Japan, pp. 215–222, 1992.

39. H. Nakahata, K. Higaki, A. Hachigo and S. Shikata, "High-frequency surface acoustic wave filter using ZnO/diamond/Si structure," *Japn. J. Applied. Phys.,* vol. 33, pp. 324–328, Jan. 1994.

40. K. Tsubouchi, H. Nakase, A. Namba and K. Masu, "Full duplex transmission operation of a 2.45-GHz asynchronous spread spectrum modem using a SAW convolver," *IEEE Trans. Ultrason., Ferroelect., and Freq. Control,* vol. 40, pp. 478–482, Sept. 1993.

—18—

Surface-Wave Oscillators and Frequency Synthesizers

18.1. Introduction

18.1.1. SURFACE-WAVE OSCILLATORS IN MOBILE AND WIRELESS COMMUNICATIONS

Advantageous features of surface-wave oscillators for fundamental frequency operation up into the gigahertz range include their high spectral purity, small power consumption, size, and ruggedness. They can be configured to perform significantly different signal-processing functions in transmitter and/or receiver stages of mobile and wireless circuitry networks communications with operational capabilities as pictured in Fig. 18.1. As delineated in Fig. 18.1, they can operate as fixed-frequency, quenched, voltage-controlled, frequency-hopping or injection-locked sources.

While Rayleigh-wave and leaky-SAW VCOs have been designed for wideband RF first local oscillator stages of mobile transceivers [1], [2] (as for the 25-MHz RF bandwidth in the AMPS 800-MHz system), their utilization in RF front-end stages still faces intensive price competition from low-cost inductance-capacitance (LC) "conventional" VCOs employing Colpitts/Clapp oscillators with silicon bipolar transistors and hyper-abrupt varactors for monotonic linear tuning [3]. The trend in analog/digital cellular and cordless communications circuitry is towards the use of VCOs with 3-V supply voltages, with 2-V tunability, for compatibility with "AA" battery (or equivalent) use. In double-heterodyne transceivers, however, surface-wave VCOs do find ready application in second local-oscillator stages to compensate for temperature drifts, and settability tolerances.

With their potential for superior far-out phase-noise performance and higher power-level capability over Rayleigh-wave oscillators (e.g., ~ 30 dBm versus ~ 12 dBm), surface transverse wave (STW) VCOs may be the preferred type in some second stages of dual-heterodyne transceivers for tight frequency control over wide temperature excursions. Coupled-mode STW resonator-filters are particularly useful for VCO use in the 1- to 2-GHz

533

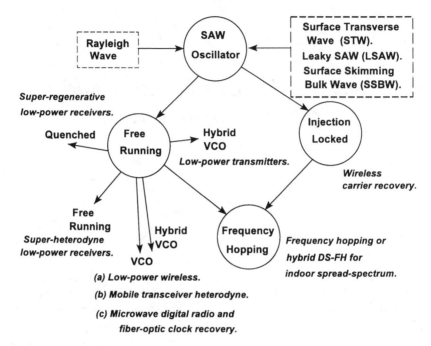

Fɪɢ. 18.1.   Classifications of surface-wave oscillators, and their mobile/wireless applications.

frequency region [4]. Injection-locked STW oscillators have been employed for carrier recovery in BPSK-modulated wireless signals. STW fixed-frequency oscillators find application in wireless security-protection circuitry.

Surface-wave VCOs are also employed in clock-recovery circuits in high-capacity (~ 600 Mb/s) microwave digital-radio links. Additionally, they are used for carrier recovery in fiber- optics communications, where data rates for current fiber networks can range from 100 to 1000 Mbits/s. Figure 18.2 illustrates an example of a compact 155.52-MHz Rayleigh-wave delay-line VCO for SONET/SDH/ATM applications[1]. This particular example of Fig. 18.2 is a highly integrated one where the SAW delay line is integrated with one single ASIC[2], which includes an on-board phase shifter. This delay line has a 3-dB Q of ~ 400, while the pull for the VCO is about ± 400 ppm.

---

[1] SONET/SDH/ATM stands for Synchronous Optical Network/Synchronous Digital Hierarchy/Asynchronous Transfer Mode, as defined in Reference [5]}.

[2] ASIC stands for Application-Specific Integrated Circuit; designed to perform a specific function.

Fig. 18.2.   A 155.52-MHz SAW VCO for SONET/ SDH/ATM optical communications. VCO employs a SAW delay line with 3-dB Q of ~ 400. VCO is about ± 400 ppm. Package incorporates a single ASIC, which includes on-chip phase shifter. (Courtesy of Vectron Technologies, Inc., Hudson, New Hampshire.)

The power output of this 155.52-MHz oscillator is ~ 0 dBm, while the SSB phase-noise at 10-kHz Fourier-frequency offset is about − 108 dBc/Hz.

Figure 18.3 shows a 130-MHz second local-oscillator module for a dual-mode IS-54 cellular radio system [6]. Here, a two-pole double-mode waveguide-coupled SAW resonator-filter is employed to provide a wide tuning range while maintaining low insertion loss. This particular oscillator module uses discrete components on a single-layer alumina substrate, and is phase-locked to a stable crystal reference. Electronic tuning compensates manufacturing tolerances and temperature variations.

To date, another major market exists for very low-power surface-wave oscillators. These are for wireless data transmission over short-range low-power unlicensed UHF links, primarily at frequencies within 200- to 900-MHz bands. These very low-power wireless systems are for a variety of consumer and commercial data-transfer applications, including those for keyless automobile entry, remote utility meter reading, wireless bar-code readers and computer-peripheral links. Transmission powers are typically less than about 1 mW (0 dBm) for coverage of up to ~ 100 m. These networks can employ a variety of Rayleigh-wave devices in feedback stages, including delay lines, resonators, and narrowband resonator-filters.

FIG. 18.3. A 130-MHz second local-oscillator module for an IS-54 dual-mode cellular radio system. Feedback element is a two-pole double-mode waveguide-coupled SAW resonator-filter. (Courtesy of NORTEL, Northern Telecom, Canada.)

18.1.2. SCOPE OF THIS CHAPTER

Before discussing SAW oscillators and frequency synthesizers for mobile and wireless communications, the chapter first reviews concepts of noise and noise measurement (and terminology) in both the frequency and time domains [7], [8]. Following this review, the chapter proceeds to illustrate the operational basics of both fixed-frequency and tunable surface-wave oscillators. Next, the principles of operation of VCOs in phase-locked loops are reviewed as related to SINAD performance specifications for mobile-radio receivers. This is followed by highlights of transmitter and receiver oscillators for very low-power fixed-frequency wireless data links, using Rayleigh-wave feedback structures.

Other surface-wave oscillators reviewed in this chapter include comb-frequency oscillators for spread-spectrum frequency-hopping circuitry, injection-locked oscillators for carrier recovery, and STW oscillators for higher-power wireless applications, as well as those calling for superior far-out phase-noise performance in phase-locked loops. Frequency synthesizer techniques are also examined. These employ Rayleigh-wave linear FM chirp filters for frequency-hopping in spread-spectrum communications systems, with capabilities of up to several thousand frequency hops per second.

18.2. Phase-Noise Spectrum of an Oscillator

18.2.1. NOISE TERMS

One of the most widely quoted models for depicting the phase-noise spectrum of an oscillator is that due to Leeson [9], [10]. His model was a heuristic one derived for the case of a single inductance-capacitance-

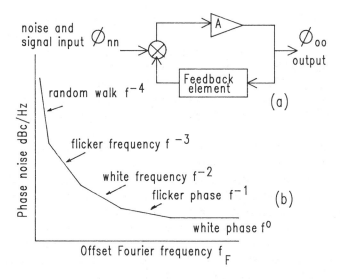

FIG. 18.4.   (a) Basic oscillator with input/output noise. (b) Noise mechanisms observed in oscillators.

TABLE 18.1

PHASE-NOISE SLOPE FOR VARIOUS OSCILLATOR PERTURBATIONS

Noise Type	Frequency Dependence	Phase Slope/Decade
White phase	f^0	0
Flicker phase	f^{-1}	−10
White frequency	f^{-2}	−20
Flicker frequency	f^{-3}	−30
Random walk	f^{-4}	−40

resistance (LCR) lumped-parameter resonator circuit in the oscillator feedback loop. As shown in his model, and as illustrated in Fig. 18.4, the overall phase-noise spectrum about the oscillator nominal carrier frequency is a composite of contributions from white phase (f^0), flicker phase (f^{-1}), white frequency (f^{-2}) and flicker frequency (f^{-3}) perturbations. The f^0 white phase-noise contribution is also known as Johnson or additive noise associated with accessory circuits. Another phase-noise term in Fig. 18.4, not included in Leeson's model [9], relates to random walk f^{-4} processes. The origins of flicker noise and random walk have never been explained satisfactorily. Flicker noise is thought to be associated with time-dependent processes causing a solid-state molecular system to reorder itself as a logarithmic function of time. This is of particular importance in SAW resonators, for example, where flicker-noise processes may be associated with time-dependent bonding strains and deformations on the piezoelectric substrate surface, following deposition of the metal-film IDTs. This can lead to frequency instability and drift in uncompensated SAW oscillators [11]. Some SAW resonator designs employ copper-aluminum alloys rather than pure aluminum for the metallization in an attempt to obtain less stressful surface bonding.

In practice, phase-noise measurements of oscillator stability usually result in less distinct noise characteristics than those shown in Fig. 18.4. "Break points" between the different regimes are usually blurred. All of these noise terms may not be dominant or evident in a particular device over the measurement period or phase-noise range.

The relations between a) white phase and white frequency noise, as well as b) flicker phase and flicker frequency noise may be demonstrated by considering a single resonator in the oscillator feedback loop in Fig. 18.4. Recall that the quality factor Q of a tuned circuit is

$$Q = \frac{\omega_o}{2B'} \qquad (18.1)$$

where $\omega_o = 2\pi f_o$ is the centre (angular) frequency, and $B' =$ the *half-bandwidth* (in radians/s) of the resonator between the 3-dB points. The Q may be expressed alternatively in terms of delay time τ_g by

$$Q = \frac{\omega_o \tau_g}{2} \qquad (18.2)$$

where the notation τ_g is used for group delay in this chapter to avoid confusion with other τ parameters. For a SAW resonator, τ_g is obtained from the phase slope $\tau_g = -d\phi/d\omega$ at resonance frequency ω_o. For a SAW delay line, τ_g is the delay between phase centres of input and output IDTs. Its Q is often termed an effective Q_e as this structure involves SAW energy transfer rather than energy storage.

The effect of input noise perturbations on oscillator stability will depend on whether or not their frequency components fall within the half-bandwidth $B' = \omega_o/2Q$ in Eq. (18.1). To understand this, recall that the closed-loop gain $G(\omega)$ of a feedback circuit is given by

$$G(\omega) = \frac{A}{[1 - AH(\omega)]} \qquad (18.3)$$

where $A =$ open-loop amplifier gain and $H(\omega) =$ (angular) frequency response of the feedback element, taken here as a SAW delay line or SAW resonator. Since phase-noise measurements relate to *power* spectra, the output power spectral density $\phi_{00}(\omega)$ of the oscillator (i.e., output phase noise) involves terms in $G^2(\omega) = G(s)G(-s)|_{s=j\omega}$ so that

$$\phi_{00}(s)\big|_{s=j\omega} = G(s)G(-s)\phi_{nn}\big|_{s=j\omega} \qquad (18.4)$$

where $\phi_{nn}(\omega)$ is the input noise-source perturbation [12], [13].

To obtain the phase-noise response at frequencies near the oscillator frequency f_o, approximations are applied to the gain term in Eq. (18.4). In this way the closed-loop power spectral response for a SAW oscillator, for example, is simplified to

$$G^2(\omega) \approx \frac{A^2}{\left[(\omega - \omega_0)\tau_g\right]^2} \approx \frac{A^2}{\left(\omega_F \tau_g\right)^2} \qquad (18.5)$$

where $\omega_F = 2\pi f_F$ is the (angular) frequency offset from the nominal carrier frequency, usually referred to as "Fourier frequencies" $\omega_F = 2\pi f_F$.

Input noise perturbations $\phi_{nn}(\omega)$ due to white phase-noise(f^0) and flicker phase-noise(f^{-1}) are given by

$$\phi_{nn}(\omega) = FkT + \frac{\alpha}{\omega_F} \qquad (18.6)$$

In the first (white phase) term F = amplifier noise figure, k = Boltzmann constant $(1.38 \times 10^{-23}\,\text{J/}^\circ\text{K})$ and T = temperature in $^\circ$K. In the second (flicker-phase) term, α = experimental constant and ω_F = angular Fourier frequency offset. Inserting Eqs. (18.2), (18.5) and (18.6) into Eq. (18.4) yields

$$\phi_{00}(\omega) \approx \frac{A^2 \omega_o^2}{4Q^2 \omega_F^2}\left(FkT + \frac{\alpha}{\omega_F}\right) + A^2\left(FkT + \frac{\alpha}{\omega_F}\right) \qquad (18.7)$$

The four terms in Eq. (18.7) have the same form as obtained by Leeson [9], except for the power-scaling factor A^2 that he set to $A^2 = 1$. Caution should be used in applying nonunity values of A^2 to flicker frequency (f^{-3}) calculations for SAW oscillators, because in many instances these are determined by the SAW device itself, independent of amplifier characteristics.

The first two terms on the right-hand side of Eq. (18.7) respectively, relate to output white frequency and flicker frequency noise within the closed-loop response. The angular Fourier frequency offset ω_{Fc} at the transition from open- to closed-loop response is given by $(\omega - \omega_o)\tau_g = \omega_{Fc}\tau_g \approx 1$. The white phase noise($f^0$) transforms into white frequency noise (f^{-2}) inside the feedback bandwidth, while the flicker phase noise (f^{-1}) transforms into flicker frequency (f^{-3}) contributions within this bandwidth in accord with the simplified phase-noise representation of Fig. 18.4. The remaining two terms in Eq. (18.7) are magnified versions of the input white phase and flicker phase perturbations outside the closed-loop bandwidth.

18.2.2. UNITS USED IN PHASE-NOISE MEASUREMENTS

Phase-noise measurements are usually normalized with respect to oscillator average output power P_o. From Eq. (18.7) this yields (in decibels)

$$\frac{\phi_{00}(\omega)}{P_o} \approx 10 \log_{10}\left(\frac{A^2\omega_o^2}{4Q^2\omega_F^2 P_o}\left(FkT + \frac{\alpha}{\omega_F}\right) + \frac{A^2}{P_o}\left(FkT + \frac{\alpha}{\omega_F}\right)\right)$$

(18.8)

Equation (18.8) is often expressed in terms of $\phi_{00}(f)$ instead of $\phi_{00}(\omega)$. Additionally, $\phi_{00}(f)$ is normally standardized to a 1-Hz bandwidth measurement. The units of Eq. (18.8) are in "dBc/Hz": meaning "decibels below the carrier in a 1-Hz measurement bandwidth." The phase noise normally refers to the "single sideband response" and is given the name "Script L" or $\mathcal{L}(f)$. An appropriate correction must be made in calculating $\mathcal{L}$ if a measurement bandwidth of other than 1 Hz is used.

Example 18.1 Oscillator Phase Noise in a 1-Hz Bandwidth. The phase noise of an oscillator is examined on a spectrum analyzer. At 10-kHz Fourier frequency offset the measured value is −94 dBm using a 100-Hz measurement bandwidth. What is the corresponding value referred to a 1-Hz bandwidth? ■
Solution. For random noise on a logarithmic display, the detected signal is smaller than the true rms value [14]. Adding 2.5 dB to correct for this gives the true measured noise level as $-94 + 2.5 = -91$ dBm. Next, apply a bandwidth correction factor of $10\log_{10}(100\,\text{Hz}/1\,\text{Hz}) = 20$ dB. The phase noise referred to a 1-Hz bandwidth is $-91 - 20 = -110$ dBm at 10-kHz offset. ■

Phase noise can be measured routinely with a modulation analyzer or a spectrum analyzer. Alternatively, a frequency stability analyzer can be employed to measure stability in the time domain or in the frequency domain [15]. An example of such a phase-noise measurement with a frequency stability analyzer is shown in Fig. 18.5 over the Fourier frequency range from 10 to 40 MHz, as measured on a 600-MHz SAW VCO on quartz. Here, the noise level is ~ −110 dBc/Hz at $f_F = 1$ kHz. Note some power-line beat interference at 60-Hz offset and above. Commercial and/or military SAW VCO designs are normally an optimization of key design parameters, including cost, size, output power-supply voltage level, input current drain, phase noise, spurious responses, set accuracy, tuning range, aging and

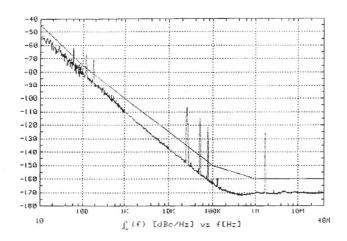

600 MHz Standard SAW VCO
Phase Noise Performance

FIG. 18.5. Lower curve is measured single-sideband (SSB) phase noise of an illustrative standard hybrid SAW VCO. Phase noise is ~–110 dBc/Hz at 1 kHz offset, with noise floor at about –170 dBc/Hz. Upper solid-line curve is the design specification. Horizontal scale: 10 Hz to 40 MHz. Vertical scale: –180 to –10 dBc/Hz. (Courtesy of Sawtek Inc., Orlando, Florida.)

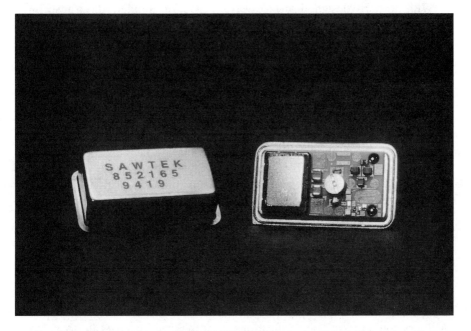

FIG. 18.6. Example of packaging for SAW fixed-frequency oscillator with low g-sensitivity. (Courtesy of Sawtek Inc., Orlando, Florida.)

541

g-sensitivity. VCOs are typically fabricated in surface-mount flatpacks, DIP packages, or ASICs. Fig. 18.6 illustrates the packaging for a fixed-frequency SAW oscillator with low g-sensitivity.

18.3.   Surface-Wave Oscillator Performance Expectations

In Rayleigh-wave oscillators on ST-quartz, phase noise below the half-loop-bandwidth is due primarily to flicker frequency, with 30 dB/decade roll-off as indicated in the example of Fig. 18.5. There is no hard and fast rule as to what the oscillator phase noise should be (it depends on the application). Because the phase noise rolls off as $1/Q^2$ in Eq. (18.8), it would be desirable to use as high a Q as possible in free-running unmodulated resonators for local oscillator applications. In this respect, SAW oscillators using high-Q resonators, and operating in the 300- to 600-MHz range, typically have loaded Qs in the range from 5000 to 15,000. Since these are difficult to tune over a frequency range greater than ≈ 100 ppm, however, SAW delay lines with Q proportional to input/output IDT separation (see Example 18.2) are routinely employed for tunable SAW oscillator applications, with typical tuning ranges of about 500 ppm for single-delay lines structures, and up to about 1400 ppm for dual differential delay-line VCOs [16].

With properly designed surface-wave VCOs for mobile-radio transceivers, phase-noise at 10-kHz Fourier-frequency offset can be low enough (-120 to -150 dBc/Hz) to avoid adjacent-channel interference. In first local oscillator stages, VCOs incorporating delay-line feedback elements do not have the tuning capability for mobile applications, such as the 25-MHz tuning requirement for 800-MHz AMPS cellular systems. This can be achieved with other surface-wave feedback elements, however, with *wideband* resonator designs on leaky-SAW or Rayleigh-wave substrates [1], [2]. Figure 18.7 illustrates the IDT for a wideband one-port resonator with slanted-finger geometry on 128° Y-X LiNbO$_3$. When incorporated into a Colpitts VCO circuit, this yielded a tunability of 5% [2].

Rayleigh-wave oscillators with high-Q resonators are restricted in their power level capability (nominally about 15 dBm maximum). At higher power levels the piezoelectric surface-wave vibrations can be violent enough for the IDTs to "self-destruct" due to migration of the metallization layer! The attainment of higher oscillator powers, up to about 30 dBm, would require the use of leaky-SAW or STW oscillator designs.

Some SAW oscillator applications for superheterodyne receivers may only require frequency stability and settability $\approx \pm 500$ ppm. More stringent usage requires a smaller frequency window on the order of $\approx \pm 10$ ppm. Military communication specifications can aim for noise floors approaching -180 dBc/Hz, with added requirements for vibrational stability in the order

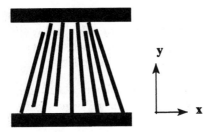

FIG. 18.7.   Example of Rayleigh-wave one-port wideband slanted finger resonator on 128° Y-X LiNbO$_3$ resonator for mobile radio local oscillator with 5% tunability. (After Reference [2].)

of 1×10^{-9} per g, at least as good as bulk-wave AT-cut devices. This is well within the capability of a precision SAW oscillator. For example, a stabilized 310-MHz SAW oscillator was reported in 1982 [17] with a noise floor of $-176\,\mathrm{dBc/Hz}$ at 20-kHz offset. This employed a two-port SAW resonator on quartz with a loaded $Q_L = 12{,}000$, encapsulated in dry nitrogen. The acoustic aperture was 148 λ_o, with an effective cavity length of 370 λ_o. The large cavity area was used to minimize the stress level in the transducer, with the oscillator running at a power level of $+20\,\mathrm{dBm}$ to reduce the noise floor (see Eq. (18.8)). Copper-doped aluminum IDT metallization was used. The peak stress level was reported as $3.4 \times 10^7\,\mathrm{N/m}^2$, which was well inside the high aging-rate level of $6 \times 10^7\,\mathrm{N/m}^2$ using pure aluminum metallization. SAW oscillators are also reported with long-term aging, attributed to random walk processes, of less than 1 ppm/year [17].

Example 18.2 Phase Noise of Tunable Rayleigh-Wave Delay-Line Oscillator.   A tunable Rayleigh-wave RF delay-line oscillator operates at frequency $f_o = 420\,\mathrm{MHz}$. The delay line has an effective $Q_e = 750$. The feedback amplifier noise figure is $F = 4.3\,\mathrm{dB}$, with open-loop gain $A = 30\,\mathrm{dB}$. The oscillator power level is $P_o = -14\,\mathrm{dBm}$. Its measured phase noise is due predominantly to white frequency (f^{-2}) roll-off over the Fourier range $1\,\mathrm{Hz} \leq f_F \leq 10\,\mathrm{kHz}$. Calculate the level of phase noise at $f_F = 10\,\mathrm{kHz}$. ∎

Solution.   From Eq. (18.8), and using f instead of ω, the phase-noise term involving white frequency noise is

$$\frac{\phi_{00}(f)}{P_o} \approx 10 \log_{10}\left(\frac{A^2 f_o^2}{4 Q_e^2 f_F^2 P_o} FkT \right) \quad (dBc/Hz) \qquad (18.9)$$

Convert dBm values ($0\,\mathrm{dBm} = 1\,\mathrm{mW}$) to dBW relative to one watt, so $P_o = -14\,\mathrm{dBm} = -44\,\mathrm{dBW}$. The thermal noise at room temperature is $kT = -204\,\mathrm{dBW/Hz}$ in a 1-Hz bandwidth, so that $FkT = 4.3 + (-204) =$

−195.7 dBW/Hz. The amplifier gain is 30 dB. The remaining terms in Eq. (18.9) yield $10\log_{10}\{f_o^2/(4Q_{ef}^2f_F^2)\} = 10 \cdot \log_{10}\{(420 \times 10^6)^2/(4 \times 750^2 \times 10{,}000^2)\} = 28.9$ dB. Summing the decibel terms yields a phase-noise $\mathcal{L}(f) = 30 + 28.9 - 195.7 - (-44) = -92.8$ dBc/Hz at $f_F = 10$ kHz. ∎

Example 18.3 Transducer Delay Time in Example 18.2. What is the time delay of the SAW transducer in Example 18.2? ∎

Solution. From Eq. (18.2), $\tau_g = 2Q_e/\omega_o = (2 \times 750)/(2 \times \tau \times 420 \times 10^6) = 568$ ns between phase centers of input and output IDTs in the delay line. ∎

18.4. Time-Domain Oscillator-Stability Measurements

18.4.1. ALLAN VARIANCE STATISTICS

It is sometimes desirable to relate the stability of an oscillator in the time domain to complement or substitute for frequency-domain characterizations. As an illustration of this point, recall from Fig. 18.4 that random walk processes in the frequency domain are dominant close in to the carrier. These may be difficult to measure for Fourier frequency offsets $f_F \ll 1$ Hz. This poses no problem, however, when random walk is measured in the time domain, as this merely requires longer counting periods. The time-domain statistics can be converted to phase-noise ones using appropriate conversion relations [8].

The time-domain definition of frequency stability uses the Allan variance parameter, which is the standard deviation of the fractional frequency deviation [8]. This variance is expressed in the notation

$$Allan\ variance = \left\langle \sigma_y^2\left(N,\ T,\ \tau,\ f_n\right)\right\rangle \tag{18.10a}$$

or

$$Allan\ variance = \left\langle \frac{1}{N}\sum_{n=1}^{N}\left(\overline{y}_n - \frac{1}{N}\sum_{k=1}^{N}\overline{y}_k\right)^2\right\rangle \tag{18.10b}$$

The bar over a "y"-term indicates that it has been averaged over a specific time interval, while the angle brackets < and > indicate a theoretically infinite time average of the enclosed quantity; $N =$ number of data values used in obtaining a sample variance; $T =$ time interval between beginnings of successive countermeasurements; $\tau =$ sampling time interval over which the frequency is averaged; and $f_n =$ high-frequency cut-off (i.e., bandwidth) of the measuring system.

In Eq. (18.10b) $y = df/f_o =$ fractional frequency fluctuation about oscillator frequency f_o, where $df = |f - f_o|$. The Allan variance with $N = 2$ and $T = \tau$ (i.e., no "dead time" between measurements) is particularly useful. This quantity is denoted by

$$\sigma_y^2(\tau) = \left\langle \sigma_y^2\left(N = 2, T = \tau, \tau, f_n\right)\right\rangle = \left\langle \frac{\left(\bar{y}_{k+1} - \bar{y}_k\right)^2}{2}\right\rangle \tag{18.11}$$

Experimentally, the forementioned infinite-time average must be estimated by a finite time average. The best estimate of this is given by

$$\sigma_y^2(\tau) \approx \frac{1}{2(M-1)} \sum_{k=1}^{M-1}\left(\bar{y}_{k+1} - \bar{y}_k\right)^2 \tag{18.12}$$

where M = number of frequency measurements made. The statistical confidence of the estimated Allan variance improves approximately as the square root of M. A common type of time-domain plot is in log-log coordinates of "sigma y of tau" (i.e., $[\sigma_y^2(\tau)]^{1/2}$) versus averaging time τ.

The terms *short-term stability*, *medium-term stability* and *long-term stability* are often used loosely when quoting the stability of an oscillator. By themselves, such terms are not significant. They should always be quoted in conjunction with the measuring period over which the averaging statistics have been carried out. "Short-term stability" values for a SAW oscillator are often quoted for a 1-s averaging period measurement.

18.4.2. COMPARISON OF TIME DOMAIN AND FREQUENCY DOMAIN OSCILLATOR NOISE

Relationships between noise-mechanism roll-off in the time and frequency domains are given in Table 18.2. Since white phase and flicker phase have the same roll-off slope τ^{-1} (i.e., 10 dB/decade), it is wise to double-check such measurements in the phase domain for proper identification.

Figure 18.8 depicts a time-domain measurement made on an injection-locked 674-MHz SAW oscillator for $N = 100$, showing 10-dB/decade roll-off in $\sigma_y(N, \tau, \tau)$ over $0.001 \leq \tau \leq 0.1$ s due to dominant flicker phase noise

TABLE 18.2

OSCILLATOR PHASE-NOISE PERTURBATION SLOPES IN TIME AND FREQUENCY DOMAINS

Noise type	Phase-Noise Dependence	Allan Variance $\sigma_y^2(N, T, \tau)$	"Sigma- y of tau" $\sigma_y(N, T, \tau)$
White phase	f^0	τ^{-2}	τ^{-1}
Flicker phase	f^{-1}	τ^{-2}	τ^{-1}
White frequency	f^{-2}	τ^{-1}	$\tau^{-1/2}$
Flicker frequency	f^{-3}	τ^0	τ^0
Random walk	f^{-4}	τ	$\tau^{+1/2}$

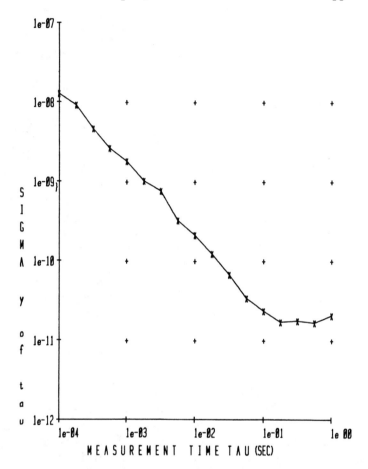

Fig. 18.8. Time-domain Allan-variance stability measurement on a 674-MHz injection—locked Rayleigh-wave single-pole resonator oscillator. For N = 100 samples, over measuring range $0.0001 \leq \tau \leq 1$ s; σ_y vertical scale is 10^{-12} to 10^{-7}. (Reprinted with permission from Campbell, Edmonson and Smith, Reference [18], © IEEE, 1985.)

(f^{-1}, τ^{-1}) in this range. The stability level transition to $\sigma_y(N, \tau, \tau) \approx 2 \times 10^{-11}$ at $\tau = 1$ s is attributed to flicker frequency (f^{-3}, τ^0) perturbations [18].

18.5. Rayleigh-Wave Oscillators

18.5.1. Fixed-Frequency Delay-Line Oscillator

On start-up, it is essential that a fixed-frequency oscillator should lock on to the desired oscillation mode and not to a spurious one. In a fixed-frequency surface-wave oscillator design incorporating a "single-mode" transversal delay-line filter or a SPUDT filter, this is achieved by proper design of the IDT path-length parameters. While this is demonstrated here for a Rayleigh-wave delay-line oscillator, similar reasoning may also be applied to a SPUDT-based design.

To this end, consider the SAW oscillator of Fig. 18.9, incorporating a delay-line feedback element. For illustrative purposes, delay-line configurations referred to here assume a narrowband input IDT and a broadband receiver. The delay time between phase centres is $\tau_g = L/v$, where L = distance and v = SAW velocity. For minimum insertion loss, the delay line is usually designed so that its input impedance is $Z_{in} \approx 1/(\omega_o C_T)$, where C_T is the total static capacitance of the input IDT.

One condition for oscillation at a desired frequency is that the amplifier gain must be greater than the total insertion loss of components in the feedback loop. Additionally, the total phase shift around the loop must be an integer number of 2π radians. Thus,

$$\tau_g 2\pi f_A + \phi_a + \phi_c = 2\pi P \qquad (18.13)$$

where ϕ_a = amplifier phase shift, ϕ_c = phase shift through cables and/or phase shifter in the feedback loop, P = an integer, and f_A is an allowable oscillation frequency. Because the amplifier phase shift may be several hundred degrees, it may be necessary to include a phase shifter or adjust the path lengths to the precise value required [19].

Because of the sidelobes of the transfer function response of either the delay-line or SPUDT-filter devices—and unless special precautions are taken—more than one oscillation condition can be satisfied within each response peak, provided there is sufficient loop gain at each frequency. This is an undesirable situation in single-mode oscillators. In the case of the delay-line oscillator It can be circumvented by designing the delay line to

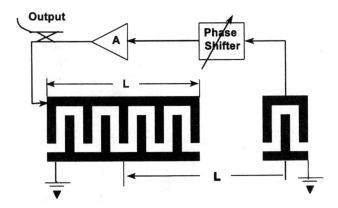

F<small>IG</small>. 18.9.   Transversal delay-line oscillator circuit configured for suppression of unwanted modes.

have input IDT length L equal to the separation between midsections of input and output IDTs, as depicted in Fig. 18.9. Under these conditions it can be shown that all of the possible oscillations but one are disallowed [19]. The single permissible oscillation is around the main peak of the $(\sin X)/X$ response over a frequency range dictated by the closed-loop gain. The exact frequency location will be dependent on the overall loop length and is fine-tuned by the value of ϕ_c.

The tuning range of a SAW delay-line oscillator is dictated by the width of the $(\sin X)/X$ amplitude response roll-off, which is dependent on the number of fingers in each IDT. On the other hand, its effective Q_e (from Eq. (18.2)) is proportional to the separation between the IDTs. This can be specified independently of the number of IDT fingers except in the mode-suppression configuration shown in Fig. 18.9.

18.5.2. SINGLE-POLE RAYLEIGH-WAVE RESONATOR OSCILLATORS

Because of the attainable high-Q values, SAW resonator oscillators are employed as highly pure signal sources at frequencies up to the 2-GHz range, normally using Rayleigh-wave structures on ST-X quartz substrates for power levels up to about +12 dBm. The temperature drift of such substrates is a parabolic function with zero temperature coefficient around 25°C. The quadratic temperature coefficient around this point of inflection corresponds to a drift of ± 15 ppm over a temperature range from 0 to 55°C [20].

Typical noise floors of precision Rayeigh-wave oscillators (mainly from the first term in Eq. (18.8)) are from −160 to −170 dBc/Hz, with ultraprecision floors of about −184 dBc/Hz. In commercial devices the set-frequency tolerances range from ± 30 to ± 200 ppm depending on cost. Vibration sensitivities are given as $\Delta f/f \approx 1 \times 10^{-9}$/g. Oscillators with two- and four-pole Rayleigh-wave feedback elements can be designed for a 1-dB bandwidth group delay which is flat over 400 ppm, to take into account 5-year aging.

Standard single-pole resonator oscillators have typical room-temperature stabilities < ~80 ppm. Temperature-compensated sources with lower phase noise are used as primary reference sources, while oven-controlled types usually exhibit the lowest phase noise at the cost of increased power consumption [21].

18.5.3. MULTIPLE-POLE RAYLEIGH-WAVE RESONATOR-FILTER OSCILLATORS

Due to their steeper—and phase-controllable—phase slopes around f_o, two- and four- pole resonator-filters are finding increased use in low-noise VCO designs for mobile communications, as in Fig. 18.3. Their 1-dB tuning bandwidths can be designed with flat group delay over 400 ppm or more, to account for the sum of: a) frequency accuracy (e.g., ±50 ppm); b) five-year aging (e.g., ±35 ppm); and c) temperature-induced drift over −40 to +70°C (e.g., ±75 ppm). This can eliminate the need for ovens in some phase-lock applications. These can exhibit phase noises as low as −80 dBc/Hz at 10-Hz Fourier-frequency offset, with noise floors from −177 to −183 dBc/Hz [22].

The most common two-port SAW resonator oscillators employ common-base circuits or Pierce-type common-emitter configurations [20]. Figure 18.10 illustrates the basic Pierce oscillator configuration. A phase-inverted resonator is employed here to handle the inversion between base and collector. Shunt inductances are employed at input and output IDTs to tune out the capacitance. Impedances Z_1, Z_2, Z_3 and Z_4 are chosen to match the transistor and resonator impedances. Maximum power transfer through the resonator is then commensurate with low phase noise.

18.6. SAW VCOs in Phase-Locked Loop Synthesizers for Mobile Radio

18.6.1. SINAD PERFORMANCE SPECIFICATIONS FOR A UHF MOBILE RADIO RECEIVER

The mobile radio receiver must satisfy specifications for in-channel and out-of-channel performance in terms of SINAD parameters [23]. A SINAD determination which can be measured using a SINAD audio analyzer) is given as the ratio of {Signal(S) + Noise(N) + Distortion(N)} at the receiver output to {Noise(N) + Distortion(D)} at the same output level so that

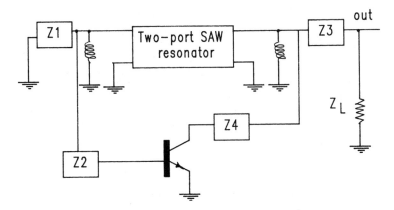

FIG. 18.10. Basic outline of one configuration of Pierce-type oscillator employing a single-pole two-port SAW resonator. (After Reference [20].)

$$SINAD = 20 \log_{10} \left(\frac{(S + N + D)}{(N + D)} \right) \quad (in \ decibels) \qquad (18.14)$$

The North American Electronics Industry Association (EIA) FM standard defines usable sensitivity as the input RF level which produces 12-dB SINAD at greater than 50% of rated audio output power. A SINAD audio analyzer first acts as a broadband voltmeter and measures the total output of the receiver. A filter in the analyzer notches out the audio modulation and the resultant noise plus distortion is measured.

Figure 18.11(a) illustrates a measurement example for obtaining the front-end in-channel specification [15]. The low-noise synthesizer frequency f_{S1} is set to $f_{S1} = (f_{LO} - f_{IF})$, where $f_{LO} = $ SAW local oscillator frequency and $f_{IF} = $ receiver IF. A 1-kHz modulation tone is employed, with the modulation index adjusted to give a ± 3-kHz frequency deviation. This

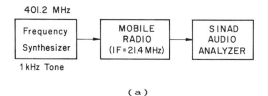

(a)

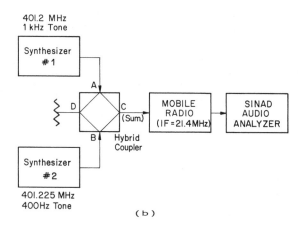

(b)

Fig. 18.11.   (a) Set-up for in-channel voice-grade SINAD measurement on mobile/wireless radio receiver employing an RF local oscillator. (b) Out-of-channel selectivity test using two low-noise frequency synthesizers for a receiver with 25-kHz channel spacing. (Reprinted with permission from Campbell, Sferrazza Papa, and Edmonson, Reference [15], © IEEE, 1984.)

signal is applied to the antenna input terminal on the receiver and the synthesizer power level is reduced until the SINAD audio analyzer reading drops to the 12-dB minimum specification. The RF input power level determines whether the performance rating is met. (In systems tested by the author, the specification was typically met with the RF signal reduced to −109 dBm, corresponding to ≈0.8 μV across 50 Ω [15].)

For adequate out-of-channel interference suppression in these systems, the selectivity should be at least 70 to 90 dB for channel separations of up to 30 kHz. One technique for measuring this selectivity employs two low-noise frequency synthesizers, as sketched in Fig. 18.11(b) for a test on a system with 25-kHz channel spacing. Frequency f_{S1}, modulation and power levels for synthesizer 1 are the same as for the in-channel test for the 12-dB SINAD reading. In this case, synthesizer 2 is applied at frequency $f_{S2} = f_{S1} \pm 25$ kHz, with 400-Hz audio modulation and the same ±3-kHz frequency deviation. Its output power is increased until the SINAD level falls from 12 to 6 dB. The difference between synthesizer power levels is the (adjacent) out-of-channel selectivity.

The S/N ratio at the input to the receiver detector will depend on the selectivity measurement employed. For example, the above 12-dB SINAD specification relates to an audio measurement, which may require 5-dB S/N co-channel rejection ratio[3] at the detector IF frequency. Different IF S/N ratio specifications would apply to different sensitivity measurements such as for the BER ones considered in Chapter 17 [24].

18.6.2. RECEIVER SELECTIVITY RELATIONSHIPS

An expression for receiver selectivity (*RS*) in the RF front-end of a multi-channel mobile/wireless radio receiver, expressed in decibels, is given by [24]

$$RS = -C - 10\log\left[10^{-\frac{IF}{10}} + 10^{-\frac{S}{10}} + BW \times 10^{\frac{L}{10}}\right]: \quad \left(in\ decibels\right) \quad (18.15)$$

where *RS* = adjacent-channel selectivity relative to nominal receiver selectivity, *C* = co-channel rejection ratio (dB), *IF* = first IF filter rejection at the adjacent channel spacing (decibels), *S* = local oscillator (LO) spurious responses in Fig. 18.12 at the channel edge offset frequency (dBc/Hz), *BW* = IF noise bandwidth (Hz), and *L* = SSB phase-noise at Fourier frequency offset, corresponding to channel spacing Fourier-frequency offset Δ.

---

[3] The *co-channel rejection ratio* (also known as *capture ratio*) is the S/N ratio required at the input to an FM detector to produce the desired baseband S/N ratio, as well as the desired operation bandwidth [24].

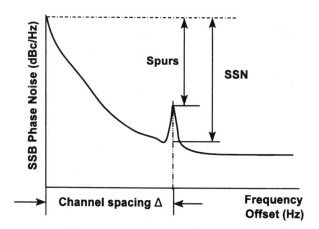

Fɪɢ. 18.12.   Spectral purity parameters of RF local oscillator used for adjacent-channel selectivity determinations in mobile/wireless receiver circuitry.

Example 18.4 Adjacent-Channel Selectivity Approximation for Cellular Receiver.   A 900-MHz base station cellular receiver with 30-kHz channel spacing is required to have a front-end adjacent-channel selectivity of 80 dB. Following the method of Example 18.1, estimate the phase-noise specification for the SAW local oscillator at a Fourier-frequency offset $f_F = 30$ kHz. ∎

Solution.   Follow Example 18.1 and employ a conversion factor for the 30-kHz channel bandwidth as $10\log_{10}(30,000/1) \approx 49$ dB. The maximum allowable oscillator phase noise at 30-kHz Fourier-frequency offset is approximated as $-80 - 49 = -129$ dBc/Hz. ∎

Example 18.5 More Exact Adjacent-Channel Selectivity Determination.
The 900-MHz base station cellular receiver in Example 18.4, with 30-kHz channel spacing, employs a SAW narrowband first IF filter with a centre frequency of 80 MHz. The receiver requires an adjacent-channel selectivity greater than 70 dB. Determine the adjacent-channel receiver selectivity from Eq. (18.15) for the following three circuit-parameter options, assuming a co-channel rejection ratio value $C = 5$ dB: (1) The IF filter has an out-of-band suppression of 80 dB, while the RF local oscillator has spurs of 110 dB and SSB phase noise of −120 dBc/Hz at $\Delta = 30$ kHz offset; (2) The IF filter selectivity is still 80 dB, while the local oscillator has spurs of 110 dB and SSB phase noise of −131 dBc/Hz at $\Delta = 30$ kHz offset; (3) The IF filter selectivity is reduced to 70 dB, while the local oscillator has spurs of 110 dB and SSB phase noise of −131 dBc/Hz at $\Delta = 30$ kHz offset. ∎

Solution. From eq. (18.15) the receiver selectivity for the three cases is obtained as: (1) RS = 68.9 dB; (2) RS = 74 dB; (3) RS = 64.8 dB. Case (2) is the only one that meets the specifications in this instance, illustrating the need for a low-noise RF local oscillator as well as for a first IF SAW filter with very good out-of-band rejection! ∎

18.6.3. Illustrative Tunable SAW Oscillator Circuit

Figure 18.13 gives the block diagram of a circuit incorporating a VCO and phase-locked loop (PLL) frequency synthesizer for use in the RF stage of a mobile/wireless communications receiver. A circuit of this type was employed by the author in a 422-MHz application, for *narrowband* VCO tuning [15]. This employed two feedback loops for control of a Rayleigh-wave local VCO. The first oscillator loop employed a Rayleigh-wave delay line in series with a voltage-controlled phase shifter. The delay-line input and output IDTs had 72 and 45 finger pairs, respectively, for 50-Ω operation. A metallization thickness of ≈ 950 Å was used to reduce mass-loading, while 5% Cu in Al was used to aid the aging response [25]. The insertion loss was 17 dB, with effective $Q \approx 750$ and group delay $\tau_g = 565$ ns. A 30-dB feedback amplifier was used with noise figure $NF = 4.3$ to give sufficient gain margin for the VCO to be tuned over about 0.5 MHz.

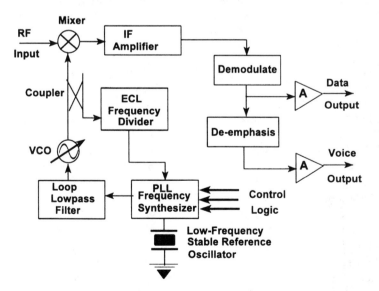

Fig. 18.13. Primitive voltage-controlled oscillator (VCO) with phase-locked loop (PLL) frequency synthesizer for data/voice recovery in mobile radio receiver.

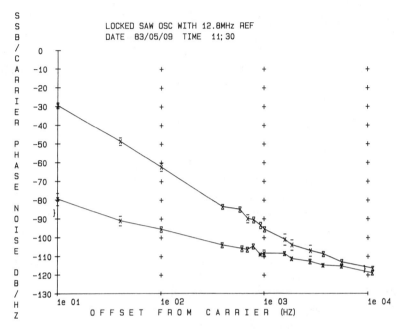

FIG. 18.14.   Upper trace: Measured phase-noiseof VCO locked at 422.60001 MHz ± 10 Hz. Lower trace: phase noise of 12.8-MHz BAW reference oscillator. Analyzer IF drift –0.157 Hz. Noise floors given by } -signs. Horizontal scale: 10 Hz to 10,000 Hz offset. Vertical scale: 10 dB/div. (Reprinted with permission from Campbell, Sferrazza Papa, and Edmonson, Reference [15], © IEEE, 1984.)

The second loop incorporated a scale-of-64 emitter-coupled logic (ECL) frequency divider plus a two-input combination PLL frequency synthesizer and phase-detector chip. A 12.8-MHz bulk acoustic wave (BAW) crystal oscillator provided the reference input to the PLL. This was divided internally by 1024 to give channel and SAW oscillator frequency increments of (12.8MHz)/1024 = 12.5 kHz. The phase-detector output to the tuning phase shifter was applied through a lowpass filter to set the loop cut-off frequency. The remaining circuitry employed a 21.4-MHz IF amplifier and demodulator/de-emphasis stage for reception of voice or data signals.

The phase-noise performance of the BAW reference oscillator in such a system must be considered carefully as it will dominate the overall phase-noise performance of the tunable local oscillator *up to* the cut-off frequency of the PLL [26]. Below the loop cut-off, the up-converted phase-noise will be increased by at least $20.\log_{10} n$, where n = multiplier ratio. In this instance this ratio was $20.\log_{10} (422\,MHz/12.8\,MHz) = 30.3\,dB$. Above the loop cut-off, the VCO phase noise will be dominated by the VCO itself.

The lower plot in Fig. 18.14 shows the phase noise of the BAW reference crystal in the forementioned circuit over the Fourier-frequency range $10 \le f_F \le 10^4$ Hz. The upper plot shows a phase-noise level for the phase-locked SAW oscillator of about -90 dBc/Hz at $f_F = 1$ kHz.

18.7. Rayleigh-Wave Oscillators for Low-Power Wireless Data Links

18.7.1. HIGHLIGHTS OF APPLICATIONS

As noted in the Introduction to this chapter, a major market currently exists for surface-wave oscillators as components for free-running or narrowband voltage-controlled oscillators (VCOs) for wireless data transmission over short-range low-power unlicensed UHF links at frequencies within 200- to 900-MHz bands. These very low-power wireless systems are for a variety of consumer and commercial data-transfer applications, including those for keyless automobile entry, remote utility-meter reading, wireless bar-code readers and computer-peripheral links. Transmission powers are typically less than about 1 mW (0 dBm) for coverage of up to about 100 m. These networks can employ a variety of Rayleigh-wave devices in feedback stages, including delay lines, resonators, and narrowband resonator filters [27]. Two of these interesting oscillator circuits for transmitters and receivers in such low-power networks are highlighted in the following.

18.7.2. EXAMPLE OF TRANSMITTER FOR AUTOMOTIVE KEYLESS ENTRY AND SECURITY SYSTEMS

The transmitter sketched in Fig. 18.15 is normally designed for low-voltage lithium-battery use, in automotive keyless entry and other security-systems, as well as for PCMCIA data-link applications at low-data rates [27]. Digital modulation is employed in these circuits for data transfer. In this example, a low-loss narrowband two-pole Rayleigh-wave resonator filter is used in the oscillator feedback loop for reduced harmonic emissions.

Restrictions on fundamental and harmonic emission in low power-levels can vary significantly from country to country. Table 18.3 illustrates examples of some regulatory levels.

18.7.3. EXAMPLE OF RECEIVER OSCILLATOR FOR LOW-POWER WIRELESS RECEIVERS

While "conventional" fixed-frequency Rayleigh-wave local oscillators are routinely employed in superheterodyne receivers for low-power wireless systems, Fig. 18.16(a) shows an unconventional one for use in a

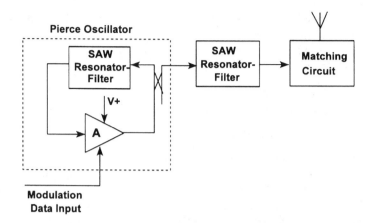

FIG. 18.15.   Hybrid low-power wireless transmitter, using Rayleigh-wave resonator-filters in oscillator and output circuitry.

TABLE 18.3

LOW-POWER WIRELESS RESTRICTIONS ON FUNDAMENTAL AND HARMONIC EMISSIONS.
(After Reference [27].)

Country	Frequency (MHz)	Fundamental Power	Second Harmonic	Third Harmonic
United States	260–470	3750–12500 μV/m	375–1250 μV/m	375–1250 μV/m
Great Britain	417.9–418.1	250 μW	4 nW	1 μW
Germany	433.05–434.79	25 mW	1 nW	30 nW
Japan	303.675–303.975	500 μV/m	35 μV/m	35 μV/m

superregenerative receiver [27]. The aim of this type of circuitry is to increase receiver sensitivity for processing very low-powered data-modulated RF signals. As sketched, this particular example employs a superregenerative SAW delay-line oscillator which is turned on and off by a quenching oscillator. In quenching oscillators, desirable operation calls for an optimum trapezoidal waveform output. It is reported that ten cycles around the loop allows the oscillation to reach maximum value, giving a maximum quench frequency for this oscillator type as $1/40\,\tau$, where τ is the group delay of the delay line in the feedback loop of the Rayleigh-wave oscillator [27]. For a delay line with $\tau = 150$ ns, this would give the maximum quench frequency as about 56 kHz.

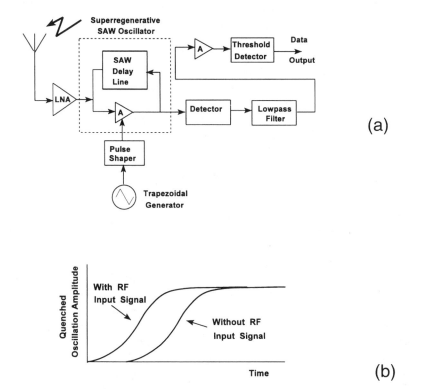

FIG. 18.16. (a) Basic low-power superregenerative receiver using a quenched SAW oscillator for enhanced sensitivity. (b) Oscillation amplitude with and without input RF signal. (After Reference [27].)

Figure 18.16(b) illustrates the build-up of the amplitude response of the quenched SAW oscillator as a function of receiver input-voltage stimulus. As illustrated, the onset of the oscillatory build-up is delayed as a function of the input-voltage amplitude. In this way, variations of the input data signal result in pulsewidth modulation of the oscillator quenched response. Following detection, filtering, and threshold-level detection, the digital data signal is restored.

The advantage of using a superregenerative receiver is, of course, the increase of the receiver sensitivity. With the SAW-based system of Fig. 18.16(a), typical receiver sensitivities of $-105\,\mathrm{dBm}$ $(1.26\,\mu\mathrm{V})$ are obtained compared with typical sensitivities of $-80\,\mathrm{dBm}$ $(22\,\mu\mathrm{V})$ for "conventional" LC-based superregenerative receivers [27].

18.8.   Multimode SAW Oscillator

18.8.1.   APPLICATION TO FREQUENCY-AGILE SYSTEMS

A multimode oscillator is defined here as one which includes a comb filter in the feedback loop, and allows stable oscillation at only one of a number of allowable comb-mode frequencies. Such oscillators, preferably with fast-hopping capability, can find application in a number of communications systems [28]–[31]. The use of a hybrid direct-sequence/frequency-hopping (DS/FH) spread-spread system was already illustrated in Chapter 17 for indoor data communications using frequency-hopping rates. As was illustrated, spread-spectrum systems are employed in indoor communications for efficient communication in the presence of severe multipath interference. Hybrid DS/FH systems offer the advantages of spectral spreading as well as the use of lower-speed PN-encoding. Hopping rates for indoor communications systems can be relatively high (e.g., 90 kb/s), since only stationary object reflections are involved.

As another example, frequency-hopped differential phase-shift keying (FH-DPSK) spread-spectrum is used to provide mobile-telephone service to a large number of users. In 800-MHz mobile systems for automobile communications, the permissible hopping rates are reduced to handle communications from moving vehicles. Typical hopping rates to/from automobiles with speeds of 160 km/h would be in the order of 5 kHz (with a waveform duration of 200 μs), since the distance travelled in this period would be small enough to provide a reliable phase reference [32].

Frequency agility of local oscillators in pulsed radars has been used for several years to provide decorrelation between radar echoes from successive transmitted pulses. Since the radar display is obtained by integrating over a period of time, relatively long-term fluctuations due to reflectivity changes are significantly reduced. For example, a 12-dB reduction in clutter intensity can be achieved over a single radar scan in $N = 15$ to 20 pulses when the transmitter is hopped with each pulse [30].

18.8.2.   PHASE CONDITIONS FOR OSCILLATION

Despite their comb-frequency response characteristics, both basic and SPUDT forms of the comb filter in Chapter 12 represent delay-line structures. As with the SAW delay line oscillator of Section 18.5.1, SAW comb-oscillator operation is critically dependent on the correct choice of filter group delay. This must be selected to ensure that the total phase shift around the loop is $2\pi n$ (n = integer) at any of the desired comb-mode frequencies. In the illustrative circuit of Fig. 18.9, consider that the

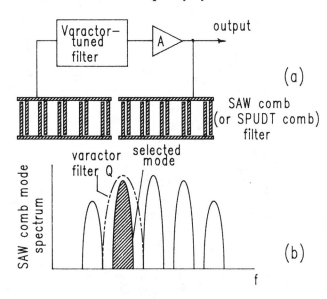

Fig. 18.17. (a) Multimode SAW oscillator using comb filter in feedback stage. Varactor-tuned filter used for mode selection. (b) Varactor-tuned filter Q must be high enough to select only one mode. (After Reference [31].)

phase shift through the amplifier, phase shifter and connecting cables is negligible compared with that through the comb filter so that the latter provides the phase specification. This is established by the separation between phase centers of input and output IDTs, just as in a conventional delay line.

In the comb-filter oscillator of Fig. 18.17, the path length between IDT phase centers is $L = (R-1)W\lambda_o + s\lambda_o$, where R is the number of "rungs" in each IDT and $W\lambda_o$ is their rung separation, while $s\lambda_o$ is the gap between the IDTs. For phase reinforcement around the loop, the total phase shift ϕ at each desired oscillation frequency must be $\phi = 2\pi n$. However, $\phi = \beta L$, where β is the phase constant (in radian/m), given by $\beta \lambda_A = 2\pi$, where $\lambda_A = v/f_A$ and f_A is an allowable oscillation frequency. If the IDT separation parameter s is set to $s = W$, the foregoing relations yield [31]

$$L = RW\lambda_o \qquad (18.16)$$

as the required spacing between phase centers of input and output IDTs for allowable oscillations at frequencies

$$f_A = \frac{nv}{RW\lambda_o} \qquad (18.17)$$

By substituting appropriate values of integer n in Eq. (18.17) it can be deduced that allowable frequencies f_A correspond to those for comb peaks. Multimode oscillator operation is thus possible when the separation between ends of the equal IDTs is arranged to be the same as the spacing between individual rungs. This spacing condition also applies to the SPUDT comb filter.

18.8.3. ACOUSTIC Q OF MULTIMODE SAW COMB FILTER

Since the forementional comb-filter structure forms a SAW tapped delay line, the acoustic Q is dictated by the separation between phase centres of input and output IDTs. Following the relations for the conventional SAW delay line, therefore, the Q of the SAW comb filter is given by

$$Q_a = \frac{\omega_A \tau_g}{2} = \pi f_A \tau_g \qquad (18.18)$$

where f_A = an allowable comb-mode frequency and $\tau_g = L/v$ = SAW propagation time between IDT phase centres at SAW velocity v. Substitution for L from Eq. (18.16) gives the acoustic Q as

$$Q_a = \pi R W \qquad (18.19)$$

Example 18.6 Comb-Q of Multimode Oscillator. The multimode oscillator of Fig 18.17 operates around a comb centre frequency of $f_A = f_o = 400\,\text{MHz}$. Each IDT is composed of $R = 50$ rungs, with $N = 2$ electrode finger pairs in each rung. The center-to-center spacing between adjacent rungs is $40\lambda_o$. Determine the approximate value of the acoustic Q of each comb mode. ∎

Solution. From Eq. (18.19), $Q_a \approx \pi R W \approx \pi \times 50 \times 40 \approx 6280$. ∎

18.8.4. AN ILLUSTRATIVE DESIGN

One reported frequency-agile VCO [31] employed a comb filter on ST-quartz, using $R = 50$ rungs in both input and output IDTs. Separation between phase centres of adjacent rungs was $W = 40\lambda_o$, with $N = 2$ finger pairs in each rung. This allowed stable oscillation at one of 21 permissible comb modes, centred around a VCO centre frequency of 400 MHz.

18.8.5. SELECTING THE DESIRED OSCILLATION MODE

One method of frequency selection sketched in Fig. 18.17 employs a varactor-tuned filter in series with the feedback loop. An extended version of a phase-locked multimode Rayleigh-wave oscillator employs two

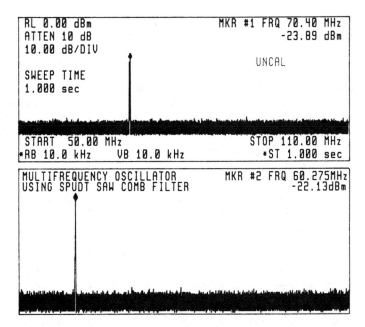

FIG. 18.18. Spectrum analyzer plots showing selection of 60- or 70-MHz oscillations within multimode SPUDT-based comb filter oscillator with 10-MHz comb-spacing. Start frequency 50 MHz. Stop frequency 110 MHz. Sweep time is 1 s, 10-kHz resolution bandwidth.

varactor diodes and two feedback loops for mode selection and stability enhancement over long dwell times [33].

Another frequency-selection technique involves the forced injection of a short signal at a desired frequency. This injection technique employs SAW linear FM chirp filter mixing, as will be described in a following section. Figure 18.18 demonstrates the spectrum of two alternately selected oscillations obtained this way in an illustrative design [34].

In measurements of mode-selection comb-frequency oscillators, using SPUDT comb filters with minimum 3.7-dB insertion loss on lithium niobate, it was demonstrated that dwell times as low as about $2\,\mu s$ could be realized [34], with the maximum dwell time dictated by the interval between input stimuli. This served to demonstrate the feasibility of such comb-oscillator techniques in circuitry for short dwell times such as for indoor DS/FH spread-spectrum communications, or for longer dwell times appropriate to FH-DPSK or similar systems for mobile communications with fast-moving vehicles.

18.9. SSBW and STW Oscillators

18.9.1. REVIEW OF SUBSTRATES AND PROCESSES

While surface skimming bulk wave (SSBW)[4] and surface transverse wave (STW) substrates and processes have previously been discussed in Chapter 2, some of their characteristics are reviewed here before proceeding to a discussion of their use in oscillator circuitry. SSBW propagation, as pictured in Fig. 1.2 of Chapter 1, involves longitudinal bulk waves with shear-horizontal (SH) polarization. These can be excited and detected by surface-deposited IDTs in similar fashion as for Rayleigh-wave or leaky-SAW devices. While Rayleigh waves are generated by the electric-field components parallel and normal to the excited IDT fingers, the SH waves are only excited by the parallel electric field. SBAW structures have essentially some of the same merits as for leaky-SAW in that:

1. Bulk waves have higher velocities than Rayleigh-waves, so that SBAW devices can be designed for operation at fundamental frequencies up to about 60% higher than with SAW IDTs of the same geometry.

2. Since the bulk waves in the SBAW devices propagate below the piezoelectric surface, the propagating wave is much less sensitive to surface contamination than for SAW (although the IDTs can still be affected).

3. SBAW substrate cuts have superior temperature coefficients to Rayleigh wave ones in some cuts of quartz and lithium tantalate (LiTaO₃).

4. SBAW filters and resonators can operate at much higher powers than their Rayleigh-wave counterparts (e.g., 25 dBm), before piezoelectric nonlinearities set in.

5. SBAW devices can yield good suppression of spurious modes.

As detailed in Chapter 2 SBAW devices may be grouped into two general categories depending on whether or not energy-trapping structures are involved. Where there is no energytrapping, the SH bulk waves are usually referred to as *surface skimming bulk waves*. When an energy-trapping grating structure is located between input and output IDTs to promote low-loss operation, the SBAW is termed a *surface transverse wave* (STW). Current prominent SSBW piezoelectric substrates are listed in Table 2.3 of Chapter 2. Desirable piezoelectric crystals and crystal cuts for SSBW de-

---

[4] *Surface skimming bulk waves* (SSBWs) are also known as *shallow bulk acoustic waves* (SBAWs).

vices are those that have: 1) large piezoelectric coupling to the SH bulk wave; 2) zero piezoelectric coupling to surface acoustic waves; 3) zero piezoelectric coupling to other bulk-wave modes; and 4) a zero or low-temperature coefficient of delay. Most of these features are to be found in singly-rotated Y-cuts of quartz.

As seen from Table 2.3, the SSBW AT-cut of quartz is a most useful one, having a high SSBW velocity and negligible temperature coefficient of delay. The SSBW cuts of lithium niobate and lithium tantalate, with propagation along the X-axis, have high values of electromechanical coupling coefficient K^2 but poorer temperature stabilities. For SSBW propagation at 90° to the X-axis, the SSBW velocity is about 60% higher than for SAW propagation along the X-axis.

18.9.2. Oscillators Using SSBW Delay Lines

SSBW delay lines and resonators are particularly suited to gigahertz-band oscillator applications. To illustrate this, 3-GHz SSBW oscillators using delay-line feedback elements on 35.5° (near AT) rotated Y-cut quartz, yielded Q-values ranging from 2200 to 2600 [35]. The frequency-time product of about 6.8×10^{12} for these AT-devices greatly exceeds those reported for oscillators using Rayleigh-wave delay lines.

BT-quartz is preferable in situations where there is a large temperature swing. For BT-based SSBW oscillators in gigahertz-range applications, the frequency drift over a temperature excursion of 100°C can be as low as 15 ppm. This is about a factor of 10 better than achievable using ST-quartz delay lines in Rayleigh-wave oscillators.

18.9.3. Surface Transverse Wave (STW) Resonators

While one- and two-port SSBW resonators can be readily fabricated with geometries comparable to those for Rayleigh-wave devices, they suffer from much higher insertion loss and lower Qs than their SAW-based counterparts. This situation can be remedied by constraining the bulk wave to propagate as a surface transverse wave (STW) mode with shear horizontal (SH) polarization. This result is achievable by using an energy-trapping grating structure between input and output IDTs. This reduces the diffraction of the shallow bulk wave into the substrate, with commensurate reduction in device insertion loss and increase in resonator Q.

Figure 18.19 outlines the geometry of a two-port STW resonator with energy-trapping grating. As with the SAW resonator, the period of the IDT fingers and the outside reflection gratings is the same. As with a multistrip coupler, however, the period of the elements in the central energy-trapping

TWO-PORT STW RESONATOR

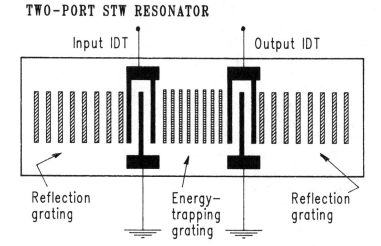

Fɪɢ. 18.19.   Location of energy-trapping grating in a two-port STW resonator.

grating is slightly less than for the IDTs. Thus, the resonator response is below the stopband of the energy-trapping grating.

Energy-trapping structures have employed periodic gratings of etched grooves on the piezoelectric surface, as well as those using open-circuit aluminum strips [36]–[38]. Both grating structures have the effect of slowing the shallow bulk wave and confining it to the surface layer. The aluminum structure can be less troublesome to fabricate. The aluminum gratings also provide additional wave-slowing in high-coupling (i.e., large K^2) materials due to the piezoelectric shorting that can effectively trap the wave in some instances [38]. Unlike Rayleigh-wave propagation, the depth of penetration of the STW wave can be controlled by the design of the energy-trapping grating. The acoustic power density can be reduced by increasing the wave-penetration depth. This is desirable in SBAW oscillator applications because it allows the power level to be increased without entering into the nonlinear operation of the piezoelectric. This is an advantageous feature because the FM noise floor of an oscillator varies inversely with the square root of the oscillator power. Additionally, it may be noted that reduction in acoustic power density can be beneficial to the aging response [25].

While theoretical values in excess of 100,000 [39] have been computed for the Q-values of STW resonators on 35.7° rotated Y-cut quartz, with propagation perpendicular to the X-axis, experimental values obtained to date are typically in the order of 10,000. For example, for 500-MHz and 632-MHz STW resonators on rotated Y cuts of quartz, with aluminum metallization and an energy-trapping grating, $Q = 10,000$ was the typical unloaded

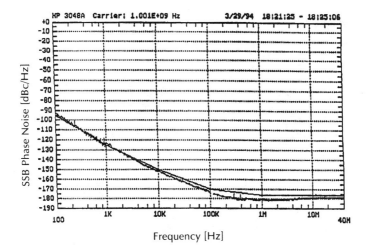

Frequency [Hz]

FIG. 18.20. The SSB phase-noise of a high-performance STW oscillator with far-out phase-noiselevel ~ −180 dBc/Hz. Lower curve is measurement, while upper curve is specification. Horizontal scale : 100 Hz to 40 MHz. Vertical scale: − 190 to 0 dBc/Hz. (Courtesy of Sawtek Inc., Orlando, Florida.)

Q obtained, with device-insertion loss less than 6 dB and measured phase-noise levels of $\mathcal{L} \approx -135$ dBc/Hz at a 1-Hz Fourier-frequency offset[5] [40].

By appropriate choice of inter-IDT lengths of the STW resonator in Fig. 18.19, multimode STW resonator-filters can be formed that comprise two STW modes plus one SSBW mode, with insertion loss ~ 7 dB, and Q = ~4000, for operation in 1-GHz STW oscillators in FM- or data-modulation wireless transmitters [41].

18.9.4. SURFACE TRANSVERSE WAVE (STW) OSCILLATORS AND APPLICATIONS

STW oscillators can have excellent far-out phase-noise characteristics as illustrated in the 1-GHz example of Fig. 18.20. Compared to typical "off-the-shelf" Rayleigh-wave oscillator performance, Fig. 18.20 serves to demonstrate that an STW oscillator could be the preferred choice in second-stage VCO phase-locked double-heterodyne transceivers if noise suppression above ~10-kHz Fourier-offset level is the important parameter. STW VCOs with output powers of up to 2 W (i.e., 30 dBm) are also employed in wireless security systems for coverage over distances of about

---

[5] Note that this result is for only the resonator!

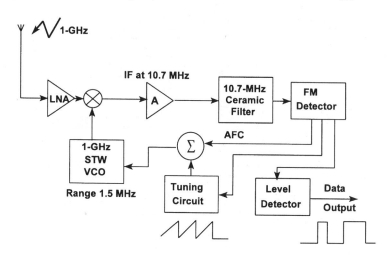

FIG. 18.21. Single-heterodyne receiver for 1-GHz STW-based VCO in a 2-W wireless security system. (After Reference [42].)

2 km [42]. In an illustrative system, the STW resonator-filter Q is low enough ($Q_L \sim 1500$) so that FSK-modulation rates of more than 100 kb/s can be applied in modulation bandwidths of about 150 kHz. The VCO tuning is carried out using a varactor-controlled phase shifter.

Figure 18.21 shows a receiver circuit for a 1-GHz, 2-W, wireless-security system. This employs a single-heterodyne receiver with a VCO employing an STW resonator-filter. Down-conversion is at 10.7 MHz for application to a standard 10.7-MHz ceramic filter. This circuit had a tuning range of about 1.5 MHz, which is much larger than transmitter frequency excursions. In this way, feedback from the FM detector can be used to track transmitter frequencies within a fabrication tolerance of about ±500 ppm [42].

18.10. Injection-Locked Oscillators for Carrier Recovery

18.10.1. INJECTION-LOCKING PRINCIPLES

Consider the free-running oscillator circuit of Fig. 18.4 with unperturbed frequency f_o and additive sinusoidal injection signal voltage V_J. It may be shown that there are three modes under which the oscillator will interact with the injected signal V_J, namely:

1. *Free-running condition* in which the power level or frequency of V_J is insufficient to perturb the oscillator frequency f_o
2. *Driven unlocked condition* in which the weak inject signals beats with the "free-running" oscillator and beat frequencies result around f_o

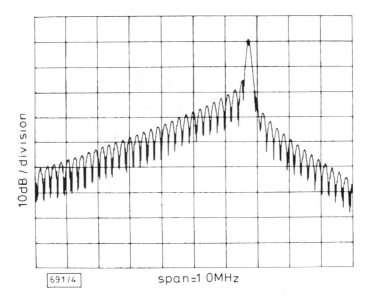

FIG. 18.22. Experimental beat spectrum of a driven-unlocked SAW oscillator with $f_o = 914.360\,\text{MHz}$, $f_J = 914.534\,\text{MHz}$, and $V_J/V_o = -21\,\text{dB}$. (Reprinted from C.K. Campbell, P.J. Edmonson, and P.M. Smith, Reference [43], Courtesy of Electronics Letters.)

3. *Injection-locked condition*, where the injection signal is within the maximum locking bandwidth Δf_o of the oscillator. In this condition, the oscillator tracks the input signal, and its output is an amplified version of V_J. Most importantly, the phase-noiseof the injected-signal output becomes the phase-noise of the injection signal itself [18].

Figure 18.22 illustrates a typical response for the driven-unlocked condition as implemented with a SAW oscillator with free-running frequency $f_o = 914.360\,\text{MHz}$, with injection at $f_J = 914.534\,\text{MHz}$, with an injection ratio $V_J/V_o = -21\,\text{dB}$. Theoretically, a driven unlocked oscillator will only exhibit a "one-sided " spectrum, with the beat-frequency sidebands on the side opposite to that for the injection. In practice, and as exhibited in Fig. 18.22, a partial double sideband results—to a degree depending on the level of saturation of the amplifier in the feedback loop [43].

For the injection-locked condition of interest here, the maximum locking bandwidth Δf_o of the oscillator may be shown to be [44], [45]

$$\Delta f_{\max} = \frac{f_o}{2Q} \frac{V_J}{V_o} \frac{1}{\sqrt{1 - \left[\dfrac{V_J}{V_0}\right]^2}} \qquad (18.20)$$

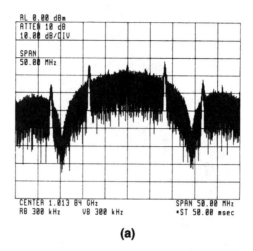

(a)

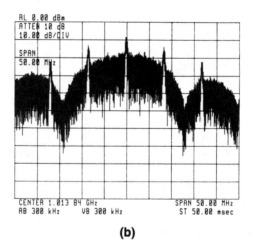

(b)

FIG. 18.23. Injection-locking carrier-recovery in a 1-GHz STW oscillator with 8 Mb/s BPSK unipolar RZ pseudorandom sequence modulation. (a) Spectrum of signal at input to locked oscillator with very weak carrier. (b) Carrier enhancement by 9 dB at output from oscillator. (Reprinted with permission from Edmonson, Smith, and Campbell, Reference [45], © IEEE, 1992.)

where V_J = injection voltage, V_o = oscillator voltage at injection point, Q = quality-factor of feedback element, and f_o = unperturbed oscillator frequency [44], [45]. An injection-locked oscillator will track a phase- or frequency-modulated input signal to a degree dependent on the injection bandwidth constraint of Eq. (18.20). For a binary phase-shift-keyed signal

(BPSK), the symbol duration T must be long enough for the oscillator to achieve a phase change of 180°. From Eq. (18.20) this minimum symbol duration that will allow this is given by

$$T_{\min} > \frac{1}{2\Delta f_{\max}} \qquad (18.21)$$

18.10.2. INJECTION-LOCKED OSCILLATOR FOR CARRIER-RECOVERY

If the forementioned bandwidth conditions for the signal are not met, the injection-locked oscillator will no longer be able to track the message modulation in the input signal. Instead, it will settle into a median-frequency state, which can serve as a carrier-recovery signal with lower noise than for a conventional PLL [45]. For carrier recovery by an injection-locked oscillator, the bandwidth of the modulated input signal must be greater than the maximum injection-locking bandwidth $\Delta f_{\max}$ in Eq. (18.20), To illustrate this capability, Fig. 18.23 shows the measured responses of a 1-GHz injection-locked STW oscillator for carrier recovery in an BPSK system with phase modulation by an 8 Mb/s BPSK unipolar return-to-zero (RZ) pseudorandom sequence. In this particular example, the input power of -15 dBm at the input to the 1-GHz STW oscillator was not strong enough to allow signal tracking. Figure 18.23(a) shows the spectrum of the signal at the input to the oscillator. The periodic peaks are separated by the bit rate, and are typical of unipolar RZ-sequences. Note the very weak carrier level at centre frequency. Figure 18.23(b) shows the carrier enhancement of about 9 dB at the output from the locked oscillator. In this instance, BER values of about 10^{-7} were obtained for a C/N ratio of 14 dB [45].

18.11. A SAW-Based Frequency Synthesizer

18.11.1. GENERAL REQUIREMENTS

Frequency hopping (FH) spread-spectrum communications techniques are used to provide low error-rate performance in the presence of jamming or high interference levels. In FH radar systems, for example, synthesizers are required that will provide a large number F of discrete frequencies with good spectral purity in the range $50 \le F \le 10,000$, with circuit bandwidths ranging from 10 to 500 MHz. Fast hopping is of particular interest in countering repeat jamming.

In FH operation, the available bandwidth is divided into a number of contiguous sidebands. Band spreading is achieved by transmitting successive constant-duration pulses on PN-selected subchannel carrier frequencies. The circuit bandwidth should also be programmable to handle single or multiple carrier frequencies as well as different hopping rates.

While multimode SAW oscillator techniques may be satisfactory for hopping over a limited number of frequencies in spread-spectrum mobile/ wireless communications, another approach is required for a very large number of hops. One technique employs SAW chirp-filter mixing, as presently described, and is an efficient and elegant technique for attaining these requirements [46].

18.11.2 Chirp Mixing Principles

To illustrate one fast FH-technique, consider the mixing circuit of Fig. 18.24(a) employing two SAW linear FM chirp filters with the same *TB*-product. One of these has dispersive slope μ_1 and starting frequency f_1. The other has dispersive slope μ_2 and starting frequency f_2. The dispersive slopes are arranged to be equal and opposite so that $\mu_1 = -\mu_2$, and the chirp slope is $\mu = B/T$. The FH operation is then based on impulsing

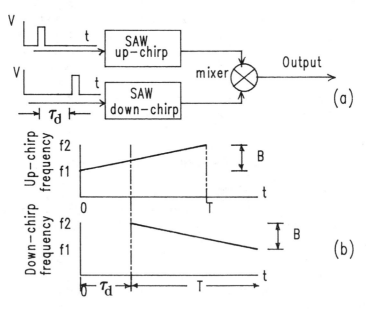

Fig. 18.24. Spectrum analyzer response of sample SAW chirp-filter FH circuit. Shows sum-frequency response at ~ 70 MHz, about 15-dB higher than adjacent sidebands. Horizontal scale: 50 to 110 MHz. Vertical scale: 5 dB/div. Sweep time 5 s with 300-kHz resolution bandwidth. (Reprinted with permission from Saw, Smith, Edmonson, and Campbell, Reference [34], © IEEE, 1988.)

the chirp filters at different times and extracting either the mixer sum or difference frequencies in the overlap period for the chirp-impulse responses. (The mixer output will, of course, include the individual chirp responses plus a host of intermodulation products.) In Fig. 18.24(b) the down-chirp filter is shown as being impulsed at time τ_d after the up-chirp one. For this case the sum frequency in the impulse response overlap interval $\tau_d \leq t \leq T$ is

$$f_{sum1} = \left(f_1 + \mu_1 t\right) + \left(f_2 + \mu_2\left(t - \tau_d\right)\right) \quad \text{for} \quad \tau_d \leq t \leq T \qquad (18.22)$$

which, for $\mu_1 = -\mu_2$, reduces to a constant value given by

$$CASE\ 1: \ f_{sum1} = \left(f_1 + f_2\right) + \mu_1 \tau_d \quad \text{for} \quad \tau_d \leq t \leq T \qquad (18.23)$$

If chirp 1 in Fig. 18.24 is impulsed after chirp 2, the sum frequency takes the constant value

$$CASE\ 2: \ f_{sum2} = \left(f_1 + f_2\right) - \mu_1 \tau_d \quad \text{for} \quad \tau_d \leq t \leq T \qquad (18.24)$$

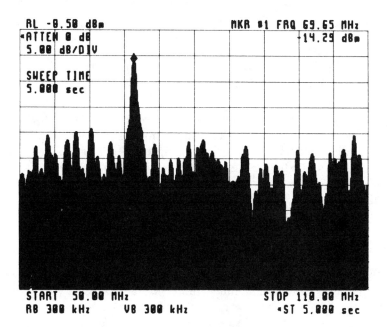

Fig. 18.25.   Sample spectrum analyzer plot of mixer-output in Fig. 18.24 due to impulse-driven SAW chirp-filter inputs. Sum-frequency response at 70 MHz is enhanced by ~15 dB. Horizontal scale: 50–110 MHz. Vertical scale: 5 dB/div. Sweep time: 5 s, with 300-kHz resolution bandwidth. Reprinted with permission from Saw, Smith, Edmonson, and Campbell, Reference [34], © IEEE, 1988.)

From Eqs. (18.23) and (18.24) the synthesizer frequency range is seen to be $2B$. The dwell time over which a frequency in this range can be synthesized varies from 0 to T, where T is the dispersion time for each chirp.

Figure 18.25 is an example of the mixer output spectrum obtained using SAW chirp filters of equal and opposite chirp slope, with bandwidth $B = 25\,\text{MHz}$ over the range 27.5 to 52.5 MHz. Here, the down chirp is impulsed after the up chirp to give a sum output at $\approx 70\,\text{MHz}$ which is about 15 dB higher than adjacent sidebands

Example 18.7 Linear FM Chirp Filter Mixing for Fast Frequency Hopping. Two linear FM chirp filters with the same TB parameters and equal and opposite chirp slopes are to be used in a chirp mixing circuit. Their centre frequency is 50 MHz, with 3-dB band edges at 75 and 25 MHz The dispersion of each filter is $T = 50\,\mu\text{s}$. Determine: (a) the chirp-slope magnitude μ; (b) the range of sum frequencies at the mixer output, and (c) the maximum and minimum dwell times. ■
Solution. (a) The chirp-slope magnitude $\mu = B/T = (75 - 25)/50 = 1\,\text{MHz}/\mu\text{s}$. (b) Start frequencies are $f_1 = 25\,\text{MHz}$ and $f_2 = 75\,\text{MHz}$. From Eq. (18.23) $f_{\text{sum1(max)}} = (25 + 75) + 50 = 150\,\text{MHz}$, while $f_{\text{sum2(min)}} = (25 + 75) - 50 = 50\,\text{MHz}$, for $2B = 100\,\text{MHz}$. (c) The maximum dwell time will be $50\,\mu\text{s}$ when both chirps are impulsed at the same time while the sum frequency is $f_{\text{sum}} = (f_1 + f_2) = (25 + 75) = 100\,\text{MHz}$. The dwell time becomes zero for $f_{\text{sum}} = 50\,\text{MHz}$ or $f_{\text{sum}} = 150\,\text{MHz}$. ■

Example 18.8 Maximum Number of Frequencies in Chirp-Mixing Synthesizer. (a) What is the maximum number of orthogonal frequencies F that can be obtained from a chirp-mixing synthesizer using SAW linear FM chirp filters with time-bandwidth parameters $T = 40\,\mu\text{s}$ and B = 50 MHz? (b) What is the minimum frequency step-size ? ■
Solution. (a) $F = TB/2 = (40 \times 10^{-6} \times 50 \times 10^6)/2 = 1000$. (b) The minimum frequency step size is $|\Delta f_{\text{step}}| = B/F = (50 \times 10^6)/1000 = 50\,\text{kHz}$. ■

18.11.3. EXAMPLE OF A HIGH-PERFORMANCE SAW-BASED SYNTHESIZER

An example of the forementioned chirp-mixing technique for spread-spectrum wireless data communications is now described [46]. This involved fast frequency-hopping at 200 khops/s between two bands under the control of a pseudonoise (PN) code. Two chirp mixers in the sum mode used four SAW chirp filters configured for operation in two bands, each 48-MHz wide. The upper band extended from 356 to 404 MHz, while the lower band ranged from 306 to 354 MHz. The output frequency hopped to one of 240 channels in either band depending on the PN code state. The two SAW chirp filters for the upper band had centre frequency $f_{ou}=190$ MHz, while the two for the lower band had centre frequency $f_{ol}=165$ MHz. The system used doubly-dispersive in-line SAW chirp filters on ST-X quartz, with dispersion $T=10\,\mu s$ and chirp slope $\mu=4.8$ MHz/μs. Dimensions of the modem were $30\times14\times7$ cm, while the power consumption was only 13 W.

In an experimental spread-spectrum link using binary FSK data modulation with this modem, bit-error rates BER $\approx10^{-1}$ were reported for signal-to-noise ratios S/N ~9 dB, as well as BER $\approx10^{-5}$ for S/N ≈18 dB.

Factors affecting the performance of such a frequency-hopping system include: a) impulse timing errors; b) chirp-slope mismatches and residual linear FM terms; c) chirp phase ripple which gives rise to a variable FM term depending on impulsing time τ_d, d) chirp centre-frequency errors which degrade frequency accuracy; and e) mixer spurious signals. Potentially, such a SAW-based technique should allow bandwidths of up to 500 MHz with up 4000 hop frequencies [46].

18.12. Summary

This chapter has reviewed a variety of surface-wave oscillator techniques for application, or potential application, to mobile/wireless circuitry. As a prelude to these circuit considerations, the chapter first reviewed oscillator noise mechanisms and their stability characterization in both the frequency and time domains. Fixed-frequency and tunable SAW oscillators were then examined together with representative examples, such as for SINAD and channel selectivity. This was followed by highlights of two Rayleigh-wave oscillator techniques for incorporation in transmit/receive circuitry for very low-power wireless systems for keyless entry and other security systems.

The principles of operation of multimode SAW oscillators were examined for application to frequency hopping in hybrid spread-spectrum communications [47]. This was followed by highlights of SSBW/STW filter and resonator techniques for use in higher-powered (30 dBm) STW VCOs for application in second-IF stages of mobile transceiver circuitry calling for enhanced narrowband tunability. STW VCO operation was also illustrated for an example of a 2-W wireless-security system with a range of 2 km.

Principles of injection-locked oscillators were presented, followed by an illustrative example of their application to carrier-recovery in BPSK-modulated data-communication systems. The chapter concluded with a description of a SAW-based frequency synthesis technique for fast frequency-hopping in spread-spectrum communications based on the mixing of responses from impulse-driven SAW linear FM chirp filters.

18.13. REFERENCES

1. M. Hikita, A. Sumioka and N. Fujiwara, "Voltage controlled oscillator and SAW resonator for use in radio communication equipments," *IEICE Technical Report*, MW-90-102, pp. 15–20, 1990.
2. K. Yamanouchi, T. Matsudo and M. Takeuchi, "Wide bandwidth SAW resonators and VCO using slanted interdigital transducers," *Proc. 1992 IEEE Ultrasonics Symposium*, vol. 1, pp. 57–60, 1992.
3. Modco Inc., "Miniature VCOs for Wireless Communication," *Microwave Journal*, vol. 39, pp. 310–312, May 1996.
4. R. Almar, B. Horine and J. Andersen, "High frequency STW resonator filters'" *Proc. 1992 IEEE Ultrasonics Symposium*, vol. 1, pp. 51-56, 1992.
5. R. C. Dorf (Ed), *The Electrical Engineering Handbook*, CRC Press, Boca Raton, Florida, p. 1446, 1993.
6. J. Saw, M. Suthers, J. Dai, Y. Xu, R. Leroux, J. Nisbet, G. Rabjohn and Z. Chen, "SAW technology in RF multichip modules for cellular systems," *Proc. 1995 IEEE Ultrasonics Symposium*, vol. 1, pp. 171–175, 1995.
7. L. S. Cutler and C. L Searle, "Some aspects of the theory and measurement of frequency fluctuations in frequency standards", *Proceedings of IEEE*, vol. 54, pp. 136–154, 1966.
8. D. W. Allan, J. H. Shoaf and D. Halford, "Statistics of Time and Frequency Data Analysis," in B. E. Blair, (ed), *Time and Frequency: Theory and Fundamentals*, National Bureau of Standards, Monograph 140, Washington, DC, pp. 151–204, May 1974.
9. D. B. Leeson, "A simple model of feedback oscillator noise spectrum," *Proceedings of IEEE*, vol. 54, pp. 329–330, February 1966.
10. G. Sauvage, "Phase noise in oscillators: a mathematical analysis of Leeson's model," *IEEE Trans. Instrumentation and Measurement*, vol. IM-26, pp. 408–410, December 1977.
11. T. E. Parker, "1/f phase noise in quartz delay lines and resonators," *Proc. 1979 IEEE Ultrasonics Symp.*, pp. 878–881, 1979.
12. Y. W. Lee: *Statistical Theory of Communication*, McGraw-Hill, New York, Chapter 13, 1960.
13. C. K. Campbell, "Z-transform analysis of a SAW delay line oscillator," *IEEE Trans. Sonics and Ultrasonics*, vol. SU-30, pp. 313–317, Sept. 1983.
14. Hewlett-Packard Note, "Spectrum Analysis . . . Noise Measurements," *Hewlett-Packard Application Note 150–4*, April 1974.
15. C. K. Campbell, J. J. Sferrazza Papa and P. J. Edmonson, "Study of a UHF mobile radio receiver using a voltage-controlled SAW local oscillator," *IEEE Trans. Sonics and Ultrasonics*, vol. SU-31, pp. 40–46, Jan. 1984.
16. T. P. Cameron, J. C. B. Saw and M. S. Suthers, "Applications of SAW technology in a SONET-compatible high capacity digital microwave radio," *Proc. 1992 IEEE Ultrasonics Symposium*, vol. 1, pp. 237–240, 1992.
17. T. E. Parker, "Precision surface acoustic wave (SAW) oscillators," *Proc. 1982 IEEE Ultrasonics Symposium*, pp. 268–274, 1982.
18. C. K. Campbell, P. J. Edmonson and P. M. Smith, "The phase noise characteristics of a driven SAW oscillator in the threshold vicinity for injection locking," *Proceedings 1985 IEEE Ultrasonics Symposium*, vol. 1, pp. 283–286, 1985.
19. J. Crabb, M. F. Lewis and J. D. Maines, "Surface acoustic wave oscillators: mode selection and frequency modulation," *Electronics Letters*, pp. 195–197, 17 May 1973.
20. P. S. Cross and S. S. Elliott, "Surface-acoustic-wave resonators," *Hewlett-Packard Journal*, vol. 32, December 1981.

21. C. K. Campbell, "SAW oscillators and resonators," *Proc. International Symposium on Surface Acoustic Wave Devices for Mobile Communication*, Sendai, Japan, pp. 171–178, 1992.

22. J. Andersen,"High-performance SAW oscillators," *SAW Scene*, Summer 1991, Sawtek Inc., Orlando, Florida.

23. Hewlett-Packard Note, "Application and measurements of low phase noise signals using the 8662A synthesized signal generator, *"Howlett-Packard Application Note #283-1*, November 1981.

24. P. Vizmuller, *RF Design Guide*, Artech House, Boston, p. 14, 1995.

25. W. R. Shreve, R. C. Bray, S. Elliott and Y. C. Chu, "Power dependence of aging in SAW resonators," *Proc. 1981 IEEE Ultrasonics Symposium*, vol. 1, pp. 94–99, 1981.

26. W. P. Robins, *Phase Noise in Signal Sources*, Peregrinus Press, London, 1982.

27. *1995 Product Data Book*, RF Monolithics, Inc., Dallas, Texas.

28. E. F. Scherer, "Compact multi-frequency STALO sources in frequency agile systems," *Microwave Journal*, pp. 41–44, March 1976.

29. P. Weissglas and S. Svensson, "Local oscillators for frequency agile systems," *Microwave Journal*, pp. 35–38, January 1977.

30. J. B. Fuller, "Implementation of radar performance by frequency agility, *IEE Conference Publication (Great Britain)*, No. 105, October 1973.

31. M. F. Lewis, "Practical frequency source for use in agile radar," *Electronics Letters*, vol. 21, pp. 1017–1018, October 1986.

32. W. C. Y. Lee, *Mobile Communications Engineering*, McGraw-Hill, New York, 1982.

33. C. G. Bailey and C. K. Campbell, "Design of a phase-locked multimode SAW oscillator," *IEEE Trans. Ultrasoics, Ferroelecttrics, and Frequency Control*, vol. 37, pp. 277–278, May 1990.

34. C. B. Saw, P. M. Smith, P. J. Edmonson and C. K. Campbell, "Mode selection in a multimode SAW oscillator using FM chirp mixing signal injection," *IEEE Trans. Ultrasonics, Ferroelectrics and Frequency Control*, vol. UFFC-35, May 1988.

35. K. V. Rousseau, K. H. Yen, K. F. Lau and A. M. Kong, "High Q, single mode S-band SBAW oscillators," *Proc. 1982 IEEE Ultrasonics Symp.*, vol. 1, pp. 279–283, 1982.

36. B. Auld. J. Gagnepain and M. Tan, Horizontal shear surface waves on corrugated surfaces," *Electronics Letters*, vol. 12, pp. 650–652, 1976.

37. A. Renard, J. Henaff and B. A. Auld, "SH surface wave propagation on corrugated surfaces of rotated y-cut quartz and berlinite crystals," *Proc. 1981 IEEE Ultrasonics Symp.*, 14–16 October 1981, Chicago, Illinois, vol. 1, pp. 123–127.

38. D. F. Thompson and B. A. Auld, "Surface transverse wave propagation under metal strip gratings," *Proc. 1986 IEEE Ultrasonics Symp.*, vol. 1, pp. 261–266, 1986.

39. A. Rønnekleiv, "High Q resonators based on surface transverse waves," *Proc. 1986 IEEE Ultrasonics Symp.*, vol. 1, pp. 257–260, 1986.

40. T. L. Bagwell and R. C. Bray, "Novel surface transverse wave resonators with low loss and high Q," *Proc. 1987 IEEE Ultrasonics Symp.*, vol. 1, pp. 319–324, 1987.

41. I. D. Avramov, "Microwave oscillators stabilized with surface transverse wave resonant devices," *Proc. 1992 IEEE Frequency Control Symp.*, pp. 391–408, 1992.

42. I. D. Avramov, Using surface transverse waves to guard your property," *Proc. 1994 IEEE Ultrasonics Symp.*, vol. 1, pp. 203–206, 1994.

43. C. K. Campbell, P. J. Edmonson and P. M. Smith, "Effect of amplifier saturation on the one-sided beat spectrum of a driven unlocked oscillator," *Electronics Letters*, vol. 28, pp. 1121–1122, June 1992.

44. R. Adler, "A study of locking phenomena in oscillators," Proc. IRE, vol. 34, pp. 351–357, June 1946. (Reprinted in *Proc. IEEE*, vol. 61, pp. 1380–1385, Oct. 1973.)

45. P. J. Edmonson, P. M. Smith and C. K. Campbell, "Injection locking techniques for a 1-GHz digital receiver using acoustic-wave devices," *IEEE Trans. Sonics Ultrasonics*, vol. SU-39, pp. 631–637, Sept. 1992.
46. B. J. Darby and J. M. Hannah, "Programmable frequency-hop synthesizers based on chirp mixing," *IEEE Trans. Sonics Ultrasonics*, vol. SU-28, No. 3, pp. 178–185, May 1981.
47. M. Kowatsch, J. Lafferl and F. J. Seifert, "Burst-communication modem based on SAW elastic convolvers," *Proc.1982 IEEE Ultrasonics Symp.*, vol. 1, pp. 262–267, 1982.

—19—

SAW Filters for Digital Microwave Radio, Fiber Optic, and Satellite Systems

19.1. Review of Coverage To This Point

Part 2 of this text has, so far, covered a variety of SAW-based techniques, devices and applications for mobile and wireless communications. Chapter 10 gave an overview of frequency bands and systems currently allocated for major analog and digital phone systems, as well as wireless bands for data and spread spectrum communications. Chapter 11 examined the design of surface wave resonators for fixed-frequency applications, including those for precision oscillators and low-power unlicensed wireless—for circuits such as used in keyless automotive entry, door openers, wireless security, bar-code readers, and other compact low-power devices. Bleustein-Gulyaer-Shimizu (BGS) resonators were also demonstrated for their use in feedback delay in DECT demodulator circuitry.

Chapter 12 covered the basics of a general type of Rayleigh-wave SPUDT filter design for either comb or single passband response implementation. Chapter 13 then dealt with applications of leaky-SAW filters for antenna duplexers and RF front-end filtering in mobile and wireless architectures. This included longitudinally coupled leaky-SAW resonator-filters, as well as ladder filters based on impedance-element structures. This was followed in Chapter 14 with discussions of some other illustrative RF filter structures of the non-resonant type for front-end and inter-stage filtering in mobile/wireless systems, including IIDT and FEUDT structures.

Chapter 15 dealt with SAW-based IF filters for mobile receivers and pagers. This included those for analog/digital cellular phones, digital cordless phones, and for PCS IF filtering. Those IF-filter structures examined were waveguide-coupled resonator-filters, longitudinally coupled resonator-filters, SPUDT-based filters, and Z-path filters. Typical IF frequency responses were given for a number of mobile phone systems, to meet channel and adjacent-channel specifications. Examples also were given for North American AMPS, IS-54, and PCS. The IF filter-response requirements, and spectral constraints, were also reviewed for European GSM and DECT requirements, while an illustrative design was presented for IF filtering in the Japanese PHP system.

Chapter 16 concerned the use of fixed-code SAW IDTs for use in spread-spectrum mobile communications. This included a review of matched-filter, direct-sequence spectral spreading, and processing-gain concepts. Some SAW-based techniques for BPSK, Barker, and quadraphase coding were also reviewed. Chapter 16 concluded with an example of an MSK-receiver for a 2.4-GHz spread-spectrum indoor link for processing Gold codes.

The coverage of SAW-based spread-spectrum techniques was expanded in Chapter 17, to incorporate real-time SAW-based convolvers and modems. Multipath interference in indoor environments was first reviewed. This was followed by the basics of correlation and convolution efficiency, before examining SAW-based real-time convolvers suitable for indoor and/ or outdoor spread-spectrum voice/data communications. Chapter 17 also included an outline of the operation of two synchronous SAW-convolvers, fabricated on lithium niobate substrates, and incorporating Kasami coding of the message signal. One of these was for packet-voice and the other was for packet-data. Chapter 17 concluded with an examination of the operation of an asynchronous full-duplex SAW convolver fabricated on a high-efficiency $ZnO/SiO_2/Si$ multilayer substrate, where the associated modems employed FSK or CSK modulation for sprectral spreading.

Chapter 18 reviewed various techniques for implementing SAW-based oscillators in mobile/wireless units. These included oscillator techniques suitable for application to superregenerative low-power receivers, VCOs for first- or second-IF stages in mobile receivers or base stations, frequency-hopping oscillators and chirp-filter synthesizers for hybrid DS-FH systems for indoor spread-spectrum modems, as well as for injection-locked oscillators for carrier recovery in data receivers.

19.2. Coverage of This Chapter

New technologies are being applied to meet ever-increasing consumer demands for voice or data communications from home or office. At the same time, current technologies are being significantly expanded or modified to handle these demands. Paging services are experiencing a major upsurge, while interactive two-way pagers are appearing on the market. Existing analog phone systems are being converted into dual analog/cellular ones, or are being replaced by digital or spread-spectrum services. To keep up with demands for instant global communications, fiber-optics SONET communications systems are under expansion, while long-haul digital microwave radio networks have been introduced to deal with SONET compatibility in regions where direct use of fiber-optic cabling is not possible or economical. Low or medium earth-orbit satellites with multibeam capability are being developed for increased flexibility in channelizing and routing of traffic at high data rates between base stations and mobile subscribers.

In concluding this coverage of SAW-based devices for mobile and wireless communications, this chapter illustrates some significant applications of SAW-filter technology to each of the three other wireless communications systems mentioned here, namely: 1) SAW Nyquist filters for digital microwave radio; 2) SAW clock-recovery filters for fiber-optic networks; and 3) SAW filters for on-board channelizing in orbiting satellites, as well as for ground stations. As a preamble to examining the use of SAW Nyquist filters in digital radio, we will first review pertinent codes and modulation schemes.

19.3. Digital Microwave Radio Concepts

19.3.1. DIGITAL MICROWAVE RADIO LINKS

With the advent of satellite communications, an increased amount of long-haul telephone and data communications traffic has been over microwave links [1]. Where digital message modulation is involved, the technology is termed "digital microwave radio," often abbreviated to "digital radio[1]." In North America the common-carrier bands for such communications are primarily at 4, 6, 8, and 11 GHz, employing bandwidths of 20, 30 and 40 MHz [2]. For enhanced spectral efficiency and high-capacity transmission of digitized voice and data signals, the trend has been to employ increasing levels of quadrature-amplitude-modulation (QAM). In the late 1970s, long-haul digital microwave radio systems in North America were operating in the 8-GHz band, with capacities of as high as 1344 voice-frequency circuits per channel, and channel bandwidths of 40 MHz [3]. In the early 1980s a 16-QAM digital system was reported, with an IF-filter bandwidth of about 50 MHz, and 140 Mb/s signalling rates [4]. This was followed by 64-QAM digital radio systems, for use in the North American digital network capability for 2016 voice channels, with a signalling-rate capability of 135 Mb/s in a 30-MHz bandwidth. As a result of the impact of SAW technology on these carrier systems, one 4-GHz North American system currently employs 512-QAM modulation (8 bits/s/Hz), with SONET

---

[1] The earliest "digital radio" is that relating to Morse-code transmissions!

compatibility, providing a transmission capacity at a standard SONET/SDH/ATM frequency of 622.08 Mb/s over two 40-MHz channels [3].

Surface acoustic wave (SAW) filters find application in two important areas of digital radio and digital communications. One area concerns their use as spectral-shaping filters for efficient data transmission with low bit-error-rate (BER), for filter fractional bandwidths of 70% or more [5]. Such filters are called *Nyquist filters*. In digital radio, SAW Nyquist filters are usually employed in 70- or 140-MHz IF stages.

19.3.2. DATA TRANSMISSION TERMINOLOGY AND UNITS

To date, a wealth of digital modulation techniques has been applied to data communications over satellite and other microwave communications links. Two such modulations referred to in this chapter concern the use of nonreturn-to-zero (NRZ) binary codes and those involving quadrature-amplitude-modulation (QAM). Binary codes are rated as power- efficient ones, where power efficiency (PE) can be defined as that value of carrier-to-noise ratio (C/N), which yields a satisfactorily low bit-error rate (BER) in discriminating between signal and noise at the decision sampling time. Typically, a modulation circuit containing additive white Gaussian noise may be considered to be power efficient for a $BER \ll 10^{-8}$ in a noisy channel with a carrier-to-noise ratio $C/N = 14\,dB$ [1]. QAM coding on the other hand is employed for efficient bandwidth utilization and spectral efficiency (SE). A modulation technique may be described as spectrally efficient for $SE > 2$ bits per s per Hz bandwidth (b/s/Hz) in a lowpass baseband filter channel, or for $S.E. > 1$ b/s/Hz in a bandpass channel.

Terms applicable to digital signal transmission are *bit rate* f_b and *symbol rate* f_s, where a symbol consists of an appropriate grouping of bits. For binary-coded transmissions $f_b = f_s$. For optimum spectral efficiency we wish to use the maximum digital signaling rate and minimum filter (channel) bandwidth, over which individual message symbols can be detected without corruption, (*i.e.* without Intersymbol Interference (ISI)). Filters designed to achieve the goal of ISI-free transmission are called Nyquist filters.

Example 19.1 Spectral Efficiency of Binary-Coded Channel. A binary-coded signal is transmitted at a data rate $f_b = 20$ Mbit/s through a baseband channel that is 12-MHz wide. What is the spectral efficiency? ■
Solution. $SE = 20/12 = 1.67$ b/s/Hz. ■

Example 19.2 Spectral Efficiency of NRZ-Coded Sequence. A 2-Mbit/s NRZ-coded sequence for a digital radio-relay system passes through a

spectral-shaping IF SAW filter with 1-MHz bandwidth. What is the spectral efficiency in this instance ? ∎
Solution. With this SAW filter SE = 2/1 = 2 b/s/Hz. ∎

19.3.3. POWER SPECTRUM OF NONRETURN-TO-ZERO (NRZ) CODES

In NRZ sequences such as the one illustrated in Fig. 19.1, successive binary "1's" are not separated by a return to the binary "0" state. It can be shown that the *power spectral density* $P_s(f)$ of an NRZ waveform has a sinc-function *squared* relationship given by

$$P_s(f) = 2A^2 T_s \left| \frac{\sin(\pi f T_s)}{\pi f T_s} \right|^2 \qquad (19.1)$$

where $|A|$ = signal magnitude and T_s = unit symbol (US) duration. This shows that the close-in spectral nulls of $P_s(f)$ occur at the signaling frequency $f = f_s = 1/T_s$. This distribution is illustrated in the spectrum analyzer plot of Fig. 19.2, as measured for a pseudo-random binary sequence (PRBS)

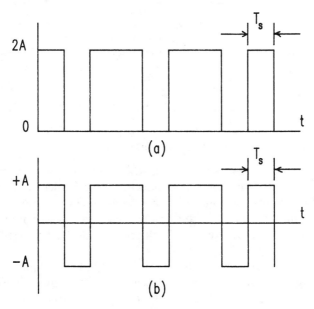

FIG. 19.1. (a) NRC sequence with a dc component. T_s is the symbol duration. (b) A balanced NRZ sequence.

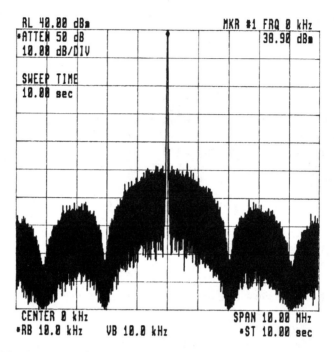

FIG. 19.2.   Spectrum analyzer measurement of an NRZ pseudo-random binary sequence at 2 MB/s signalling rate. Peak at center is the zero-reference line. Close-in adjacent nulls are at ±2 MHz. Horizontal span: 10 MHz. Vertical scale: 10 dB/div. Sweep time is 10 s.

at a signaling rate of 2 Mbit/s. In Fig. 19.2 the central peak is at 0 Hz while the adjacent spectral nulls are at ±2 MHz, as required by Eq. (19.1).

19.3.4. Quadrature-Amplitude Modulation

Quadrature-amplitude-modulation (QAM) codes are used in microwave digital radio for transmission with enhanced spectral efficiency over limited channel bandwidths. This modulation system allows two independent orthogonal signals to be transmitted in the same channel without interfering with each other. The required bandwidth reduces as the message bit stream is converted to higher QAM levels.

A basic outline for a transmitter employing QAM modulation is shown in Fig. 19.3; Fig. 19.4 outlines the receiver segments. Two controlling signals— termed the in-phase (I) and quadrature (Q) components—are required to implement complex carrier amplitude and phase states for quadrature modulation. The number of QAM states is 2^N, as determined by the number of binary bits per symbol. Thus a 16-QAM system, ($N = 4$), is one for which

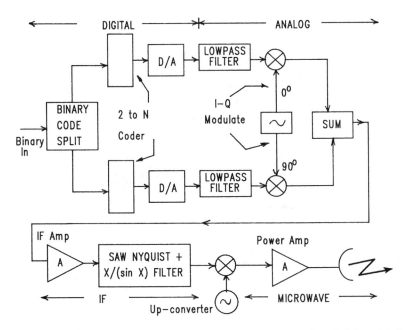

FIG. 19.3. Basic outline of digital microwave radio transmitter with QAM modulation, showing location of Nyquist filter.

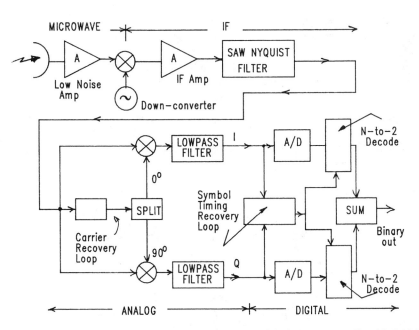

FIG. 19.4. Basic outline of corresponding receiver for digital microwave radio with QAM modulation, showing location of Nyquist filter.

Techniques, Devices and Mobile/Wireless Applications

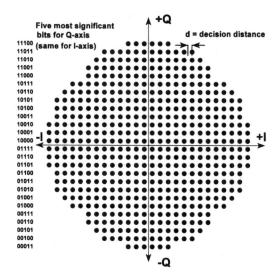

FIG. 19.5.   Constellation for 512-QAM modulation (After Reference [3].)

the (microwave) carrier is modulated into any one of sixteen different amplitude and phase states [6], [7]; the carrier of a 512-QAM system has 512 different amplitude and phase states. Both QAM circuit complexity and constraints increase drastically with the level of the QAM code. Figure 19.5 illustrates an idealized constellation for a 512-QAM modulation. In operation, however, the size of each cluster in a constellation will depend on the carrier-to-noise (C/N) ratio. As the system degrades with decreasing C/N ratio, the size of the individual clusters will increase as demonstrated in the experimental responses shown in Figure 19.6 for a 16-QAM system. Figure 19.7 illustrates a vector signal analyzer measurement [8] on a 64-QAM European Digital Video Broadcast (DVB) system. Crosses mark ideal symbol locations. Square dots show actual carrier values at symbol times.

19.4.   Nyquist Theorems and Filters

19.4.1.   RESPONSE OF AN IDEAL LINEAR-PHASE FILTER TO A SINGLE IMPULSE

Before considering the response of a transmission filter to a binary-coded signal sequence such as of the NRZ type, let us examine the impulse response behavior of an ideal lowpass (or bandpass) filter of Fig. 19.8 when subjected to a single delta-function input voltage stimulus as shown in

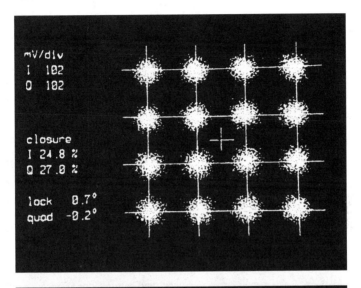

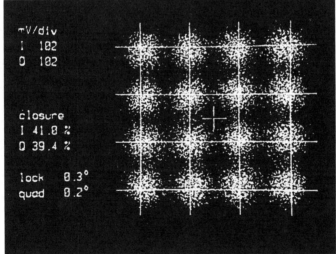

Fig. 19.6. (a) Constellation clusters in a 16-QAM digital microwave radio link with carrier-to-noise ratio C/N = 20 dB. (b) Cluster sizes increase as C/N ratio decreases to C/N = 15 dB. As measured on a Constellation Display. (Courtesy of Hewlett-Packard Canada Ltd.)

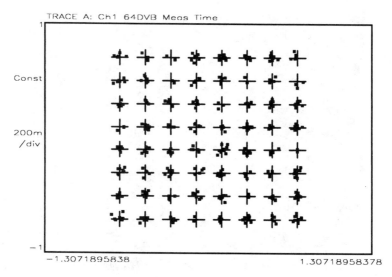

FIG. 19.7. Vector signal analyzer measurement and constellation display for a 64-QAM European Digital Video Broadcasting (DVB) system. Crosses mark ideal symbol locations. Square dots show actual carrier values, shown only at symbol times (Courtesy of Hewlett-Packard Company.)

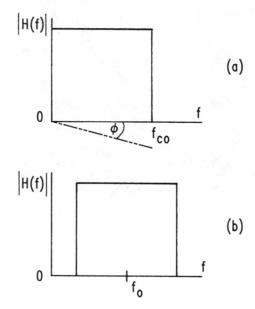

FIG. 19.8. (a) Frequency response of ideal linear phase lowpass filter. (b) Ideal bandpass filter response.

Fig. 19.9 and a sequence of delta-function impulse voltages of uniform time separation and arbitrary polarity as shown in Fig. 19.10.

With reference to Fig. 19.9, application of a single delta-function voltage to an ideal lowpass filter will yield impulse response $h(t)$, with a $|(\sin X)/X|$ amplitude response. For the bandpass filter, this amplitude modulation would appear as a modulation envelope on the carrier signal. To

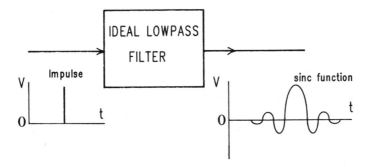

FIG. 19.9. Sinc-function impulse response $h(t)$ of ideal linear lowpass filter.

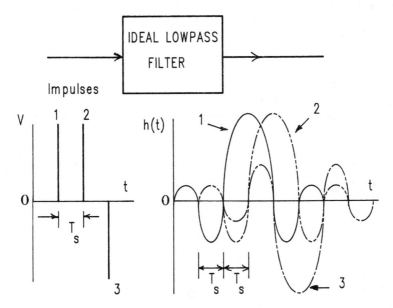

FIG. 19.10. Impulse response of idealized lowpass filter for three input voltage impulses at synchronous rate $f_s = 1/T_s$. This is free from intersymbol interference (ISI) at peaks of each sinc-function.

demonstrate this for the lowpass filter impulse response, first define the cut-off frequency f_{co} of the lowpass filter in terms of a Nyquist cut-off frequency, $f_{co} = f_N = 1/2T_s$, where T_s is the unit symbol (US) duration for the binary sequence, as in Fig. 19.1. The impulse response $h(t)$ is obtained as the inverse Fourier transform of the lowpass transfer function $H(f)$, namely

$$h(t) = \int_{-\infty}^{+\infty} H(f)e^{j2\pi ft}df, \qquad (19.2)$$

giving

$$h(t) = \frac{\sin(2\pi f_N t)}{(2\pi f_N t)} = \frac{\sin(\pi t/T_s)}{(\pi t/T_s)} \qquad (19.3)$$

in terms of Nyquist frequency f_N and unit symbol duration T_s. The delay through the filter is omitted for convenience, so that $h(t)$ has maximum amplitude at $t=0$, together with zero-crossings for $t=nT_s$ (n = integer).

19.4.2. NYQUIST BANDWIDTH THEOREM

Now consider the intermingled sinc-function impulse responses of the ideal lowpass filter when it is subjected to a sequence of uniformly spaced delta-function impulse voltages of random polarity, as sketched in Fig. 19.10. From this example, demonstrated for three input impulses uniformly spaced at intervals T_s, it can be deduced that it should be possible to detect any one of the sinc function responses without interference from any of the others. (When one sinc function response is maximum, all others have zero amplitude.) This is the result implicit in *Nyquist's Bandwidth Theorem* [1] for impulses applied at the synchronous rate $f_s = 1/T_s = 2f_N$ and represents the condition for complete freedom from ISI. While these considerations have tacitly assumed zero group delay through the filter for ease of illustration, the inclusion of a linear group delay term would only cause each of the sinc function impulse responses to experience the same delay. This would allow the maintenance of ISI-free transmission under ideal conditions.

19.4.3. NYQUIST VESTIGIAL SYMMETRY THEOREM

The idealized filters of Fig. 19.8 cannot be realized in practice. As a result — and to maintain ISI-free data transmission — the rectangular lowpass filter of Fig. 19.8(a) is modified in a particular way; it is handled through the addition of a skew-symmetric transfer function to the ideal filter $H_1(f)$ as illustrated in Fig. 19.11. This approach forms the basis of Nyquist's *Vestigial Symmetry Theorem*. The addition of this skew-symmetric response does not

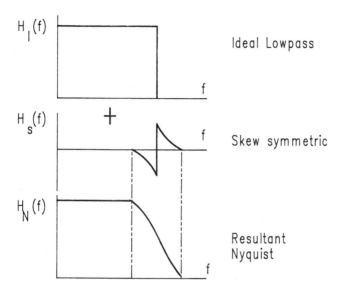

$H_I(f)$ — Ideal Lowpass

$H_s(f)$ — Skew symmetric

$H_N(f)$ — Resultant Nyquist

FIG. 19.11. Application of a skew-symmetric response to ideal lowpass filter. Yields practical filter design for realizing ISI-free transmission.

corrupt the timing of the signal responses at zero crossover points, thereby maintaining the necessary condition for ISI-free data transmission at the synchronous rate $f_s = 2f_N$.

A particular filter function that satisfies the Vestigial Symmetry Theorem is the *raised cosine function* $H_{RC}(f)$ given in lowpass form

$$H_{RC}(f) = \begin{cases} 1 & \text{for } 0 < f < 1(1-\alpha)/2T_s \\ \cos^2\left(\dfrac{T_s\pi}{4\alpha}\left[2f - \dfrac{(1-\alpha)}{T_s}\right]\right) & \text{for } \dfrac{(1-\alpha)}{2T_s} < f < \dfrac{(1+\alpha)}{2T_s} \\ 0 & \text{for } f > 1(1-\alpha)/2T_s \end{cases}$$

(19.4)

where α is defined as the *channel roll-off factor*. The condition $\alpha = 0$ corresponds to a physically unrealizable filter. A 50% excess bandwidth is employed for $\alpha = 0.5$, while $\alpha = 1$ results in the channel bandwidth being twice that for $\alpha = 0$.

In digital microwave radio, the filter appropriate to the bandpass form of Eq. (19.4) can theoretically be partitioned in an arbitrary manner between transmitter and receiver, provided the product function equals that of Eq. (19.4). For matched filter performance, however, the function must be split equally between transmitter and receiver. The individual Nyquist filter

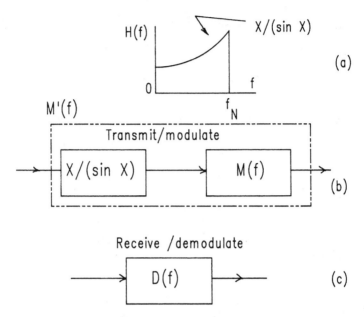

FIG. 19.12. (a) Amplitude equalizer compensates for power spectral response of NRZ sequence. (b) Equalizer and Nyquist filters in transmitter can be combined. (c) Nyquist-matched filter in digital microwave radio receiver.

response $M(f)$ for the transmitter (modulator) stage and that for the receiver (demodulator) stage, $D(f)$, are

$$M(f) = D(f) = \sqrt{H_{RC}(f)} \tag{19.5}$$

as shown in Fig. 19.12.

The Nyquist filter relations given in Eqs. (19.4) and (19.5) are for lowpass filters. The results may be applied to SAW bandpass filter realization and implementation by a simple change of variable from frequency f to $(f-f_o)$ at bandpass centre frequency f_o.

19.4.4. Nyquist Filters and Matched Filters

Previous chapters of this text have examined the design and use of SAW matched filters for attainment of optimal signal-to-noise (S/N) performance. In this respect, however, some analysts consider that such optimal performance is strictly valid only for (a) wideband systems where ISI is negligible or (b) for the correlation of single pulses, as in pulse compression radar. As a result the matched filter and the Nyquist filter are sometimes considered to be equivalent only in the event that the matched filter has additionally been designed for ISI-free data transmission performance when a sequence of synchronous pulses or bits is transmitted.

19.5. Illustrative SAW Nyquist Filter Response

19.5.1. DESIGN REQUIREMENTS AND RESTRICTIONS

Many factors can degrade the BER performance and increase the Intersymbol Interference within a digital microwave radio system. These include amplitude and group delay ripple in the Nyquist filters, as well as misaligned modulator and demodulator circuitry, phase noise in the timing recovery circuit, power amplifier nonlinearities, and noise in the transmission channel.

All SAW filters for digital radio must meet stringent specifications if the BER within the system is to be minimized. In terms of circuit performance, this means minimizing all second-order degradation due to diffraction, bulk waves, circuit loading, and IDT end-effects. The IDT end-effect by itself can yield unacceptable passband amplitude and group delay ripple, thereby degrading the ISI performance. In some designs stray coupling is reduced by using offset IDTs in conjunction with a multistrip coupler. Other designs employ in-line IDTs with offset apodization, such as the "V-line" structure. With the V-line structure, fingers with short overlaps near the ends of the IDTs have very little coupling to ground [9], [10].

In implementing the Nyquist channels in both the transmitter and receiver stages, two needs have to be met in employing signaling waveforms such as the NRZ type. First of all, the transmission system must include a spectral shaping (amplitude equalizer) filter to compensate for the $|(\sin X)/X|^2$ power spectral distribution associated with NRZ waveforms in Eq. (19.1). And ISI-free transmission would not be attainable without this compensation, because the zero crossover points of the filter impulse response would be corrupted. This compensation is readily attained using a SAW filter with the reciprocal amplitude response $H_s(f)$,

$$H_s(f) = \frac{\pi(f - f_o)T_s}{\sin[\pi(f - f_o)]T_s} \tag{19.6}$$

for bandpass implementation.

Second, for matched filtering the Nyquist raised-cosine filter function is equally partitioned between transmitter and receiver as in Eq. (19.5). The two SAW filters required in the transmitter may be combined into a single SAW filter with the transfer function $M'(f) = M(f)H_s(f)$, to accommodate

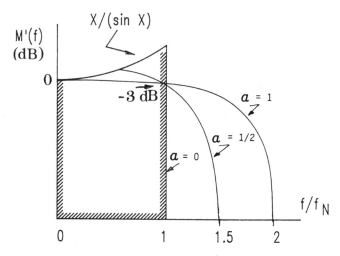

Fɪɢ. 19.13. Lowpass prototype of Nyquist raised-cosine response combined with X/(sin X) amplitude equalizer in digital microwave radio transmitter, for various values of roll-off factor α. Note that $\alpha = 0$ represents an unrealizable design.

both $H_S(f)$ and $(H_{RC}(f))^{1/2}$. Figure 19.13 shows the lowpass prototype frequency response of $M'(f)$ in the transmitter for various values of roll-off factor α.

As well as minimizing ISI, the SAW Nyquist filter must also be designed to minimize adjacent channel interference (ACI) from channels on *each* side of the desired message channel. This means rigid control of its shape factor and roll-off response at passband edges. One technique for obtaining the finger apodization patterns for the IDTs involves an adaptation of the Remez optimization algorithm, originally applied to digital filter synthesis [11]–[13]. For a given data rate, the filter roll-off will vary *inversely* with the number of modulation levels. In a design reported for a 512-QAM system [3], the carrier-to-interference ratio (C/I) was specified as 50 dB, requiring a channel roll-off factor $\alpha = 0.12$. Figure 19.14 shows the experimental amplitude and phase responses of a 140-MHz SAW Nyquist filter in a 512-QAM digital microwave radio.

Another important factor in SAW Nyquist filter design relates to its insertion loss. In considering its effects on circuit performance, the reader will recall from Table 4.1 and Eq. (4.13), repeated here, that an estimate of triple-transit interference (TTI) level in filters with bidirectional IDTs, when the loss is greater than about 10 dB, is

$$TTI \; suppression \approx 6 + 2(insertion \; \text{loss}) \quad (dB) \qquad (4.13)$$

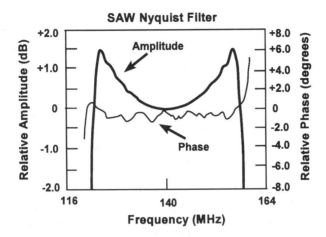

Fig. 19.14. Amplitude and phase responses of illustrative 140-MHz SAW Nyquist filter in a 512-QAM digital microwave radio. Horizontal scale : 116 to 164 MHz. Vertical scales: relative amplitude is −2 to +2 dB, while phase deviation from linearity is over a −8 to +8 degree range. (After Reference [3].)

where the factor of 6 in Eq. (4.13) relates to the minimum insertion loss of the SAW filter with matched source and load impedances. Moreover, SAW Nyquist filters are often significantly mismatched for suppression of TTI, with resultant insertion losses in the range of 20 to 30 dB. The noise figure of a matched SAW filter is approximately equal to its matched insertion loss. To minimize the detrimental effect of such noise figure levels on system performance, it is most desirable to drive the SAW Nyquist filter with a high-gain low-noise preamplifier; this serves to minimize noise degradation.

Example 19.3 SAW Nyquist Filter Using NRZ Transmission. A SAW Nyquist filter is to be designed for a digital radio system using NRZ transmission. The center frequency of the SAW bandpass filter is to be $f_o = 140$ MHz, with an NRZ signaling rate of $f_b = 140$ Mbit/s. What would be the minimum (unrealizable) bandwidth of a raised-cosine channel with roll-off factor $\alpha = 0$? ∎

Solution. For $f_b = 140$ Mb/s, the corresponding Nyquist cut-off frequency is $f_N = f_s/2 = 70$ MHz, so that the minimum unrealizable bandwidth (with $\alpha = 0$) is $2 f_N = 140$ MHz. ∎

Example 19.4 Digital Radio Using 16-QAM Coding. The digital radio system in the previous example is replaced by a 16-QAM system.

Determine (a) the required minimum bandwidth of the spectral shaping demodulator SAW filter for this system for an unrealizable filter roll-off factor $\alpha = 0$, and (b) the spectral efficiency SE. ∎

Solution. (a) In binary transmission systems, the signaling frequency f_s = the bit rate f_b and the unit symbol duration $T_s = T_b$. In multilevel systems, however, we must consider the symbol duration $T_s = T_b \cdot \log_2 L$ where L = number of signalling levels. Here $L = 16$, giving $f_s = 140/4 = 35$ Mbit/s for 4-bit symbols. The corresponding Nyquist frequency is $f_N = 35/2 = 17.5$ MHz in the lowpass prototype. The minimum IF bandwidth is $2 \times 17.5 = 35$ MHz. (b) The (unrealizable) spectral efficiency is SE = $140/35 = 4$ b/s/Hz. ∎

Example 19.5 SAW Nyquist Filter with Roll-Off Factor $\alpha = 0.5$. The SAW Nyquist filter in Example 19.4 is to be redesigned for a practical roll-off factor $\alpha = 0.5$. Determine (a) the required minimum bandwidth and (b) the resultant spectral efficiency. ∎

Solution. (a) The cut-off frequency in the lowpass filter prototype is $f_{co} = f_N + \alpha \, f_N = 17.5 + (0.5 \times 17.5) = 26.25$ MHz. The required IF bandwidth is $2 \times 26.25 = 52.5$ MHz. (b) The IF spectral efficiency is reduced to SE = $140/52.5 \approx 2.4$ b/s/Hz. ∎

Example 19.6 Digital Radio Using 512-QAM Coding. The digital radio system in Example 19.3 is replaced by a 512-QAM system. Determine (a) the required minimum bandwidth of the spectral-shaping demodulator SAW filter for this system for an unrealizable filter roll-off factor $\alpha = 0$ and (b) the spectral efficiency SE. ∎

Solution. (a) The signaling frequency f_s = the bit rate f_b and the unit symbol duration $T_s = T_b$. The symbol duration $T_s = T_b \cdot \log_2 L$ where L = number of signaling levels is now L = 512. Consider this as a reduced 1024-QAM scheme, with 8-bit symbols. This gives $f_s = 140/8 = 17.5$ Mbit/s for 8-bit symbols. The corresponding Nyquist frequency is $f_N = 17.5/2 = 8.75$ MHz in the lowpass prototype. The minimum IF bandwidth is $2 \times 8.75 = 17.5$ MHz. (b) The (unrealizable) spectral efficiency is SE = $140/17.5 = 8$ b/s/Hz. ∎

19.5.2. A Basic 16-QAM Digital Radio System

Figure 19.15 outlines the main features of an illustrative 140 Mb/s digital radio system, using 16-QAM transmission [4]. In the transmitter, an input digital signal with a bit rate $f_b = 140$ Mbit/s is fed into the coder circuit. This causes a serial-to-parallel conversion of 4 bits into 1 symbol at a rate $f_s = 140/4 = 35$ Mbit/s, as in Example 19.4. In the modulator stage each of these 4-bit symbols is mapped into one of 16 possible states of the 140 MHz carrier, as illustrated in the constellation display of Fig. 19.6. The circuitry for achieving this is based or the use of ring mixers and quadrature carriers supplied

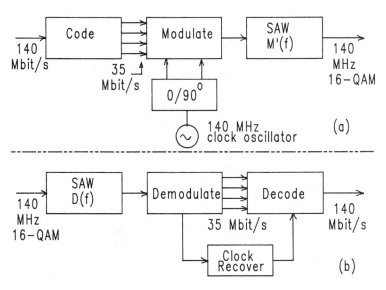

FIG. 19.15. Outline of a 140-Mb/s 16-QAM digital microwave radio with (a) a combined SAW Nyquist and $X/(\sin X)$ amplitude equalizer in the transmitter, and (b) a SAW Nyquist matched filter in the receiver. (After Reference [4].)

by a 140-MHz crystal oscillator and 90° phase-shift network. The single SAW spectral shaping filter in the transmitter includes a raised-cosine frequency response, together with an $X/(\sin X)$ spectral-shaping filter. The Nyquist filter is partitioned equally between transmitter and receiver, for matched filtering. The SAW filters were fabricated on 128° YX-lithium niobate. The filter bandwidth between -60 dB points was or the order of 50 MHz. (Compare Example 19.5.) ISI was found to be negligible, while the simulated degradation of the carrier-to-noise (C/N) ratio due to the SAW filters was only 0.2 dB at a BER of 10^{-8}.

19.6. IF Filters for Digital Radio Employing Slanted-Finger IDTs

As introduced in Chapter 8, slanted finger IDTs find application for IF filtering in digital microwave radio systems. Figure 19.16(a) shows the fabrication of a 70-MHz slanted-finger IDT on a SAW substrate, as designed for a 3-dB bandwidth of 10%. This was designed for use in a nonregenerative 4-GHz band digital multiplex radio transmission system operating at about 6 Mb/s and requiring an IF filter with very-low passband ripple. Figure 19.16(b) shows its flat passband response and high out-of-band suppression.

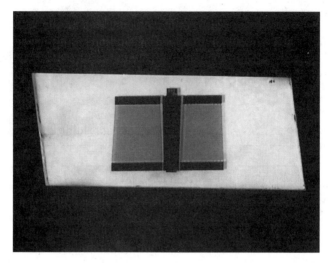

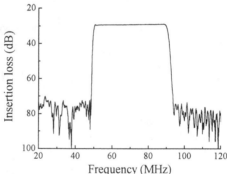

Fig. 19.16. (a) 70-MHz slanted-IDT IF filter, with 3-dB fractional bandwidth of 10%, for nonregenerative 4-GHz digital multiplex radio transmission system operating at ~6 Mb/s. (b) Frequency response shows required flat passband and high out-of-band suppression. Horizontal scale: 50 to 90 MHz. Vertical scale: 20 to 100 dB. (Courtesy of Japan Radio Co., Tokyo, Japan.)

19.7. SAW Filters for Clock Recovery in Optical Fiber Data Systems

19.7.1. CLOCK RECOVERY CIRCUITS

The second important area of application highlighted in this chapter relates to the use of SAW transversal filters in clock-recovery circuits for regenerative repeaters in optical-fiber data-communications systems. These SAW filters operate at center frequencies up to the 2 GHz range, commensurate

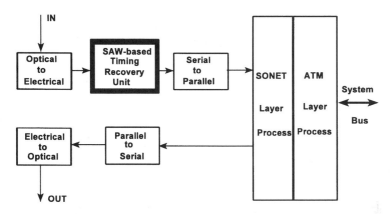

FIG. 19.17. Outline of functional components for fiber-optic SONET/ATM network interface card. (After Reference [14].)

with the signaling rate, and with effective Qs in the approximate range $800 \leq Q_e \leq 1500$. These filters can play a vital part in the attainment of low BER performance, aimed at $BER < 10^{-11}$ per repeater, in conjunction with exceedingly stringent demands pertaining to long-life and reliability.

In long-haul optical-fiber digital-communications systems employing pulse-code-modulation (PCM), a sequence of optical fibers is connected to digital regenerative repeater circuits, which "refreshes" the incoming digital waveforms that have been corrupted by attenuation and dispersion [4]. Both PLL-based clock and data-recovery modules are available, as well as transceiver chip sets with built-in clock and data-recovery functions. The SAW-based timing-recovery modules can have the best jitter performance in many instances, for example, for those ATM/SONET/SDH network interfaces[2] illustrated in Fig. 19.17 [14]. (The abbreviation SDH refers to synchronous digital hierarchy for trunk transmission.) This is a byte-oriented system, known as STM-1[3], with a basic bit rate of 155.52 Mb/s. (See Table 19.1 on page 598.) A main difference between SDH and SONET is that the latter has a basic bit rate of 45 Mb/s, of which three fit into one STM-1 frame. Higher bit rates are obtained in both SONET and SDH systems by byte interleaving of the STM-1 frames [15], which yield bit rates of 622.08 Mb/s (STM-4) and 2488.32 Mb/s (STM-16), as given in Table 19.1. The advantage of SDH is that it is backwards compatible with current

---

[2] ATM/SONET/SDH stands for *Asynchronous transfer mode/Synchronous optical network/Synchronous digital hierarchy.*

[3] STM stands for *synchronous transfer mode.*

TABLE 19.1

SPECIFICATIONS FOR ILLUSTRATIVE SAW FILTERS FOR SONET/
SDH/ATM CLOCK RECOVERY (After Reference [14])

Frequency (MHz)	155.52	622.08	2488.32
3-dB Q-value	420	800	750
Insertion Loss (dB)	17	15.5	19.5
Phase slope (degrees/kHz)	−0.72	−0.33	−0.07

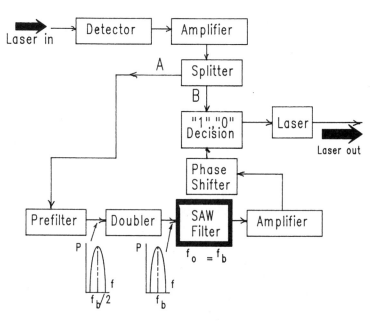

FIG. 19.18. Basic functional blocks for clock-recovery circuit in typical regenerative repeater for fiber-optic communications link employing NRZ modulation. Employs precision narrowband SAW filter, centered at clock frequency $f_o = f_b$ for jitter suppression.

American, European and Japanese transmission standards. Currently, most new trunk connections employ the SDH standard. The ATM mode offers flexibility in trunk capacity and throughput, because it can vary the transmission byte-rate [15].

Figure 19.18 outlines the circuit basics of an optical regenerator employing NRZ modulation at transmission clock rate f_b. The input optical signal is first detected and amplified. The output from the amplifier is applied to two circuits, along paths labeled A and B. That along A proceeds to the clock-recovery stage, incorporating a SAW transversal filter at centre frequency $f_o = f_b$. Proceeding to the next stage of waveform processing, recall from Eq. (19.1) and Fig. 19.2 that the power spectral nulls of an NRZ sequence occur at the signaling rate f_b. Before application to the SAW filter, this signal passes through an optional prefilter stage, which peaks the spectral response at $f_b/2$. The prefilter output goes to a squaring (i.e., frequency-doubling) circuit, which can be readily implemented using a nonlinear transistor amplifier, followed by a bandpass filter centered at the second harmonic [16]. The output from the doubler circuit has its spectral peak at the signalling clock rate f_b, for application to the SAW transversal filter with $f_o = f_b$.

Clock-extraction frequencies for SONET/SDH/ATM boards include 155.52 MHz, 622.08 MHz, and 2488.32 MHz. Those SAW-based filters for the 2488.32-MHz clock-recovery circuits have been fabricated using surface skimming bulk wave (SSBW) IDTs to reduce restrictions on electrode geometries and tolerances [14]. Figure 19.19 illustrates a commercial 622.08-MHz clock-recovery filter, incorporating the central components in Fig. 19.18 into a single ASIC.

Composite layered substrates have also been applied to the fabrication of SAW retiming filters for 2.5-GHz circuitry. Figure 19.20 shows a jig for

FIG. 19.19. Highly integrated SAW-based 622.08-MHz timing module in ASIC package, for SONET/SDH/ATM applications. (Courtesy of Vectron Technologies, Inc., Hudson, New Hampshire.)

FIG. 19.20. Jig for testing individual 2.5-GHz SAW filters fabricated on two-inch wafer of composite layered SiO₂/ZnO/Diamond/Si. (Courtesy of Sumitomo Electric Industries, Itami, Japan.)

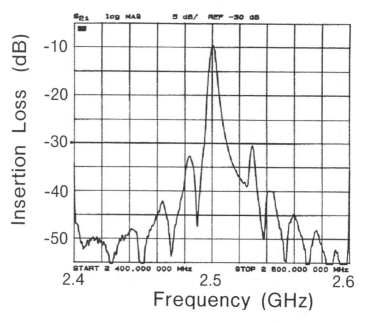

FIG. 19.21. Measured frequency response of a 2.5-GHz SAW filter chip on layered wafer in Fig. 19.20. Insertion loss is 10 dB, with Q = ~ 650. Horizontal scale: 2.4 to 2.6 GHz. Vertical scale: 5 dB/div. (Courtesy of Sumitomo Electric Industries, Itami, Japan.)

testing individual 2.5-GHz SAW filters of SiO$_2$/ZnO/Diamond/Si on a two-inch wafer. Figure 19.21 shows the measured frequency response of one of the SAW filter chips on the two-inch wafer in Fig. 19.20, with insertion loss of 10 dB and $Q = {\sim}650$. These 2.5-GHz layered structures can have temperature coefficients of frequency (TCF) of less than 2 ppm/°C by optimizing the thickness of each layer. Operation is in the first Sezawa Rayleigh-wave mode with a velocity of 9000 m/s. This allows electrode line-widths of 0.9 μm, which is more than twice the width for conventional designs on quartz substrates [17], [18].

19.7.2. BANDWIDTH OF CLOCK-RECOVERY FILTER

The bandwidth of the SAW clock-recovery filter is always chosen to be much greater than the spectral response due to the clock at f_b. As a result the SAW filter "sees" a strong spectral component at its center frequency f_o, with some residual message spectrum. The output from the filter, which is essentially sinusoidal, is amplified and passed through a phase shifter, to gate the decision circuit in path B. The phase shifter is required to adjust the zero-crossings of the sinusoidal signal to the optimum times for the "1," "0" decision gating circuits that control the modulation of the output laser.

Extremely high precision is required in the decision circuitry, requiring BER levels to be kept to BER $\approx 10^{-11}$ in each repeater circuit [4]. Additionally, the timing recovery filter must have an impulse ringing time (i.e., group delay) τ, which is long enough to maintain a sufficiently large sinusoidal output over the largest number of anticipated strings of continuous "1" or "0" occurrences. It is reported that the BER in such optical repeater circuits is most affected by detuning due to static phase errors [4]. In its implications to the SAW filter components of interest here, this means that detuning of the SAW filter by aging or temperature drifts must be severely curtailed.

19.7.3. RESTRICTIONS ON THE Q OF THE SAW CLOCK FILTER

In considering the drifting of the SAW clock-filter response, a fractional detuning parameter η has been employed and is given by

$$\eta = \frac{\left[f_o - f_b\right]}{f_b} \tag{19.7}$$

where f_b = message signal rate and f_o = centre frequency of the clock-recovery filter, with effective Q_e

$$Q_e = \frac{f_o}{\Delta f} = \pi f_o \tau \tag{19.8}$$

where $\Delta f = 3$-dB bandwidth of the SAW filter and $\tau =$ group delay or ringing time.

Studies of the sources of static detuning over the entire assembly of repeater links for worst-case detuning have placed a limit on the maximum usable value of Q_e in the range $Q_{eu} < \approx 800$. At the other extreme, requirements on filter ringing time place a lower one $Q_{el} > \approx 165$ [4]. In support of these, Table 19.1 gives the bandwidth specifications for three such commercial SAW filters for clock recovery on SONET/SDH/ATM boards.

Whereas one-pole two-port SAW resonators generally have much higher Q values than the upper limit of 800 considered in the foregoing, the use of SAW transversal or delay-line filters appears to be favored in such circuitry. Recall that the 4-dB filter fractional bandwidth of a SAW delay line is BW_4 % $\approx 1/N_p$ where $N_p =$ number of finger pairs in input IDT. Estimating the 3-dB bandwidth as BW_3 % $\approx 0.9 BW_4$ %, it is seen that for $Q_e = 900$, the number of required finger pairs in the SAW IDT would be $N_p \approx 1000$. Even with the use of split-electrode IDT geometry to minimize spurious reflections at center frequency, it can be appreciated that such SAW clock-filter designs at frequencies up into the 2-GHz range can be exceedingly exacting ones.

19.8.   SAW Filters for Satellite Systems

19.8.1.   FIXED AND MOBILE SATELLITE SERVICES

Satellite systems may generally be classified as those involving fixed satellite services (FSS) and also mobile satellite services (MSS). The FSS systems have provided wideband communications between major urban areas. Transmissions for the FSS geostationary satellites [19] normally are based around C-band (with uplink from 5.9 to 6.4 GHz, and down-link from 3.7 to 4.2 GHz), and Ku-band (with up-link from 14.0 to 14.5 GHz, and down-link from 11.7 to 12.2 GHz).

A basic outline of a satellite repeater is sketched in Fig. 19.22. (Although it is illustrated here for a single frequency conversion, dual frequency conversion can also be employed in some systems.) In its operation, the repeater of Fig. 19.22 receives up-link signals, into low-noise amplifiers (LNAs). This is followed by down-conversion to the lower return frequency. The input frequency band is divided into channels (also called *transponders*) by input multiplexers (MUXs). The signals in each channel are amplified by either a solid-state power amplifier or by a travelling-wave tube amplifier, before being recombined in output multiplexers for retransmission to the down-link. While frequency modulation has been widely used to date, digital modulation techniques are being applied to digital-

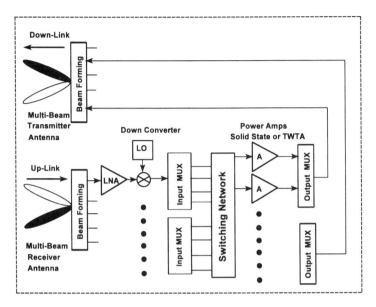

Fɪɢ. 19.22.   Outline of basic on-board processor for satellite up-link and down-link.

voice and video compression [19]. Multiple antenna beams allow for enhanced frequency reuse, as well as higher antenna gains that give higher levels of effective isotropic radiated power (EIRP) for users.

19.8.2.   SAW Fɪʟᴛᴇʀs ɪɴ FSS ᴀɴᴅ MSS Sᴀᴛᴇʟʟɪᴛᴇs

In typical FSS satellite processors, and with reference to Fig. 19.22, the 500-MHz input band for the C- or Ku-transmissions is divided into channels which can be 36, 54, or 72 MHz wide, with guard bands of 4, 6, and 8 MHz, respectively. In allocating customer traffic to transponders, each channel can be subdivided into smaller segments of SAW filter banks [20].

The MSS systems include the INMARSAT-3 system of geostationary satellites, which provide links between mobile users and base stations, where the base station link is at C-band and the user link is at L-Band (1.6-GHz band). Each satellite can employ one global beam and seven spot beams [20]. On-board SAW-based processors are employed for the up-link and down-link transmissions. The number of SAW filters in a forward processor will depend on the system design. One system employs down-conversion to L-band, following which the signals are divided between 15 filter modules, with a further down-conversion to an IF of 160 MHz, for a total of 168 SAW filters on ST-quartz substrates.

19.8.3. SATELLITES FOR PERSONAL COMMUNICATION NETWORKS

While geostationary commercial-traffic satellites at orbital altitudes of 35,785 km can also be used as personal communications systems, they would require higher-power levels to compensate for the small antenna apertures of small mobile terminals. Because of their altitude, they could also suffer from reduction in acceptable signal performance due to relatively long delays between satellite and ground terminals.

These problems can be overcome by the use of low earth orbit (LEO) satellites (with orbital altitudes ~ 765–1389 km) as sketched in Fig. 19.23, or medium earth orbit (MEO) satellites (with orbital altitudes ~ 10,354 km). There would, of course be a trade-off in their use. As they would no longer be in geostationary orbits, this would mean that a number of spaced

LEO Satellites

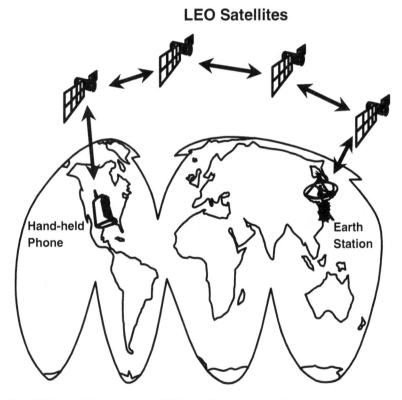

FIG. 19.23. Artistic impression of LEO satellites for personal network communications at L-band.

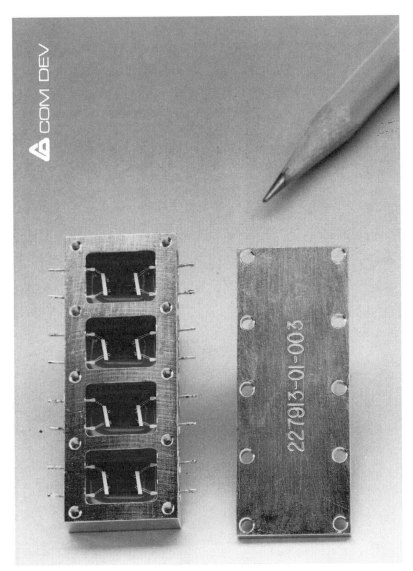

FIG. 19.24. A satellite on-board L-band (1.6 GHz) SAW filter bank, with four filters of different bandwidths. Intended for routing traffic in upcoming satellite systems for personal network communications. (Courtesy of COM DEV, Cambridge, Ontario, Canada.)

satellites would be required to give continuous coverage to a particular location. There are a number of such LEO/MEO satellite systems under consideration at this time, including ODYSSEY, GLOBALSTAR, IRIDIUM, and ELLIPSO. Up-links from users to satellites would be at L-Band (1.6 GHz); down-links from satellite to user would be at S-band (2.5 GHz). The number of satellites required by these systems would be as high as 66, depending on the concept [21].

Figure 19.24 illustrates a bank of four L-Band (1.6 GHz) SAW filters intended for use in such satellite-based personal communications networks. These SAW filters have different bandwidths, and are used to route traffic

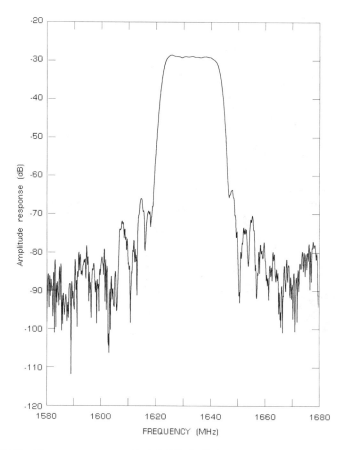

Fig. 19.25. Frequency response of a 1.6-GHz SAW filter for on-board satellite processing in upcoming satellite systems for personal network communications. Horizontal scale: 1580–1680 MHz. Vertical scale: −120 to −20 dB. (Courtesy of COM DEV, Cambridge, Ontario, Canada.)

to various satellite beams. The switching of satellite bandwidths to various locations thereby ensures the optimal usage of satellite capacity. The frequency response of one such L-band SAW filter is shown in Fig. 19.25.

19.8.4. IF FILTERS FOR SATELLITE EARTH STATIONS EMPLOYING SLANTED-FINGER IDTS

Wideband SAW IF filters employing slanted-finger IDTs also find application in satellite communications systems. Figure 19.26 illustrates a 70-MHz SAW IF filter with a 3-dB fractional bandwidth of 50%. This is used in a

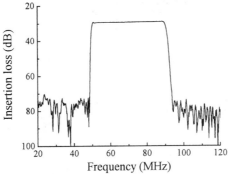

FIG. 19.26. (a) Lithographic patterns for slanted-IDT 70-MHz SAW IF filters with fractional bandwidth of 50%, for digital data terminal in mobile earth station for INMARSAT-C satellite system. (b) Frequency response showing flat passband and large out-of-band suppression. Horizontal scale: 20 to 120 MHz. Vertical scale: 20 to 100 dB. (Courtesy of Japan Radio Co., Tokyo, Japan.)

digital data terminal for a mobile earth station (MES), in conjunction with an INMARSAT-C satellite communications system. Again note the flat passband and large out-of-band suppression required for these filters.

19.9. Summary

This chapter has highlighted some applications of SAW filter technology in digital microwave radio, fiber optic and satellite systems. The principles of Nyquist filter design were presented for digital radio communications, with emphasis on the design and use of SAW filters in such systems. Two modulation schemes, one relation to NRZ sequences and the other to 16-QAM modulation, as well as examples were given to illustrate the parameter values involved.

Following this, SAW delay-line transversal filters were illustrated for application as timing recovery filters in SONET-compatible long-haul optical fiber systems for digital communications. The restrictions on the permissible Q-values of the SAW delay lines in the clock-recovery circuits were also examined.

The chapter concludes with a brief review of some satellite frequencies, on-board processor systems for FSS and MSS satellite systems, and the role of SAW filters in on-board processors. In addition, the use of LEO/MEO satellites for personal communications networks was also introduced, together with the use of on-board SAW filter banks for routing traffic to various satellite beams as a way of enhancing satellite capacity. The use of SAW IF filters employing slanted-finger IDTs was also illustrated for digital microwave radio, as well as for digital data terminals in mobile earth stations for satellite communications networks.

19.10. REFERENCES

1. K. Feher, *Digital Communications — Satellite/Earth Station Engineering*, Prentice-Hall Inc., Englewood Cliffs, 1983.
2. M. S. Suthers, T. Cameron, P. Kennard and J. McNicol, "SAW Nyquist filters for digital radio,"*Proc. 1987 IEEE Ultrasonics Symposium*, vol. 1, pp. 117–122.
3. J. C. B. Saw, T. P. Cameron and M. S. Suthers, "Impact of SAW technology on the systems performance of high capacity digital microwave radio, *Proc. 1993 IEEE Ultrasonics Symp.*, vol. 1, pp. 59–65, 1993.
4. E. Ehrmann-Falkenau, H. R. Stocker, C. Ruppel and W. R. Mader, "SAW-filters for spectral shaping in a 140 Mbit/s digital radio using 16 QAM," *Proc. 1983 IEEE Ultrasonics Symp.*, vol. 1, pp. 17–22, 1983.
5. C. Ruppel, E. Ehrmann-Falkenau and H. R. Stocker, "Compensation algorithm for SAW second order effects in multistrip coupler filters," *Proc. 1985 IEEE Ultrasonics Symp.*, vol. 1, pp. 7–10, 1985.

6. M. J. McKissock, "Constellation measurement: a tool for evaluating digital radio,"*Hewlett-Packard Journal*, vol. 38, pp. 13–17, July 1987.

7. G. Waters, "A digital radio noise and interference test set,"*Hewlett-Packard Journal*, vol. 38, pp. 19–26, July 1987.

8. Hewlett-Packard Note, "Using Error Vector Magnitude Measurements to Analyze and Troubleshoot Vector-Modulated Signals," Product Note 89400-14, 1996.

9. M. Suthers, G. Este, R. Streater and B. McLaurin, "Suppression of spurious SAW signals,"*Proc. 1986 IEEE Ultrasonics Symp.*, vol. 1, pp. 37–42, 1986.

10. A. Vigil, B. P. Abbott and D. C. Malocha, "A study of the effects of apodized structure geometries on SAW filter parameters,"*Proc. 1987 IEEE Ultrasonics Symp.*, vol. 1, pp. 139–144, 1987.

11. J. H. McClellan, T. W. Parks and L. R. Rabiner, "A computer program for designing optimum FIR linear phase digital filters," *IEEE Trans. Audio and Electroacoustics*, vol. AU-21, pp. 506–526, Dec. 1973.

12. P. M. Smith and C. K. Campbell, "The design of SAW linear phase filters using the Remez Exchange Algorithm, *IEEE Trans. Ultrason., Ferroelec., and Freq. Control*, vol. UFFC-33, pp. 318–323, May 1986.

13. S. F. Yuen, "Design and synthesis of surface acoustic wave pulse-shaping filters for digital radio," *M.Eng. Thesis in Electrical Engineering*, McMaster University, Hamilton, Ontario, Canada, 151 pages, April 1988.

14. C. S. Lam, D. S. Stevens and D. J. Lane, "BAW- & SAW-based frequency control products for modern telecommunications systems and their applications in existing & emerging Wireless communications equipment," *Proc. International Meeting on the Future Trends of Mobile Communication Devices,"* Tokyo, Japan, Jan. 1996.

15. P. W. Hooijmans, *Coherent Optical System Design*, Wiley, New York, ch. 1, 1994.

16. L. W. Couch II, *Digital and Analog Communications Systems*, Third Edition, Macmillan Publishing Co., New York, p. 263, 1990.

17. H. Nakahata, H. Kitabayashi, S. Fujii, K. Higaki, K. Tanabe and S. Shikata, "Fabrication of 2.5 GHz SAW retiming filter using SiO2/ZnO/Diamond structure," *Proc. 1996 IEEE Ultrasonics Symp.*, vol. 1, 1996.

18. S. Shikata, H. Nakahata and N. Fujimori, "High frequency bandpass filter using polycrystalline diamond," *Diamond and Related Materials*, vol. 2, pp. 1197–1202, 1993.

19. R. C. Dorf (ed.), *The Electrical Engineering Handbook*, CRC Press, Boca Raton, Florida, p. 1543, 1993.

20. R. C. Peach, "SAW based systems for communications satellites," *Proc. 1995 IEEE Ultrasonics Symp.*, vol. 1, pp. 159–166, 1995.

21. J. G. Schoenenberger, "Satellite personal communications networks," *Proc. IEE Third European Conference on Satellite Communications (ECSC-3)*, pp. 128–132, Nov. 1993.

—20—
Postscript

20.1. Trends in Mobile/Wireless Systems

The goal of mobile/wireless communications in the coming years will be to allow user access to the global network at any time, without regard to mobility or location. Cellular and cordless telephone communications have begun this process, but do not yet allow total communications. Continued developments in satellite communications, fiber-optic and digital micro-wave radio communications are aimed at this goal. In Europe, the long-term aim is for a Universal Mobile Telephone System (UMTS). The intention is to exploit the potential of GSM and DECT, as well as to unify cellular, cordless, LAN, paging, Radio Local Loop (RLL), and Private Mobile Radio (PMR) [1].

In contrast, the Federal Communications Commission (FCC) in the U.S. has allocated 140 MHz of spectrum near 2 GHz to PCS communications. This allows licensed spectral blocks for 51 major trading areas (MTAs) and 492 Basic Trading Areas (BTAs). A minimum of 20 MHz has been allocated for unlicensed applications for voice and packet data. Licenses awarded by the FCC are on an "auction" basis, where the winner of each license is free to use any desired air interface and system architecture, provided that these comply with FCC regulations on power levels and performance specifications.

Automobile installations of position indicators using the Global Positioning System (GPS)—now routinely used in Japan—should see significant worldwide adoption. Mobile broadband systems (MBS) are also under investigation for quasi-mobile services at 155 Mb/s in the 60-GHz band [2].

20.2. Implications for Surface Wave Device Technology

As a result of increasing demands by the communications industry for SAW products, many North American, European, and Japanese manufacturers of SAW devices are expanding their facilities. It is anticipated that the

611

demand for SAW-based antenna duplexers, RF filters, and IF filters will continue to expand. As the price of dielectric resonators has dropped below the $1 level (United States currency), some SAW manufacturers are offering RF filters costing between $1 and $1.50. As prices continue to fall, it has been predicted [3] that dielectric resonator filters will be displaced in most portable circuitry, and that SAW prices for RF filters between 800 MHz and 1.5 GHz will fall to between 50¢ and $1 in the next few years [3].

Two-chip zero-IF (direct conversion) techniques recently developed for DECT telephones eliminate the use of an IF filter in superheterodyne-based systems [4]. For efficient direct conversion, however, the input signal level must fall to within a relative-small amplitude range. Additionally, because the mixer LO input is at the input frequency, leakage between mixer and antenna must be kept to a minimum. It is noted, however, that the superheterodyne method is currently still the easiest circuitry for implementation and incorporation of components [3], [5].

As consumer adoption of mobile/wireless communications technology continues to expand, It is anticipated that RF frequencies for some systems will extend into higher GHz-band regimes. One example of this relates to the recent U.S. availability of 300 MHz of spectrum in the 5-GHz band for unlicensed radio communications. This would include short-range high-speed WLAN communications, with bandwidths supporting up to 20 Mb/s data transmission rates at power levels of 10 mW EIRP [6]. This should offer new challenges in the development of surface wave technologies and techniques at these higher frequencies that will include fabricational and power-density limitations imposed by smaller SAW device sizes. It is anticipated by this author that this will include increased focus on the efficient harmonic-operational capability of SAW IDTs and reflection gratings.

20.3. REFERENCES

1. J. E. Padgett, C. G. Günther and T. Hattori, "Overview of Wireless Personal Communications," *IEEE Communications Mag.*, pp. 28–41, Jan. 1995.
2. J. S. daSilva and B. E. Fernandez, "The European research program for advanced mobile systems," *IEEE Personal Communications Mag.*, vol. 2, pp. 14–19, Feb. 1995.
3. C. S. Lam, D-P Chen, B. Potter, V. Narayanan and A. Vishwanathan, "A review of the applications of SAW filters in wireless communications," *International Workshop on Ultrasonics Application*, Nanjing, China, Sept. 1996.
4. "New zero IF chipset from Philips," *Electronics Engineering*, p. 10, Sept. 1995.
5. C. S. Lam, D. S. Stevens and D. J. Lane, "BAW- & SAW-based frequency control products for modern telecommunications systems and their application in existing and emerging wireless communications equipment," *International Meeting on the Future Trends of Mobile Communications Devices*," Tokyo, Japan, Jan. 1996.
6. *Applied Microwave and Wireless*, p. 10, April 1997.

Glossary

Phone Standards (with some representative countries)

ARDIS	Advanced Radio Data Information Service
AMPS	Advanced Mobile Phone Service (North America, Central America, South America, Australia, New Zealand, China, South Korea, Switzerland, Hong Kong, Thailand, Malaysia, Singapore)
CDPD	Cellular Digital Packet Data
CT1/CT1+	Cordless Telephone (analog) (European countries)
CT2/CT2+	Cordless Telephone (digital); (European countries, China, New Zealand, Singapore)
DCS	Digital Cellular System (European countries)
DECT	Digital European Cordless Telephone (European countries, China, Hong Kong)
EAMPS	Extended Advanced Mobile Phone Service (See AMPS)
EGSM	Extended GSM (See GSM)
ETACS	Extended Total Access Communication System (See TACS)
GSM	Global System for Mobile Communications (Europe, Eastern Europe, Mid-East, South Africa, India, Singapore, Hong Kong, Malaysia, Australia, New Zealand)
IS54	North American Digital Cellular (North America, Mexico)
IS95	North American Digital Cellular (North America, Mexico)
JCT	Japanese Cordless Telephone (Japan)
JDC	Japan Digital Cellular (Japan)
JTACS	Japan Total Access Communication System
NADC	North American Digital Cellular
NTACS	Nippon Total Access Communications System

613

NMT	Nordic Mobile Telephone (Norway, Sweden, Finland, other European countries, Mid-East, India, Malaysia)
NTT	Nippon Telephone and Telegraph (Japan)
PACS	Personal Access Communications Services (North America, Mexico)
PCS	Personal Communications Services (North America, Mexico)
PCN	Personal Communications Network (Europe)
PHS (PHP)	Japanese Personal Handy Phone System
PDC	Personal Digital Cellular (Japan)
RTMS	Radio Telephone Mobile System (Italy)
TACS	Total Access Communication System (UK, Ireland, China, Japan, Hong Kong, Singapore, Malaysia, Spain)
UMTS	Universal Mobile Telecommunications Service (in Europe)
USDC	U.S. Digital Cellular (USDC) system
WAN	Wide-area network
WLAN	Wireless local area network

General Communications

ACI	Adjacent channel interference
AFC	Automatic frequency control
AGC	Automatic gain control
ASIC	Application specific integrated circuit
ATM	Asynchronous Transfer Mode
B_c	Coherence bandwidth
BER	Bit-error rate
BPSK	Binary phase-shift keying
BW	Bandwidth
C/N	Carrier-to-noise
CDMA	Code division multiple access
CPSM	Continuous phase-shift modulation
CSK	Code-shift keying
dB	Decibel
dBm	Decibels relative to $1\,mW$
DVB	Digital Video Broadcast (in Europe)
dBW	Decibels relative to $1\,W$
DPSK	Differential phase shift keying
DQPSK	Differential quadrature (or quaternary) phase shift keying
DS	Direct sequence
DSSS	Direct sequence spread-spectrum

EIA	Electronics Industry Association
EIRP	Effective isotropic radiated power
F	Amplifier noise figure (usually expressed in dB)
FDD	Frequency division duplexing
FCC	Federal Communications Commission (in USA)
FDMA	Frequency division multiple access
FH	Frequency hopping
FIR	Finite impulse response
FM	Frequency modulation measured in Hz peak frequency deviation
FSK	Frequency shift keying
FSS	Fixed satellite services
GFSK	Gaussian-filtered frequency-shift-keying
GHz	Gigahertz (10^9 Hz)
GPS	Global Positioning System
IF	Intermediate frequency
IM	Intermodulation distortion
ISI	Intersymbol Interference
ISM	Industrial, Scientific, and Medical (frequency bands)
k	Boltzmann constant (1.38×10^{-23} J/$^\circ$K)
K	Frequency-reuse factor
kHz	Kilohertz (10^3 Hz)
LCR	Inductance-capacitance-resistance
LEO	Low earth orbit
LFSR	Linear-feedback shift register
LNA	Low-noise amplifier
LO	Local oscillator
MBS	Mobile broadband systems
MEO	Medium earth orbit
MES	Mobile earth station
MWC	Microwave ceramic
MCF	Monolithic crystal filter
MHz	Megahertz (10^6 Hz)
MSK	Minimum-shift keying
MSS	Mobile satellite services
MUX	Multiplexer
NF	Noise figure (usually expressed in dB)
NRZ	Non return to zero
PE	Power efficiency
PCM	Pulse-code modulation
PCMCIA	Personal Computer Memory Card International Association

PLL	Phase-locked loop
PMR	Private Mobile Radio
PN	Pseudo-noise
PRBS	Pseudo-random binary sequence
QAM	Quadrature amplitude modulation
QPSK	Quadrature phase shift keying
RF	Radio frequency
RLL	Radio Local Loop
RS	Receiver selectivity
rms	Root mean square
RAM	Random Access Memory
Rx	Receiver
SDH	Synchronous Digital Hierarchy
SE	Spectral efficiency
SHF	Super-high frequency
SINAD	Signal, Noise and Distortion
SNR	Signal-to-noise ratio (normally expressed in dB)
SONET	Synchronous Optical Network
STM	Synchronous transfer mode
TDMA	Time division multiple access
TEM	Transverse electromagnetic
Tx	Transmitter
UHF	Ultra-high frequency
US	Unit symbol
VCO	Voltage-controlled oscillator
VSWR	Voltage standing wave ratio
α	Channel roll-off factor
Δ	Delay spread

SAW Devices

BGS	Bleustein-Gulyaev-Shimizu (wave)
BWC	Beam width compressor
COM	Coupling-of-modes
ETL	Effective transmission loss
EWC-SPUDT	Electrode-width controlled single-phase unidirectional transducer
B_a	Unperturbed radiation susceptance of IDT
$C(f)$	Circuit factor
C_o	IDT capacitance/finger pair/unit length (F/m)
C_S	IDT Capacitance/finger pair (F)
C_T	Total IDT capacitance (F)
FEUDT	Floating electrode unidirectional transducer

G_o	Acoustic characteristic conductance (mho)
$G_a(f)$	Unperturbed radiation conductance of IDT (mho)
$G_{am}(f)$	Radiation conductance of IDT, incorporating velocity shift (mho)
$G_{amf}(f)$	Radiation conductance of IDT with velocity shift and IDT finger reflections (mho)
h/λ	IDT film-thickness ratio (dimensionless)
IDT	Interdigital transducer
IIDT	Interdigitated interdigital transducer
k_o	Wave vector (*i.e.*, phase constant β_o in rad/m)
K^2	Electromechanical coupling constant
MSC	Multistrip coupler
N	Number of IDT finger pairs
NGFEUDT	Narrow-gap floating electrode unidirectional transducer
NSPUDT	Natural single-phase unidirectional transducer
PG	Processing gain
RAC	Reflective array compressor
SAW	Surface acoustic wave
SBAW	Shallow bulk acoustic wave
SPUDT	Single-phase unidirectional transducer
SF	Shape factor (of SAW bandpass filter)
SSBW	Surface skimming bulk wave
STW	Surface transverse wave
T_b	Bit period
T_c	Chip period
TB	Time-bandwidth product
TCD	Temperature coefficient of delay (ppm/°C)
TCF	Temperature coefficient of frequency (ppm/°C)
TDL	Tapped delay line
TTI	Triple-transit interference
v_a	Average shift velocity (m/s)
v_o	Free-surface SAW velocity (m/s)
W	Acoustic aperture (m)
Z_o	Acoustic characteristic impedance (Ω)
Å	Angstrom (1 Å $= 10^{-8}$ cm)
α	Attenuation coefficient (m^{-1})
β	Phase constant (rad/m)
η	Metallization ratio (dimensionless)
θ	Electrical transit angle (rad)
δ	Detuning parameter; also phase-mismatch parameter
λ	Acoustic wavelength (m)
κ_{11}	Self-coupling coefficient (m^{-1})

κ'_{11}	Frequency-normalized self-coupling coefficient (dimensionless)
κ_{12}	Mutual-coupling coefficient (m^{-1})
κ'_{12}	Frequency-normalized mutual-coupling coefficient (dimensionless)
ε_r	Relative permittivity, or dielectric constant
ε_o	Permittivity of free space (8.85×10^{-12} F/m)
ρ	Reflection coefficient (dimensionless)
τ	Group delay (s)
ζ	Transduction coefficient
$\chi(\tau,\phi)$	Ambiguity function

Index